Transforming food systems

The quest for sustainability

It is widely recognised that agriculture is a significant contributor to global warming and climate change. Agriculture needs to reduce its environmental impact and adapt to current climate change whilst still feeding a growing population, i.e. become more 'climate-smart'. Burleigh Dodds Science Publishing is playing its part in achieving this by bringing together key research on making the production of the world's most important crops and livestock products more sustainable.

Based on extensive research, our publications specifically target the challenge of climate-smart agriculture. In this way we are using 'smart publishing' to help achieve climate-smart agriculture.

Burleigh Dodds Science Publishing is an independent and innovative publisher delivering high quality customer-focused agricultural science content in both print and online formats for the academic and research communities. Our aim is to build a foundation of knowledge on which researchers can build to meet the challenge of climate-smart agriculture.

For more information about Burleigh Dodds Science Publishing simply call us on +44 (0) 1223 839365, email info@bdspublishing.com or alternatively please visit our website at www.bdspublishing.com.

Related titles:

Climate change and agriculture
Print (ISBN 978-1-78676-320-4); Online (ISBN 978-1-78676-322-8, 978-1-78676-323-5)

Developing circular agricultural production systems
Print (ISBN 978-1-80146-256-3); Online (ISBN 978-1-80146-257-0 , 978-1-80146-258-7)

Key issues in agricultural ethics
Print (ISBN 978-1-80146-313-3); Online (ISBN 978-1-80146-314-0, 978-1-80146-315-7)

Pesticides and agriculture
Print (ISBN 978-1-78676-276-4); Online (ISBN 978-1-78676-278-8, 978-1-78676-279-5)

Chapters are available individually from our online bookshop:
https://shop.bdspublishing.com

BURLEIGH DODDS SERIES IN AGRICULTURAL SCIENCE
NUMBER 99

Transforming food systems

The quest for sustainability

Dr Dave Watson

Published by Burleigh Dodds Science Publishing Limited
82 High Street, Sawston, Cambridge CB22 3HJ, UK
www.bdspublishing.com

Burleigh Dodds Science Publishing, 1518 Walnut Street, Suite 900, Philadelphia, PA 19102-3406, USA

First published 2024 by Burleigh Dodds Science Publishing Limited

Library of Congress Control Number: 2024935134

British Library Cataloguing in Publication Data
A catalogue record for this book is available from the British Library

ISBN 978-1-78676-455-3 (print)
ISBN 978-1-78676-458-4 (PDF)
ISBN 978-1-78676-457-7 (ePub)
ISSN 2059-6936 (print)
ISSN 2059-6944 (online)

DOI: 10.19103/AS.2024.0156

Typeset by Deanta Global Publishing Services, Dublin, Ireland

Contents

Series list

Title | Series number

Achieving sustainable cultivation of maize - Vol 1 001
From improved varieties to local applications
Edited by: Dr Dave Watson, CGIAR Maize Research Program Manager, CIMMYT, Mexico

Achieving sustainable cultivation of maize - Vol 2 002
Cultivation techniques, pest and disease control
Edited by: Dr Dave Watson, CGIAR Maize Research Program Manager, CIMMYT, Mexico

Achieving sustainable cultivation of rice - Vol 1 003
Breeding for higher yield and quality
Edited by: Prof. Takuji Sasaki, Tokyo University of Agriculture, Japan

Achieving sustainable cultivation of rice - Vol 2 004
Cultivation, pest and disease management
Edited by: Prof. Takuji Sasaki, Tokyo University of Agriculture, Japan

Achieving sustainable cultivation of wheat - Vol 1 005
Breeding, quality traits, pests and diseases
Edited by: Prof. Peter Langridge, The University of Adelaide, Australia

Achieving sustainable cultivation of wheat - Vol 2 006
Cultivation techniques
Edited by: Prof. Peter Langridge, The University of Adelaide, Australia

Achieving sustainable cultivation of tomatoes 007
Edited by: Dr Autar Mattoo, USDA-ARS, USA and Prof. Avtar Handa, Purdue University, USA

Achieving sustainable production of milk - Vol 1 008
Milk composition, genetics and breeding
Edited by: Dr Nico van Belzen, International Dairy Federation (IDF), Belgium

Achieving sustainable production of milk - Vol 2 009
Safety, quality and sustainability
Edited by: Dr Nico van Belzen, International Dairy Federation (IDF), Belgium

Achieving sustainable production of milk - Vol 3 010
Dairy herd management and welfare
Edited by: Prof. John Webster, University of Bristol, UK

Ensuring safety and quality in the production of beef - Vol 1 011
Safety
Edited by: Prof. Gary Acuff, Texas A&M University, USA and Prof. James Dickson, Iowa State University, USA

Ensuring safety and quality in the production of beef - Vol 2 012
Quality
Edited by: Prof. Michael Dikeman, Kansas State University, USA

Achieving sustainable production of poultry meat - Vol 1 013
Safety, quality and sustainability
Edited by: Prof. Steven C. Ricke, University of Arkansas, USA

Achieving sustainable production of poultry meat - Vol 2 014
Breeding and nutrition
Edited by: Prof. Todd Applegate, University of Georgia, USA

Preface

This book is the culmination of my spare time (evenings and weekends) over the past 5 years. I dedicated my time to write this book for two important reasons. First, I was driven by the need to capture decades of experience in food system transformation in both developing and developed country contexts. Secondly, to the exclusion of alternative pathways and approaches, much of the contemporary writing on food system transformation focus on the promotion of relatively narrow pathways to sustainability. This book broadens discussion by comparing a number of competing approaches to achieving sustainable food systems. Each approach contains innovative science, technologies, practices, and polices that have something to contribute to the development of more sustainable food systems.

The book is divided into three parts. Comprising three chapters, Part 1 of the book provides an overview of the global food system. Chapter 1 provides a potted history of the evolution of food systems from the advent of settled agriculture to the dominant industrial food systems that exist today. Chapter 2 provides a relatively detailed overview of both successes and failures of the industrial food system. Whilst Chapter 3 focusses on the contemporary drivers of food system change.

Comprising four chapters, Part 2 focusses competing paradigms of production. Chapter 4 showcases some of the technologies and approaches proffered by the dominant Neo-Productivist food system paradigm. These include development of the bioeconomy, cultured meat, plant-based meat analogues, genetically modified and gene-edited plants and animals, and the emergence of vertical farming. Chapter 5 showcases some of the technologies and approaches proffered by the so-called Reformist food system paradigm. These include examples of sustainable intensification and land sparing brought about via the application of Precision Agriculture, Climate Smart Agriculture, Conservation Agriculture, and recent advances in nanotechnologies and biopesticides. Chapter 6 showcases some of the technologies and approaches proffered by the so-called Progressive food system paradigm. These include examples of ecological intensification and land sharing brought about via the application of Regenerative, Organic, and Agroecological food system approaches. Chapter 7 showcases some of the technologies and approaches proffered by the so-called Radical food system paradigm. These include examples of degrowth, food sovereignty, Slow Food, Permaculture, food sharing, and the localisation and regionalisation of food systems.

Part 3 comprises two chapters. Chapter 8 of the book evaluates paradigms vis-à-vis their ability to address, food and nutrition security, environmental sustainability, economic sustainability, social sustainability, and political sustainability. Drawing on a multi-level perspective framework, and notions of lock-ins and path dependency, Chapter 9 attempts to characterise likely food system trajectories over the foreseeable future.

I hope that you enjoy reading this book as much as I have enjoyed writing it. The opinions expressed in this book are solely those of the author and do not reflect the views of either employer or publisher.

Dr Dave Watson

Part 1

The global food system

Chapter 1

Emergence of the global food system

The global food system encompasses 'the entire range of actors and their interlinked value-adding activities involved in the production, aggregation, processing, distribution, consumption, and disposal of food products that originate from agriculture, forestry, or fisheries' (FAO 2018a). In turn, the global food system itself is couched within wider economic, social, political, and environmental systems. The global food system is not static, it is continually influenced by biophysical, socio-cultural, economic, and political landscape-level drivers (Sperling et al. 2020; Geels 2002), as well as socio-technological innovations that emerge within the food system itself. Whilst the term 'global' is widely used, the global food system is, in fact, a series of interconnected, heterogeneous, and often nested local, regional, and national food systems. In turn, these nested food systems are often woven together by global food value chains (OECD 2021). Figure 1 depicts a stylised version of the food system concept.

1.1 The journey from the dawn of agriculture to the Second Food Regime

This section of the book provides a concise history of the global food system's evolution from the emergence of hunter-gathering *Homo sapiens* through to the early twenty-first century. It aims to highlight the interconnectivities and interdependencies between people, food, nature, and the modern world in which we live.

Similar to other living organisms, humans exist within ecosystems and form part of intricate food webs. Initially, our ancestors hunted or trapped prey and gathered edible and nutritious foods (plants and fungi) within these food

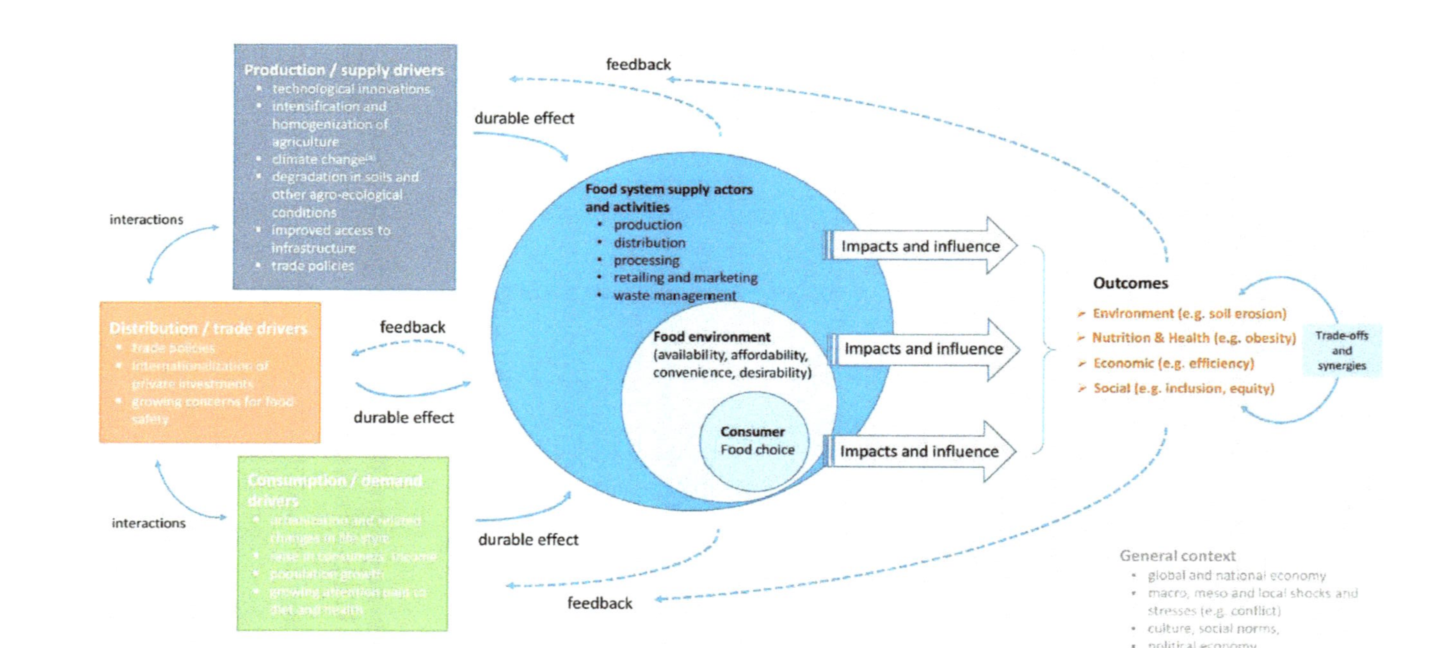

Figure 1 Conceptual framework of food system. Source: Adapted from: Béné (2019b).

webs. Indeed, a few isolated communities around the world still do. When hunting and gathering were bountiful, and disease pressure was low, family groups of humans were able to survive. However, when animals did not return when expected, or the opportunities for gathering food were scarce, family groups struggled to survive. Approximately 10 000 years ago, in attempts to better control the erratic nature of relying on foraging and hunting for food supplies, groups of humans around the world began to experiment with sedentary agriculture.

Over the centuries, sedentary agriculture slowly became the mainstay of food systems (Mabhaudhi et al. 2019) in most parts of the world. However, the precarious balance between food production, availability and price remained, and still remains, a principal concern of individual rural families, urban communities, and both local and national leaders. To main stability and order, leaders needed to ensure the supply of adequate volumes of food at affordable prices. Understanding this link between food availability and social stability, most early civilisations, including the Egyptians, Greeks, and Romans, ensured that adequate food stocks were established to buffer consumption during lean times. Many leaders also attempted, usually unsuccessfully, to enforce price controls on key staple foods to ensure sufficient access to food and prevent social and political unrest. Food trade remained limited until the invention of the wheel, oxen/horse-drawn carts, and boats capable of carrying cargo. Even then, food was rarely transported more than 10 miles to the local village or town. This 10-mile radius struggled to feed a town population of more than 10 000 people (Lumen 2021). However, food trade did increase to some extent within vast empires such as Egypt and Rome. For example, as a province of the Roman Empire, Egypt supplied significant amounts of grain to Rome. Ultimately, however, human population growth was kept at bay due to a combination of limited agricultural land, low levels of agricultural productivity (output of crop/livestock per unit of land), regular pest and disease outbreaks, and protracted periods of conflict. For example, a combination of the Black Death, animal murrains (diseases), and poor harvests (linked to a period of cooler and wetter weather in the early fourteenth century), together with war, decimated the population of Europe in the 1300s. It took almost 500 years for the European population to reach pre-1300 levels.

From the mid-seventeenth century, this delicate and precarious balance between people, food, and nature underwent a significant step change. This change initially occurred across Western Europe and the USA. In the UK, the Agricultural Revolution (circa 1650–1850) underpinned an unprecedented increase in labour productivity and agricultural output. This was brought about by a number of key factors, such as (Watson 2018)

1 improved drainage, land reclamation, and the conversion of permanent pasture to arable land;
2 mechanisation, including the development of the horse-drawn seed drill, corn reaper, and threshing machine, and improvements to the traditional plough;
3 introduction of new crops (such as turnips, clover, and potatoes) and animal breeds (mainly cattle and sheep);
4 introduction of the Norfolk Four Course Rotation with reduced fallows;
5 enclosure of the medieval open-field system, allowing more efficient utilisation of land;
6 the application of science brought about through the Period of Enlightenment in the 1600s and 1700s;
7 increased and more effective utilisation of organic pesticides on high-value horticultural crops;
8 improved and more cost-effective transportation of agricultural produce; and
9 development of sophisticated market infrastructure (roads and inland waterways) and institutions.

The embrace of science across Western Europe and the USA evolved through the process of 'Enlightenment' from the late seventeenth through to the eighteenth century. The enlightenment period can be characterised by a focus on the power of rational, empirical enquiry to explain the world (rather than the accumulated dogmas of the Catholic and other established religions). Through the application of scientific methods, gentlemen farmers quickly developed a rudimentary scientific understanding of agricultural production and its agroecosystems, especially the basic science around plant nutrition, crop rotation, and pest identification, biology, and ecology (Watson 2018). By the early 1800s, the individual work of gentleman farmers, educators, journalists, and farm input producers in the UK, Western Europe, and the USA had led to the establishment of agricultural societies in pursuit of increased productivity, profitability, and professionalism. In 1843, this culminated in the establishment of the world's first agricultural research station at Rothamsted in the UK which was set up to conduct field trials for the first synthetic fertilisers (Rothamsted 2023).

The Agricultural Revolution led to a doubling of cereal yields in the UK during the 1700s (Krausmann and Langthaler 2019) and almost tripled yields and farm-labour productivity by 1870. This allowed agricultural production to outstrip population growth (Britannica 2015). By the mid-1800s, English agriculture was by far the most productive in Europe, with yields as much as '80% higher than the Continental average' (Britannica 2015). Lower unit costs of food production, brought about by rapid increases in the productivity of land, labour, and capital, backed by government protection for agriculture (particularly the Corn Laws 1815–1846), combined to create veritable halcyon days for farming in countries such as the UK. This profited both landed elites and yeoman farmers alike (NUS 2003).

Ultimately, whilst the Agricultural Revolution in the UK provided somewhat of a blueprint for Western European nations and their New World colonies, it also paved the way for the Industrial Revolution, which began in England (circa 1765–1840). The Agricultural Revolution underpinned the Industrial Revolution in several ways:

1. Rapid population growth in England and Wales, from 5.5 million in 1700 to over 9 million by 1801 (Britannica 2015). This was primarily based on regular national food production surpluses;
2. The displacement of farm labour to towns and cities, due to productivity-enhancing mechanisation, farmland enclosure, and farm consolidation;
3. Animal traction for farm work, transportation (horse-drawn carts and barges), and industrial work (mining, etc.); and
4. Agricultural outputs, such as animal skins and timber, could be used as inputs for industrial processes.

However, by the mid to late nineteenth century, despite significant productivity gains, it became increasingly difficult for the UK and other European nations to compete with cheap food imports (especially wheat) from the USA (Krausmann and Langthaler 2019).

1.2 The First Food Regime (1870–1930s)

As the Industrial Revolution progressed, it became evident to the increasingly powerful nouveau riche industrialists in Britain and Western Europe that, to maintain high and stable profits, it was necessary to secure physical access to regular supplies of cheap, yet nutritionally adequate, food for their increasingly urbanised workforce (Watson 2018). The prosperity of industrialists was heavily dependent on a highly productive workforce. In the 1800s, factory employees still spent most of their income of food. The price of basic staples was, therefore, used to determine baseline industrial wages (Watson 2018). It was during this time that, industrialists, in partnership with national governments, used their political and economic power to reorganise food production, distribution, and consumption to serve their needs. Indeed, industrialists were the principal architects of what was latterly termed the First Food Regime, or the diasporic-colonial Food Regime (Friedmann and McMichael 1989), established during the so-called extensive period of capitalism (Watson 2018). Orchestrated by Britain, the First Food Regime was established during a time of British global imperialism and hegemony when Britain was the 'workshop of the world' (Watson 2018).

Key components of the First Food Regime were (Watson 2018) a sharp increase in international food trade; and the reorganisation and industrialisation of food production, processing, distribution, and consumption in nationally organised economies.

These are discussed here.

1.2.1 Increased international food trade

The First Food Regime was underpinned by two key trade-related components. First, after forcing repeal of the Corn Laws, powerful industrialists in Britain, and latterly in France, Holland, and Spain, secured cheap imports of unprocessed and semi-processed foods and materials, principally wheat and meat (wage foods), from white settler countries, namely, Australia, Canada, and the USA, with tropical fruits and vegetables being imported from other colonies (Pereira et al. 2020). This created the foundation for the contemporary global trade in foodstuffs. Figure 2 illustrates current trade in wheat between the USA and major wheat importers across the world.

Cheap imports from settler countries were aided by two key factors:

1. Cheap labour and inputs, and greater economies of scales in settler countries, which led to the production of cheap grain; and
2. Low transportation costs.

By the early nineteenth century, with the advent of low-cost trans-Atlantic shipping based on cheap coal, it cost as much to ship a ton of wheat 3000 miles

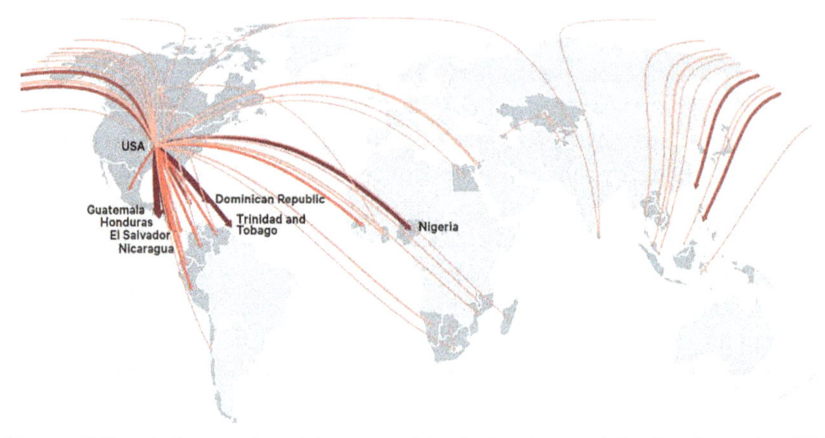

Figure 2 Twenty-first century international trade in wheat originating from the USA. Source: Adapted from: Adams et al. (2021).

across the Atlantic as it did to transport a ton of wheat 32 miles by wagon over unimproved roads in the UK (Lumen 2021). At the same time, the introduction of refrigerated ships allowed more perishable food products, such as meat, butter, and tropical fruits, to be exported to Europe (Peet 1969). In the USA, the Government developed an export-led agriculture policy and infrastructure interventions. The two most important interventions were the Homestead Act of 1862, which gave free land in the American Prairies to anyone who would farm it, and the 1862 Pacific Railroad Act, which provided a low-cost connection between emergent agricultural production areas in the west and expanding US urban markets in the East, as well as Eastern docks for exportation (Agriculture Outlook 2000).

Second, the First Food Regime instigated a new division of labour, namely, the bilateral exchange of imported foods for exported manufactured goods, labour, and capital from Europe (Watson 2018; Mabhaudhi et al. 2019). This quickly evolved into multilateral trade based on comparative and competitive economic advantage, which deconstructed historic trading monopolies and promoted a global market economy (Watson 2018).

1.2.2 Reorganisation of food systems

The move towards a global market economy necessitated the reorganisation of food systems into urban-centred industrial agro-food complexes within nationally organised economies. The term 'urban-centred' acknowledges two things (Watson 2018):

1 The transfer of power over food systems from rural elites to urban elites.
2 Populations across developed countries were increasingly urbanised.

The growing capitalist marketplace increasingly insisted on increased competitiveness of key activities along agricultural value chains. Reductions in the unit cost of production were generally achieved through greater economies of scale, increased productivity of land, labour, and capital via the specialisation of labour, and more efficient and effective application of off-farm industrial inputs, such as, machinery, fertilisers (such as Guano, Chile Saltpeter, and nitrates) from South America (Krausmann and Langthaler 2019), and crop protection products. The reorganisation of food systems increasingly fostered a more business-like and commercial approach to farming (Watson 2018).

At this time, the USA began to emerge as an agricultural powerhouse. Access to cheap food (grains and meat) not only fed a growing industrialised workforce and underpinned competitive industrial growth, but competitive food exports facilitated significant foreign exchange from the agriculture sector. The American Government ramped up investments in agriculture through the establishment of the United States Department of Agriculture (USDA) in 1862, Land Grant agricultural colleges and experimental stations, an adult education system (via the Hatch Act of 1887), and the creation of the USDA's Cooperation Extension Service in 1914 (Watson 2018).

Reorganisation of food production, distribution, and consumption within the framework of the First Food Regime substantially increased the capacity of farmers and farming systems in both the Old and the New World to provide greater volumes of marketable food commodities at increasingly competitive prices. Growing agricultural surpluses continued to facilitate the release of agricultural labour to supply the growing demand for industrial workers, predominantly in urban areas, and to supply adequate quantities of reasonably quality food to this growing non-agricultural workforce (Watson 2018; Mabhaudhi et al. 2019).

1.3 The Second Food Regime (1947–73)

The First Food Regime faced a crisis after the food shortages of the First World War and economic depressions of the 1920s and the 1930s. Food production capacity in Europe was further decimated by the Second World War. Indeed, many European governments were obliged to maintain wartime food rationing several years after the war ended (Watson 2018). The food shortages and the political and economic upheaval associated with the Second World War signalled the need for a restructuring of the global economy and, within it, food systems. After the significant loss of British hegemony, the United States of America quickly filled the power void and moulded the global economy to suit US interests, just as the First Food Regime had been moulded by the UK (Hough 1998). The spirit of the time is aptly captured in President Truman's inaugural speech in 1949..... 'Economic recovery and peace itself depend on increased world trade.... Our aim should be to help the free peoples of the world, through their own efforts, to produce more food, more clothing, more materials for housing, and more mechanical power to lighten their burdens' (Benton and Bailey 2019).

The USA orchestrated several pivotal policy and institutional changes. First, the declaration by 26 'United Nations' in 1942 'to continue their united fight against the Axis Powers'. Second, 2 years later, the USA backed the United Nations Monetary and Financial Conference at Bretton Woods in 1944. The resultant Bretton Woods Agreement of 1945

1. stabilised exchange rates between national currencies (based on the dollar/gold standard);
2. underpinned the national model of economic growth, primarily in the USA's favour; and

3 led to the establishment of the World Bank (International Bank for Reconstruction and Development) and the International Monetary Fund (IMF) in 1945.

Third, the General Agreement on Tariffs and Trade of 1947 (which evolved into the World Trade Organisation in 1995), which set the rules of liberalised international trade, except for most agricultural commodities. Fourth, the creation of the Food and Agriculture Organisation (FAO) in 1945 (Watson 2018; Benton and Bailey 2019; Krausmann and Langthaler 2019). Further efforts quickly led to the established the United Nations (UN) in 1945, and, in 1949, international military stability was achieved via the North Atlantic Treaty Organisation (Watson 2018). Ultimately, America's dominance in the Second Food Regime led to US surpluses of wheat and maize being offered as food aid to postcolonial states across the globe, and the sharing of Green Revolution technologies, to bolster strategic alliances with the USA during the Cold War (Pereira et al. 2020).

Whilst remnants of the First Food Regime remained, the Second Food Regime, also known as Mercantile-Industrial Food Regime (1947–73) (Friedmann and McMichael 1989), replaced the free market philosophy with market regulation based on the goal of food self-sufficiency (Watson 2018). The stage was now set for the evolution of a fossil fuel-based industrialisation of agriculture 'capable of delivering adequate supplies of low-cost quality foods' (Hansen 2019). Indeed, the Second Food Regime was responsible for restructuring the world's principal farming systems, 'firstly in developed countries during the 1950s and 1960s and latterly in developing countries during the late 1960s and the early 1970s' (Watson 2018). This restructuring produced 'totally new and distinctive agro-food complexes' (Watson 2018).

Ultimately, the Second Food Regime was built on five key tenets (Symes and Marsden 1985; Watson 2018):

1 National food self-sufficiency;
2 Increased commoditisation of production;
3 Increased industrialisation of production;
4 Increased subsumption of farming systems within deepening urban-centred agro-food complexes; and
5 Evolution of productivist high-input/high-output farming systems.

1.3.1 National food self-sufficiency

Immediately after the Second World War, the US Government achieved significant increases in productivity and output through the direct provision of income support to farmers and the application of modern industrial inputs and technologies – discussed later (Agriculture Outlook 2000; Krausmann and Langthaler 2019).

In Europe, the cause of national food self-sufficiency was initially taken up by the UK through the 1947 Agriculture Act (Watson 2018). This Act had three principal aims:

1 Increase domestic food production;
2 Decrease food imports; and
3 Support rural areas.

The Act achieved this through the provision of:

1 guaranteed prices for most farm produce;
2 provision of free advice to farmers; and
3 government-funded agricultural R&D.

In 1962, the newly established European Economic Community (EEC) launched the Common Agricultural Policy (CAP). In a similar vein to the UK, the CAP aimed to (1) 'increase agricultural productivity by promoting technical progress and by ensuring the rational development of agricultural production and the optimum utilisation of all factors of production, particularly labour; (2) ensure a fair standard of living for the agricultural community, by increasing the individual earnings of persons engaged in agriculture; (3) stabilise markets; (4) guarantee supplies; and (5) ensure reasonable prices for consumers' (Wrathall 1988).

1.3.2 Agricultural commoditisation

Agricultural commoditisation refers to the transformation of subsistence to market forms of production, namely, agricultural inputs and outputs became marketable commodities. The term commodity refers to objects (including agricultural inputs and outputs) that are produced for the market rather than one's own use (Watson 2018). US and European governments encouraged farmers to increasingly purchase off-farm inputs to maximise food production for the market (Benton and Bailey 2019).

Evolving agribusinesses capitalised on agricultural support by extracting surplus value through the sale of off-farm seeds, fertilisers, machinery, and pesticides (Bowler 1992). Agribusiness, which refers to 'vertically integrated businesses involved in the supply of agri-inputs (e.g. pesticides), the processing of agri-outputs and the distribution of food products' was first coined at the start of the Second Food Regime (Wallace 1985).

1.3.3 Agricultural industrialisation

The increased industrialisation of agriculture refers to the application of modern technologies, innovations, and practices that emerged from non-agricultural sectors of the economy. The process of industrialisation attempted to decouple agricultural production from the expansion of agricultural land (Krausmann and Langthaler 2019). Key industrial innovations applied to the modernisation of agriculture include (Symes and Marsden 1985; Swagemakers et al. 2019):

1 adoption of economies of scale;
2 production intensification;
3 production specialisation; and
4 production concentration.

The term 'economies of scale' refers to the concept that as the volume of production increases, the unit cost of production declines. For example, the average cost of producing a tonne of wheat will diminish as each additional tonne is produced since the fixed costs are spread over more grain. Widespread belief in the concept of economies of scale led to the rapid expansion of farm size across much of the developed world (Watson 2018).

Production intensification is defined as increasing agricultural output per area of land through increased agricultural productivity. Agricultural output was increasingly achieved through the utilisation of industrial off-farm inputs, such as new crop varieties and improved animal breeds, machinery, inorganic fertilisers, fossil fuels, credit, and inorganic pesticides. Efficient combinations of industrial inputs significantly enhanced the productivity of land, labour, and capital (Watson 2018).

Production specialisation refers to a narrower range of crop and livestock commodities produced on the farm. Production concentration refers to both spatial and size concentration of farm businesses. Spatial concentration refers to increasing concentration of farm production in areas with a comparative advantage for producing a particular agricultural commodity. For example, by the end of the twentieth century, almost 80% of Europe's agricultural production came from just 20% of farms located in agri-industrial hotspots: Paris Basin (wheat), East Anglia (wheat and barley), Emilia-Romagna (milk), and the Netherlands (hydroponic crops) (Byé and Font 1991; Benton and Bailey 2019). Size concentration refers to the consolidation of small farmers into large farms. Spatial and size concentration were usually combined in the search for economies of scale (Watson 2018). Globally, small farms (<2 ha), which comprise 84% of all farms worldwide, produce only around 35% of the world's food on just 12% of the world's agricultural land (Lowder et al. 2021). Figure 3 illustrates the spatial and size concentration of farms across the globe, showing the predominance of large-scale farming in Europe, North America, and South America.

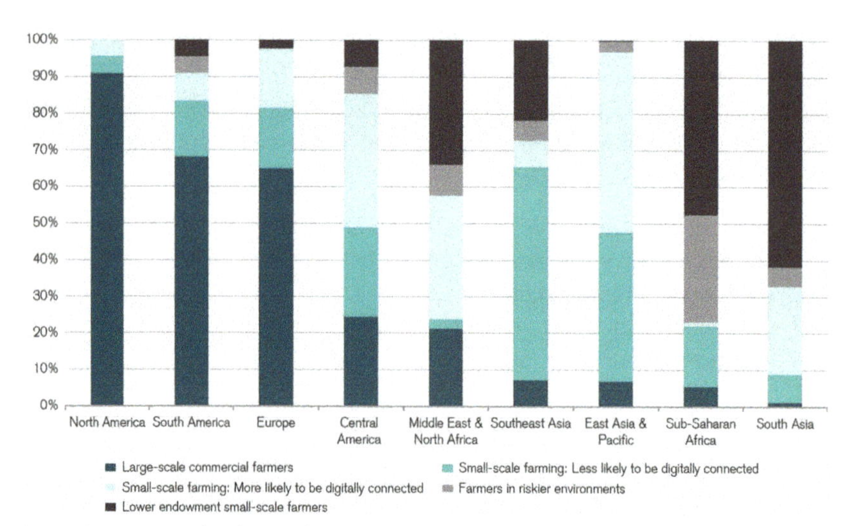

Figure 3 Farm size distribution by region. Source: Adapted from: Credit Suisse (2021).

1.3.4 Agriculture's increasing subsumption within urban-centred industrial agro-food complexes

Although urban-centred industrial agro-food complexes were present during the First Food Regime, it was not until the Second Food Regime that private non-agricultural companies committed significant and prolonged investments in food production, processing, distribution, and retail sectors. This interest often took the form of subsumption and appropriation. Direct or real subsumption is the process whereby traditionally non-farm businesses take direct control of the agricultural production process. This usually took form of backward integration of farm businesses by food processing companies. Exceptionally, the whole value-chain from food production through to retail were integrated. CWS Agriculture, which is a direct subsidiary of the UK-based Co-operative Wholesale Society and dates back to 1890s, is a good early example of direct subsumption (Sauer 1990; Bowler 1990). On the other hand, indirect or informal subsumption refers to processes where non-farm businesses increasingly extract higher amounts of value added from agricultural production, without the need for actual farm ownership. Examples include the sale of farm machinery, pesticides, fertilisers, and credit to farmers (Sauer 1990; Bowler 1990).

Throughout the Second Food Regime, food systems became increasingly subsumed within broader capitalist structures so that industrial agriculture became the norm (or conventional) across the developed world as well as many developing countries (Watson 2018). Non-farm businesses increasingly appropriated elements of farm production, namely industrial inputs replaced organic inputs. A good example of this is the replacement of organic plant nutrients with inorganic plant nutrients, which are supplied by fertiliser manufacturers, or the replacement of crop rotations (used for disease control) with pesticides (Cloke and Goodwin 1992). Non-farm businesses also increased value added through the process of substitutionism, whereby farm outputs are increasingly substituted by non-agricultural substitutes (Watson 2018).

Productivist agriculture is the term used to conceptualise both the development and institutionalisation of state-supported intensive (high-input/high-output) industrialised food systems based on science and technology (Lowe et al. 1993; Marsden et al. 1993). Indeed, the Second Food System is often referred to as the Productivist Period (Watson 2018). During 1950s, 1960s, 1970s, and 1980s, governments of most developed countries helped develop 'broad-based consensus and support from agricultural scientists, farmers, the public and emerging agribusinesses for the so-called techno-policy model of productivist agriculture and the associated processes of agricultural modernisation and commercialisation' (Watson 2018). Governments achieved increased farm productivity and output through (Ilbery and Bowler 1998):

1. minimum price guarantees for priority farm outputs;
2. protection from cheap imports via tariffs; and
3. subsidised fertiliser, pesticides, farm equipment, field drainage, and agricultural research and development.

Furthermore, the techno-policy model was also guarded by the establishment of exclusive agricultural policy communities, e.g. by establishing or substantially expanding multi-stakeholder platforms that supported national agricultural policy-making processes (Perkins 1997; Agriculture Outlook 2000).

The public's endorsement of the techno-policy model of agricultural development can be attributed to:

1. hunger and malnutrition during the global economic recessions of the 1920s and 1930s and the Second World War;
2. the promise of regular supplies of a wide range of quality foods at reasonable prices; and
3. unquestioned acceptance of modern, science-based, technologies, and innovations.

Apart from a few, so-called laggards, the new approach to farming was slowly taken on board. Likewise, support from agribusinesses, such as plant and animal breeders, and machinery, fertiliser, and pesticide manufacturers, was also assured, as they stood to profit from a rapid expansion of input markets and the sale of advice (Dent 2000). Even food processors, wholesalers, and retailers stood to profit from the larger volumes of a more diverse range of food products (Watson 2018).

1.3.5 Applying the urban-centred industrial agro-food system model to developing countries

After more nearly two decades of food security and prosperity in the developed world, the USA was keen to extend the industrial agro-food system model to the developing world (Bowler 1992), especially in countries where communism was taking a foothold. In the words of President H. Truman, 'more than half the people of the world are living in conditions approaching misery. Their food is inadequate; their poverty is a handicap and a threat both to them and to more prosperous areas. The USA is pre-eminent amongst nations in the development of industrial and scientific techniques. Our imponderable resources in technological knowledge are constantly growing and are inexhaustible. I believe that we should make available to peace-loving peoples the benefits of our store of technical knowledge to help them realize their aspirations for a better life' (Perkins 1997).

With both food security and national security objectives in mind, this statement institutionalised political and economic commitment initially from the US Government, and major US philanthropic foundations (e.g. Rockefeller and Ford), and latterly most OECD member states, to the modernisation of food systems in developing countries by promoting a 'Green Revolution' (Perkins 1997; Krausmann and Langthaler 2019).

To this end, the Rockefeller and Ford Foundations established the Consultative Group on for International Agricultural Research (CGIAR) in 1972 (Perkins 1997). Together with the FAO, the CGIAR was charged with transferring the industrial agro-food system model to the developing world. Much of the initial CGIAR research focussed on crop breeding, especially, wheat, rice, and maize, and crop agronomy (Benton et al. 2021).

Realising the financial opportunities, multinational corporations quickly jumped onto the Green Revolution bandwagon (Perkins 1997). Companies such as Coca-Cola, Del Monte, Heinz, Kellogg's, Nabisco, Pepsi, and Unilever quickly leveraged connections with former colonies to draw developing countries into the global agro-food system (Atkins and Bowler 2001). However, contrary to most developed countries, where agriculture was seen as a means to national food security or for strengthening economic and social stability, agriculture in developing countries was primarily seen as an engine of economic growth and taxed to provide rents and revenues for urban and industrial development (E-council 2003). Indeed, during the 1960s, 1970s, and 1980s, agriculture in developing countries was taxed an average of 30%. Indeed, when overvalued exchange rates were factored in, the combined direct and indirect tax on agriculture could be as high as 75% (E-council 2003). This necessitated developing country governments to subsidise both agricultural credit and agricultural inputs, especially fertilisers, irrigation, and pesticides, and develop national agricultural development plans.

1.4 Neo-productivism: the Third Food Regime?

Academics suggest that the Second Food (or Fordist) Regime of accumulation slowly collapsed during the early 1970s (Ward 1993; Levidow and Bijman 2002; Friedmann 2005; McMichael 2009b; González Esteban 2017). To put it another way, the unwritten compact between nation-states, businesses, and workers/consumers had been fractured.

It is suggested that, due to a slowdown in productivity and reduced profitability in most sectors of the economy, the Fordist Regime slipped into crisis in the late 1960s in the USA and early 1970s in Western Europe (Atkins and Bowler 2001). Whilst there is agreement amongst academics that the Fordist Regime had fallen into crisis by the mid-1970s, there remains plenty of disagreement regarding what new organisational paradigm replaced it.

In what Watson (2018) calls the Divergent Period (1974 onwards), claims are made that several competing organisational paradigms emerged, and that, just like the First Food Regime, elements of the Second Food Regime and productivism remained. For example, in what he terms as Residual productivism, Watson (2018), in agreement with (Pritchard 2009), claims that the Second Food Regime did not actually collapse but managed to adjust to the new global realities.

Alongside the remnants of a weakened Second Food Regime, Watson (2018) argues for the emergence of three competing paradigms:

- Neo-productivism;
- Post-productivism; and
- Sustainability.

Aligned with theorisations developed by McMichael (2005, 2016) on a neo-Fordist or a Corporate Food Regime, and Goodman and Watts (1997) on the emergence of the Bioeconomy Economy and the Financialisation of economies, Watson (2018) argues for the emergence of neo-productivism, which refers to a deepening of corporate control (appropriationism, substitutionism, market liberalisation, globalisation, and financialisation) of the global food system.

Aligned with theorisations developed by Friedmann (2005) on post-Fordism in the form of a Corporate-Environmental Food Regime, post-productivism refers to attempts to commoditise nature, landscapes, and ecosystem services, etc. For example, this includes payments to farmers for producing or conserving habitats and biodiversity, or payments for carbon sequestration in the form of reafforestation or efforts to increase soil organic carbon. Lastly, the sustainability paradigm refers to the broad array of sustainability transitions within food systems.

In food systems literature, the crisis of Fordism and the Second Food Regime was referred to as the international farm crisis. This crisis was attributed to two key events. First, the sudden and sharp increase in oil prices in 1973, which arose due to a strategic tightening of oil supplies by the Organisation of Petroleum Exporting Countries (OPEC). Given that the global agro-food system was dependent on cheap oil, the actions of OPEC quickly led to increased agricultural input prices, particularly fuel, fertiliser, pesticides, and machinery, as well as increasing food transportation and processing costs further down the food chain (Watson 2018). The second world event that shook the global agro-food system was the virtual elimination of surplus international wheat stocks, due to US–USSR grain deals (which resulted in the USSR becoming a major importer of wheat). When combined, these two factors plunged the global economy into recession, destroyed the Bretton Woods System of exchange rate management, and caused a significant spike in world grain prices, spreading real concern over international and domestic food security (Watson 2018).

If the Third Food Regime has in fact emerged, the new regime could only be defined as highly differentiated and full of contradictions and tensions (Watson 2018). Indeed, Peck and Tickell (2000) suggest that, amidst significant experimentation between competing organisational models, it is way too early to proclaim the arrival of a stable post-Fordist regime of accumulation. According to Watson (2018), 'what seems increasingly likely is that, throughout the last 50 years, agrofood capitals have explored a range of strategies that sought to maximise accumulation over heterogeneous social, economic, and political landscapes.' That said, of all the competing global, regional, national, and sub-national organisational models that have been subject to experimentation over the past 50 years, the so-called Neo-Productivist, Corporate Food Regime, or Food from Nowhere Regime (described more fully in Chapter 4), which is part of the broader Neo-Fordist/Neo-Liberal Regime, has by far gained the most traction and has become the most institutionally embedded.

In summary, the 'Corporate Food Regime' or 'Food from Nowhere Regime' describes a range of key social, political, economic, organisational, and technological characteristics that underpin a 'new and stable regime of accumulation and mode of social regulation' McMichael (2009a,b). The concept encompasses the processes of agro-industrialisation, sustainable intensification of production, international sourcing of raw commodities, and international trading of processed foods (McMichael 2016). Whilst most of the key defining characteristics of neo-Fordism were identical to the previous Fordist Regime, two new and critical differences allegedly emerged:

1 Restructuring of the labour process; and
2 Reduced agency of nation states.

These differences are discussed in more detail here.

Restructuring of the labour process centres on what has been termed as 'lean production', 'Toyotism', or the 'Just-in-Time' (JIT) system, which eliminates anything in the production process that fails to contribute to added value. Namely, this entails cracking down on over-production, time delays, multiple handling, unnecessary steps or work/procedures, unnecessary transportation, excess inventory (inputs and outputs), over-processing, production of a defective part, or something that needs reworking, and by implementation of the concepts of continuous flow and consumer pull (Watson 2018). Neo-Fordist principles have also led to the reduction of the unit costs of production. To a large extent, this has been achieved by dismantling the labour process, primarily by moving parts of the production process to developing countries, both in manufacturing and agriculture. Indeed, this re-location of production has been a key to the rapid rise in international trade in manufactured goods (or parts).

As noted earlier, multinational agri-food corporations played a key role in linking lucrative food markets in developed countries with low-cost production and processing platforms in developing countries. This reflected a wider shift which has come to be known as 'globalisation'. The term describes the growing interdependence of national economies brought about by increasing cross-border flows of raw materials, labour, investment, information, goods, and services. This has shifted the balance of power from national governments to multinational corporations able to source raw materials and production from almost anywhere in the world, potentially playing one country off against another in getting favourable terms for investing in a particular location. Currently, around 80% of world trade now comes from supply chains dominated by multinational corporations, giving them significant bargaining power and influence in trade flows. This reduced role of nation states has been reinforced by the neoliberal shift to deregulation by many governments in favour of market liberalisation, strengthening the role of key market actors such as corporations.

According to McMichael (2009a), the Corporate Food Regime created a situation where staple grains from developed countries are traded for high-value agricultural products such as fruits and vegetables from developing countries. Figure 4 illustrates the growth in international food trade.

McMichael (2013) argues that under the emergent Corporate Food Regime, agri-food corporations were able to orchestrate the rapid expansion of high-value horticultural and agricultural value chains, linking lucrative food markets in developed countries with low-cost production and processing platforms in the developing countries. McMichael (2013) also argues that this new regime orchestrated the rapid expansion of land grabbing (by both corporations and governments) to secure bulk staples for human food, animal feed, or biofuel.

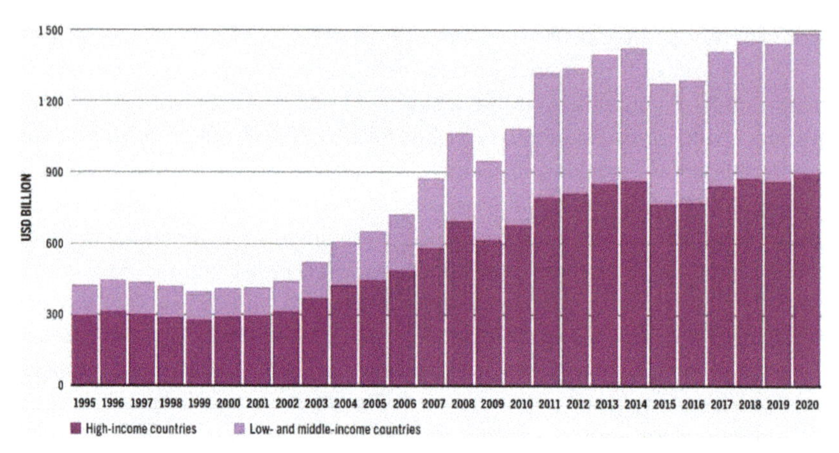

Figure 4 Increased international trade in food 1995 to 2020. Source: Adapted from: FAO (2022).

In attempts to rekindle economic growth during the late 1970s, nation states increasingly deregulated markets through the adoption of free market neoliberal policies. Two countries, the USA and UK, were pivotal in leaving behind the Keynesian policies of the post-war period and embracing Milton Friedman's supply-side monetarist policies. The political hegemony of Margaret Thatcher (UK prime minister 1979–90) and Ronald Reagan (US president 1981–89), backed by New Right economic strategists, became the vanguard through which laissez-faire thinking re-entered the realm of development thinking (Watson 2018). Galvanised by the debt crisis of 1982, the Thatcher and Reagan administrations aligned behind trade liberalisation strategies, which were a cornerstone of the IMF's and World Bank's Structural Adjustment Programmes. These Programmes directly linked moves towards deregulation and market liberalisation with loan disbursements. This linkage became known as the Washington consensus (Benton and Bailey 2019). The New Right theorised that liberalised markets and individual choice would efficiently distribute resources and wealth.

By the late 1980s, neoliberal economic doctrine and policies had been adopted across much of the world and underpinned both the theory and practices of the WTO, the IMF, and the World Bank (Watson 2018). According to the World Bank (2000), with over 122 member states, the WTO has grown to be the most powerful institution in the world. When WTO trade liberalisation talks stalled in Cancun, the corporate and US-backed Transatlantic Trade, and Investment Partnership (TTIP) picked up the trade liberalisation gauntlet, with the aim to dismantle residual market protection, increasingly protect investors, and restrict use of ad hoc public policy attempts to manage food systems (Hansen-Kuhn and Suppan 2013). TTIP's ambitions include relaxation of the EU's Precautionary Principle with regard to pesticides, food packaging and additives, and use of nanotechnologies, biotechnologies (GMOs and gene editing) and removal of 'localization barriers to trade' (Hansen-Kuhn and Suppan 2013). Under neoliberalism, many governments cut social services and rescinded regulations that protected labour, consumers, and the environment.

1.4.1 Neo-productivism/the Second Food Regime and food production

Since the advent of the Second Food Regime, global food production has increased almost exponentially (Benton et al. 2021; OECD 2021). Significant increases in productivity in the developed world have seen output not only keep pace with rapid population growth but surpass it, leading to the so-called grain mountains and milk lakes of the 1980s and 1990s (Benton et al. 2021). Figure 5 illustrates the relative changes in agricultural production and agricultural land against population growth. Figure 5 shows that agricultural production has surpassed population growth, even though population growth rates have soared since the 1960s. And, whilst there has been a modest increase in agricultural land of approximately 10–15% (OECD 2021), most of the increase in food production has arisen through increased productivity of existing agricultural areas due to the adoption of industrial agro-food systems (Benton et al. 2021; OECD 2021).

Global food surpluses and national food security, at least for the majority, were primarily achieved through increased farmer productivity, both in the developed and developing world (Benton et al. 2021). For example, in the 1940s, US farmers produced enough food to feed themselves and nine others. By the 1970s, US farmers produced enough to feed themselves and 31 others (Green 1976).

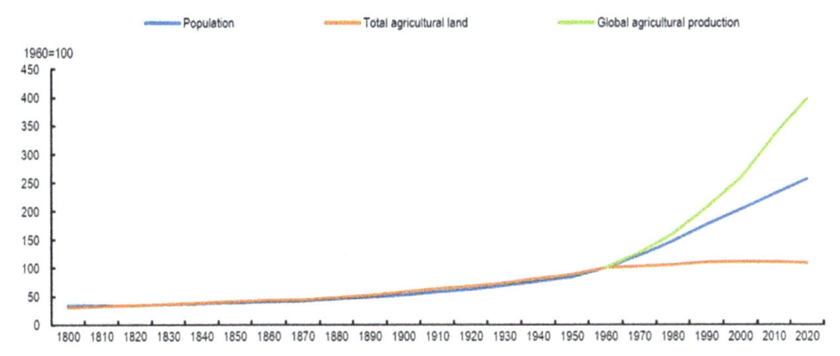

Figure 5 Global food production, agricultural land expansion, and population growth. Source: Adapted from: OECD (2021).

Since the 1960s, increased productivity has been achieved primarily through the development of high-yielding crop varieties, application of synthetic fertilisers and pesticides, mechanisation, and improved irrigation (Benton et al. 2021). During this period, the area of irrigated cropland increased from 12% to 21%, and blue water (fresh surface and groundwater) use increased from 1500 km^3 to 2800 km^3 per year (OECD 2021).

During the 1960s, 1970s, and 1980s, annual gains in cereal yields were around 1.7% per annum (Krausmann and Langthaler 2019). Similarly, Britton (1990) estimated annual yield increases in the UK during the 1950s and 1960s in the following crops: oats (2.3%), wheat (2.2%), barley (1.9%), potatoes (1.7%), and sugar beet (1.6%). Whilst global population grew by 142% (1961–2016), cereal yields grew between 193% (Benton and Bailey 2019) and 300% (Forster et al. 2013). According to the (OECD 2021), between 1930 and 2011, improved varieties accounted for '59–79% of the seven-fold increase in US maize yields.'

During the 1970s, 1980s, and 1990s, the Green Revolution increased wheat, rice, and maize yields in South Asia and Latin America between 100% and 200%, leading to increased net calorie availability and poverty alleviation (IPES-Food 2016). Between 1981 and 2000, improved crop varieties accounted for 40% of this growth (OECD 2021). By the 1990s, 70% of wheat and rice varieties were modern and high-yielding (IPES-Food 2016). In the case of countries such as India, this not only helped achieve national food security and provided surplus production for exports but was also the principal driver in India's transition to a modern market economy (Kesavan and Swaminathan 2018).

Investments in yield-increasing crop breeding have been targeted at food staples such as wheat, rice, maize, and cassava, major oil crops, and high-value fruits and vegetables. Figure 6 illustrates the yield growth of major crop types since the 1960s. Aside from pulses (such as peas and beans), all major crop types have witnessed a significant increase in yields since the 1960s.

Similar yield growth can also be seen in livestock and livestock products. Figure 7 illustrates the rapid growth in milk yields, from around 2 tons per year to over 9 tons per year, in both Denmark and the USA between 1900 and 2020. In the case of the US dairy production, compared to 1944, improved genetics, feed, housing, and care have made it possible to produce the same amount of milk with only 21% of the number of cows, 23% of the feed, 35% of the water, 10% of the land, 24% of the manure output, 43% of the methane emissions, and 56% of the nitrous oxide emissions (OECD 2021). The carbon footprint/kg of US milk produced in 2007 was just 37% of that produced in 1944 (OECD 2021).

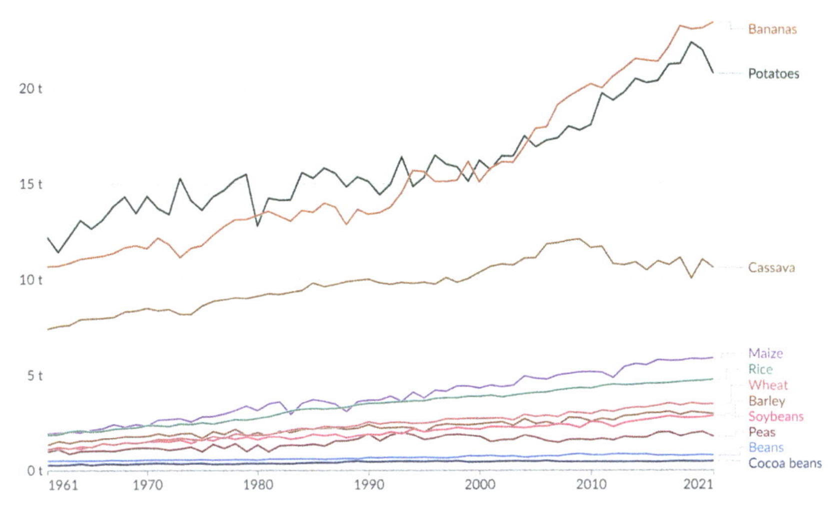

Figure 6 Growth in yields of major crops 1961 to 2021 (NB yield is measured in tonnes per hectare, based on FAO data). Source: Adapted from: Our World in Data (2022).

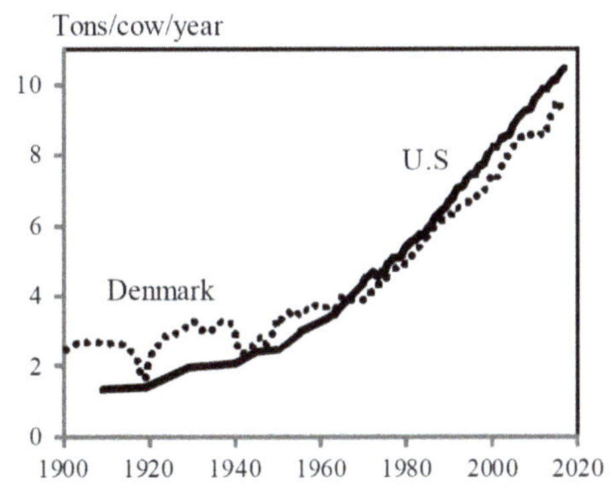

Figure 7 Increasing milk yields in Denmark and the USA between 1900 and 2020. Source: Adapted from: Hansen (2019).

1.4.1.1 Fertilisers

The use of synthetic fertilisers was also central to the significant increases in farmer productivity observed during the Second Food Regime (Hayes 1991). Of the key nutrients supplied through synthetic fertilisers, nitrogen, potassium, and phosphorus were the most important. Of these, nitrogen, which traditionally came from leguminous crops and animal manure, has probably been the most pivotal nutrient in attaining advances in crop yields. Synthetic nitrogen fertiliser, which could now be fixed from the atmosphere using the Haber process, meant that, if farmers had money in their pockets, nitrogen need no longer limit yield.

Indeed, synthetic fertilisers were so cheap and accessible that many farmers often over-applied them to realise potential yields. Since 1961, the global consumption of nitrogen-based fertilisers increased by 900%.

Phosphate and potassium fertilisers increased by 400% (OECD 2021). Current nitrogen fertiliser applications range between farms and farming regions. Rates remain extremely high in parts of China, India, and other East Asian countries at more than 170 kg/ha. In other parts of the world, due to the increasing cost of fertilisers, nitrogen applications have reduced to more economically sound levels (150 kg/ha) in Europe and remain around 73 kg/ha in North America (OECD 2021). Figure 8 illustrates the growth in fertiliser use per hectare from just over 40 kg/ha in the 1960s up to just under 140 kg/ha in 2021.

1.4.1.2 Pesticides

During the Second Food Regime, farmers quadrupled pesticide application from 8.5 million tons in 1940 to 37 million tons in 1968 (Pimentel and Pimentel 1979; Hayes 1991). Globally, production of pesticides expanded from approximately 100 000 tonnes in 1944 to 450 000 tonnes in 1955, 1 000 000 tonnes in 1965, and 1 500 000 tonnes in 1970 (WHO 1990).

By the 1950s and 1960s, pesticide use had become routine in most developed countries. On average, farmers used 1.5 kg/ha of pesticides in the USA, 1.9 kg/ha in Europe, and 10.8 kg/ha in Japan. In Japan, the production and use of pesticides doubled between 1965 and 1970 and again between

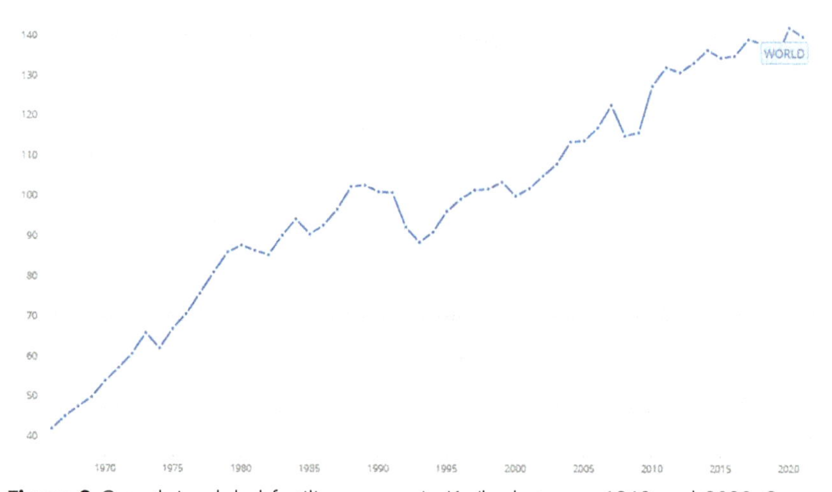

Figure 8 Growth in global fertiliser usage in Kg/ha between 1960 and 2020. Source: Adapted from: The World Bank: https://data.worldbank.org/indicator/AG.CON.FERT.ZS; published under a CC BY Licence.

1972 and 1974 (Boardman 1986). Between 1962 and 1976, synthetic pesticide production in the USA increased 124% from 730 million pounds to 1630 million pounds, and herbicide sales grew by 487% (Pimentel and Pimentel 1979). According to the OECD (2021), 'between 1990 and 2015, global pesticide use increased from 2 700 000 tonnes of active ingredients to 4 000 000 tonnes.' Pesticide use in the UK was still increasing until 2013. In the year 2000, pesticides were being applied to 59 100 000 ha of land. This increased to 78 200 000 ha in 2013, mainly through increased numbers of applications per hectare rather than pesticides being applied to more land (Williamson 2019). Since 2015, pesticide use has slowly begun to decrease. Figure 9 illustrates the global growth in pesticide use between 1990 and 2020.

1.4.1.3 Cheap food

During the time of the Second Food Regime, productivity growth in agriculture has made food not only more abundant but also cheaper. In real terms, the overall price of food has fallen by 37% (Benton and Bailey 2019). In the case of the major world staple foods (rice, maize, and wheat), prices have fallen by 60% between 1960 and 2000 (FAO 2020). Figure 10 illustrates the relationship between increases in wheat yields over time and the relative fall (adjusted for inflation) in the price of wheat.

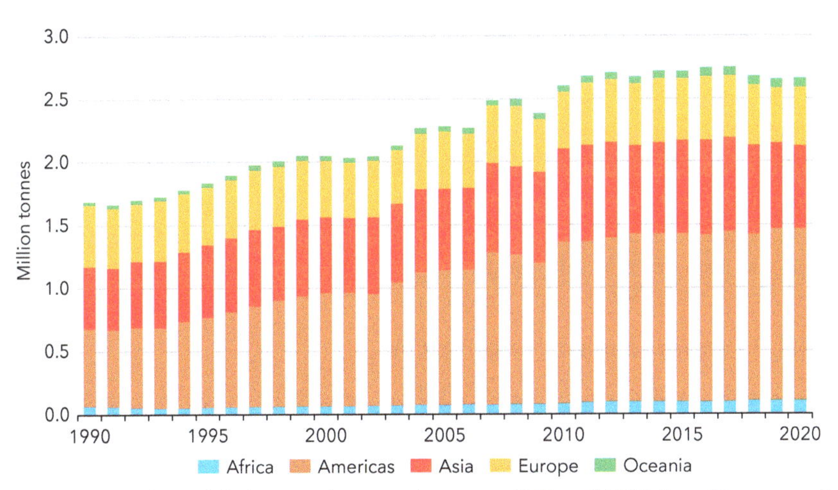

Figure 9 Growth in global pesticide usage between 1990 and 2020 (in million tonnes). Source: Adapted from: FAO – http://www.fao.org/faostat/en/#data/RP).

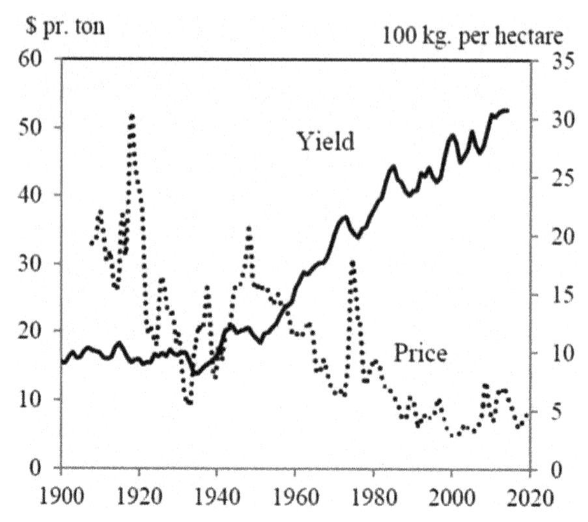

Figure 10 Growth in wheat yields versus price between 1900 and 2020. Source: Adapted from: FAO (2020).

1.4.1.4 International food trade

During the Second Food Regime, the central logic of global food trade policy has been that trade will lower food prices, increase food availability and food choice, and generally increase global food security (Swagemakers et al. 2019; Benton and Bailey 2019). Driven by cheap fossil fuels and surplus food production, agricultural exports grew substantially during the Second Food Regime. The USA increasingly dominated net exports of agricultural commodities, expanding from 4% of global production in the pre-war years to 12% of global production by the mid-1980s. In terms of volumes, US agricultural exports expanded from 20 000 000 tonnes per year in the pre-war years to 140 000 000 tonnes per year by the mid-1980s. Indeed, total global per capita exports of cereals doubled from below 20 kg in the pre-war years to 40 kg per year in 1973 and 50 kg per year in the late 1970s (Krausmann and Langthaler 2019). Figure 11 illustrates the growth of cereal exports from the 1850s to 2010, mapped against fossil-fuel-based energy consumption.

1.4.1.5 Increased meat consumption

The production of cheap commodities in volume has been central to increasing global meat consumption, allowing many more global citizens to enjoy both red and white meats. Figure 12 illustrates the rapid growth in meat consumption between 1961 and 2022. This rapid growth has only been possible due the relatively cheap price of maize, wheat, and soybean as sources of animal feed. Indeed, some have coined this process as the 'meatification' of Western diets,

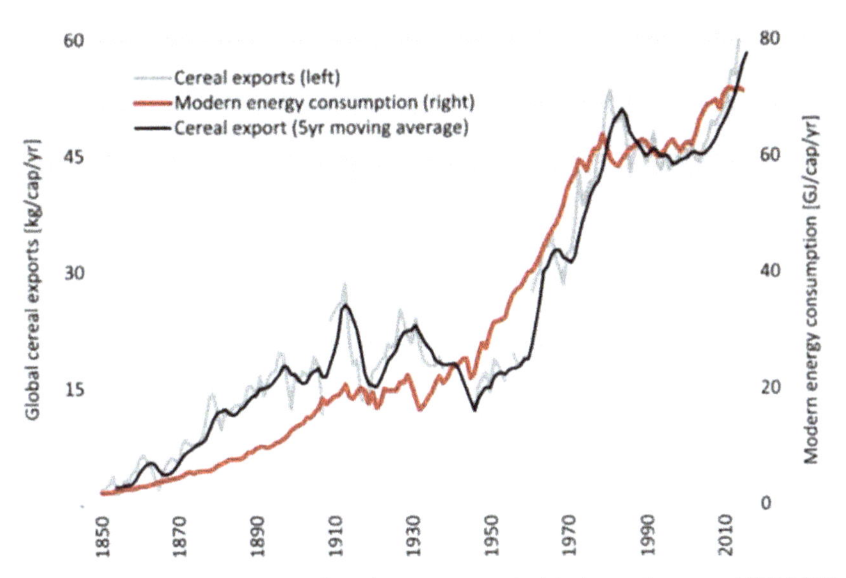

Figure 11 Global consumption of modern energy and global cereal exports 1850-2015. Source: Adapted from: Krausmann and Langthaler (2019).

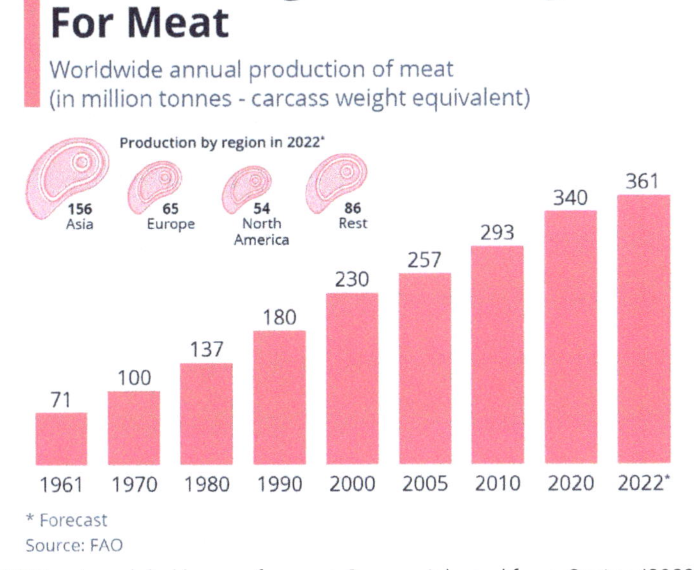

Figure 12 Growing global hunger for meat. Source: Adapted from: Statista (2023a).

i.e. a massive increase in the per capita consumption of meat and other animal products in industrialised countries after the Second World War (Krausmann and Langthaler 2019). Indeed, given the reliance on cheap and nutritious animal feeds, especially soybean-derived oil cakes, and the diversion of soybean oil into human consumption, the world also witnessed an associated 'oilification' of Western diets (Krausmann and Langthaler 2019).

By the end of the productivist period, the Fordist Food Regime had institutionalised regular aggregate global food surpluses and assured national food security in most countries, even though access to regular and sufficient quantities of nutritionally adequate food was not always assured in some developing countries (Watson 2018).

Chapter 2

Trouble at t'mill

Despite the phenomenal success of the Second and (so-called) Third Food Regime in delivering cheap and plentiful food, the global food system faces a growing list of problems. The title 'Trouble at t'mill' refers to a colloquial phrase, which is still used in my home county of Yorkshire, England, to infer 'a problem'. The phrase arose during the 1800s due to running disputes between the industrial mill workers and mill owners, hence, trouble at the mill (shortened to trouble at t' Mill).

Depending on one's understanding of the problems and their potential solutions, food system academics and practitioners rationalise these problems in one of three ways:

- First, there are those who deem food system problems as normal, ongoing challenges that can be fixed by a combination of policies, technologies, or finance (Béné et al. 2019b).
- Second, there are those who are convinced that the global food system is in crisis, has failed, and needs to adopt radical solutions to fix the problems (De Schutter 2014; FAO 2016a; Haddad and Hawkes 2016; UNEP 2016).
- Third, there are those who are convinced that the global food system, and the overarching capitalist world, have already passed planetary limits and must now face a terminal or near-terminal crisis before being replaced (if at all) by something completely new (Ingram 2011; Rockström et al. 2009a,b; Pereira et al. 2018).

Whilst some may find valuable insights in all three perspectives, many academics and practitioners tend to favour one rationalisation above the others, often resulting in quite polarised and entrenched opinions. Chapter 2 focusses on the problems and concerns that have been raised around industrial food systems. The principal problems associated with the global food system are outlined below.

2.1 Natural resource exploitation and pollution

There are 16 mineral nutrients that are essential for healthy crop growth. Amongst these, nitrogen, phosphorus, and potassium are the major nutrients. In industrial food systems, phosphorus and potassium are mined (UNEP 2016). Nitrogen, on the other hand, is either fixed from the atmosphere (i.e. Haber process) or by nitrogen-fixing nodules on the roots of leguminous plants (biological nitrogen fixation-BNF) (UNEP 2016).

2.1.1 Nitrogen and phosphorus fertilisers

Nitrogen remains an essential, and often limiting, input into agricultural production. However, excess nitrogen, unused by growing crops, and released into the atmosphere and hydrological systems, can cause significant environmental pollution. Since the 1960s, the global consumption of nitrogen fertilisers increased almost nine-fold (Holden et al. 2018). Figure 13 illustrates the current global consumption of both nitrogen and phosphate fertilisers and the expected range of growth in utilisation up to 2050.

One of the challenges of nutrient management, especially nitrogen and phosphorus, is the potentially high rate of loss to the environment (Amate and de Molina 2013). In some cases, only 15–20% of the nitrogen and phosphorus fertiliser applied to crops is eventually consumed by consumers, with the remaining 80–85% lost to the environment (UNEP 2016). Average losses of nitrogen, for example, are around 50% (Zhang et al. 2021).

The loss of both nitrogen and phosphates to the environment occurs via a range of modalities. For example, nitrogen can be lost directly to the atmosphere (as ammonia gas) through a process known as volatilisation (UNEP 2016), and into groundwater, rivers, lakes, and the ocean via water-based nitrate leaching (UNEP 2016). Intensive livestock production and nitrogen fertiliser application are key contributors to ammonia emissions (Frison and Clément 2020). High levels of ammonia in the atmosphere have been associated with respiratory illnesses, heart disease, and lung cancer (Lelieveld et al. 2015). Phosphates are also lost via leaching but can also be lost via direct wind erosion of topsoil (Sutton et al. 2013; Steffen et al. 2015a).

It must be noted that fertiliser leaching and run-off are spatially distributed. In developed countries, significant nitrate and phosphate pollution from intensive/large-scale feedlots can occur if contaminated water runoff is poorly

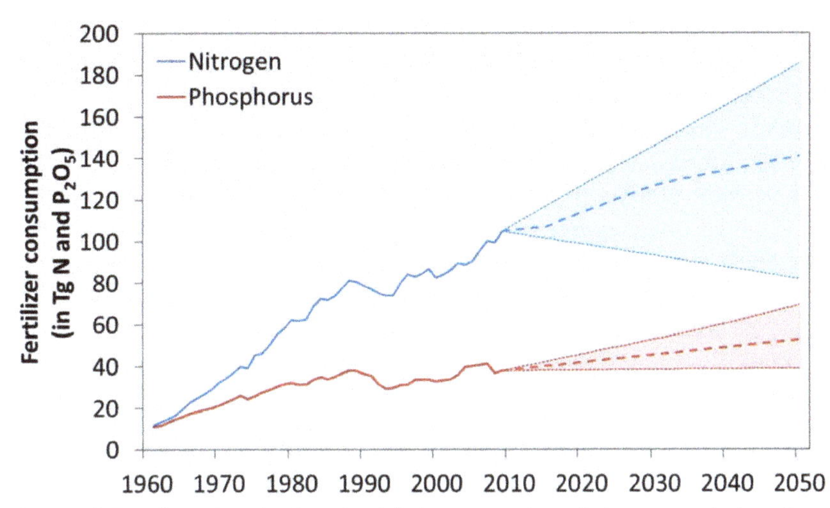

Figure 13 Trends and projections in global consumption of nitrogen and phosphorus fertiliser. Source: Adapted from: Sutton et al. (2013).

managed. In addition to nitrates and phosphates, runoff can contain copper, zinc, arsenic, as well as disease-carrying pathogens such as *Escherichia coli* and other zoonotic diseases (Anderson and Sobsey 2006; Liverani et al. 2019). In many developing countries, nitrate and phosphate run-off occurs due to the over-application of fertilisers. For example, it is estimated that nitrogen leaching and runoff in Asia is as high as 15 million tonnes per year or, to put this into context, 64% of the global total (Liu et al. 2010). Conversely, low-input/low-output traditional food systems in sub-Saharan Africa (SSA) apply, on average, just 8 kg of fertiliser per hectare (UNEP 2016). With such limited inputs of key fertilisers, harvests often remove more nutrients in the grain and stover (fodder for livestock) than have been applied - thus constantly mining nutrients from the soil (OECD 2021).

2.1.2 Eutrophication

Since the 1970s, the loss of nitrates and phosphates has had both a direct cost to the farmer and an impact on the environment and human health (EC 2002; EEA 2003). One of the most significant impacts of nitrate and phosphate leaching is the eutrophication and acidification (nitrogen) of rivers, lakes, sea, and oceans (Seitzinger et al. 2010; Sutton et al. 2013; Carpenter 2005; Moss 2008; Camargo and Alonso 2006; Conijn et al. 2018). For example, the eutrophication of the Baltic Sea by agricultural fertiliser leaching and run-off has been especially problematic (Wezel et al. 2018b). Other examples include run-off from the Mississippi River into the Gulf of Mexico (Pimentel et al. 2005; Holt-Giménez 2019). Indeed, phosphorus run-off into Lake Erie (USA) led to the closure of public water supplies (Chung 2014).

Whilst small amounts of nitrates and phosphates can enhance freshwater and marine ecosystems, eutrophication can lead to problematic algal and phytoplankton blooms (Heisler et al. 2008). In turn, these blooms lead to high levels of decomposing organic matter that remove significant amounts of dissolved oxygen from water bodies. This then leads to hypoxia (low levels of oxygen) which, in turn, causes the death of fish and invertebrates and can create lifeless/dead zones of water (OECD 2021). Nitrates and nitrites (a step in the breakdown of nitrogen fertilisers) also impact on human health. Excessive concentrations of nitrates and nitrites need to be removed from drinking water supplies (Sutton et al. 2013; Withers et al. 2014). Studies in both Europe and North America have linked nitrite and nitrate contamination in drinking water to various cancers and non-Hodgkin's lymphoma (Dubrovsky et al. 2010; Lawniczak et al. 2016; Munné et al. 2015; Ward et al. 2018; Wheeler et al. 2015). Nutrients are also lost from the food system through food processing and human sewage. It is estimated that global annual losses of nitrogen and phosphorus, which were 6.4 and 1.3 Tg (teragram) respectively in 2000, could be as high as 12–16 and 2.3–3.1 Tg by 2050 (Van Drecht et al. 2009; Bouwman et al. 2013). This rapid increase is linked to the over-application of synthetic fertilisers, concentrated application of manure, and increased urban emissions (Seitzinger et al. 2010; OECD 2018; Carpenter 2005; Moss 2008).

2.1.3 Energy-intensive production of nitrogen

Both direct and indirect energy inputs into crop and livestock production account for over 34.1% of overall energy consumption. Of this, nitrogen fertiliser, animal feed, and fuel (mainly diesel) require over 85% of the 480 PJ of primary energy used in agricultural production (Amate and de Molina 2013). According to the ETC Group (2017), the current energy required to produce nitrogen fertiliser is equal to 3–5% of the world's annual natural gas. To put it another way, for every hectare of fertilised land, 62 L of fossil-fuel are required to produce and distribute nitrogen fertilisers. To put it even another way, 50% of the energy used to grow a crop of wheat is used to manufacture fertilisers and pesticides (ETC Group 2017).

The demand for fossil-fuel-powered nitrogen is significant, both in the developed and developing world. For example, in developed countries, nitrogen accounts for between 32% (Jasinski et al. 1999) and 40% of the total energy required for agricultural production. In developing countries, this can be as high as 70% (Amate and de Molina 2013).

2.1.4 Depletion of phosphorus and potassium reserves

Agriculture is the world's principal user of mined phosphorus and potassium (UNEP 2016). Based on current consumption, supply from known phosphate and potassium reserves will be able to meet demand for between 50 and 500 years. Unknown reserves could extend the supply of these key fertilisers (UNEP 2016). However, depletion of these non-renewable sources of fertiliser is of growing concern. Holden et al. (2018) provides one of the most pessimistic accounts of phosphate supply, insisting that the world peak for phosphate fertilisers (namely, when global phosphate supply begins to fall short of global supply) could be reached as early as 2027 (Mohr and Evans 2013) or 2033 (Cordell et al. 2009). Others suggest that, combining both known supplies and estimations of unknown reserves, it is unlikely that the world will face shortages of either phosphorus or potassium this century (IPNI 2008; Heckenmüller et al. 2014). Currently, in most modern and transitional economies, minerals flow from rural to urban areas, with as little as 4% being recycled back to rural areas as fertilisers (Morée et al. 2013). Unless something is done to address this, the situation will probably become worse over time (Bouwman et al. 2013; Neset and Cordell 2012; Sutton et al. 2013). According to the FAO (2012), total fertiliser consumption of nitrogen, phosphorus, and potassium could increase from 166 million tonnes in 2006 to 263 million tonnes in 2050. Fertiliser consumption is expected to increase in SSA, South Asia, and Latin America (UNEP 2016). There is an increasing need to stimulate the recycling of phosphorus and other key nutrients from wastes and human excreta (Conijn et al. 2018).

Additionally, many observers are increasingly concerned about the growing stranglehold that fertiliser companies have over food systems (Holden et al., 2018; ETC Group 2019b). For example, according to the ETC Group (2019b), 25% of rock phosphate is currently mined and distributed from just three countries. Over 50% of the world's production of ammonia (nitrogen), phosphate, and potash (potassium) fertiliser comes from Canada, China, the USA, India, and Russia (Hernandez and Torero 2013; IPES-Food 2017b). In the future, these countries could significantly increase the cost of phosphates, via supply restriction, potentially pushing fertiliser out of reach for many smallholder farmers. In line with other corporate sectors of the food system, the global fertiliser industry is constantly consolidating, passing control into fewer and fewer hands (ETC Group 2019b). For example, with the (current) exception of China, just four fertiliser companies control more than 50% of fertiliser production and distribution. In the case of North America, just three companies (PotashCorp, Mosaic, and Agrium) control the lion's share of potash sales (IPES-Food 2017b).

2.1.5 Pesticides

2.1.5.1 Human health effects of pesticides

Concerns over the human health impacts of pesticides have existed since the development of early synthetic pesticides. Many of these early pesticides were highly toxic by-products of industrialisation. For example, sold as the pesticide Paris Green, the toxic dye copper acetoarsenite was developed in the mid-1800s (Watson 2018). Whilst the dangers of this arsenic-based pesticide were well known, it became widely used in the late 1800s due to its cost-effectiveness in controlling key agricultural pests (Holmes 2001). In the USA alone, Paris Green was used as the first line of defence against the economically important Colorado beetle on potatoes, gypsy moth on fruit trees (Boardman 1986; Hough 1998), and codling moth (Bailey 1909). Introduced in 1860, the Colorado beetle spread rapidly across the USA and caused significant economic damage to potato crops. Just to put Colorado beetle and Paris Green into perspective, by the late 1800s, total economic losses caused by insect pests in the USA were US$700 000 000. This equated to more than the entire expenditure of the US Government (Bailey 1909). By the late 1800s, sales of Paris Green quickly rose to £2000 per year, and by 1910, combined sales of lead arsenate and Paris Green were £10 000 000 per year (Bailey 1909). Concerns over the human health implications of pesticides resulted in policy interventions. These were the Federal Food and Drugs Act (1906), creating the Food and Drug Administration (FDA), and The Federal Insecticide Act (1910) (Watson 2018). However, it must be noted that early policy interventions aimed to control the misuse and adulteration of pesticides. In general, as long as they were used properly, pesticides themselves were deemed safe (Watson 2018).

Whilst concerns over the use of pesticides are not a recent phenomenon (see above), they increased significantly throughout the 1950s, 1960s, and early 1970s across the most intensive food systems, notably the USA, Western Europe, and Japan (Watson 2018). It was during this time that the focus of concern turned to the pesticides themselves and not the way they were misused or adulterated. Initially, attention was focussed on specific groups of pesticides, such as organochlorines, organophosphates, and carbamates, which were the principal groups of pesticides (PANNA 1995).

Originally formulated in 1874, it was not until 1938 that the Geigy Chemical Company recognised the effectiveness of Dichlorodiphenyltrichloroethane (DDT) as an insecticide (Hoechst 2003). DDT was the first, and most infamous, organochlorine pesticide. Indeed, it was the focus of Rachel Carson's seminal work 'Silent Spring' in 1963, in which she exposed its highly bioaccumulative nature in ecological food chains (BCERF 1998). Later work also indicated a probable link between organochlorine insecticides and cancer (Tarjan and

Kemeny 1969) and type-2 diabetes (Arrebola et al. 2013), but it has been almost impossible to definitively prove the link to cancer (Efron 1985; Kolata 1997).

DDT was initially developed as an insecticide for use on malaria and yellow fever-carrying mosquitoes and insect vectors linked to typhus and the plague (Dunlap 1981). It was used extensively during the Second World War to control malaria amongst US troops serving in tropical war zones. It was such a scientific breakthrough that Dr Muller, the scientist who recognised the pesticidal nature of DDT, was awarded the Nobel Prize in Medicine and Physiology (Watson 2018). By the end of the War, 140 000 tonnes/year of DDT were being produced annually. After the War, stockpiles of DDT were used both in food systems, as a cheap and highly effective insecticide, and in the worldwide battle against malaria. Indeed, DDT has been acclaimed as saving more lives than penicillin (Boardman 1986). DDT's role in the World Health Organisation's (WHO) Global Malaria Eradication Programme was truly remarkable (Hough 1998). According to Pampana and Russell (1955), malaria kills 2.5 million people every year. The application of DDT virtually eliminated malaria wherever it was used. For example, in Sardinia, DDT single-handedly reduced malaria cases from 78 000 cases per year in 1942 to nine cases in 1951 (McEwen and Stephenson 1979). In Sri Lanka, DDT was also responsible for reducing malaria from 2 800 000 cases and 7300 deaths per year in the 1940s to 17 cases and no deaths in the 1990s (Hicks 1992).

Ultimately, growing public concerns surrounding organochlorine insecticides and constant lobbying by anti-pesticide groups led to the banning in developed countries of most uses for a range of organochlorine insecticides. Six out of the ten best-selling pesticides in the USA in 1968 are now banned, including DDT (Phillips McDougall 2018). The organochlorine insecticides Heptachlor and Chlordane were both banned in the USA in 1974 due to their probable links to cancer (EPA 1975).

Unlike organochlorines, organophosphate (OP) pesticides are based on nerve gases (cholinesterase inhibitors) developed by the military in preparation for the Second World War during the 1930s and 1940s (Berenbaum 1995). Whilst the initial OPs were highly toxic to both people and animals, later formulations, such as tetraethyl pyrophosphate (TEPP) released in 1946 and ethyl-parathion released in 1947, showed themselves to be cheap and effective insecticides with lower persistence in the environment and less risk to users. However, they were still deemed as toxic chemicals and could lead to irreversible OP poisoning. Agricultural workers applying the insecticide and non-target animals were the most at risk from OP poisoning. As time went by, OPs became linked to a range of medical problems, including acute poisoning (FAO 1997), respiratory illnesses (William et al. 1993), overt renal insufficiency with massive proteinuria (Albright et al. 1983), depression, anxiety, confusion, impaired concentration, fatigue, headaches, and leucopenia (Davignon et al. 1965). Organophosphate pesticides were estimated to generate health costs of US$121 000 000 000 per annum (Trasande et al. 2016). Organophosphates have been banned in most developed countries and many developing countries.

First discovered by Geigy Chemical Company scientists in 1947, carbamates were released as insecticides in the late 1950s. Carbamates act in a similar manner to OPs by inhibiting the cholinesterase enzyme, but, unlike OPs, their effect is reversible. Early carbamate insecticides include: Carbaryl (1959) (EPA 2003a); Methomyl (1968) (EPA 2003b), and Aldicarb - better known as Temik (1970). Some carbamates also have herbicidal and fungicidal properties. As with OPs, carbamates have also been linked to acute poisonings (Schonfield et al. 1995; Hicks 1992; Siedenburg 1991).

Today, pesticides remain a major threat to farmers and farmworkers (WHO 2010; (Mostafalou and Abdollahi 2017), especially in developing countries. Annually, there are an estimated 200 000 acute poisoning deaths due to pesticides, of which, 99% occur in developing countries (Elver 2017). Pesticide poisoning primarily occurs during pre-application of pesticides (splashes and spills), inappropriate handling and application, and poor disposal of pesticide containers, such as the reuse of pesticide containers for water or storage, and the dumping of empty pesticide containers near water courses (Eddleston et al. 2002; Konradsen et al. 2003). Part of the blame has been apportioned to pesticide manufactures, especially with regard to badly formulated products and poor labelling practices. In developing countries such as India, pesticide residues in drinking water have become a major health risk (Van Drecht et al. 2009). Concerns have also been raised regarding pesticide residues in food products, especially fruits and leafy vegetables (Aktar et al. 2009; Fountain and Wratten 2013; Kumar 2012). Indeed, according to (IDDRI 2018), pesticides, including organophosphates and organochlorines, have been detected in more than half of the food consumed in Europe.

2.1.5.2 Impact of pesticides on non-target species

Growing concern has also been expressed regarding the impact of pesticides, especially organochlorines, organophosphates, and carbamates on non-target species (Anderson et al. 1999; PAN 2002; EPA 2003a; Aktar et al. 2009; Fountain and Wratten 2013; Kumar 2012). The impact of pesticides on non-target species and ecosystems was formally recognised as early as 1966 by the International Union for Conservation of Nature and Natural Resources (IUCN) (Boardman 1986). Increasingly, risk assessments for highly toxic pesticides began to include non-target species. For example, the USA's Environmental Protection Agency found that the organochlorine Endosulfan posed 'risks for terrestrial and aquatic organisms and can cause reproductive and developmental effects in non-target animals, particularly birds, fish, and mammals' (EPA 2003c).

Many of the early insecticides were non-selective and killed many insects other than the target pests (Geiger et al. 2010; Benton et al. 2021). According to the ETC Group (2017), as little as 1-5% of pesticide acts on the target pest. Many early pesticides were persistent and remained active on the leaves, stems, and soil surface, sometimes for many years. Pesticides also act on soil microbiota (Duah-Yentumi and Johnson 1986; Hartmann et al. 2015; Komorowicz et al. 2010) and can reduce soil fertility and plant growth (Malik et al. 2017). Just like nitrogen, many pesticides are mobile in the soil and reach water bodies via run-off and leaching (Chaplain et al. 2011; Chowdhury et al. 2008).

The impact of pesticides on non-target species has been widely documented. The use of insecticides has been linked to a dramatic reduction in insect populations in both terrestrial and aquatic ecosystems (Seibold et al. 2019; Sánchez-Bayo and Wyckhuys 2019). Pesticides have also been associated with declining bee populations, both wild and farmed (Brittain et al. 2010), and various insect-eating birds (Hallmann et al., 2014). According to Dainese et al. (2016), with current applications of pesticides, up to 40% of insect species, especially dietary specialists, could become extinct within the next few decades.

2.1.5.3 Growing biological resistance to pesticides

Biological resistance is growing to a wide range of old and new pesticides (DEFRA 2000; Aktar et al. 2009; Fountain and Wratten 2013; Kumar 2012; Bernardes et al. 2015; Chowdhury et al. 2008; Kaur and Garg 2014). Whilst initial resistance to insecticides was first noted in the early twentieth century, it was not until the 1940s, and the introduction of synthetic organic insecticides, that the problem became serious (Georghiou 1986). Indeed, the first case of resistance to DDT occurred as early as 1947 (Georghiou 1986), just 8 years after it was released. By the 1950s, resistance to modern synthetic insecticides had developed in approximately 50 insect species (Georghiou and Taylor 1976), a number that doubled during the 1960s and 1970s (Allen 1980). According to Bull (1982), 432 species of arthropods had developed resistance to one or more pesticides by 1980, rising to 480 species of insects, mites, or ticks by the late 1980s, many of which were major agricultural pests (Georghiou 1986).

Whilst insecticide resistance was the biggest problem, resistance to both fungicides and herbicides also slowly began to emerge in the 1940s (Georghiou 1986). By the late 1970s, more than 70 species of fungal pathogen had developed resistance to commonly used fungicides, including Benomyl (Georghiou 1986). By 1990, 150 fungal and bacterial pathogens had developed resistance (WRI 1994). Herbicide resistance took a little longer to materialise and was not seen as a major problem until the 1970s, 1980s, and 1990s. By the late 1980s, at least 107 weed species had developed resistance to commonly used herbicides (DEFRA 2000). By the early 1990s, 113 weed species had developed resistance to at least one herbicidal formulation (Pretty 1995). By 2014, 210 weed species had developed resistance to herbicides (Heap 2014).

2.1.6 Genetically modified organisms

Interest in the new biotechnologies, especially genetically modified organisms (GMOs), evolved during the early 1970s when

- the pesticide industry faced growing biological resistance to pesticides; and
- the discovery and development of pesticides were becoming increasingly expensive, and as public hostility towards pesticides grew.

This coincided with emergent molecular biology knowledge and products from publicly funded universities. Blue sky research, initiated in the 1960s, began showing commercial potential in the early 1970s, first in the USA and then in Western Europe (Goodman and Redclift 1991). The world's first biotechnology company, Genentech, was founded in 1976 (Genentech 2003), followed by Calgene in 1980 (University of Barcelona 2003). Both the companies and the IP and commercial regulatory framework were supported by the US Government (NTTC 2003). By the early 1980s, these and other emergent biotechnologies companies were increasingly bought out or bought into by pesticide corporations (Thompson 1971). The same pattern followed in Western Europe with the emergence of Plant Genetic Systems (PGS) in 1982, which, in 1983, successfully generated a GMO plant (Bijman 2001).

The initial tranche of GMOs aimed to simplify crop management, reduce pesticide use, increase crop yields, and increase profits for all food system actors (UNESCO 2003). Monsanto led the commercialisation of GM crops with the release of herbicide-tolerant (Roundup Ready) soya beans, maize, and cotton, as well as insect-resistant maize and cotton, in 1996 (Deutsche Bank 1999). Figure 14 illustrates the adoption of GM crops (herbicide-tolerant and insect-resistant maize, soya bean, and cotton) in the USA between 1996 and 2017.

Other developed countries, such as Canada, Argentina, and China, also adopted GM crops (Clapp 2002). GM crops have also been marketed in developing countries. According to ISAAA (2016), the area of GM crops in developing countries expanded from 7.1 million ha in 1999 to 100 million ha in 2016. Initially, the production of Bt cotton led to an 80% decrease in pesticides used and a decrease in pesticide-related illnesses from 22% to 5% in farmers growing Bt cotton (New Scientist 2002).

However, whilst the rapid expansion and initial agronomic performance of selected GM crops was remarkable, GM crops have faced several challenges.

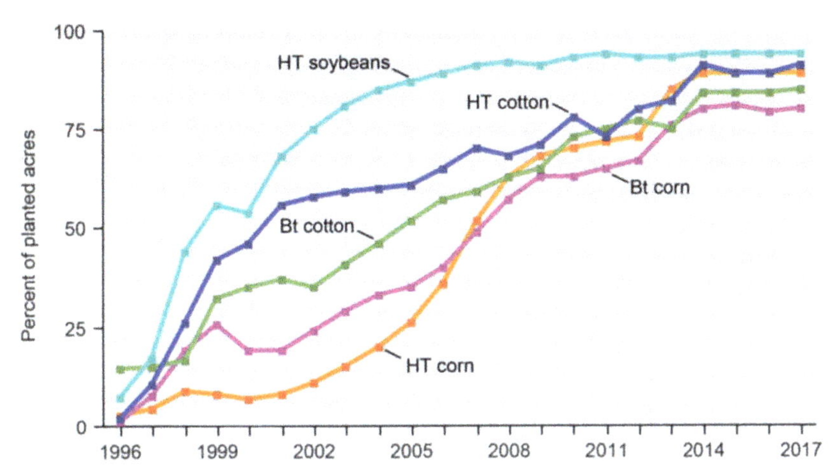

Figure 14 Area expansion of GMO soya bean, maize, and cotton between 1996 and 2017. Source: Adapted from: USDA.

One of the most significant challenges continues to be the limited acceptance of GM technologies by the public. Since their release, GM crops have increasingly exposed industrial food systems to combative press around 'Frankenstein Foods' in national newspapers, public protests, and even destruction of GM fields, especially in Europe (Ilbery and Maye 2005a; Renting et al. 2003). Concerns have also arisen regarding genetic pollution (Sirohi et al. 2014) and the impact of GM crops on non-target organisms, including the creation of herbicide-resistant superweeds (Moore 2010). The evolution of superweeds has been widely documented in the USA (Moore 2010), Argentina, and Brazil (Villar and Freese 2008), especially in soya bean-growing areas. In turn, this has led to the application of greater quantities of (different types of) herbicides (Moore 2010).

2.1.7 Water use

Given that water pollution by fertilisers and pesticides has already been addressed, the focus will be on concerns around the supply and demand of fresh water. Demand for fresh water comes from several sources: domestic (drinking, cooking, washing, bathing, etc.), crop and livestock production, aquaculture, food processing, and a range of other industrial uses (UNEP 2016). The supply of water comes from two interconnected sources:

- Precipitation and water held in soil and plants (green water); and
- Water held in rivers, lakes, and groundwater aquifers (blue water).

Globally, the food system (from production to consumption) is the principal user of fresh water. It is estimated that agriculture uses between 70% (UN Water 2015; Clapp et al. 2018; Holden et al. 2018) and 80% (Holt-Giménez 2019) of global freshwater (blue water) reserves, mainly for crop irrigation. In the Middle East, and several developing countries, such as Afghanistan, Bangladesh, Bhutan, India, Nepal, Maldives, Pakistan, and Sri Lanka, 95% of extracted blue water is used for agricultural purposes (FAO 2017a; Babel and Wahid 2008; OECD 2021).

Globally, since 1961, irrigated cropland has increased from 12% to 21%, and water usage from 1500 km^3 to 2800 km^3 per annum (Hertel et al. 2014). These irrigated lands are responsible for producing approximately 40% of global crop production (Gleick et al. 2002). In Asia, approximately 40% of the cropland is irrigated, but this produces 60–80% of the region's food (Yadvinder-Singh et al. 2014). According to Hoekstra and Mekonnen (2012), nearly 4000 L are required per capita for food production. Due to high industrial and domestic demand for blue water in developed countries, the proportion used by agriculture is usually less than 50% (OECD 2010).

As demand for water continues to increase, both for domestic, agricultural, and industrial purposes, the supply of blue water is coming under increasing pressure (Wheeler and von Braun 2013). For example, whilst the world population has increased roughly 4.4 times in the past 100 years, global water extractions have increased 7.3 times in the same period (AQUASTAT 2016). Linked to the Green Revolution, irrigated cropping has more than doubled during the past 50 years in Asia (UNEP 2016). To meet future food needs, agricultural demand for blue water is projected to increase at 0.7% per year (Doll and Siebert 2002; Rosegrant et al. 2009). Between 2000 and 2080, blue water demand is expected to increase by 50% in developing countries and 16% in developed countries (Yadvinder-Singh et al. 2014).

Currently, at least 20% of the world's groundwater aquifers are overexploited (Gleeson et al. 2012). Excessive extraction, and depletion, of groundwater reserves has already led to severe water shortages, and the situation is projected to worsen (UNEP 2016; Bellamy and Ioris 2017; OECD 2020; Foley et al. 2011; Gomiero et al. 2011; Gomiero 2016; Rees 2019; Dardonville et al. 2020). Figure 15 illustrates the regions of the world that are increasingly at risk of crop production losses due to irrigation water shortages. According to the OECD (2012), 'the number of people living in severely water-stressed river basins is projected to increase from 1.6 billion in 2000 to 3.9 billion by 2050, or over 40% of the world population of 2050'.

Whilst providing food security for Asia's growing population, irrigated agriculture, especially for rice and wheat, has led to excessive groundwater extraction (Balwinder-Singh et al. 2015). Excessive groundwater extraction affects key agricultural regions, such as North-Western India, the North China

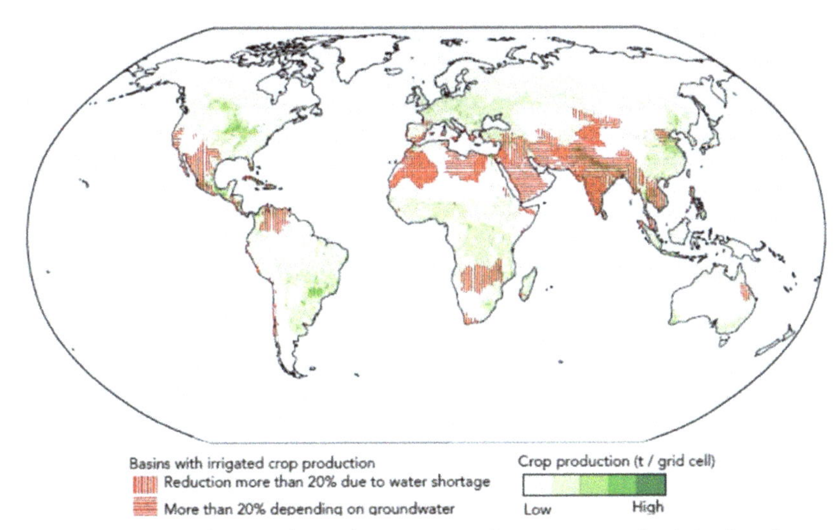

Figure 15 Regions that are dependent on groundwater sources for irrigation. Source: Adapted from: FAO.

Plain, Central USA, and California (Dalin et al. 2017). In South Asia, excessive blue water extraction has led to rapidly declining and unsustainable water tables, as well as increased costs associated with extracting water from deeper underground (Yadvinder-Singh et al. 2014). In many areas, the quality of extracted water has also deteriorated due to salinisation (Balwinder-Singh et al. 2015). Without adjustments, by 2050 more than 50% of global irrigated arable land may suffer from some degree of salinisation (Jamil et al. 2011).

Depletion of groundwater in many Asian countries reduced per capita availability of water by 40–60% between 1955 and 1990. Water availability is expected to decline by a further 15–54% over the next 35 years (Lampayan et al. 2015). According to Tuong and Bouman (2003), 15–20 million ha of rice land will suffer from water scarcity by 2025.

Key hotspots of water scarcity have already been reported in the USA, Mexico, Iran, China, Philippines, Bangladesh, northern China, the northwest Indo-Gangetic Plains, India's Cauvery delta, and the Chao Phraya delta in Thailand (Lampayan et al. 2015; Holden 2018; Gleeson et al. 2012). India is projected to run out of extractable groundwater within 300 years (Holden et al. 2018). According to OECD (2012), one in three people in the world are living in areas of medium to high water stress. Projections suggest that by 2050, over 40% of the world's population will be living in severely water-stressed river basins. In absolute terms, this is an increase from 1.6 billion in 2000 to 3.9 billion by 2050 (OECD 2012), and 50% of the world's population growth will be in these dryland areas (Holden et al. 2018). The demand for fresh water is expected to surpass its supply by at least 40% by 2030 (Guardian 2023). Lastly, due to globalised food trade, water depletion, just like habitat destruction, etc., is often exported from developed countries to developing countries. For example, according to Hoekstra (2012), 75% of UK food-related water use occurs outside the UK.

2.1.8 Habitat and species loss

Whilst concern over habitat and species loss is probably higher today than at any time in history, it is not a new phenomenon. Concerns over the loss of natural habitats and species have been of growing concern since pre-industrial times but significantly increased during the Industrial Revolution of the mid 1800s, especially in North America and Western Europe (Benton et al. 2021). According to the Convention on Biological Diversity, biodiversity or biological diversity means 'the variability among living organisms from all sources including, inter alia, terrestrial, marine and other aquatic ecosystems, and the ecological complexes of which they are part; this includes diversity within species, between species, and of ecosystems' (Engels 1995). Indeed, the so-called bioethics principle, promoted by visionaries such as John Muir, was mobilised in the nineteenth and twentieth centuries to protect wilderness areas, national parks, and wildlife habitats from the pressures of exploitation (O'Riordan 1981).

Support for species and habitat conservation emerged in the UK in the late 1880s with the establishment of the Royal Society for the Protection of Birds (RSPB), the National Trust in 1907, and the Society for the Promotion of Nature Reserves (SPNR) in 1912, which, amongst other things, aimed to protect and conserve important habitats and landscapes (Watson 2018). The first global body charged with the conservation of nature was the International Union for the Protection of Nature (IUPN), founded in 1948 (IUCN 1991), which was later to become the International Union for Conservation of Nature and Natural Resources (IUCN) in 1956, and the World Conservation Union (IUCN) in 1990 (IUCN 1991). As public concerns about biodiversity grew, the agribusiness community also established initiatives to reconcile the need for food production and the need to conserve biodiversity. In the UK, initiatives such as the Farming and Wildlife Advisory Group (FWAG), established in 1970, and the Linking Environment and Farming (LEAF) initiative, established in 1991, were the industry's vanguard instruments.

Biodiversity is attributed to a diverse range of values, including (Selman 2000)

- aesthetic (for natural beauty, variation, and inspiration);
- recreational;
- existence (the value/sense of well-being associated with the continued existence of biodiversity);

- bequest (the value the current generation places on preservation of biodiversity for future generations);
- scientific (e.g. as a source of inquiry and new knowledge);
- genetic variation in wild species (specifically for agricultural and industrial uses);
- bioremediation (neutralising pollution and buffering other environmental changes);
- evolution (e.g. in promoting organismal evolution to meet changing conditions);
- medicines (e.g. as a source of potential new pharmaceutical products); and
- ecosystem services (including water filtration, as CO_2 sinks or indicators of ecological change or stress)

Costanza et al. (1997) estimated the value of 17 key ecosystem services to be in the range of US$16–54 trillion/year.

Whilst human-influenced extinctions have occurred over the past 50 000 years (Tilman et al. 2017), species extinctions have risen quickly in recent history. Over the past 200 years, species diversity is estimated to have declined by around 14% (Benton et al. 2021). Whilst there has been recognition of the value of biodiversity, since the 1960s, the world has witnessed the highest rate of habitat fragmentation and species loss (including extinction) (Pimm et al. 1995; Holden et al. 2018; UNEP 2016; Tilman et al. 2017; IPES-Food 2016). Species extinction rates are suggested to range from 5000 to 150 000/year (Goodland 1991). According to Ramankutty et al. (2008), cropland expansion led to the clearance/conversion of more than '28% of tropical forests, 40% of temperate forests, 50% of shrub land and 58% of savannah and natural grassland'. Key habitats are continually being lost, and many of them as part of the expansion of agricultural land (Marchetti et al. 2020). Indeed, agriculture is responsible for 80% of land use change, and habitat loss (Tilman et al. 2017; Benton et al. 2021). Currently, approximately 80% of extinction-vulnerable terrestrial birds and mammals are threatened by the conversion of natural habitats for agricultural land and logging (Tilman et al. 2017).

Whilst most of the increase in food production since the 1960s has been derived from a tripling of yield on existing farmland through the application of industrial farming practices, the limited expansion of cropland and rangeland of 10–15% was still significant. At least an additional 450 000 000 ha of land, mainly biodiverse tropical forests, have been lost since the 1960s (Gibbs et al. 2010). This is an area equal to the combined size of India and South Africa. Between 2000 and 2010, it is estimated that conversion to agricultural land was responsible for 73% of tropical and sub-tropical deforestation (Hosonuma et al. 2012).

Agriculture has likely been responsible for the loss of 60% of terrestrial biodiversity (UNEP 2016). Between 1980 and 1995, over 200 million ha of land were deforested (FAO 1997). Most deforestation occurs in poor developing countries, principally in the high species-diverse tropical zones of Africa and Asia. Conversion is primarily for soy, cattle, and palm oil (IPBES 2019). According to IPBES (2019), between 1980 and 2000, over 42 million hectares of tropical forest were converted to cattle ranches in Latin America. An additional 6 million ha were converted to palm oil plantations in Southeast Asia. Between 2001 and 2015, over 27% of deforestation across the globe was for soy, cattle, and palm oil (Curtis et al. 2018). In the Amazon Rainforest alone, at least 75 million ha of forest have been cleared between 1978 and 2020, equivalent to 1.5 times the land area of Spain. It is estimated that most of this clearance has been for cattle ranching (Butler 2021). Deforestation in the tropics has doubled during the past 20 years, with approximately 70% of this loss being due to the conversion of forests into agricultural land (Feng et al. 2022). Figure 16 illustrates the key deforestation hotspots across the tropics.

Crucially, the loss of key habitats also leads to species extinctions. Habitat loss and fragmentation are especially dangerous for megafauna (large herbivores and carnivores) (Tilman et al. 2017; Benton et al. 2021). Habitat destruction also significantly affects small animals and insects (Benton et al. 2021). It is estimated that global deforestation between 1990 and 2020 could equate to the extinction of between 15 000 and 50 000 species (UNEP 1992). Conversion to agricultural land and other habitat losses threaten 25% of all mammal species and 13% of all bird species, as well as more than 21 000 other species of plants and other animals, with extinction (IUCN 2016). According to UNEP (2016), by 2050, global cropland could have increased by 300 million ha. Large-scale cropland expansion is predicted in Africa (+121 million ha) and Latin America (+57 million ha). Between 1997 and 2011, the loss of ecosystem services linked to cropland expansion cost an estimated €3.5–18.5 trillion per annum year in ecosystem services (EU 2020).

Habitat and species loss are also occurring in non-biodiversity hotspots, including the European Union (EU). Recent studies suggest that more than 75% of wildlife habitats are deteriorating (Wezel et al. 2018b). Key indicator species, including both rare and common birds, have declined, and many are at risk of disappearing (Wezel et al. 2018b). Hallmann et al. (2017) measured a 75% reduction in flying insect biomass across Germany over a period of 27 years. Overall, terrestrial and freshwater insect numbers have reduced by approximately 9% and 11%, respectively, every decade (van Klink et al. 2020). As the area of natural habitats declines and the area of agricultural lands slowly expands, concerns are increasingly turning to declining agrobiodiversity (Woomer and Swift 1992; Greenland and Szabolcs 1994;

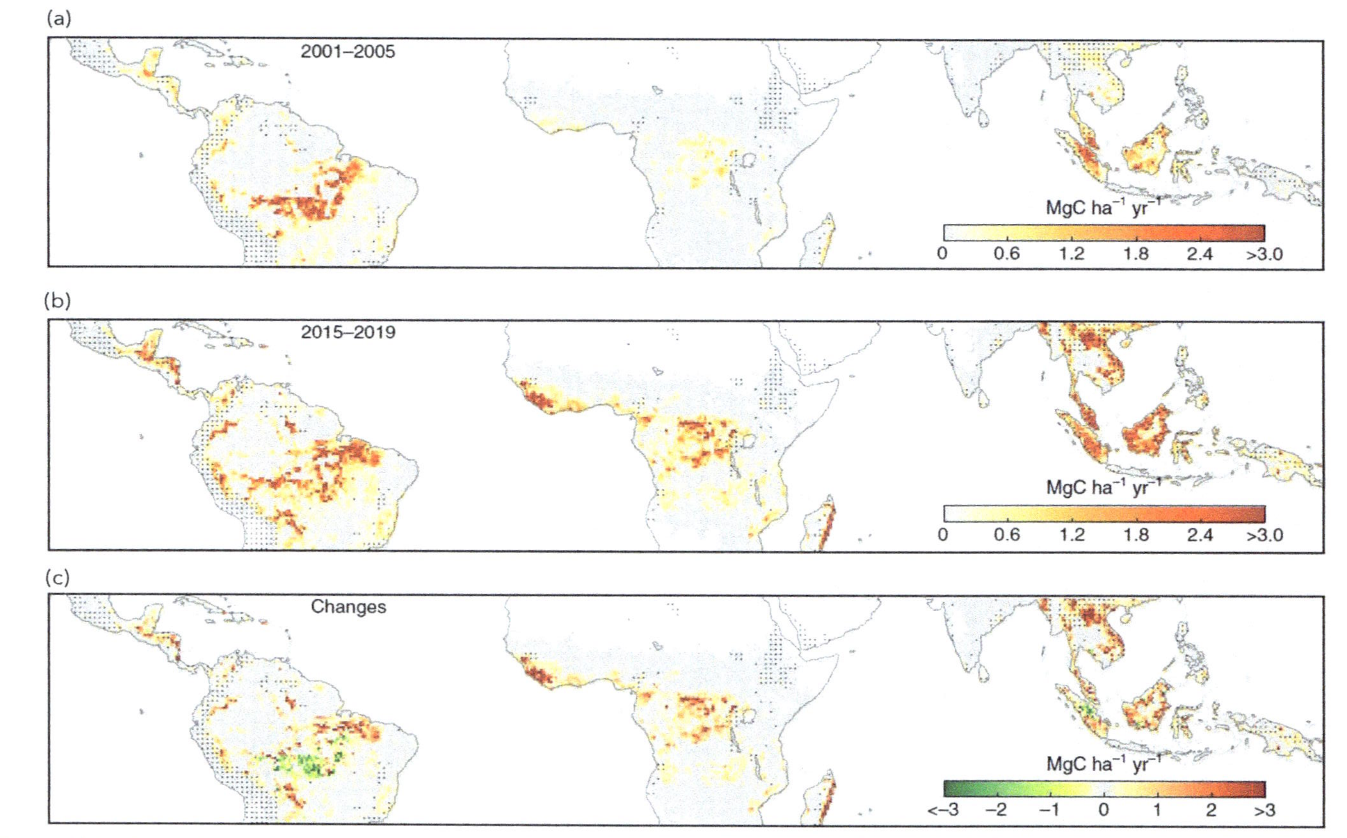

Figure 16 Spatial pattern of forest carbon loss across the tropics. Source: Adapted from: Feng et al. (2022).

Altieri 1995; Marsden et al. 1993; Morris and Bate 1999; Chagnon et al. 2015; Benton et al. 2021).

In farmed areas (both arable cropping and grasslands), there is generally a negative relationship between crop yields and biodiversity. Increasing yields, larger farm sizes, mechanisation, use of synthetic inputs (pesticides and fertilisers - especially nitrogen), etc. under industrial food systems lead to reduced biodiversity (Gabriel et al. 2013). Species diversity in intensively farmed land has declined by between 75% (Newbold et al. 2015) and 90% (Holt-Giménez 2019). Key European indicators species, such as wild bees (Brittain et al. 2010) and insectivorous birds (Hallmann et al. 2014) have declined. If arable land continues to expand into all suitable remaining natural areas, this could lead to a further reduction in biodiversity (species richness 30% and species abundance 31%) in the tropical zones of South America and Africa (Kehoe et al. 2017).

Climate change, in part a consequence of fossil-fuel-based industrial food systems, is also a key driver of biodiversity loss (Newbold et al. 2015). Increasing global temperatures directly affect habitats, slowly relocating habitats and species - towards the poles and higher elevations, or deeper/cooler waters for aquatic species (Pecl et al. 2017). Figure 17 illustrates the threats to species extinction from agriculture and other pressures.

Mega biodiversity hotspots, where cropland expansion and intensification threaten significant loss of biodiversity, are expected to be areas of Central and South America (affecting an area of 1.097 million km^2), SSA (773 375 km^2), and Australia (79 490 km^2). Cropland expansion is likely to affect biodiversity hotspots along the tropical Andes, the Brazilian Atlantic Forest, and both West and East Africa. Key biodiversity hotspots expecting significant losses due to production intensification are likely to be in SSA (122 702 km^2) and Brazil (3560 km^2) (Zabel et al. 2019).

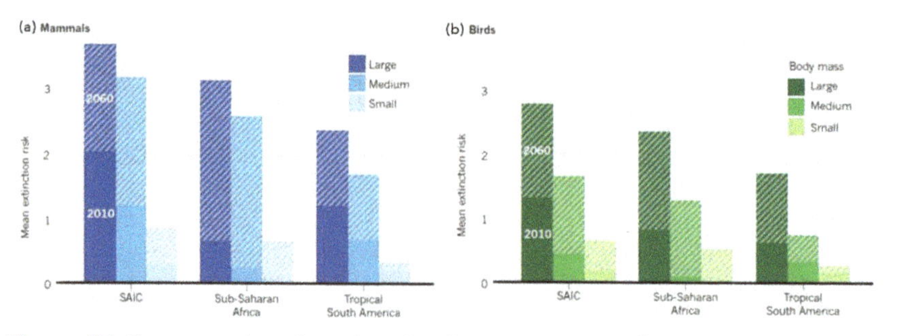

Figure 17 Current and projected regional extinction rates for mammals and birds. Source: Adapted from: Tilman et al. (2017).

2.1.9 Land degradation

Land degradation has been a key concern of food system actors for millennia. Indeed, land degradation undermined many ancient civilisations, including the Sumerians (4000 BC) and the Mayans (AD 900) (Kesavan and Swaminathan 2018). But what exactly is meant by land degradation and why is it considered a problem in the twenty-first century?

Land degradation has many definitions (Eswaran et al. 2001). Here, I use the GEF (2021) definition 'the deterioration or loss of the productive capacity of the soils for present and future'. Whichever definition is used, land degradation is a growing and significant phenomenon that undermines food system capacity to continue feeding the world (Migliorini et al. 2018; Bellamy and Ioris 2017; Rees 2019; IPES-Food 2016; Frison and Clément 2020; Khurana and Kumar 2020; Benton et al. 2021; Benton and Bailey 2019; Reganold and Wachter 2016; Rosa-Schleich et al. 2019; Mabhaudhi et al. 2019; Marchetti et al. 2020). EU (2020) estimates that the costs associated with land degradation are between €5.5 and 10.5 trillion per year.

Whilst the figures are fiercely contested (EU 2020; OECD 2021), it is estimated that between 20% (UNCCD 2012) and 33% (FAO 2015a) of the world's land mass suffers from some kind of soil degradation (Springmann et al. 2018; Sachs et al. 2019). Within this figure, up to 40% of agricultural land is degraded to some extent (UNEP 2016). As much as 12 000 000 additional hectares of agricultural land are degraded each year (ELD Initiative 2015). According to the WEF (2012), the world only has 'about 60 years of topsoil left'. This claim is fiercely contested (OECD 2021). However, what can be agreed is that soil degradation in many places is occurring much faster than soil regeneration (Bindraban et al. 2012). A significant proportion of degraded land is in developing countries. Indeed, it is estimated that soil degradation affects up to 500 million ha in the tropics (FAO 2015b).

Ultimately, when land is degraded and, by default, less productive, it becomes more likely that the remaining forests, natural grasslands, and wetlands will be converted to agricultural production to compensate. For land currently under production, this is likely to be farmed more intensively to provide food for a growing global population, thus potentially causing even further degradation (UNEP 2016). In addition, more than 40% of the world's poor live in these degraded areas (Conway 2012). In many instances, the poor lack the finance and technologies to undertake land restoration to tackle the problem of degradation. Many poor farmers and herders therefore face a downward spiral of land degradation and yields (Tittonell and Giller 2013). This situation is especially worrying when set against the projected expansion of dryland areas to around 50% of the world's total land surface (Holden et al. 2018).

Land degradation is often the result of a culmination of factors, such as loss of organic matter, soil erosion, nutrient depletion, acidification, salinisation, compaction, and chemical pollution (UNEP 2016; Amede 2003; Henao and Baanante 2006; Folberth et al. 2014; Omotayo and Chukwuka 2009). The loss of organic matter from the topsoil has been recognised as a key issue. The most important drivers of organic matter loss are

- regular cultivation of land for cropping (Adamtey et al. 2016); and
- net off-take of biomass and nutrients due to overgrazing of grassland and continuous removal of grain and stover without replenishment (Henao and Baanante 2006; Sileshi et al. 2010; Omotayo and Chukwuka 2009).

Loss of organic matter is important because of its role in soil nutrient cycling and balance, soil structure, water-holding capacity, subterranean biodiversity (soil-dwelling organisms), and others. The loss of organic matter occurs in both industrial and traditional food systems.

The loss of topsoil, and the nutrients contained therein, is also accelerated by frequent cultivation and breakdown of soil structure, especially the breakdown of stable soil aggregates (Henao and Baanante 2006; Sileshi et al. 2010). Brought about by over-cultivation of arable lands, drought, and high temperatures, the Dust Bowl of the 1930s, covering 400 000 km^2 across Texas, Oklahoma, and parts of New Mexico, Colorado, and Kansas, is, to date, the worst example of large-scale soil erosion (Shannon et al. 2015). According to Thaler et al. (2022), up to 1.8 mm/year of topsoil is being lost in the Mid-Western states of the USA. During the past 40 years, 33% of the world's arable land has been lost due to soil erosion, degradation, or chemical pollution (FAO 2015a; Holden et al. 2018). Globally, estimates suggest that approximately 75 000 000 000 tons of topsoil are eroded each year at a cost of between US$5 500 000 000 000 and US$10 600 000 000 000 00 (Holt-Giménez 2019; EU 2020). More importantly, the loss of topsoils through over-cultivation is estimated to be occurring at more than twice the natural rate of regeneration (Montgomery 2007). Salinisation is also becoming a major issue. Jamil et al. (2011) estimate that, if current trends continue, by 2050, 50% of irrigated arable land will be salinised.

2.1.10 Industrial agriculture and climate change

Industrial agriculture of the Second Food Regime contributes to climate change in several ways. The principal way is through its reliance on fossil-fuel energy for agricultural mechanisation (especially tractors and combines), production of synthetic fertilisers and pesticides, storage, processing, and transportation of agricultural commodities and final food products (Jasinski et al. 1999; Kucukvar and Samadi 2015; FAO 2017b; Oosterveer and Sonnenfeld 2012; UNEP 2016). Figure 18 illustrates the proportional contribution of different components of the food system to greenhouse gas emissions.

The production of synthetic nitrogenous fertilisers consumes approximately 5% of the global annual natural gas demand (UNEP 2016) and is responsible for generating 575 Mt CO_2/year UNEP (2016). Food refrigeration consumes approximately 15% of total global electricity, generating around 490 Mt CO_2/year (Vermeulen et al. 2012). Nearly 80% of the energy used in industrial food systems is non-renewable (i.e. fossil-fuel-based). Crop production and intensive livestock systems (requiring energy for heating, cooling, ventilation, and production of feed (Koneswaran and Nierenberg 2008)) account for nearly 33% of the energy consumed by food systems (FAO 2011c; Jasinski et al. 1999), and 86% (12 000 Mt CO_2/year) of the GHG generated by food systems (Vermeulen et al. 2012). Much of this energy is currently derived from fossil fuels (Zou et al. 2001; Kudo and Miseki 2009; FAO 2016d). Industrial food systems are so energy-reliant that, in 2013, they accounted for approximately 30% of the world's total energy consumption (Monforti-Ferrario 2015; Bajželj

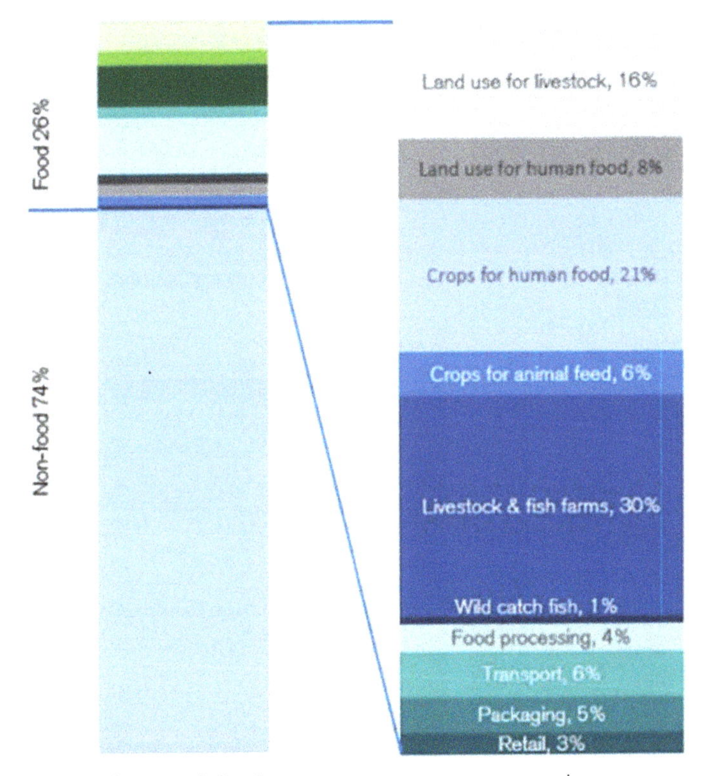

Figure 18 Contribution of food system components to greenhouse gas emissions (GHGEs). Source: Adapted from: Credit Suisse (2021).

et al. 2013; FAO 2013a), between 12% (IPCC 2019) and 33% (Tilman 1999) of anthropogenic GHG emissions, and 56% of non-CO_2 (N_2O and CH_4) GHG emissions (Vermeulen et al. 2012).

OECD (2019a) estimates that agriculture is responsible for approximately 44% of methane (CH_4) and 82% of nitrous oxide (N_2O) emissions. Figure 19 illustrates food system GHG emissions by food type and source in the EU. The conversation of forests and peatlands to agricultural production (cattle ranching, and maize and soya bean feed production), and clearing native forest for oil palm production, generate significant amounts of greenhouse gas (GHGs) (Tilman et al. 2011; Smith et al. 2014; Willett et al. 2019; Pendrill et al. 2019), probably over 22% of total food system emissions (UNEP 2016).

Approximately 66% of agricultural GHG emissions (including deforestation) can be attributed to livestock production. The digestion system of ruminant livestock contributes approximately 40% of direct emissions from agriculture

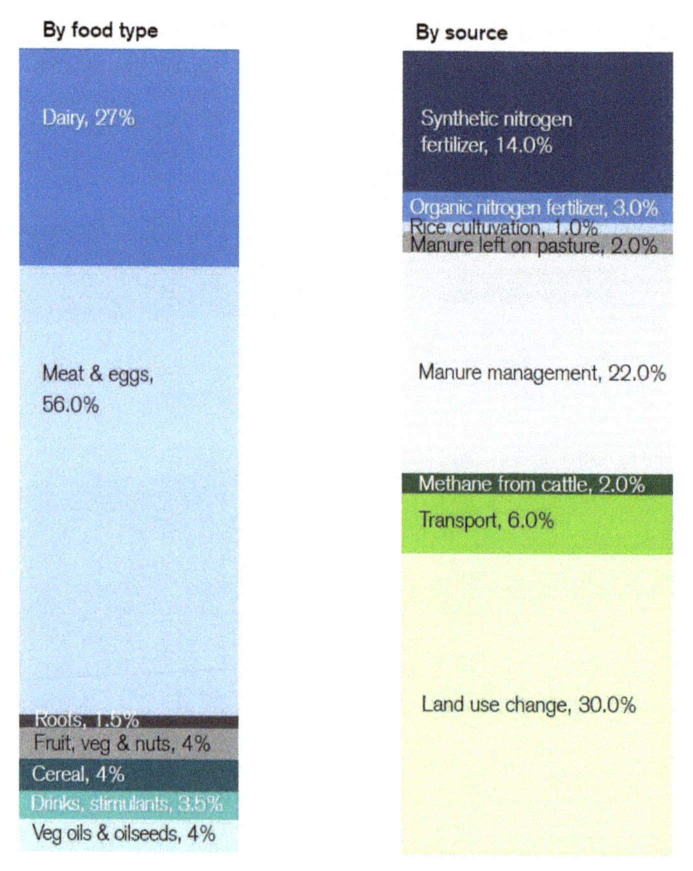

Figure 19 Food system greenhouse gas emissions by food type and source in the EU. Source: Adapted from: Credit Suisse (2021).

(FAO 2020). In 2005, the global livestock sector emitted 7.1 giga-tonnes CO_2, which equates to between 14.5% and 16.5% of total human-induced GHG emissions (FAO 2014; FAO 2018h). Of this number, beef production was responsible for 41% of livestock emissions. Milk production accounted for 20% of livestock-related GHG emissions (Gerber et al. 2013). The production of animal-based protein generates higher overall carbon emissions compared to the production of plant-based protein (Nijdam et al. 2012; Eberle and Fels 2014; Treu et al. 2017; Röös et al. 2017; Meier et al. 2014; Davis et al. 2010; Berners-Lee et al. 2012; Risku-Norja et al. 2009; Saxe et al. 2012; Westhoek et al. 2014). Figure 20 illustrates the combined contribution of GHG emissions by food types, clearly demonstrating the significant contribution of ruminant livestock species to GHG emissions.

Land cultivation for annual crops, in both industrial and traditional food systems, also leads to the direct loss of organic matter from soils which, in turn, leads to increased atmospheric carbon dioxide (CO_2) (Montgomery 2007; Willett et al. 2019; IPCC 2019; Blandford and Hassapoyannes 2018; Smith et al. 2014). Between 2007 and 2016, land use change, especially for agriculture, generated approximately 44% of agriculture-related GHG emissions (IPCC 2019; OECD 2019a,b). Dwivedi et al. (2017) suggest that, by 2050, GHGE from food production and land clearance could increase by 80%.

In addition to the generation of CO_2, the production of rice, ruminant livestock, and farm animal manure also releases methane (CH_4) (Vermeulen

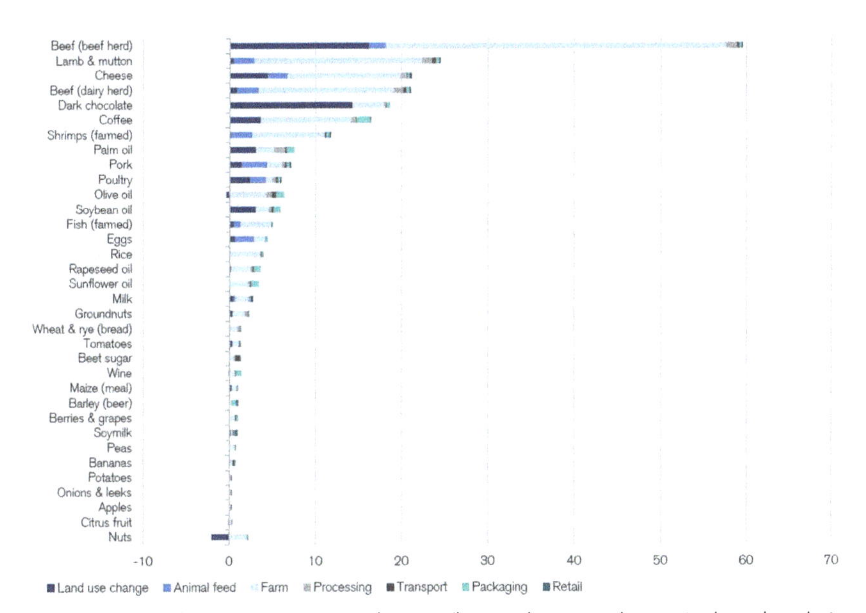

Figure 20 Greenhouse gas emissions by crop/livestock type and stage in the value chain. Source: Adapted from: Credit Suisse (2021).

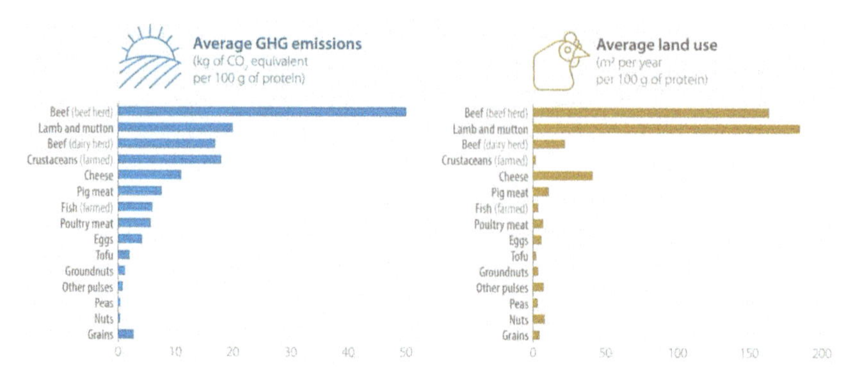

Figure 21 Average greenhouse gas emissions and land use associated with the production of 100 g of protein . Source: Adapted from: IGS (2019).

et al. 2012; FAO 2014; HLPE 2016), approximately 17% of total food system emissions (UNEP 2016). A small amount of nitrous oxide (N_2O) is also lost to the atmosphere after nitrogen fertiliser application and via animal manure (Willett et al. 2019; UNEP 2016). The food system (primarily livestock production) is responsible for 44% of global CH_4 emissions and 53% of N_2O emissions (Gerber et al. 2013).

According to GRAIN and IATP (2018), 'the world's top five meat and dairy corporations – JBS, Tyson, Cargill, Dairy Farmers of America and Fonterra – combined are responsible for more annual greenhouse gas emissions than ExxonMobil, Shell or BP'. Whilst absolute amounts of agricultural GHGs have increased since the 1960s, increased productivity of both livestock and crop production systems has led to a reduction in GHG emissions per unit of output (Blandford and Hassapoyannes 2018). Indeed, the low levels of milk and meat yields and the longer time required until slaughter mean that, in many cases, livestock production by small-holder farmers and pastoralists often generates more GHGEs than livestock in industrial units, which have high-quality feed, improved genetic productivity and controlled living environments (Herrero et al. 2013). Figure 21 illustrates the GHG emissions associated with the production of a range of protein-based foods.

Food loss and food waste also make a significant contribution to GHG emissions. Figure 22 illustrates the GHG emissions associated with food loss and waste. According to Credit Suisse (2021), if food loss and waste were a country, it would be the third-largest contributor to GHG in the world.

2.2 Food loss and food waste

Food loss refers to the loss of agricultural produce before it is transformed into food product. Losses occur during the production, storage, processing, and distribution nodes of food systems. This includes the pre-harvest loss of

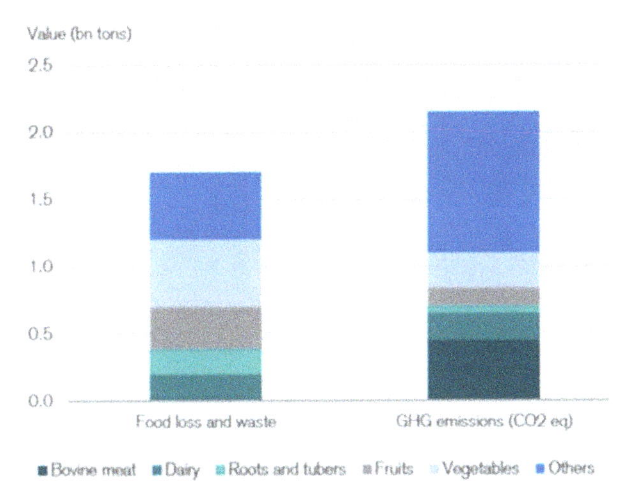

Figure 22 Greenhouse gas emissions associated with food loss and waste. Source: Adapted from: Credit Suisse (2021)

quantity and quality (due to pests, diseases, and weed competition), losses caused by pests and disease whilst in storage, physical damage/losses during transportation, as well as loss during processing (Lipinski et al. 2013). Figure 23 visualises the different stages during which food loss and waste occur.

Whilst food loss in developed countries tends to be during food processing, food loss in developing countries tends to occur pre-harvest (due to pests and diseases) and post-harvest (due to vermin, insect pests, moulds, etc., predominantly linked to inadequate drying and storage) (UNEP 2016; Credit Suisse 2021). Whilst accurate estimates are difficult to come by, post-harvest food loss in developing countries can be as high as 40%, and this does not account for losses linked to pre-harvest yield reductions due to pests and diseases. For example, cereal loss in SSA generally ranges from 25% to 40% UNEP (2016), whilst root crops, fruit, and vegetable losses range between 30% and 40% (UNEP 2016), or even 80 % (HLPE 2014).

Food waste refers to the loss of final food products (sold in retail outlets, served in restaurants, or cooked in the home). The primary causes of food waste are over-buying, poor storage, loss during cooking, plate waste by consumers, uneaten food in restaurant/catering/hospitality, overstocking/under-purchasing (Stuart 2009), poor storage, technical malfunction, damaged packaging, and rejection on cosmetic grounds by food retailers and shoppers (Griffin et al. 2009; Stuart 2009; Parfitt et al. 2010; Gustavsson et al. 2011; BCFN 2012; Principato et al. 2015; UNEP 2016; Credit Suisse 2021; OECD 2021).

Figure 24 visualises the sheer scale of food loss and waste. Approximately 33% of global food production is wasted (Nature Editorial 2019). According to UNEP (2021b), the global food system wastes 931 million tonnes of food each

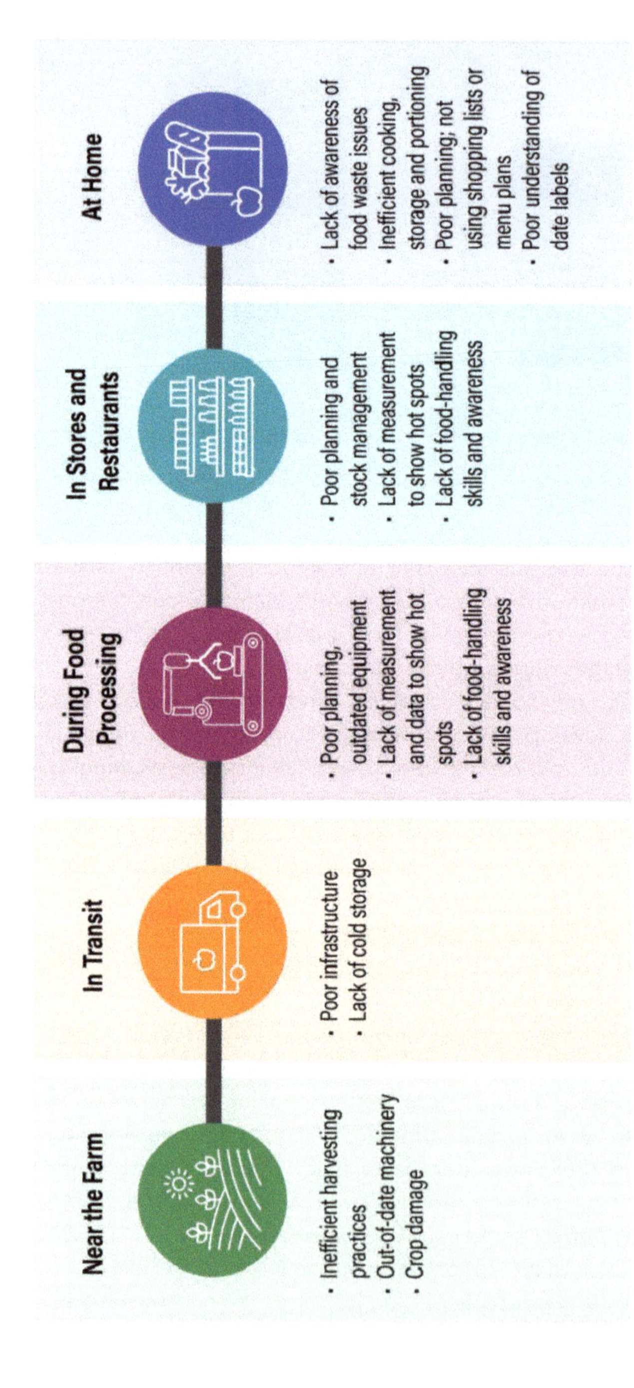

Figure 23 Main drivers of food loss and waste throughout the supply chain. Source: Adapted from: WRI (2023).

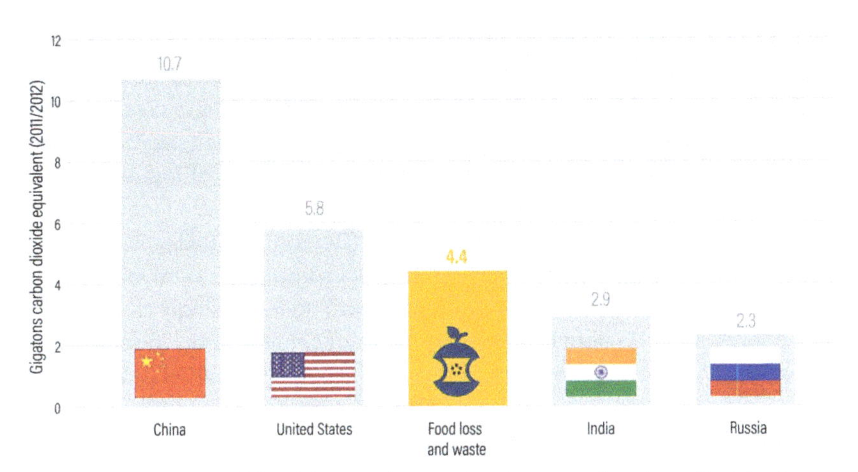

Figure 24 The scale of food loss and waste in gigatons of carbon dioxide. Source: Adapted from: Searchinger et al. (2019).

Figure 25 Food loss and waste by region and value chain component. Source: Adapted from: Searchinger et al. (2019).

year (61% households, 26% food service, and 13% retail). Between 30% and 50% of food is wasted between the farm gate and consumption (Holt-Giménez 2019). Figure 25 clearly illustrates significant food loss in developing regions versus significant food waste in developed regions.

Despite recent efforts, data on food waste remain relatively sparse (UNEP 2021b). According to UNEP (2021a), direct food waste by consumers is considerably higher in Europe and North America (95–115 kg/year). In total, the EU wastes approximately 153 m tonnes of food each year (Guardian 2022a). Conversely, direct food waste by consumers in SSA, South, and Southeastern

Asia is as little as 6–11 kg/year. This general relationship between poorer and richer consumers was also noted over a decade ago (Monier et al. 2010). However, analysis by Dou and Toth (2020) concluded that there was 'no clear relationship between the level of household food waste per capita and the GDP of a country'. Lastly, as part of the capitalist expansion of industrial food systems, increasing yields and decreasing food prices relative to incomes have also been blamed for increases in food waste (Benton and Bailey 2019; Holt-Giménez 2019). Whilst it is difficult to make direct comparisons and infer trends in food waste over time (UNEP 2021b), it is becoming increasingly clear that, in a time of growing food demand, the sheer scale of contemporary food loss and food waste can no longer be ignored (Irfanoglu et al. 2014; FAO 2019).

2.3 The triple burden of malnutrition

The triple burden of malnutrition refers to the coexistence of three nutrition-related concerns, namely, undernutrition (underweight, stunted, or wasted), micronutrient deficiency, and overweight and obesity. Whilst trade-dependent industrial food systems produce and distribute regular quantities of relatively low-cost food and feed for both people and livestock, there are a growing number of concerns around the uneven nature of food security and several important health externalities.

2.3.1 The uneven nature of food security

Although the industrial food system produces surplus calories, the distribution of these calories is highly uneven. In part, this is due to physical access to food. This might be due to the limited amounts or diversity of food produced locally, and the lack of access to markets - often linked to the lack of infrastructure (especially roads). Whilst affected by other factors, such as conflict and social unrest, displacement, and climate change (i.e. drought), it can also be argued that poor physical access to food is due to underdevelopment. Economic access to food, which is often the principal barrier to food security, is, however, often a much more complex issue, as it occurs in both developed and developing countries.

2.3.2 Economic access to food

Despite the global food system producing enough calories for the current global population, approximately 700 million people, around 11% of the global population, remain undernourished (UNEP 2016; FAO, IFAD, UNICEF, WFP and WHO 2017, 2020; Credit Suisse 2021). Figure 26 illustrates that, whilst the number of undernourished people in the world had been falling in the longer term, in recent years the number of undernourished people in the world

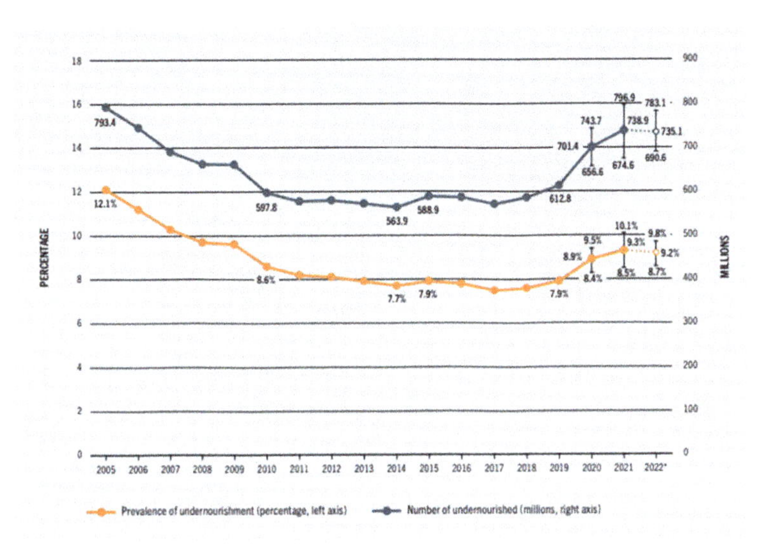

Figure 26 Global hunger 2005-22+. Source: Adapted from: FAO, IFAD, UNICEF, WFP, and WHO (2023).

has increased again. COVID-19 increased numbers even further, possibly by an additional 132 million people (FAO, IFAD, UNICEF, WFP, and WHO 2020).

Whilst most food insecurity is restricted to developing regions, such as SSA and Central and South Asia, where food and nutrition insecurity can affect more than 60% of the population, people in developed regions, such as North America and Europe, also face food insecurity (FAO, IFAD, UNICEF, WFP and WHO 2023). Figure 27 illustrates the percentage of undernourishment (moderate and severe food insecurity) across the globe.

Even when food prices were at all-time lows in the early 2000s, more than 800 million people around the world remained undernourished (OECD 2021). Whilst low food prices are important for economic access to food, it is the relationship between the cost of food and an individual's/family's income that makes the difference between being able to put enough food on the table each day (Hirvonen et al. 2019). This is particularly pertinent in developing countries. Studies undertaken during the 2008 Food Crisis, as well as recent COVID-19-related studies, have revealed coping strategies based on the restriction of diets to cheap starchy staples and highly processed foods (FAO-AU 2020; Mercy Corps 2020; WFP 2020; Watson 2021).

In many instances, families were restricted to eating just one meal, or less, per day (Olwande and Ayieko 2020). Even in developed countries, the increase in food prices since the year 2000 has noticeably affected the purchasing patterns of low-income families. When food prices increase, poor families tend to purchase cheaper meats, such as chicken and pork, rather than

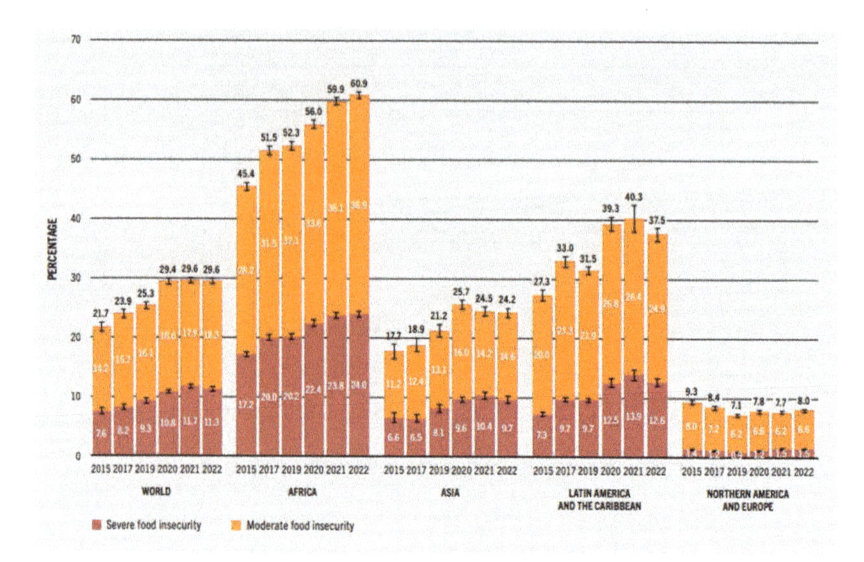

Figure 27 Food insecurity by region 2015–22. Source: Adapted from: FAO, IFAD, UNICEF, WFP, and WHO (2023).

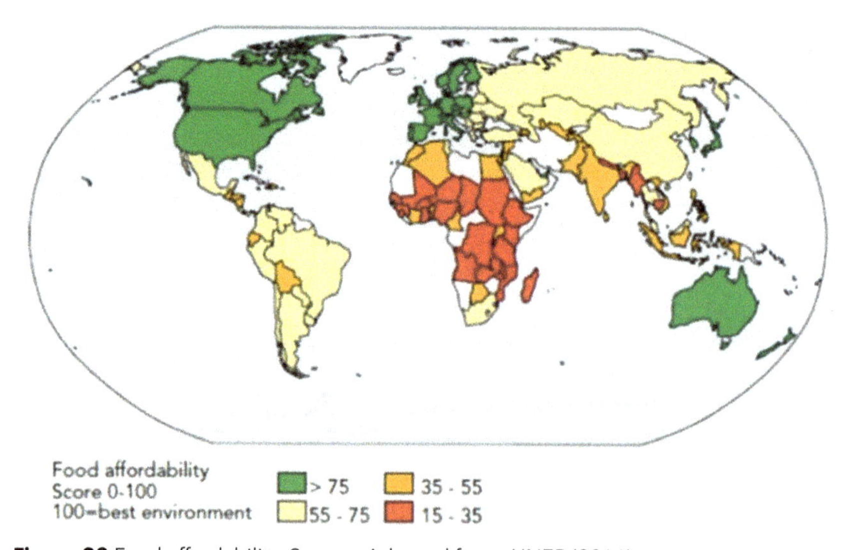

Figure 28 Food affordability. Source: Adapted from: UNEP (2016).

purchasing fruits, fish, and more expensive meats (DEFRA 2014). According to Gentilini (2013), in 2013, 60 000 000 people used food banks. This equates to approximately 7% of the global population. Figure 28 illustrates the spatial distribution of food affordability.

2.3.3 Food systems and health

According to FAO (2002a), food security is where 'all people, at all times, have physical, social, and economic access to sufficient, safe, and nutritious food to meet their dietary needs and food preferences for an active and healthy life'. Both physical and economic access to food has far-reaching implications on health. This goes further than the capacity to purchase adequate amounts of food calories. Indeed, up to 3 000 000 000 people worldwide have neither physical nor economic access to a healthy diet containing adequate amounts of calories and nutrition (FAO, IFAD, UNICEF, WFP and WHO 2020).

Aside from a few notable exceptions (such as pulses and dairy products in India and dark leafy greens in Western and Central Africa), economic access is restricted for nutrient-rich and healthy diets (GPAFS 2016; GPAFS 2020; FAO 2018h). According to FAO, IFAD, UNICEF, WFP, and WHO (2020), healthy nutrient-rich diets are approximately five times more expensive than diets that simply meet satisfactory calory intake. Figure 29 illustrates the percentage of the population who can afford a healthy diet, whilst Figure 30 illustrates the cost of healthy diet versus a calorie-sufficient diet across developed and developing regions. This suggests limited economic access to healthy diets across the world.

In less developed countries, many people suffer from what has been termed as hidden hunger (limited intake of micro-nutrients), lacking iron, zinc, vitamin A, iodine, calcium, and folic acid in their diets (Muthayya et al. 2013; Popkin 2017; FAO 2015d, 2016b; Tilman and Clark 2014; Haddad and Hawkes 2016; Popkin 2009; Popkin and Reardon 2018; Siegel et al. 2014). Several studies highlight that poor families have less variation in their diets, which, along with health and sanitation challenges, reduces the body's stock of vital vitamins and micro-nutrients (Credit Suisse 2021; IPES-Food 2017b; Frison and Clément 2020; Khoury et al. 2014). In turn, this affects the health status of individuals and families, leading to a range of health problems, such as poor-quality breast milk available for babies (Credit Suisse 2021), stunting in children and cognitive disorders, and reduced capacity to fight disease (Headey 2013). Malnutrition is not just a problem for developing countries. Driven by a combination of poor individual food choices and financial constraints, roughly 50% of Europeans have some form of micronutrient deficiency (SAPEA 2020).

Zoonotic diseases are becoming increasingly prevalent and have especially been linked to intensive livestock production, where significant numbers of genetically similar, and immune-suppressed animals - due to the prophylactic use of antibiotics in animal production - are kept in confined spaces (Jones et al. 2013; Otte et al. 2007). Public health problems, such as COVID-19, often originate from the consumption of poultry, livestock, and other sources of meat (Doyle et al. 2015; Painter et al. 2013). In the case of poultry, 58% of all

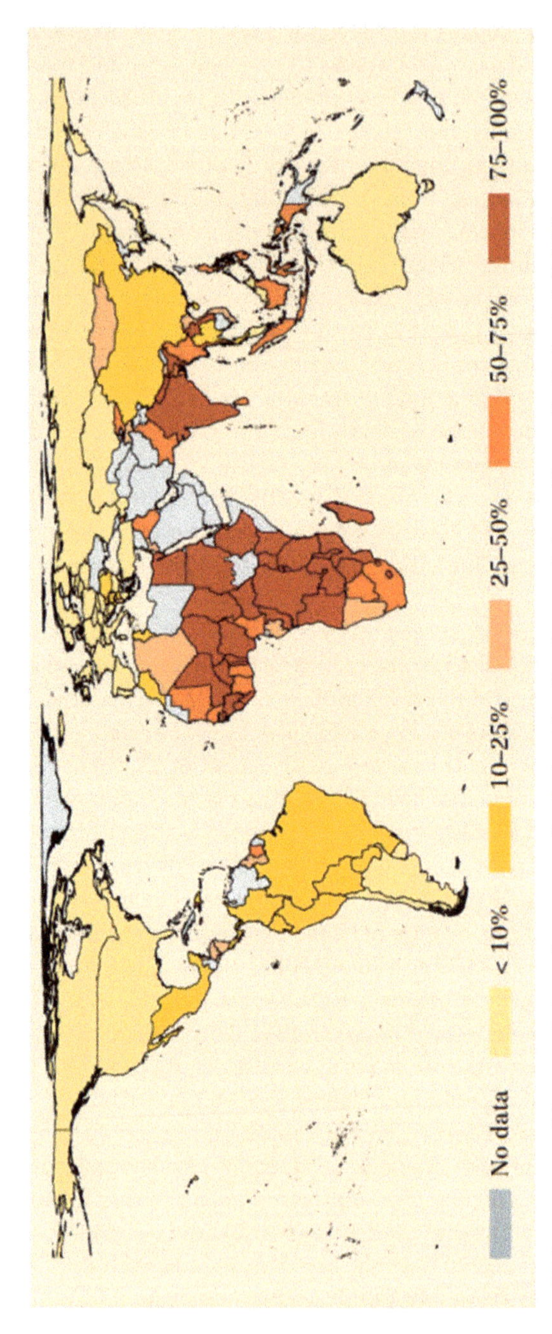

Figure 29 Percentage of the population who cannot afford a healthy diet. Source: Adapted from: Herforth et al. (2020).

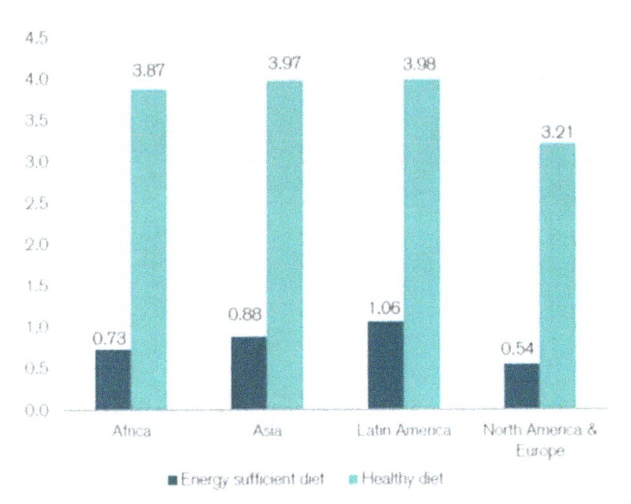

Figure 30 US$ cost/capita/day of a healthy diet compared to an energy-sufficient diet. Source: Adapted from: Herforth et al. (2020).

Salmonellosis cases in Europe were from intensive egg production and an estimated 29% of all campylobacteriosis cases were linked to the consumption of poultry meat (Pires et al. 2010).

2.3.4 Changes in diets

During the past 100 years or so, diets have changed in two important ways. Firstly, the range of foods consumed has reduced considerably. Whilst diets have been affected by changing consumer tastes, the principal reason for this change can be attributed to the increased corporate control of the food system (discussed later). Secondly, as income per capita has increased, diets have changed from starch-rich to refined sugar, protein, and refined fat/oil-rich foods, livestock-based products, and vegetables (UNEP 2014; Tilman and Clarke 2014; Popkin 1994). Meat consumption has become more accessible due to reduced unit cost of production (due to cheap cereals and oilseeds) in large/efficient production facilities which house sometimes tens of thousands of animals/birds or through the large-scale conversation of forest to grassland (Benton et al. 2021). Figure 31 illustrates the regional breakdown of meat consumption.

As countries become richer, they generally undergo a 'nutrition transition', which comprises an increased intake of calories and then increased intake of animal protein as well as, nuts, fruits, and vegetables (Popkin and Gordon-Larsen 2004; Popkin 2017). This change has been witnessed in both developed and developing countries. For example, China's and Brazil's consumption of meat jumped from 14 kg and 40 kg/year/per capita in the early 1970s to 52 kg

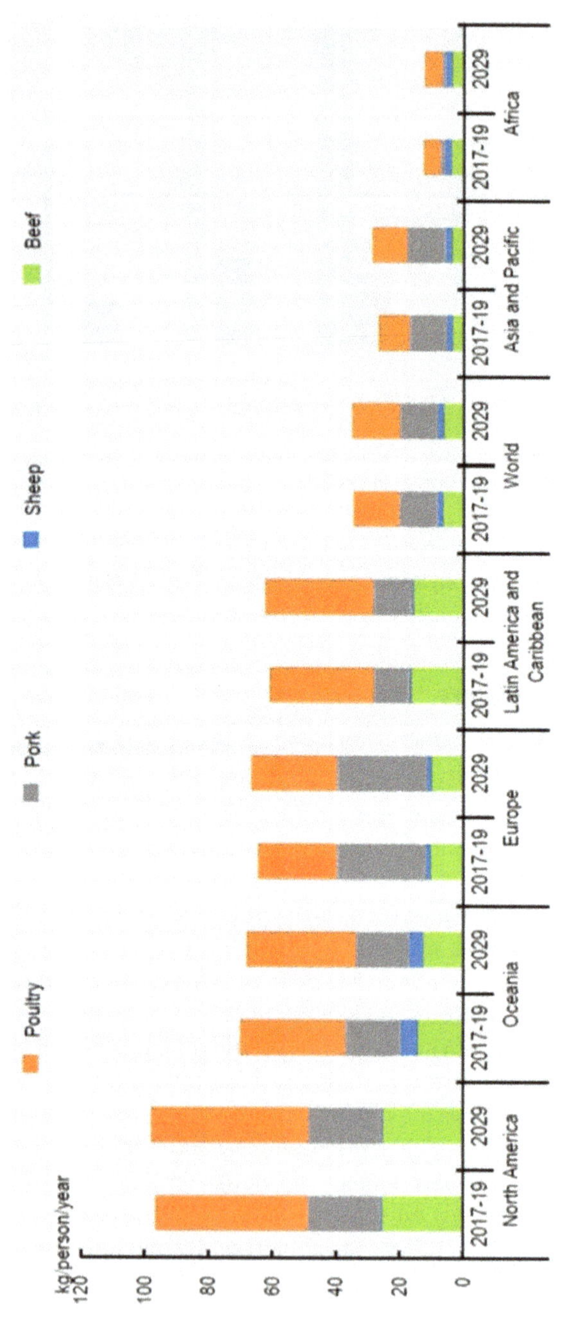

Figure 31 Per capita meat consumption by region and type. Source: Adapted from: OECD (2021).

and 78 kg/year/per capita in 2010, respectively (Alexandratos and Bruinsma 2012).

In 2013, it was estimated that the combination of livestock rangeland and land used to grow cereals and oilseeds for livestock feed amounted to one-third of the earth's total land surface (Herrero et al. 2013).

In China, the transformation of livestock-based food systems during the past 35 years has led to the pollution of soil and water, overgrazing, and land degradation (Garnett and Wilkes 2014). Currently, cereals and oilseeds for livestock feed are being grown on approximately 560 million ha of land, which, alternatively, could be used to produce food for direct human consumption (OECD/FAO 2019). In total, more than 3 000 000 000 ha of land are used for grazing. Of this number, 685 000 000 ha could have been used to produce food for human consumption (Mottet et al. 2017). Closing all feedlot beef production and restricting ruminant (cattle, sheep, goats) grazing to land that could not otherwise produce food crops would increase global land area for food crops by 50% (OECD/FAO 2019). Even the European Union does not produce enough feed for its livestock and relies on countries such as Brazil and Argentina for soya bean and maize for livestock feed (EU 2013; Amate and de Molina 2013). Figure 32 illustrates the growth (actual and predicted) in livestock numbers across different parts of the world.

Changes have also occurred in the way foods are being consumed. During the past 60 years or so, there has been growth in consumption of (fast) foods which are heavily processed, pre-packaged, chilled, and transported over long distances (UNEP 2016; Khoury et al. 2014).

The causes of unhealthy eating are complex, especially given both physical and economic factors affecting access to healthy foods (OECD 2019a,b; Alston et al. 2016). However, several studies have shown that the rapid change in diets is driven by factors such as global trade, urbanisation (Reardon et al. 2021), cultural change - including a growing proportion of women working outside the home and the exponential growth in eating out (Bleich et al. 2008; Seto and Ramankutty 2016; Popkin 2006; Hawkes 2006; Popkin et al. 2012), reduced exercise, and more sedentary lifestyles (Hawkes et al. 2005; Popkin et al. 2012; Graf and Cecchini 2017).

These changes have led to the overconsumption of calories, especially sugar and fats, and underconsumption of nutrients (Benton et al. 2021). According to Frison and Clément (2020), 'poor diets are responsible for 49% of the burden of cardiovascular disease, which remains the leading cause of death in the EU'. Afshin et al. (2019) found that, in 2017, the consumption of unhealthy diets caused 11 million deaths and 255 million disability-adjusted life years (DALYs). It is estimated that poor diets now account for 10% of the total global disease burden (UNEP 2016), and to more than 20% of adult deaths (GPAFS 2020; Credit Suisse 2021).

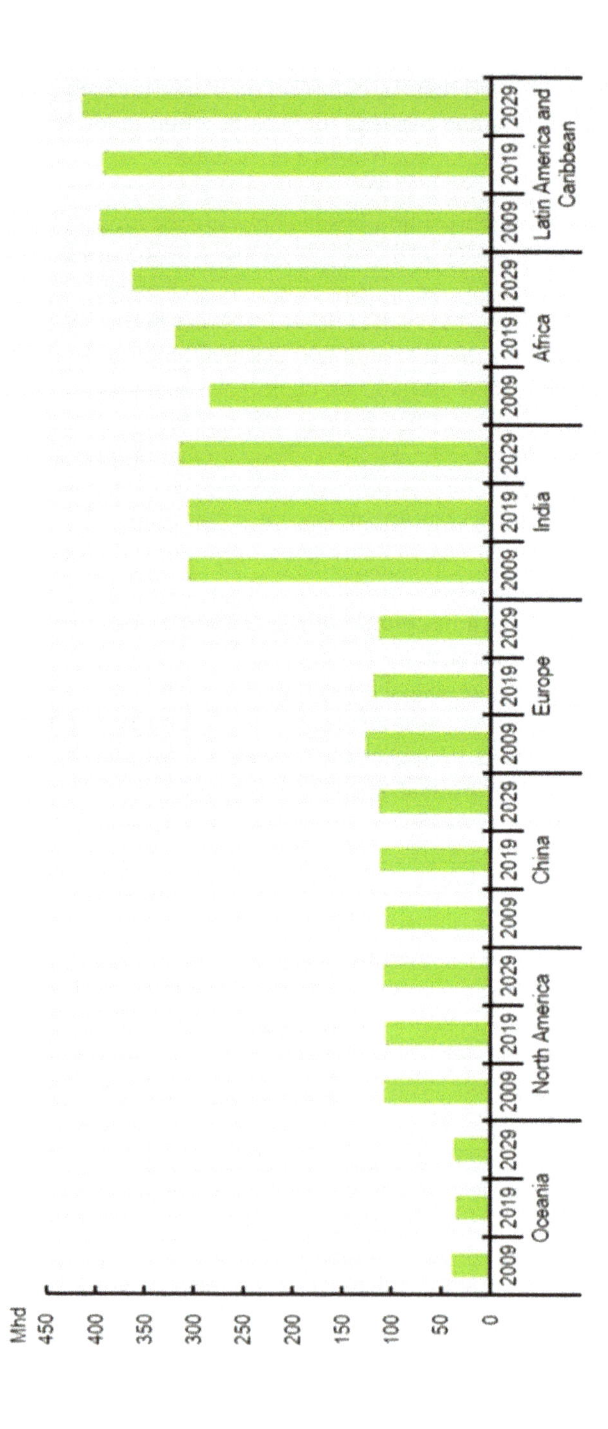

Note: Mhd = Million head.

Figure 32 Cattle inventory by region – 2009, 2019, 2029. Source: Adapted from: OECD (2021).

2.3.5 Obesity

Overconsumption, especially of sugars and fats, and underconsumption of fruits, vegetables, pulses, whole grains, and unsaturated fats (GLOPAN 2016; Gonçalves et al. 2021) have resulted in increased cases of obesity, chronic diseases, and non-communicable diseases (Butland et al. 2007; Wang et al. 2011), such as diabetes (NCD-RisC 2016; Ley et al. 2014; Hall et al. 2019), dementia (Hugenschmidt 2016), hyper-tension, strokes (Hall et al. 2019), and cardiovascular disease (Brandt and Erixon 2013; Omran 1971; Popkin 1993), a range of cancers (including breast and colon cancer), (Wagner and Brath 2012), osteoporosis and auto-immune conditions (Pimbert and Moeller 2018; Pangaribowo and Gerber 2016; Reardon et al. 2021; GPAFS 2020; Credit Suisse 2021; Butland et al. 2007; Wang et al. 2011; Moorsom et al. 2020). Even the children of obese mothers stand a higher chance of developing diabetes (Cordero et al. 2016; Tilman and Clarke 2014). Overconsumption of unhealthy diets has also been linked to increased absenteeism, reduced worker productivity, and early retirement (Credit Suisse 2021).

The scale of overconsumption is significant. Currently, approximately 2 000 000 000 people are either overweight or obese, a figure that is predicted to rise to 3 280 000 000 by 2030 (GPAFS 2016). According to Pereira et al. (2018), the prevalence of obesity has more than doubled since the 1980s to about 600 million people, leading to a situation, especially in developing countries, where people carry the dual burden of being both obese and malnourished. Figure 33 illustrates the percentage of overweight adults per region. More than

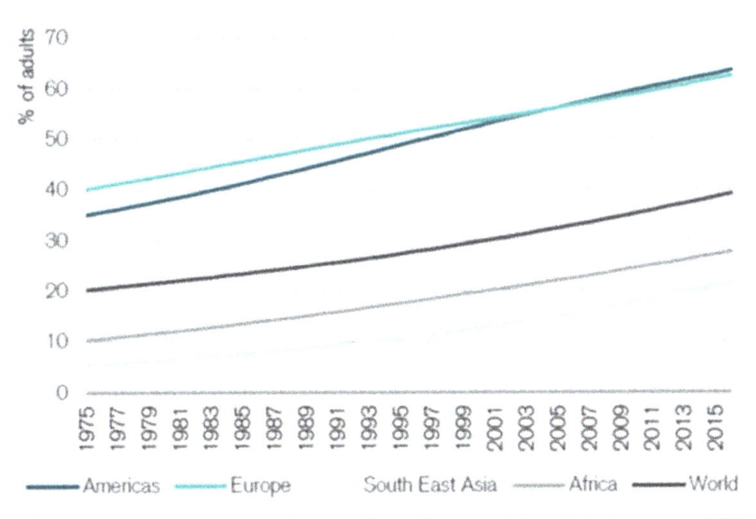

Figure 33 Percentage increase in the number of overweight adults between 1975 and 2015. Source: Adapted from: Credit Suisse (2021).

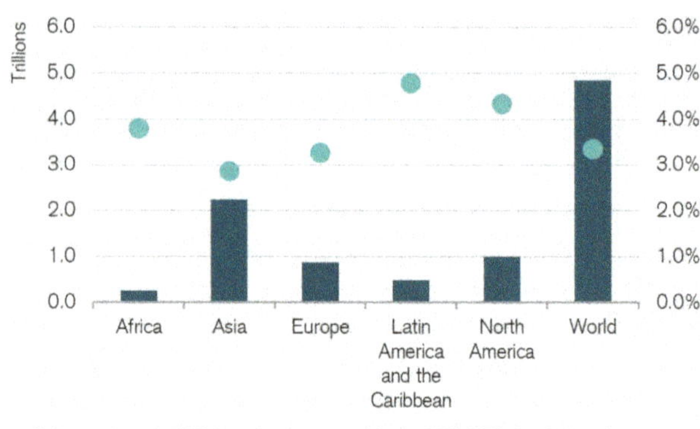

Figure 34 Estimated yearly GDP loss due to overweight, average 2020-35. Source: Adapted from: Credit Suisse (2021).

39% of the world's population is overweight or obese (FAO 2019). Obesity is also rapidly increasing amongst high-income urban families in SSA (Popkin et al. 2019; NCD Risk Factor Collaboration 2019). In 2013, it was estimated that 64% of the population of England was overweight, and that this was particularly prevalent amongst poor households (HSE 2013).

According to Brandt and Erixon (2013), obesity has been linked to the emergence of 80% of type 2 diabetes cases, 55% of hypertensive diseases, and 35% of heart disease in Europe. Benton and Bailey (2019) estimate that the global cost of type 2 diabetes alone in 2015 amounted to 4–5% of GDP, which was higher than the global contribution of agricultural GDP (3.79%). The annual global cost of obesity has been estimated at between US$2 trillion (De Schutter et al. 2020) and US$4.8 trillion (Credit Suisse 2021).

Figure 34 illustrates the regional costs of obesity in US$ trillions. In 2013, approximately 20 000 000 people in the EU suffered from disease-related malnutrition, which costs EU governments around €120 billion per year (Freijer et al. 2013).

In the EU, non-communicable diseases (NCDs) - often linked to obesity and unhealthy diets - are estimated to cost €700 billion annually, representing 70–80% of healthcare spending by Member States (Seychell 2016). According to HLPE (2017), obesity now kills more people than starvation. OECD (2019a) estimates that if the calorie content of energy-dense foods were cut by 20%, at least 1.1 million cases of NCDs could be avoided with associated healthcare savings of US$13 200 000 000.

2.4 Corporatisation of food systems

According to Reardon et al. (2019, 2021), food systems evolve through three stages:

- First, the traditional stage is characterised by flows of unprocessed or mildly processed foods along short supply chains coordinated by fragmented micro and small enterprises.
- Second, the transitional stage is characterised by flows of mildly to moderately processed foods along medium-length supply chains (linking rural to urban areas) by fragmented small- to medium-sized enterprises (SMEs).
- Third, the modern stage is characterised by flows of moderately and ultra-processed foods along long supply chains by increasingly concentrated processors and retailers, i.e. supermarkets.

A more recent classification is that of 'ultra-processed' foods, generally defined as foods with five or more ingredients, particularly additives associated with industrial food production such as artificial colouring and flavouring agents, sweeteners, and preservatives. As described earlier, the corporate concentration of food systems initially began in the early twentieth century, especially in the agricultural inputs sector. By the late twentieth century, corporate concentration had begun to accelerate in downstream parts of the food system, namely food aggregation, processing, wholesale, and retail. Concentration has been especially prolific in the processing and retail sectors in high-income regions, such as the USA, Europe, and Latin America (Lang 2003; Wilkinson 2009; Hawkes 2008; Chopraand Darnton-Hill, 2004), a process better known as the 'supermarketisation' of food (UNEP 2016). Concentration in food manufacturing has tended to focus around the ultra-processed market sectors, especially soft drinks, biscuits, and snack foods (Hawkes 2010; Wei and Cacho 2000). This concentration has resulted in two principal changes:

- Food system efficiencies increased, which, in turn, led to a reduction in the relative cost of ultra-processed foods (Baker and Friel 2016); and
- Market power, exerted via aggressive advertising and promotional campaigns, resulted in increased desirability and consumption of ultra-processed foods (FAO 2011a; Kearney 2010; GPAFS 2016; Hawkes 2007; Hawkes et al. 2009; Gonçalves et al. 2021; Monteiro et al. 2013; Frison and Clément 2020).

Concentration in food processing and retailing is also rapidly expanding across developing regions. Global food corporations have entered markets in middle, as well as low-income, countries through buyouts of domestic food companies, foreign direct investment (FDI), and aggressive sales campaigns (Schmidhuber and Shetty 2005). For example, in the 1990s, supermarkets had captured approximately 15% of retail food sales in Brazil. By the turn of the twentieth century, supermarket sales had reached more than 60% of retail food sales. A similar picture emerged in South Africa, which by 2003, 55% of retail food sales were through supermarkets (Reardon et al. 2003). Fast-food restaurants and street foods have grown rapidly in both high, middle, and low-income countries (Baker and Friel 2016; GPAFS 2016). For example, in 2012, McDonalds and Yum! Brands, respectively, which are based on 'cheap, convenient, and addictively tasty meals' (Schlosser 2001), had more than 35 000 and 39 000 restaurants, respectively, and outlets in over 100 countries (Mikler 2014). High levels of physical and economic access to fast foods have been a major cause of obesity (Stuckler et al. 2012; Moodie et al. 2013; Popkin 2006; Monteiro and Cannon 2012; Hawkes 2005; Lang 2003; De Vogli et al. 2011; Offer et al. 2010).

The food and beverage manufacturing industry now ranks amongst the top three economic activities in terms of value added in 27 OECD countries (Parsons and Hawkes 2019), generating over US$539 billion in revenues (Vanheukelom et al. 2018). In recent years, attempts by governments to influence customer purchasing habits and lifestyles have met with limited success (Marteau et al. 2012). It is argued that, due to the immense political power of global food corporations, questions around food insecurity and poor diets have not been framed in such a way as to directly challenge industrial food processing and retailing practices (IPES-Food 2016; Credit Suisse 2021).

2.4.1 Increased production and consumption of highly processed foods

Traditionally, food processing has been seen as benefiting consumers. Transformations, such as the boiling of rice, baking of bread, pasteurising of milk, and turning various fruits into jams and preserves, have improved both the quality, safety, and shelf-life of food products (Ludwig 2011; Wrangham 2013; Pollan 2013; IPES-Food. 2016). Several crops (such as cassava and yam) cannot be eaten unless they are processed. Indeed, if one were to take the US Institute of Food Technologists (IFT) definition of processed food, which is any food product that has undergone change from its natural state – including washing, cleaning, cutting, freezing, packaging, storing, filtering, and fermenting, and inclusion of preservatives, flavours, and nutrients, then virtually all food is processed to one degree or another (Credit Suisse 2021). During the past 50 years or so, however, the growth in production and consumption of (low-cost, energy-dense, and nutrient-poor) ultra-processed foods has grown exponentially (Bahadur et al. 2018; OECD 2021).

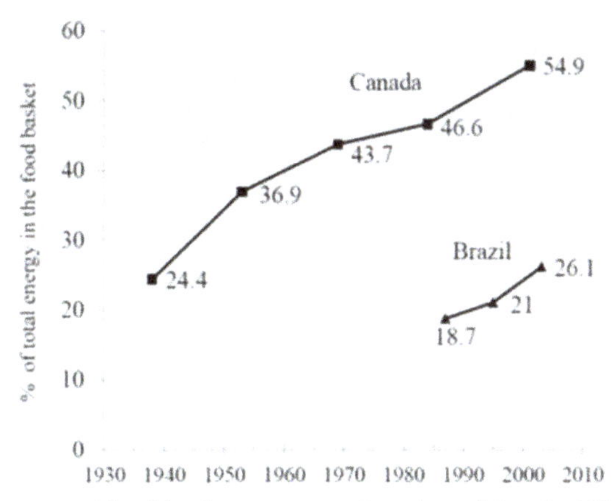

Figure 35 Increase total food basket energy in Canada and Brazil – 1930–2010. Source: Adapted from: Monteiro et al. (2013).

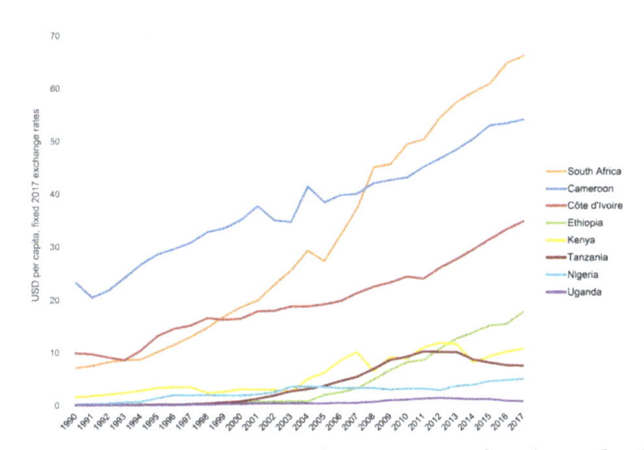

Figure 36 Increase in per capita expenditures on away-from-home food and non-alcoholic beverages. Source: Adapted from: Reardon et al. (2021).

Figure 35 illustrates the rapid growth in consumption of ultra-processed food and drink products in Canada and Brazil between 1930 and 2010.

Figure 36 illustrates the rapid growth in ultra-processed food and drinks in eight countries across SSA. Ultra-processed foods are basically foods that have primarily been produced from constituents extracted or synthesised from primary food sources such as oils, fats, sugar, starch, proteins, or flavour enhancers and colours (Baker and Friel 2016).

In Reardon's Traditional Stage (or pre-industrial stage) of food system transformation, fresh foods (generally grown locally) are prepared at home and consumed during regular mealtimes. Conversely, in Reardon's Modern Stage of food system transformation, ultra-processed foods (such as carbonated and soft drinks, burgers, frozen pasta, pizza and pasta dishes, nuggets and sticks, crisps, biscuits, confectionery, meats, cereal bars, and snacks products) (Monteiro et al. 2013), are prepared and/or sold in restaurants or by the roadside and purchased by busy consumers anytime as ready-to-eat or ready-to-heat 'convenience foods' (Monteiro et al. 2011). Often high in saturated (trans)fats, free sugars, and salt and low in dietary fibre, micronutrients, and phytochemicals, these ultra-processed convenience foods dominate the food systems of high and middle-income countries, to such an extent that many traditional diets have effectively been eroded (Monteiro et al. 2013; Swinburn et al. 2007; WHO 2003; Asfaw 2011; Monteiro et al. 2013; Moodie et al. 2013; Baker and Friel 2014; Zhou et al. 2015; IPES-Food 2016). For example, in the USA, 61% of food energy comes from ultra-processed foods (Poti et al. 2015).

Whilst fine when taken in small amounts, regular consumption of ultra-processed foods has been linked to both obesity and premature death (Hall et al. 2019; Cunha et al. 2018; Mendonça et al. 2016, 2017; Popkin 2019; Rico-Campà et al. 2019, Rohatgi et al. 2017; Vandevijvere et al. 2019; Demmler et al. 2018; Khonje and Qaim 2019; Khonje et al. 2020; Wanyama et al. 2019). Where the molecular composition of macronutrients and micronutrients in dairy products, eggs, wheat, nuts, and shellfish has been altered due to irradiation, homogenisation, thermal processing, or hydrolysis, resultant food products have also been associated with increased carcinogenic and gastro-enteric risks (Chaudhry et al. 2008; Shi et al. 2013; Verhoeckx et al. 2015). Consumption of both processed and ultra-processed foods, such as sugar-sweetened beverages (French and Morris 2006; Popkin and Hawkes 2016; WHO/FAO 2002), trans fatty acids (Stender et al. 2016), products high in sodium (NRC 2015; WHO 2014), and animal products (EPHA 2016; Feskens et al. 2013; Oggioni et al. 2015) has been closely associated with the increase in both obesity and non-communicable diseases such as cancer and cardiovascular disease (Ludwig 2011; Monteiro and Cannon 2012; Moodie et al. 2013; Moreira et al. 2015; Stuckler et al. 2012). According to Donnelly (2018), a 10% increase in the consumption of 'ready meals, sugary cereals, and salty snacks is linked to a 12% rise in cancer risks'.

2.4.2 Trade and diets

Trade can, and often does, result in more healthier and diverse diets. For example, in temperate countries, trade can provide access to a wide range of competitively priced fruits, vegetables, and nuts, which would not otherwise be available to consumer, and can do this all-year-round (GPAFS 2016), smoothing out both seasonal and climate-related price fluctuations (Hawkes 2015; Camara 2013; Jacks et al. 2011).

However, trade can also either increase the availability of cheap, nutrient-dense cereals, processed sugars, and saturated oils – for transformation into unhealthy convenience foods (Popkin 2002; Hawkes 2005) or provide access to already ultra-processed convenience foods at highly competitive prices (Reardon et al. 2021). Significant evidence exists to support the role of increasingly liberalised global food trade in population weight gain in low- and middle-income countries (LMICs) due to the importation of cheap, obesogenic food and drink products, such as high-sugar content drinks and packaged foods (Bleich et al. 2008; De Vogli et al. 2014; Friel et al. 2013a,b; Hawkes 2005; Hawkes et al. 2009; Lopez et al. 2017; Ng'ang'a et al. 2013; Stuckler et al. 2012; Wagstaff and van Doorslaer 2000). This exportation of developed country diets, known as 'cultural globalization', 'Westernization', and 'McDonaldisation', is leading to the replacement of local diets with western-styled diets (Khoury 2014; Ram 2004; Pingali 2006; Watson and Caldwell 2017). The length and opaqueness of food supply chains, and the often-complex processes that foods are exposed to, increasingly dislocate consumers from producers and reduce consumers' ability to trace the origins of the food they consume and its safety (Ercsey-Ravasz et al. 2012; Puma et al. 2015).

2.5 Financialisation of food systems

It is suggested that the financialisation of food systems came about due to overaccumulation in the global economy in the early 1970s. That is, the supply of goods and services became out of sync with demand. Or, to put it another way, over-production and the supply of products and services depressed both prices received, wages earned, and more importantly, return on capital invested. Financialisation allowed investors to make higher profits from stocks and shares, derivatives, insurance, and provision of credit (Arrighi 2009).

In its broadest sense, financialisation refers to the increased importance of financial motives, financial actors, financial markets, and financial institutions in both national and international economic and political decision-making (Epstein 2001; Epstein 2005; Engelen 2008; Montgomerie 2008). Financialisation, as defined by Epstein, refers to the 'increasing importance of financial markets, financial motives, financial institutions, and financial elites in the operation of the economy and its governing institutions, both at the national and international levels' (Epstein 2005). Putting it another way, financialisation is the extraction of added value through financial mechanisms compared to productive activities (Krippner 2011; Fairbairn 2014), namely, where liquid capital is exchanged in 'expectation of future interest, dividends or capital gains from the real economy where the actual production and trade of commodities occurs' (Isakson 2014).

Numerous studies have acknowledged the increasing share of financial profits since the late 1970s (Epstein and Jayadev 2005; Krippner 2005, 2011; Palley 2007; Orhangazi 2008). The process of financialisation is comprised of four interrelated trends:

- Domestic profits earned by, and economic and political dominance of, financial companies such as banks, mutual funds, hedge funds, pension funds, and private equity funds have increased;
- Other nonfinancial firms have crowded into the financial activities space, earning significant revenues from consumer credit lines and other financial activities;
- The so-called shareholder revolution of the 1990s placed increasing pressure on company CEOs to satisfy the demand for dividends from shareholders, and to consider ever shorter investment horizons (Parenteau 2005); and
- The switching of resources from productive activities to financial activities has led to reduced economic growth, increased volatility, stagnated wages for workers, and increased income equalities (Palley 2007; Krippner 2011).

In many respects, agriculture, and the natural world that supports it, has been subject to pressures from financialisation. The type and speed of change have accelerated quickly across Western Europe and North America since the Agricultural and Industrial Revolutions. Underpinned by fossil fuels and wholesale financial and political commitment in developed countries, agriculture experienced a significantly higher level of investment, innovation, and development since the Second World War. Much of this development, whether it be in mechanisation, pesticide, or GMO development, has been underpinned by either corporate profits or fluid and speculative venture capital investments. The fact that, against all odds, there are enough calories to feed the world, at a price that most can afford, and that leaves significant amounts of purchasing power to drive global manufacturing and service sectors, is, in a large part, due to the power of investment. The difference today is the footloose, often speculative, and unregulated role that money/capital generally plays in increasingly neoliberal food systems and the broader global economy (Palley 2007; Crotty 2009; Krippner 2011; Burch and Lawrence 2009; Daniel 2012). Whilst many see the process of financialisation as both natural and progressive, others see financialisation of natural resources (land, food, nature, water, etc.), as highly regressive and as a major threat to both society and the planet (Isakson 2014; McMichael 2016). The proceeding paragraphs outline a few of the major concerns that are being raised regarding the nature of footloose capital in food systems, especially since the Food Crisis of 2007–08.

2.5.1 Farmland

Investment in farmland by corporate investors, especially pension-fund managers, is not new. For the last 60 years or so, farmland has fallen in and out of vogue as both a medium to long-term financial and productive investment. During the late twentieth century, land was often purchased by pension companies and then rented out to farm management companies or independent farmers. In this way, investors received both a regular rental income (3-7% ROI), and significant capital gains on the land itself (~6-7% per annum) (Allison 2005). More recently, however, especially since the sub-prime property bubble implosion and the Food, Finance, and Fuel Crisis of 2008, agricultural land (especially cheap, yet productive, land in developing countries) has been increasingly viewed as a much safer and more tangible investment and has prompted 'land grab' investments (Fairbairn 2014; Cotula 2012; Daniel 2012; McMichael 2012; White and Dasgupta 2010; Clapp and Helleiner 2012; HighQuest 2010; Fairbairn 2014; Cotula 2012; McMichael 2012; White et al. 2012). Based on expected rising incomes from agricultural production, during the early twenty-first century, investments in land are increasingly being made with a longer-term view to own and operate (Cotula 2012; Grantham 2011; Kolesnikova 2011). Many of these farms produce 'flex crops' such as maize, soya bean, and sugar that can be sold into food, feed, and fuel markets depending on the strength of each market (McMichael 2012, 2014; Borras et al. 2012; Baines 2015).

Investment in farmland has been facilitated by the development of tailored land-specific financial derivatives (Burch and Lawrence 2009; McMichael 2012). Indeed, several agriculture funds now specialise in farmland acquisition, and more than 66 funds, including BlackRock, the world's largest asset manager, include land in their portfolio (Buxton et al. 2012; BlackRock 2012). In 2014, the estimated value of land owned by investment funds was US$30-40 billion and could easily increase to US$1 trillion (Wheaton and Kiernan 2012).

2.5.2 Agrofood enterprises

Whilst the financialisation of farmland may have received the greatest attention, the process has also impacted agricultural input companies (ICIS 2007; Davis 2011; Friedland 2011; IFC 2011), food processing, commodity trading (Murphy et al. 2012), and food retailing (Burch and Lawrence 2009). Indeed, according to Isakson (2014), agricultural input companies are earning higher revenues from selling inputs on credit than from the inputs themselves. In the case of food manufacturers, against a backdrop of exacting shareholder demands, processors have concentrated on the production of cheap and unhealthy food containing high fat, sugar, and salt content, often outsourced to developing countries with weak labour and environmental regulations, rather than focussing on healthier but less profitable processed foods produced locally (Moss 2013). In the USA, the world's largest grain traders - Archer Daniels Midland (ADM), Bunge, Cargill, and Louis Dreyfus have established investment vehicles that allow third-party investors to speculate on agricultural commodities and even land (Isakson 2014). For example, Calyx Agro, Louis Dreyfus's investment fund, buys, converts, and sells farmland in Latin America (Murphy et al. 2012). Cargill has also been offering financial services, such as commodity index funds, asset management services, and insurance, to third-party investors since 2003 (Isakson 2014).

2.5.3 Agricultural derivatives

The use of derivatives (in simple terms, forward contracts) and micro-insurance in food systems is also experiencing rapid financialisation (Clapp 2014). Whilst futures exchanges for agricultural commodities have existed for nearly 300 years (Bryan and Rafferty 2006), speculative investments in agricultural commodities, similar to those in agricultural land, have sky-rocketed since the 1990s and financial deregulation, and increased ten-fold after the food price spikes between 2007 and 2012 (Spratt 2013), and even recently during the 2020/2021 COVID-19 pandemic, when food speculation has been able to generate significant returns (Clapp 2014; Isakson 2014). For example, taking gold and precious metals aside, speculative financial investments in commodities reached US$5.85 trillion by June 2006, US$7.05 trillion by June 2007, and US$12.39 trillion by June 2008 (BIS 2009). Financial speculation in the USA wheat futures market increased from 12% in the mid-1990s to 61% in 2011 (Worthy 2011). In 2008, in the USA, index investors owned at least 35%, 42%, and 64% of maize, soya bean, and wheat futures contracts, respectively (Ghosh 2010). For the big four grain traders, ADM, Bunge, Cargill, and Louis Dreyfus, speculation over futures markets, including legitimate insider trading, has been part and parcel of their operations since they were established between 100 and 200 years ago (Morgan 1979). Indeed, they even established specialised financial subsidiaries, such as Cargill's Black River Asset Management, to sell Over the Counter (OTC) derivatives to third-party customers (Murphy et al. 2012).

The financialisation of agricultural derivatives has been blamed for exacerbating price spikes in wheat, maize, soya beans (IATP 2008, 2009; Robles et al. 2009; UNCTAD 2009; Wahl 2009; Ghosh 2010; Hernandez and Torero 2010; Ghosh et al. 2012; Spratt 2013), especially during 2007 and 2008, when more than 50 000 000 consumers were unable to afford adequate amounts of food (FAO 2013c). Figure 37 illustrates the rapid rise in rice, wheat, and maize prices in 2007 and 2008.

2.5.4 Food retailing

The financialisation of food retailing occurred much later than the food traders outlined above and has been closely linked to what has been termed as the supermarket revolution of the twenty-first century. This revolution had its roots

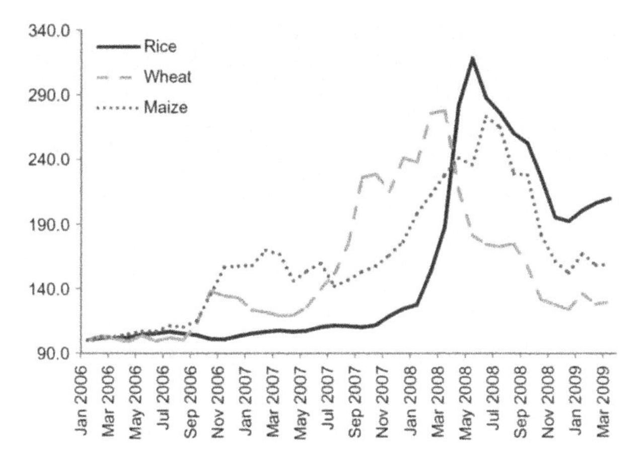

Figure 37 Index numbers of world trade prices of food grains. Source: Adapted from: Ghosh (2010).

in the establishment of the one-stop grocery shops of the 1800s (Sainsbury in 1869 in London, Albert Heijn in 1887 in Holland, and the first Casino store in France in 1898). These grocers were the precursors of the giant food retail oligopolies, such as Tesco (established in 1952), Carrefour (established in 1958), and Wal-Mart (established in 1962). Accelerated by easier foreign direct investment (FDI), as part of the World Bank's Structural Adjustment Programmes, and often close connections with post-colonial states, US and European-based supermarkets spread the supermarket revolution rapidly across most of the world (Weatherspoon and Reardon 2003; Reardon et al. 2003; Reardon and Hopkins 2006). For example, in 1990, in Latin America, supermarkets represented 10–20% of total food sales. By 2001, the five largest supermarket chains represented more than 60% of food sales (Clapp 2012). This phenomenon was replicated in India, China, and Vietnam and is currently underway across much of sub-Saharan Africa, where international supermarkets promote western-styled diets (Reardon et al. 2009).

The enhanced concentration and rapid geographical spread of food retail have made food retailers the most powerful actors in the global food system (Burch and Lawrence 2013), exercising power both over consumers and their immediate suppliers (traders, processors, wholesalers), right down to farmers that produce most of the raw material. This oligopolistic power endows retailers with significant power over what food (type and quality) is available for consumers (Friedmann 1994), how and where food is produced (Princen 2002; Kneen 1995), and the profit margins of farmers through to wholesalers (Busch and Bain 2004; Kneen 1995; Princen 1997; Dauvergne 2008). Just as with traders and processors, finance capital has infiltrated food retail and initiated structural changes such as rationalising the number of product lines and suppliers, instigating of 'just in time' inventory and supplies, lengthening payment times for suppliers (sometimes up to 90 days), direct contracts with farmers, reducing employee numbers and increasing the workload of remaining employees, downgrading commitments to labour and environmental standards, and rationalising land and property holdings – similar to the banking sector's closure of rural banks (Baud and Durand 2012; Reardon and Berdegué 2002). Retailers have also entered the financial sector offering bank accounts and credit, insurance and even mortgages (Juhn 2007; Werdigier 2009).

2.5.5 Impacts of food systems financialisation

Whilst financialisation of the global food system may have brought greater efficiencies and generated higher profits and shareholder dividends, this process has generated a growing number of concerns. Food systems are increasingly focussed on efficiency-driven and speculative profits. For example, land is increasingly seen as an asset (World Bank 2007, 2010; Bush 2012). It is something to be bought and sold, and to appreciate or depreciate. It's no longer seen as a home to, and a livelihood for, people and wildlife, or as a living, breathing, ecosystem.

The objective of market speculation is not to stabilise or raise overall output or to increase sustainability, it is to make a quick buck with little care for the economic, environmental, socio-cultural, or political fallout (Clapp 2014; Clapp and Fuchs 2009; Mittal 2009; Ghosh 2010; Friedmann 1994; Kneen 1995; Princen 1997, 2002; Clapp 2012). For example, traders (even countries) who restrict the supply of wheat, when traded volumes are already tight, to create and benefit from rapidly increasing wheat prices, have little or no concern for the mixed market signals that they create for other food systems actors (especially farmers) or the price volatility they create (Ghosh 2010; Breger Bush 2012; Spratt 2013; De Schutter 2010; Daviron et al. 2011; FAO 2011a; IFPRI 2011; UNCTAD 2011; Russi 2013). Retailers' growing control over both physical and economic access to food is especially pertinent for the poor in both developed and developing countries, where increased food prices and food price volatility induced by market speculation can mean the difference between eating dinner and going to bed hungry (Dauvergne 2008, 2010). In more extreme cases, the struggle to afford to eat can lead families to withdraw their children from school, work longer hours, and enhance emotional stress, domestic violence, and crime (Clapp 2012; Heltberg et al. 2013).

Chapter 3

Drivers and outcomes of change

In search of more sustainable food systems, there seem to be three central challenges that different approaches to 'sustainability' need to address. These are

- reconciliation of food supply and demand;
- redressing climate change (mitigation and adaptation), and depletion of biophysical resources; and
- addressing inequality in food systems.

These are explored in more detail below.

3.1 Reconciliation of food supply and demand

When it comes to providing nutritious food for all, in a manner that balances prudent management of the planet's environment and biophysical resources, population growth is the most pressing challenge. Given its centrality vis-à-vis the planet's carrying capacity, there is surprisingly little recent discussion of population growth and food. Where the topic is broached, it tends to form part of doomsday narratives that report the planet has either exceeded its carrying capacity or is coming close to it. There are some ongoing efforts, such as those of Bill Gates, that attempt to both reduce population growth, whilst, at the same time, investing in technological fixes that aim to increase food production. Namely, a meeting in the middle between food production and projected food consumption.

Those tackling population growth from a planetary boundaries perspective tend to draw on neo-Malthusian models (named after the economist Thomas

Malthus) of the population-limiting relationship of food production. Namely, the Earth's capacity to supply food, fuel, and fibre only goes so far and will, at one stage, restrict the human population (Malthus 1798; Ehrlich 1968; FAO 2013d; IFAD 2011; Ehrlich and Holdren 1971; Hardin 1968, 1993). Couched in humanity's inherent capacity to fix whatever problems materialise, those adhering to a more Boserupian interpretation (named after the economist Ester Boserup) of the Earth's carrying capacity, tend to place faith in the capacity of technologies and finance to constantly produce more food and ensure that the supply of food matches the demand.

Since the 1960s, the Boserupian approach to feeding the world has prevailed (OECD-FAO 2017). As previously mentioned, this belief in technology and finance, supported by strong political and economic infrastructure, has moved the world from aggregate food shortages to aggregate food surpluses (IPCC 2019). For the past 30 years, global food production has increased by more than 2% per annum, faster than population growth, which has grown by a little over 1% per annum (UN 2019a). Indeed, between 1961 and 2017, global food production increased by 240%. Whilst area expansion did occur during this period (Martin et al. 2018), much of the growth came from the increasingly intensified industrial agri-food system (IPCC 2019). However, the distribution of food is another matter (FAO, IFAD, UNICEF, WFP, and WHO 2017).

Whilst there has been significant progress in addressing undernutrition in developing countries since the 1970s, reducing rates of undernutrition from 37% in 1969–71 to less than 15% today (Steiner et al. 2020), in the contemporary period, exacerbated by the COVID-19 pandemic, approximately 820 000 000 people on the planet do not have access to sufficient calories (Willett et al. 2019; FAO 2019), and 3 000 000 000 do not have access to a healthy diet (FAO, IFAD, UNICEF, WFP, and WHO 2020). And, in the future, the population is predicted to grow further from 8 000 000 000 in 2023 to 9 700 000 000 by 2050 (UN 2019a, 2020; UNFPA 2023), necessitating an increase in food production of 50–70% (UN 2019b).

According to Lipper et al. (2014), most of this population growth (2.4 billion) is expected to be in developing countries, especially South Asia and sub-Saharan Africa, where more than 20% of the population are currently malnourished. Regarding Africa, more than 250 000 000 people, 19.1% of the population, are currently undernourished. By 2030, this figure is expected to increase to 412 000 000, 25.7% of the population (FAO 2020). Set against the predicted doubling of population to 2 000 000 000 by 2050, SSA needs to increase crop production by 260% to meet its food needs (UN 2019b). According to the UN (Wikipedia 2021a), using a high-growth scenario, the global population may have reached 15 600 000 000 by 2100. Many already determined that it will be either hard or impossible to produce enough food for the UN's median forecast of 11.2 billion people by 2100 (Rees 2019). Conversely, if population growth, especially in developing countries, can be arrested, the global population could decrease; perhaps to 7 300 000 000, or even 6 000 000 000 (Dixson-Declève 2022) by 2100 (Fig. 38).

If the absolute growth in population was not enough to contend with, the world faces a worrying convergence in diets as incomes rise, with those currently eating plant-based foods increasingly consuming more western-styled diets of animal-based proteins, especially red meat from ruminants. Figure 39 clearly illustrates the relationship between income growth and consumption of animal-based proteins.

Combined with the predicted population increase, this convergence of diets is likely to lead to nearly a 50% increase in land required for food production (Fig. 40).

Given the limited agricultural land available, concerns are also being raised regarding the poor feed conversion to meat of cattle, which can be anywhere from 6:1 to 20:1 (Mottet et al. 2017). To put this another way, it takes between 6 and 20 tonnes of feed (e.g. maize and soybean) to produce 1 tonne of live weight. Pigs have a better feed conversion rate of ≤3.5:1. Poultry has the best feed conversion rate of ≤2:1 (MAFRA 2021). Currently, approximately 560 million ha of the world's cropland is being used to produce feed crops of cereals and oilseeds instead of crops consumed directly as food (OECD/FAO 2019). Furthermore, more than 3 billion ha of land globally is used for grazing,

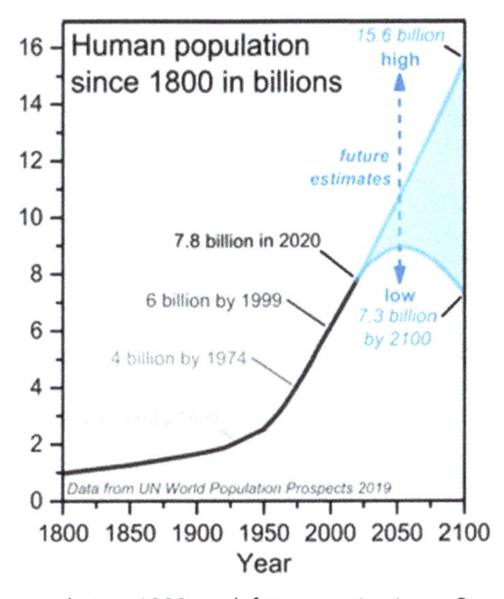

Figure 38 Human population 1800 and future projections. Source: Adapted from: Wikipedia (2021a).

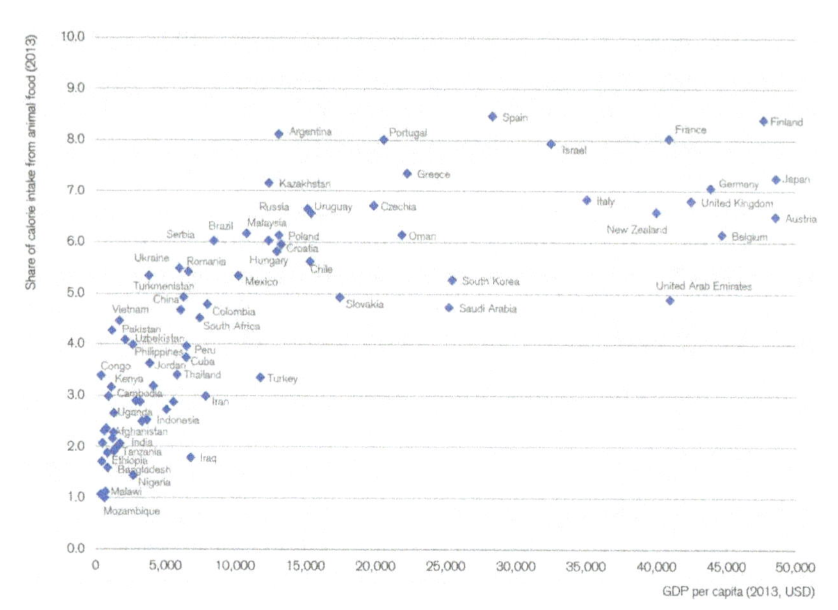

Figure 39 Increase in animal-based protein consumption in relation to GDP. Source: Adapted from: Credit Suisse (2021).

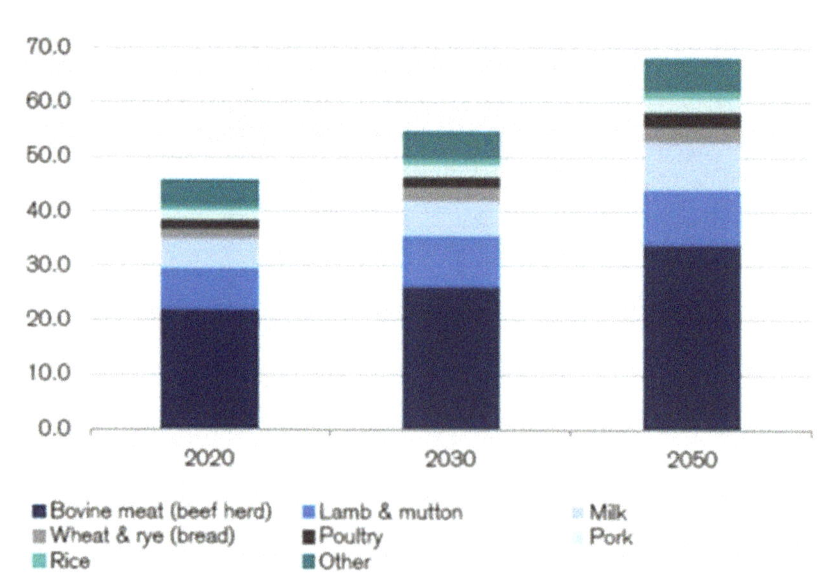

Figure 40 Total land use needed to feed the global population. Source: Adapted from: Credit Suisse (2021).

of which 685 million ha could be used for producing crops for food (Mottet et al. 2017). The inclusion of this land would increase global arable land by half (OECD/FAO 2019). The remaining 2 315 000 000 ha of land producing approximately 4 560 000 000 tonnes of fodder (grass, leaves, crop residues, etc.) could not have been directly consumed by people (Mottet et al. 2017).

In addition to population growth (including the expansion of towns and cities) and dietary convergence, there is a need to consider the growth of the bioeconomy, where land may be increasingly used for non-food purposes, such as biofuels and other plant-based products (nutraceuticals and other industrial uses). Given existing opportunities associated with the production of green energy (solar, tidal, wind, geothermal, hydro, etc.) and the nascent nature of plants being used to produce nutraceuticals and other high-value organic products, it is difficult to predict the growth in biofuels. However, to date, even the diversion of crops, such as maize, wheat, sugar cane, and oilseed rape, for the production of biofuels has been linked to the volatility of crop commodity markets and food price rises. Figure 41 illustrates the recent increase in the use of maize and wheat for bioethanol.

When all the conditions above are factored together, economic growth and increased living standards across the globe signify increasingly unsustainable levels of natural resource consumption and negative impacts on the environment (Garnett and Dodfray 2012; Pereira et al. 2018; Béné et al. 2019a). Figure 42 illustrates the relationship between high levels of consumption and humanity's unsustainable use of the planet.

Figure 41 Use of maize and wheat to produce bioethanol 2005/2006 to 2018/2019. Source: Adapted from: Green Peace (2020b).

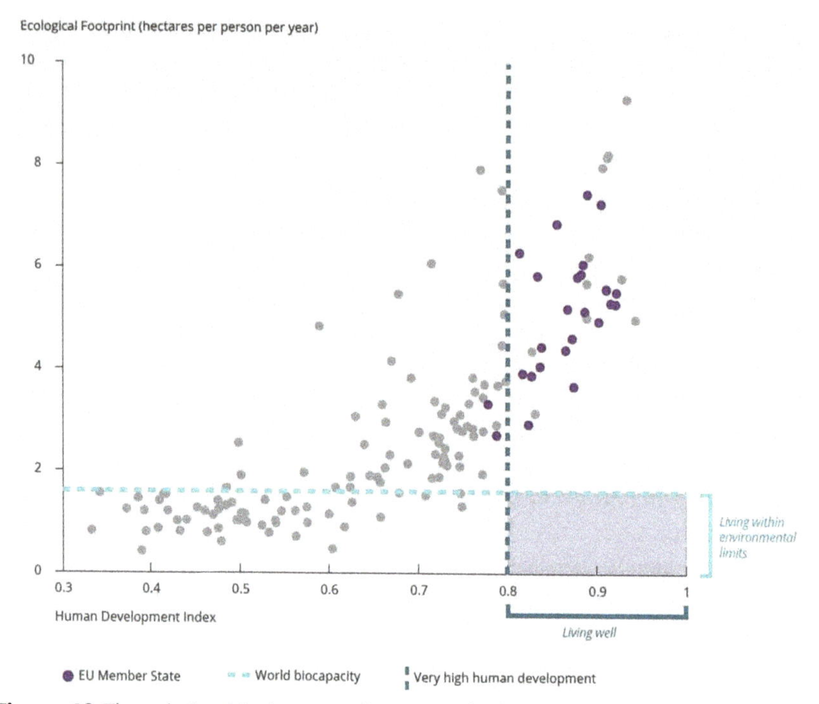

Figure 42 The relationship between living standards and ecological footprint in ha/person/year. Source: Adapted from: EEA (2018).

3.1.1 *Improving diets*

Concerns around both the quantity and quality of diets have taken centre stage in recent years. This is due to the central role that diets play in both human health and nutrition, as well as the impacts that the production of food has on climate, natural resource use (water, nutrients, soil), and biodiversity (Benton et al. 2021; Herrero et al. 2014; Tilman and Clark 2014; Ranganathan et al. 2016). Given the growing prevalence of both undernourishment, overnourishment (obesity), and undernutrition across the globe, local and national governments, as well as inter-governmental bodies and NGOs, have been exploring ways to improve diets (Benton et al. 2021; Willett et al. 2019).

3.1.1.1 *What needs to be done?*

There is growing support from governments, inter-governmental bodies, academics, and NGOs for radically changing diets (Willett et al. 2019). Indeed, many have called for a Great Food Transformation (Willett et al. 2019). A broad consensus has been reached around the need for a significant reduction in the consumption of meat (for those currently over-consuming) (Shepon et al. 2016; Alexander et al. 2017), especially red meat and dairy products, (Nugent 2011; Popkin et al. 2012; van Dooren et al. 2014; UNEP 2016), and ultra-processed foods (Reardon et al. 2021), and to increase the consumption of a diverse range of healthier and more land and resource-use efficient foods, such as a fruits, vegetables, vegetable oils, pulses, whole grains, nuts, and fish (Nugent 2011; Popkin et al. 2012; FAO, IFAD, UNICEF, WFP, and WHO 2020; WHO 2018; Benton et al. 2021; OECD 2021; Dwivedi et al. 2017; Frison and Clément 2020).

The production of red meat generates significant greenhouse gas emissions and uses significant amounts of cropland, water, and fertilisers (especially nitrogen and phosphorus) compared to crop production (IFPRI 2017). Poore and Nemecek (2018) highlight the huge difference in environmental footprint between red meat and tofu consumption. They contend that producing 100 g of beef protein requires approximately 164 m^2 of land. This compares with only 2.2 m^2 of land to produce 100 g of tofu. However, despite the growing consensus for the consumption of animal-sourced foods, evidence for the abandonment of livestock farming is far from conclusive. Whilst a few studies (Poore and Nemecek 2018) call for a planetary transition to vegan diets, most food system analysts still see a key role for farm animals. Key examples include the nutritional benefits of consuming small quantities of meat in otherwise plant-based diets, and the role of livestock in the circular food economy (de Boer and van Ittersum 2018; Van Zanten et al. 2018) through the recycling biomass unsuitable for human consumption, and in recycling nutrients and building soil organic matter through the production of manure (Schader et al. 2015; Van Kernebeek et al. 2016; Röös et al. 2017; Van Zanten et al. 2018; Van Hal et al. 2019). Parallel calls have also been made to increase investment in research on nutrient-dense orphan crops (Dwivedi et al. 2017; Sanchez 2020), and the cost-effective and nutrient-preserving production (including irrigation for all-year-round production), processing and storage, transportation (including feeder-road construction), and trade of nutrient-dense crops and food products (FAO, IFAD, UNICEF, WFP, and WHO 2020; GPAFS 2020). Figure 43 illustrates Harvard University's interpretation of what a healthy diet should look like.

The Harvard Healthy Eating Plate model represents a significant change to current diets. Globally, if everyone embraced the change, consumption of red meat would need to decrease by 67%, whilst consumption of fruit, vegetables, nuts, and whole grains would need to increase by 146%, 44%, 462%, and 500%, respectively (Credit Suisse 2021). However, the high cost of this diet would be beyond the financial reach of more than a quarter of the world's current population (Herrero et al. 2020a).

For many years, governments have used public funds to undertake diet-related research and develop and pilot elaborate healthy eating advertising campaigns, promotion of healthy eating in schools, and compulsory labelling

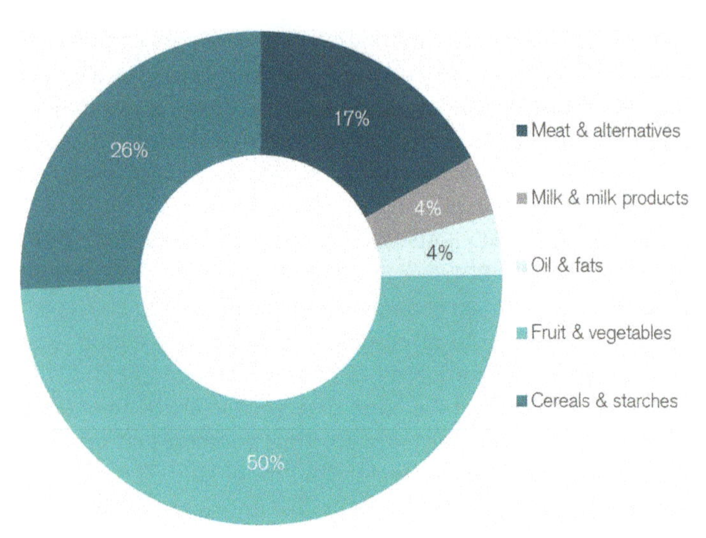

Figure 43 Harvard University's healthy eating plate model. Source: Adapted from: Credit Suisse (2021).

schemes (De Schutter et al. 2020; Baragwanath 2021; OECD 2019a). To date, these campaigns have had only a marginal impact on household diets. According to the FAO, IFAD, UNICEF, WFP, and WHO (2020), nutrition-sensitive social protection policies are needed to provide economic access to nutrient-dense diets for the most vulnerable populations.

Incentivisation of healthy and environmentally friendly eating has already been piloted. For example, in the USA, the USDA-Food and Nutrition Service (FNS) has been piloting a means-tested subsidised healthy eating programme called the Supplemental Nutrition Assistance Program (USDA 2021). This programme provides electronic bank cards to qualifying families who can then purchase healthy foods at participating food outlets. Economic simulations and controlled field trials conclude that financial incentives for the purchase of healthy foods (such as fruits and vegetables) and cash disincentives for the purchase of unhealthy foods can promote the consumption of more healthy diets. However, other studies have concluded that money saved from purchasing subsidised healthy foods is often used to purchase unhealthy products (Nnaoaham et al. 2009; Thow et al. 2014; Dharmasena and Capps 2012).

The sugar tax in Mexico has often been quoted as a relatively successful example of disincentivising sugar consumption. To redress a growing pandemic of obesity amongst its populace, in 2014, the Mexican Government introduced a levy of 1 peso/L on soda containing sugar. Within the first year of enactment, the sugar tax reduced the consumption of sugar-sweetened soda by 5.5%, followed by a 9.7% reduction in the second year. Following Mexico's lead, approximately 40 countries now implement a sugar tax. Recent analyses have shown that a 10% tax on sugar can lead to a reduction of 10% in the consumption of sugar-sweetened drinks (Credit Suisse 2021). However, sugar taxes have met with significant resistance from food manufacturers (Credit Suisse 2021).

Another example of food taxation is the so-called chip tax, which was introduced by the Hungarian Government in 2011 to curb the consumption of processed foods high in salt, sugar, or caffeine. The government also passed associated food industry regulations that formally restricted trans fats to a maximum of 2% of total fat content (SAPEA 2020). In some circumstances, even the threat of legislation can result in significant changes. For example, in the UK, the threat of the soft drinks levy prompted several leading companies to reformulate their products (Davies 2019).

Lastly, whilst civil society organisations such as DebateWise in the USA are advocating for a ban on junk food sales (DebateWise 2021), due to the significant political and economic power of food manufacturers, few attempts have been made to date to ban unhealthy foods. The World Health Organization continually lobbies national governments to legislate to ban industrially produced trans fats. Indeed, bans on trans fats are in place in a growing number of countries (Health Canada 2018). Other government interventions are moving ever closer to banning unhealthy foods. For example, according to the BBC (2021), to reduce exposure of children to advertisements promoting the consumption of unhealthy snacks and drinks, the UK Government proposed a UK-wide ban on TV adverts for foods high in sugar, salt, and fat before 9:00 pm. Due to pressure from food processing and retail lobbies, this ban has yet to come into force. However, as part of the UK Government's campaign to reduce obesity, as of 1 October 2022, unhealthy snacks were banned from prominent locations, such as checkouts, at supermarkets in England (Independent 2022). Products affected include chocolate, burgers, soft drinks, cakes, sweets, ice cream, biscuits, sweetened juices, crisps, chips, and pizzas (Guardian 2020c). Ultimately, healthy diets do not always correspond to sustainable diets (Béné et al. 2019a).

3.1.2 Changes in food consumption

It has been widely recognised that, alongside the reduction of food losses and food waste, changing human diets away from meat and towards plant-based foods is by far the most impactful way to ensure food security, improve consumer health, and reduce many of the negative externalities (including habitat and species loss and climate change) (Bryngelsson et al. 2016; Dwivedi et al. 2017, Gill et al. 2015; Hedenus et al. 2014; Greenpeace 2020a; Steiner et al. 2020). Poore and Nemecek (2018) suggest that, if natural regeneration was promoted on surplus farmland, an additional 8.1 billion metric tons of atmospheric CO_2 could be captured and stored each year for the next 100 years. Indeed, Springmann et al. (2018) are more optimistic, suggesting that the move towards plant-based diets could reduce food system emissions by up to 80%. Poore and Nemecek (2018) project that the transition to plant-based diets could reduce the area of land required for food production between 2.8 and 3.3 billion ha, which equates to a 76% reduction in land. In turn, this would reduce the global food systems' GHGEs between 5.5 and 7.4 billion metric tons of CO_2, which equates to a 49% reduction.

However, as Erb et al. (2016) point out, it is important to note that a transition from predominantly cereal-based farming systems to healthier crops, such as nuts and legumes, will result in both lower yields and reduced water use efficiency. Indeed, according to Willett et al. (2019), much of the expected reduction in the agricultural land area needed to feed a growing population would be lost due to these low yields. It should also be noted that a balance also needs to be struck regarding, on the one hand, ensuring that trade increases the availability of healthy foods and provides income stability for producers, against, on the other hand, that food miles associated with international trade, especially trade in unhealthy ultra-processed foods, are minimised. IGS (2019) stresses that only a combined approach can significantly reduce food system emissions. (see Fig. 44).

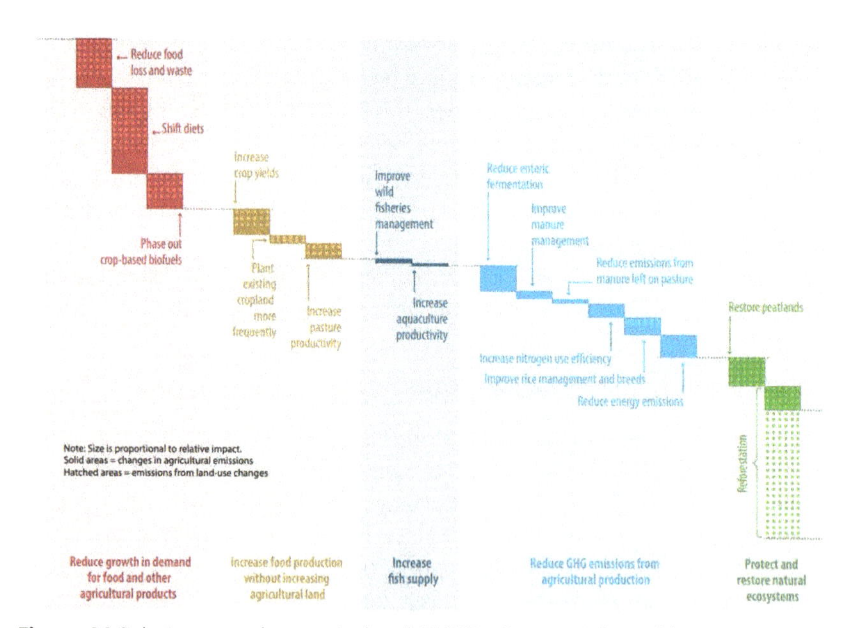

Figure 44 Solutions to reduce agricultural GHGEs. Source: Adapted from: IGS (2019).

3.1.3 Reducing food loss and food waste

Ultimately, whilst changes in diet can reduce the food system's impact on both climate and natural resources, changes in diet must go together with other measures. If humanity is to live within planetary boundaries, reducing food loss (quantity and quality) and food waste is an equally imperative challenge (FAO, IFAD, UNICEF, WFP, and WHO 2020; Poore and Nemecek 2018; Harwatt et al. 2017; Bahadur et al. 2018; Shepon et al. 2018). It is estimated that reducing food loss and food waste by 50% would lead to a 16% reduction in the global food system's GHGE, cropland needed to feed the population in 2050, and natural resource use (water, nitrogen, and phosphorus). If, for example, food loss and waste were reduced by 75%, this could lead to a 24% reduction in GHGEs. If high-level reductions in food loss and waste could be achieved in tandem with the dietary changes outlined above, GHGEs could be reduced by up to 50% (Benton et al. 2021).

As mentioned previously, farm-level and immediate post-farm food loss is most acute in low and middle-income countries (Willett et al. 2019; SAPEA 2020). Investments in food harvesting, threshing, drying, storage and packing, and preservation, road infrastructure, and transportation (including cold chain technologies for fruit and vegetables), and good access to the market are required (HLPE 2013; Credit Suisse 2021; Godfray et al. 2010).

On the food waste side, up to 50% of losses tend to occur at the household level in the developed world (Willett et al. 2019; Stenmarck et al. 2016). The economic cost of food waste is estimated to be around €900 billion per annum, with an additional €800 billion in social costs (SAPEA 2020). Along with the disposal of waste, this generates between 8% and 10% of global GHGEs (Mbow et al. 2019). Food waste and food losses fall into two broad categories: preventable and non-preventable. Preventable loss and waste refer to that occurring along the food chain between harvest and consumption, which can be reduced or eliminated by improved harvesting, primary processing (threshing, drying, grading, etc.) and storage, better packaging and labelling, and more sustainable consumer choices/behaviour (SAPEA 2020). Non-preventable waste includes leftover meals in the home and in restaurants. According to Cornell University's Food and Brand Lab, approximately 17% of meals in restaurants remain uneaten. Less than half of uneaten food is taken away to be eaten at home (Credit Suisse 2021).

Three approaches have been developed to address non-preventable waste (including unfinished meals in the home). These are (i) prevention (at source), (ii) recovery (food sharing – Morone et al. 2018; Davies and Legg 2018), and (iii) cost-efficient recycling (animal feed and compost, etc.) (Mourad 2016; Morone 2019).

Several non-profit organisations have been established, especially across the developed world, that are dedicated to 'fighting food waste' (Mourad

2016). In line with Target 12.3 of the UN Sustainable Development Goals, many governments, such as the USA, France, Mexico, and South Africa, have signed up to reduce food loss and food waste by 50% by 2030 (Bloom 2015; Willett et al. 2019). To date, food corporations have been the most active in reducing food waste. This enthusiasm is driven by both cost savings that make sound financial sense and corporate social responsibility. However, significant opportunities remain to reduce food waste at the consumer level. Whilst they should not be viewed as a panacea, public information campaigns and improved nutritional guidelines have been found to reduce school canteen waste by up to 28% (Reynolds et al. 2019). Altering plate type and size, known as 'nudging', has also been shown to reduce food waste in restaurants by up to 57% (SAPEA 2020). Unless food losses and food waste are tackled head-on, it is estimated that by 2050, the demand for food, and its incumbent environmental footprint will have increased by 50% (SAPEA 2020). Figure 44 illustrates the likely effect of reducing food loss and waste, as well as changing diets, increasing productivity, and rewilding agricultural lands, on agricultural GHGEs.

3.2 Redressing climate change (mitigation and adaptation) and depletion of biophysical resources

3.2.1 Improving resource use

The previous chapter described growing concerns over unsustainable resource use, especially soil (Holden et al. 2018), nutrients such as phosphorous, water (Recanati et al. 2019), and fossil fuels (UNEP 2016). Whilst water is renewable (via the water cycle), only green water (via rainfall) has a relatively quick, albeit erratic cycle. Recharging of groundwater reserves has a much longer cycle of perhaps hundreds or hundreds of thousands of years and is probably best treated as a non-renewable resource (OECD 2002). Known sources of phosphate are more finite and even less renewable, as phosphate, once used in agriculture and industry, tends to end up dispersed in rivers, lakes, and oceans (Steffen et al. 2015a), and given current technologies, pretty much lost to mankind (Molden 2007; Bindraban et al. 2012; Van Vuuren et al. 2010; UNEP 2016). Rock phosphate deposits are expected to be mined out within the next 50–100 years (Cordell et al. 2009). Calls, therefore, have increasingly come to improve the resource-use efficiency of these vital resources, and, where possible, to recycle (UNEP 2016).

There are a range of positions regarding how serious resource-use concerns are. The more Boserupian opinions believe ardently in technological solutions to resource-use problems and in the power of the market, via price signals (driven by supply and demand dynamics), to balance the use of key agricultural resources. At the other end of the spectrum, many are convinced that we have

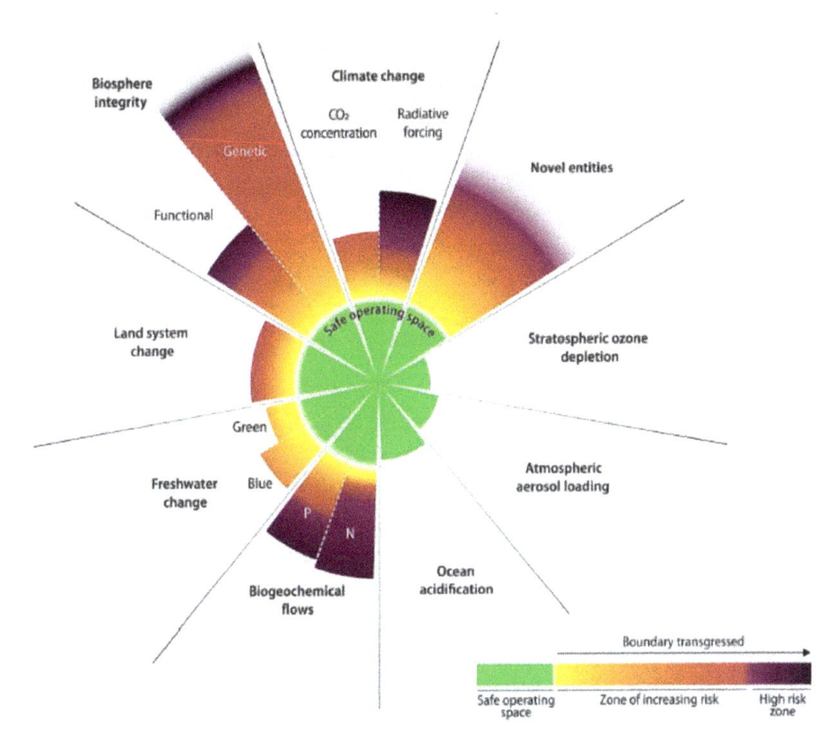

Figure 45 Status of control variables for all nine planetary boundaries. Source: Adapted from: Richardson et al. (2023).

already, or will soon, reach planetary boundaries for certain resources (such as demand for cropland, water, and phosphorus) (Rockström et al. 2009a; Raworth 2017; Springmann et al. 2018; Ehrlich 1968; Meadows et al. 1972). According to Richardson et al. (2023), the planet has already surpassed six of the nine planetary boundaries (Fig. 45).

In the case of water, it has been estimated that the global freshwater planetary boundary for human use is approximately 2800 km³/year, of which between 1800 and 2100 km³/year is currently being used. Of this amount, the food system consumes approximately 1400–1800 km³/year (Wada et al. 2011; Willett et al. 2019). By 2050, agriculture's share of global water use is likely to have increased from the current level of 75–84% (Rosegrant et al. 2009) to 90%, approximately 2500 km³/year (Willett et al. 2019).

Significant scope currently exists for the more efficient use of water. Industry's water use efficiency has already improved, and there are many existing opportunities to improve agricultural water use efficiency through improved crop breeding, new irrigation technologies (Jägermeyr et al. 2015), and the move towards more plant-based diets (Jalava et al. 2014). Improved production practices already have the potential to reduce water use by approximately 30% and reducing food waste by 50% could reduce water use by 13% (Springmann et al. 2018). However, changes in diets (including eating more water-intensive products such as legumes and nuts) would reduce savings made from reducing food waste and improving water management (Willett et al. 2019). Increasingly, concerns around resource use have driven demands to adopt a more circular approach to resource management.

3.2.1.1 Circularity in the food system

There is a growing need to move from the current linear approach of 'extract-produce-consume-discard' towards a more circular food system (Termeer et al. 2017; O'Sullivan et al. 2018; SAPEA 2020; Bauwens et al. 2020; Ruben et al. 2021). The concept of circularity is derived from industrial ecology and espouses reduced consumption of resources and polluting emissions by either preventing/reducing the use of non-renewable resources and reusing and recycling of materials and substances (Ghisellini et al. 2016; Jurgilevich et al. 2016). In food systems, circularity has significant traction with regard to closing nutrient loops, improving nutrient-use efficiency, improving food quality, and reducing food waste, or bio-refining unpreventable losses (de Boer and van Ittersum 2018; SAPEA 2020). The principle of circularity is especially pertinent with regard to the management of finite resources such as land and rock phosphate. Circularity promotes the prevention of food system leakage, such as nitrogen and phosphates into rivers, lakes, and oceans, and the extraction and reuse/recycling of phosphates as well as nitrogen from human sewage treatment plants, food processing, and composting operations, and intensive livestock production facilities (Willett et al. 2019). This looping back helps to keep finite resources in the food system and reduces the release of potentially environmentally damaging pollutants (Ghisellini et al. 2016; Jurgilevich et al. 2016). Using a business-as-usual scenario, it is estimated that feeding a global population of 10 billion or more people by 2050 will require application rates of nitrogen and phosphorus fertilisers that will exceed planetary boundaries (Mueller et al. 2012). As such, the EU is planning to reduce nutrient losses (mainly nitrogen and phosphorus) by 50%, which translates to a reduction in fertiliser use by more than 20% by 2030 (Credit Suisse 2021). Figure 46 illustrates the European Commission's interpretation of circularity.

Companies, such as Yara, are already considering using 'greener' hydrogen-related technologies to produce ammonia-based fertilisers (Credit Suisse 2021). Due to the inevitable leakage of nutrients, proponents of circularity advocate for the production of synthetic nitrogen fertiliser using renewable energy and the use of livestock to recycle nutrients and build organic matter via the consumption of biomass that is inedible to humans (UNEP 2016; Van Zanten

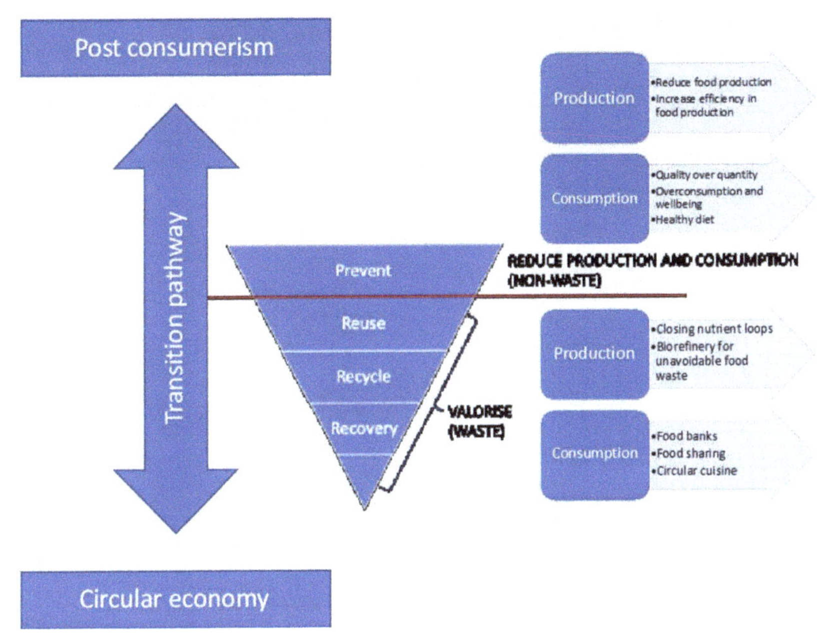

Figure 46 The circular economy concept. Source: Adapted from: SAPEA (2020).

et al. 2019). It is estimated that, by using inedible biomass from grasslands, which are unable to produce arable crops, and recycling by-products, this approach can provide approximately 33% (9-23g) of humanity's protein needs (~50-60 g/day) (Van Zanten et al. 2018). Adoption of circularity principles not only increases resource-use efficiency and reduces environmental pollutants; it can also make sound financial sense. For example, according to the Ellen MacArthur Foundation (2015), the introduction of circularity principles could save as much as €420 billion in Europe by 2030; €60 billion of food industry savings, and €360 billion in reduced environmental and health costs.

3.2.2 Addressing climate change

Climate change is now acknowledged as one of the most serious challenges facing food systems, and indeed, society and economies in general. Without the unprecedented adjustment of our current activities, humanity could cause irreversible changes to our global climate and suffer the catastrophic whirlwind of climate breakdown (Greenpeace 2020a). There is a clear and present need to significantly reduce the amount of greenhouse gases (e.g. carbon dioxide, methane, and nitrous oxide) released into the atmosphere from the agriculture sector (IPCC 2014, 2019). Currently, GHGEs from food systems contribute between 8·5 and 13·7 Gt of carbon dioxide equivalent per year. Of this number,

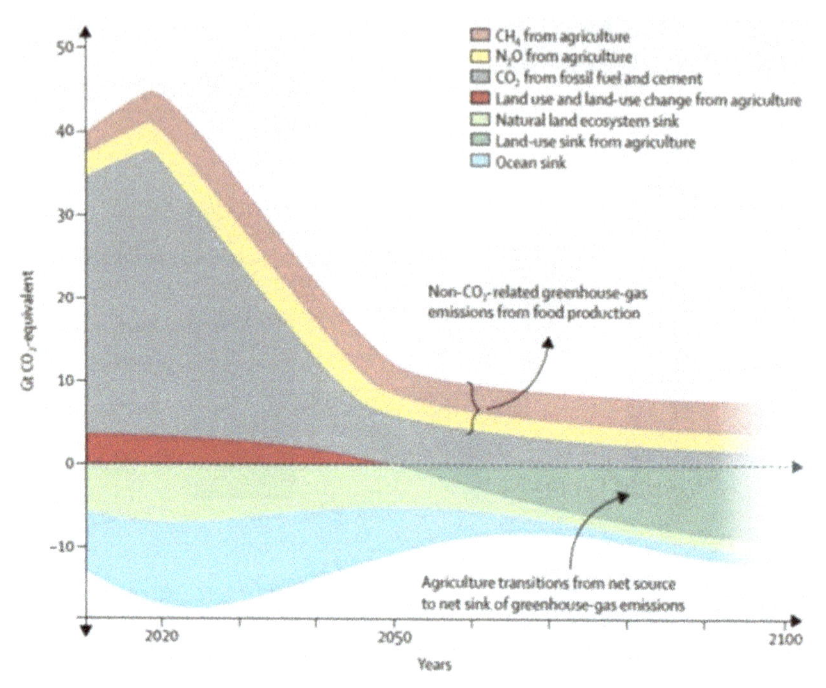

Figure 47 Projections of global GHGEs to keep global warming below 2°C. Source: Adapted from: Willett et al. (2019).

non-carbon dioxide gases (methane and nitrous oxide) are currently estimated at between 5·0 and 5·8 Gt of carbon dioxide equivalent per year (Smith et al. 2014). Food systems need to move towards becoming carbon positive at best, and carbon neutral at worst. Figure 47 illustrates the kind of proportional changes that are required in food systems to reduce agricultural emissions sufficiently to keep global warming below 2°C.°

3.2.2.1 Reducing agricultural greenhouse gas emissions

There is little disagreement that agriculture must play its role in reducing GHEs. However, with population growth likely to increase food demand by 50–70% by 2050 (IPCC 2019), the debate around how to achieve sizeable cuts remains fierce and involves difficult trade-offs. Indeed, in a business-as-usual scenario, with expected population growth and dietary convergence, it is suggested that food system GHGEs could increase between 30% (SAPEA 2020) and 52% (Greenpeace 2020a). Figure 48 highlights the projected impact of global dietary convergence (around the Western-styled diet) on agricultural GHEs.

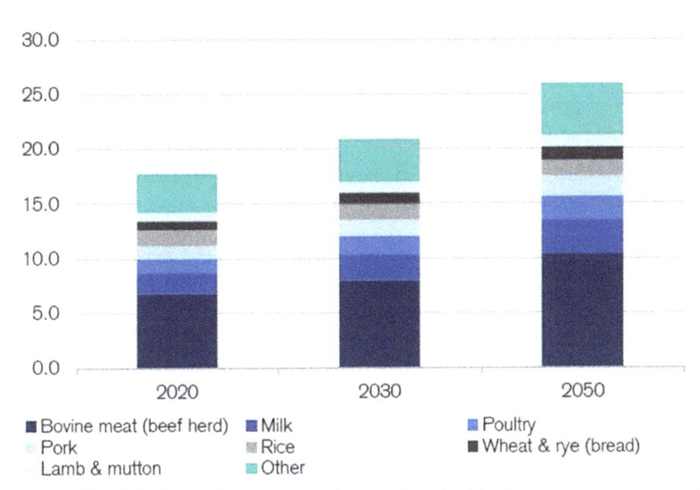

Figure 48 Total GHGEs (Gt CO_2 equivalent) associated with dietary convergence. Source: Adapted from: Credit Suisse (2021).

Whilst the prognosis appears quite dire, analysts have identified that changes in both food production and consumption have the potential to significantly increase food security as well as reduce the negative externalities associated with food production, including the production of GHGEs.

3.2.2.2 Changes in food production

Significant reductions in CO_2 generated by the global food system can only be achieved through a massive reduction in reliance on fossil fuels (Pereira et al. 2018; Benton and Bailey 2019; Stefanovic et al. 2020). Industrial forms of agriculture rely heavily on fossil fuels (especially diesel) as the principal power source for agricultural machinery involved in crop and livestock production. Agricultural machinery alone generates approximately 1·0 Gt of carbon dioxide equivalent per year (Bennetzen et al. 2016). The transportation of agricultural inputs and outputs and the production of synthetic fertilisers (especially nitrogen-based fertilisers) and pesticides rely heavily on fossil fuels as does the energy used to process, store, and transport food products. However, recent technological advances made in the generation and supply of renewable energy, such as solar, wind, and hydro, mean that the cost of green energy is now broadly competitive with fossil fuels. In concert with parallel advances in the development of electric and/or hydrogen-powered tractors (Robotics and Automation News 2019) and lorries, the scope for the replacement of fossil fuels as the primary energy source of industrial agriculture looks promising.

In addition to the opportunity to reduce CO_2 emissions from fossil fuels, a growing range of new pipeline climate-mitigation technologies offer opportunities to reduce non-CO_2 GHGs. These technologies include nitrogen fixation in cereal crops, biological nitrification inhibition (BNI) (O'Sullivan et al. 2016), reduced methane emissions in rice production via alternative wetting and drying (AWD) cultivation techniques (Lampayan et al. 2015), low-emitting manure storage (Montes et al. 2013; Bryngelsson et al. 2016), improved feed quality and feed conversion (both pasture and concentrate) - that increase emission-efficient growth in pigs and poultry, and feed additives - that reduce enteric fermentation in livestock (Poore and Nemecek 2018, Beach et al. 2015).

Increased agricultural productivity and efficiency, through the sustainable intensification of production on existing arable land, also offers an effective solution to the pressing need to produce more food whilst reducing environmental externalities (including lower GHGEs) and reducing the expansion of agriculture into biodiverse forests and wetlands (Tilman et al. 2011; Valin et al. 2013; Wirsenius et al. 2010). Indeed, according to Smith et al. (2014) and Bennetzen et al. (2016), whilst global food production has continued to grow, total emissions have remained relatively stable due to decreasing emissions per unit of food produced. According to Springmann et al. (2018), increased crop productivity and efficiency could potentially reduce food system GHGEs by approximately 10% by 2050. Sustainably increased production per unit of land, in turn, can also potentially reverse the conversion of biodiverse forests and wetlands into farmland (Poore and Nemecek 2018), which themselves act as significant carbon sinks (Benton et al. 2021). According to Vermeulen et al. (2012), the conversion of forests to arable and pastureland releases between 2·2 and 6·6 Gt of carbon dioxide equivalent per year. Willett et al. (2019) estimate that if actual crop yields could reach 75% of attainable crop yields and food loss and waste were reduced by 50%, cropland expansion would not be necessary to feed the growing world population.

Whilst fiercely contested, other commentators are convinced of the capacity of cropland to capture and store carbon (Rockström et al. 2017; Griscom et al. 2017). They suggest that soil organic carbon can be significantly increased through incorporating farm organic wastes into the soil, applying reduced or zero tillage (such as conservation agriculture), using nitrogen-fixing crops, and incorporating perennial crops on farms.

3.2.2.3 Catalysing change

Whilst there is a growing consensus that, in combination, the transitions outlined above can deliver much more sustainable food systems that operate within planetary boundaries, there is less agreement on how to achieve these transitions. In some respects, compelling scientific evidence around climate change and its far-reaching impacts, combined with growing support for change from the public and politicians, have already prompted initial market-based responses from the private sector. These include the rush to promote regenerative agriculture (discussed in Chapter 6), the reduction of food waste through increased marketing of irregularly shaped and superficially blemished fruit and vegetables, and the production of artificial meat (discussed in Chapter 4).

Without government intervention, however, market-based responses to climate mitigation will most likely be too little and too late. In this regard, food systems have increasingly become an important sub-component of ongoing climate change agreements. However, as yet, clear scientific targets for the global food system do not exist (Willett et al. 2019). Poore and Nemecek (2018) suggest that by sharing both information on the climate impact of supply chains and good practices for reducing climate impacts (Cui et al. 2018), both processors and retailers will be encouraged to instigate protocols to reduce food losses and food waste in their supply chains. According to Segerson (2013), maximum flexibility promotes least-cost mitigation through market-based innovation. However, to date, the encouragement of procurement from low climate impact farms has met with limited success. Waldman and Kerr (2014) cite the case of the Roundtable on Sustainable Palm Oil (RSPO), which, despite significant efforts, has only managed to certify 20% of oil palm production. Poore and Nemecek (2018) are also convinced that consumer education is a worthwhile endeavour but conclude that any change in consumer behaviour would be too late to meet the climate mitigation targets set by the Paris Agreement.

The use of taxes and reorientation of existing agricultural subsidies, worth more than US$0.5 trillion/annum globally, may offer more timely and scalable solutions for climate mitigation (OECD 2017; Bryngelsson et al. 2016). Ultimately, depending on the rate of progress in meeting climate change targets, the use of formal regulation may begin to play a more dominant role. For example, the European Union's Farm to Fork (F2F) and EU Biodiversity strategies seek to reduce both nitrogen and phosphorus fertiliser pollution by 50% by 2030, which relates to a reduction in fertiliser use of approximately 20% (Credit Suisse 2021). Yara, one of the world's largest fertiliser manufacturers, has already announced investments in the production of green ammonia. Green ammonia uses 100% renewable energy and produces carbon-free nitrogen-based fertiliser (Yara 2021).

Whilst climate change mitigation is essential and will hopefully reduce the scale of climate change, food systems must be capable of adapting to more erratic rainfall patterns, increases in temperature, and extreme climatic events (IPCC 2019; SAPEA 2020). Figure 49 highlights the geographies most likely to be severely affected by climate change.

Figure 49 Areas of extreme climate vulnerability. Source: Adapted from: Steiner et al. (2020).

According to Brown (2008), by 2050, climate change is likely to have contributed to the displacement of up to 200 million people. Most of these people are located in developing countries, which have limited capacity to cope with climate change (Clapp et al. 2018). The food system needs to adapt to climate change (Branca et al. 2021). Part 2 of this book discusses the different food systems approaches being developed, adopted, and adapted to address the challenges posed by climate change.

3.2.3 Reducing habitat and species loss

Pressure to reduce habitat and species loss has been growing since the days of Rachel Carson's *Silent Spring*, published back in 1963. Recent studies have identified that planet Earth is approaching or even has already exceeded, an ecological tipping point from which many proclaim there is no turning back (Rockström et al. 2009b; EEA 2018; Conijn et al. 2018; Campbell et al. 2018; Steffen et al. 2015b). Because of this, habitat and species conservation has increasingly become a political priority, especially for governments in developed countries. Whilst there are several drivers of change, agriculture (including perennial crops such as oil palm), because of its dependence on, and growing need for, land, has been identified as the principal cause of habitat and species loss (EEA 2015, 2017; Tilman et al. 2017; Díaz et al. 2019; Cherlet et al. 2018; Stefanovic et al. 2020). Indeed, according to Conijn et al. (2018), by 2050, an additional 3.5 Gha will be required to feed the world's human population, thus reducing the world's afforested area from 4.0 to 0.5 Gha. However, given humanity's biological need for food, the 'how' to create a world where food and bioeconomic products from agriculture are balanced with sustainable habitat and species conservation remains hotly debated. Two fundamental concepts are

- land sparing; and
- land sharing.

Coined by Waggoner (1996), the concept of land sparing is based on increasing yields on prime agricultural land and allowing less productive land to be set aside for biodiversity conservation. It has been juxtaposed against land sharing (Green et al. 2005), where both agricultural production and biodiversity conservation objectives are sought in the same fields.

3.2.3.1 Land sparing

The land-sparing approach promotes high-input agricultural production on highly productive land with low levels of biodiversity, namely, much of the existing industrially farmed areas (Salles et al. 2017; Balmford et al. 2018). The resultant high yields reduce the need for agricultural expansion into new areas, especially biodiversity hotspots. Land sparing works best at a landscape scale (such as the prairies of North America or steppes of Russia) but can be operationalised in more heterogeneous mixes of zoned agricultural land and sustainable/self-sustaining natural habitats, such as a large (unpolluted lake or unmanaged woodland, etc.), perhaps connected by riparian buffer strips, flower field margins, hedges, or grassy strips (Salles et al. 2017), which constitute 10% of more of the land area (Tscharntke et al. 2005). According to Knapp and Řezáč (2015), habitat heterogeneity is essential for conserving both biodiversity and maintaining ecosystem functionality.

If successful, application of the land-sparing approach would, at worst, reduce the speed and scale of agricultural encroachment onto remaining wilderness areas of the planet and probably buy enough time to ensure the expansion of agricultural land into managed ecosystems, pasture/rangelands, or secondary forests, which minimises species loss (Willett et al. 2019). At best, in combination with closing existing crop yield gaps (Dorward and Chirwa 2011; Druilhe and Barreiro-Hurlé 2012; Cui et al. 2018), minimisation of food loss and waste, mass uptake of predominantly plant-based diets (Hawkes 2015; Vallgårda 2015; Colchero et al. 2017), and adoption of integrated land management practices (Phalan et al. 2016), land-sparing, could, in theory, not only halt agricultural expansion but even release some of the existing agricultural land for rewilding - conversion back to biodiverse forests, wetlands, and grasslands (Benton 2012). Indeed, according to Williams et al. (2020), if the above conditions were met, agricultural land required to feed an expanded global population in 2050 could decline by nearly 3.4 million km^2 compared to 2010, and by a staggering 6.7 million km^2 compared to the 'business as usual' scenario.

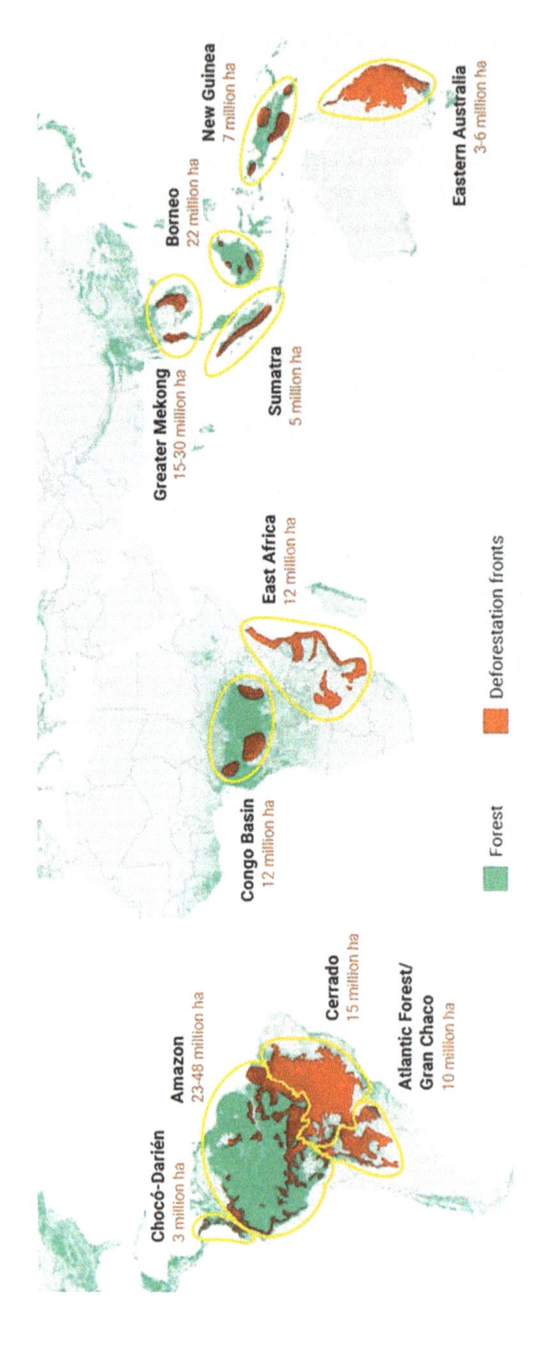

Figure 50 The 11 deforestation fronts with projected losses – 2010 to 2030. Source: Adapted from: Steiner et al. (2020).

Conversely, if agricultural expansion continues at its current rate, habitat and species loss will be catastrophic (Tilman et al. 2017; Brondizio et al. 2019). Williams et al. (2020) predict that by 2050, population growth combined with dietary convergence (i.e. more people eating meat, especially red meat), and continued food loss and waste, could result in the relatively uncontrolled expansion of agriculture into between 2 and 10 million km^2 of natural habitat. Figure 50 illustrates the key areas of forest under threat from logging and uncontrolled expansion of agricultural lands.

Williams et al. (2020) suggest that this will reduce the habitat of nearly 90% of the world's terrestrial invertebrates, with 1280 species losing more than 25% of their habitat area. According to the International Union for the Conservation of Nature (IUCN), 20% of terrestrial vertebrates or now threatened with extinction and are categorised as either vulnerable, endangered, or critically endangered (IUCN 2018). Additionally, it is estimated that up to 347 species would lose at least 50% of their remaining habitat by 2050, 96 species would lose at least 75% of their remaining habitat, and 33 species would lose at least 90% of their remaining habitat. Sub-Saharan Africa and South America are likely to be particularly hard hit (Tilman 2017).

Figure 51 illustrates the estimated loss of global biodiversity between 1750 and 2010. According to IPBES, more than 1 000 000 animal and plant species are threatened with extinction, many within decades (IPBES 2019).

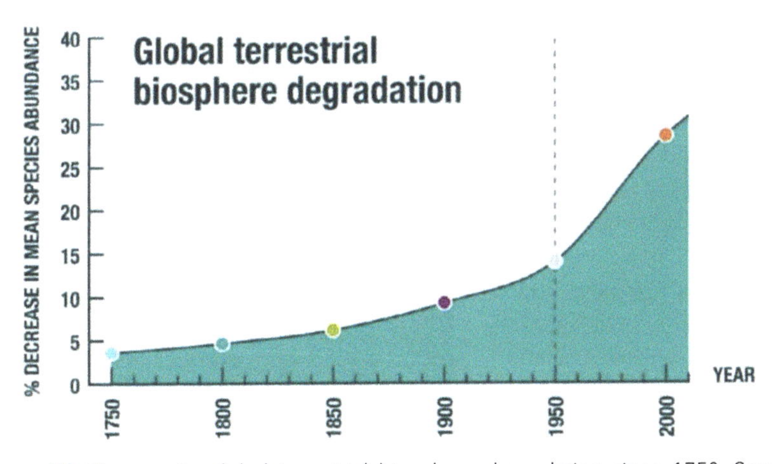

Figure 51 Changes in global terrestrial biosphere degradation since 1750. Source: Adapted from: Dasgupta (2020).

3.2.3.2 Land sharing

Conversely, the land-sharing approach promotes low-input and environmentally sensitive agriculture, such as High Nature Value farmland (Paracchini et al. 2008), agroecological, and organic agriculture, across fine-scale heterogeneous landscapes, with the aim to produce food and conserve both biodiversity and ecosystem services (Salles et al. 2017). Land sharing also relies on increased agricultural biodiversity, namely, the use of a wider range of crops, with significant genetic diversity, which can deliver multiple harvests around the year (Powell et al. 2015), increased incomes, and dietary diversity (IPES-Food 2016). Traditionally, humanity relied on more than 7000 different plant varieties for their food requirements. Today, consumers rely on rice, maize, and wheat for more than 50% of their food intake (FAO 1995).

Proponents of land sharing stress the importance of increased biodiversity for the biological control of weeds, pests, and diseases and increased pollination of crops (Ricketts et al. 2004; Klein et al. 2003; Chaplin-Kramer et al. 2011), as well as its capacity to increase carbon sequestration and enhance resilience to both climatic and economic shocks (FAO 2013b; Willett et al. 2019). However, it is becoming increasingly accepted that this approach would result in the need for significantly more agricultural land to feed the growing global population (Green et al. 2005; Benton and Bailey 2019).

3.2.3.3 Land sparing versus land sharing

The balance of these two approaches depends on the exact relationship between crop yields and biodiversity, which can vary considerably from farm to farm, region to region, based on soil type, rainfall, ecosystem services, etc. (Hodgson et al. 2010). Ultimately, if a minimal reduction in crop yield results in a significant increase in biodiversity, there is likely to be a concave relationship (Salles et al. 2017). In this situation, a land-sharing approach should be considered (Gabriel et al. 2013). However, if a significant reduction in crop yield is required to provide small increases in biodiversity, there is likely to be a convex relationship and a land-sparing approach should be taken (Gabriel et al. 2013). Often, the decision is not so clear-cut and both strategies can be simultaneously deployed (Fischer et al. 2008; Brady et al. 2009; Swinnen 2010; Salles et al. 2017; Phalan 2018).

Current evidence seems to support a more convex relationship between agricultural intensification and biodiversity, namely, as agricultural production intensity increases, biodiversity decreases (Trewavas 2001a; Green et al. 2005; Balmford et al. 2005; Salles et al. 2017; Tilman et al. 2017). However, this has not prevented the continuing land-sparing versus land-sharing debate (Fischer et al. 2008; Perfecto and Vandermeer 2008; Tscharntke et al. 2012). In practice, this has spawned the evolution of patchworks of single and multi-functional landscapes. Single-function landscapes focus primarily on either high input/ output agricultural production or nature conservation, such as nature reserves or sites of special scientific interest. Multi-functional landscapes promote sustainable food production, conserve important habitats and biodiversity, enhance ecosystem services, and provide other services such as landscape and amenity value – including tourism (Horlings and Marsden 2011; Kirwan et al. 2013; Odegard and Van der Voet 2014; Benton et al. 2021; Solazzo et al. 2016). The degree to which humanity can change diets towards more plant-based foods, whilst simultaneously reducing food losses and waste, will determine the balance that can be struck between land sparing and land sharing (Willett et al. 2019; Benton et al. 2021).

3.2.4 Reducing environmental pollution

Reducing environmental externalities of agricultural production is a central tenet of subsequent chapters in this book and will only be summarised here. However, what can be said as a kind of introduction to the potential ways forward is that virtually all actors involved in food systems, from input providers, farmers, processors, retailers, and consumers have acknowledged the impact that pollution of all kinds has on our planet and the need for change (Bush and Martiniello 2016; Maye and Duncan 2017; Kirwan et al. 2017; EEA 2018; Gomiero 2018; Springmann et al. 2018; Rockström et al. 2009; Stefanovic et al. 2020; Benton et al. 2021; Credit Suisse 2021). In this sense, change can mean anything from the more judicious and targeted use of inputs (fertilisers, pesticides, genetically modified of gene-edited organisms), to the development of safer, biodegradable, and short-lived inputs, to the replacement of many off-farm and synthetic inputs with natural processes (recycling of nutrients, biocontrol of pests and diseases, and diversification of crop species and cropping systems).

3.3 Addressing inequality in food systems: more equitable food systems

There is no doubt that the Second Food Regime, based on industrial forms of agriculture, has delivered unimaginable volumes of relatively cheap food. However, increased concentration and centralisation (discussed in Chapter 2) and evolving oligopolistic control (Bush and Martiniello 2016) have increasingly skewed benefits towards the large global corporations involved in non-farm components of the food system (Béné et al. 2019b). Many farmers and small-scale input providers, food aggregators, processors, wholesalers, and retailers involved in industrial food systems struggle to remain profitable. Additionally, small-scale, predominantly self-sufficient farmers - selling small volumes of surplus production in local communities, especially indigenous peoples, find themselves struggling to survive on often marginal agricultural lands and selling produce in local markets often flooded with, and dependent on, cheap basic food and grains spilling over from industrial food systems (Bailey et al. 2015; Pereira et al. 2020).

The marginalisation of small and indigenous farmers, both within and on the periphery of industrial food systems, has ignited a proliferation of tens of thousands of grassroots protest groups (Hart et al. 2015; Ingram 2015), often banding together into local, national, regional, and international associations, such as the 'Food Sovereignty' movement, to fight against what is seen as the overdominance of the global industrial food system and all the challenges associated with it, especially increased price volatility (Bush and Martiniello 2016), and to insist on land reform, fair incomes for farmers and farmworkers (www.europeanfooddeclaration.org), food justice (Moore 2014; SAPEA 2020; Gottlieb 2009), and improved local conditions in rural areas in both developed and developing countries (Horlings and Marsden 2011; Kirwan et al. 2013; Odegard and Van der Voet 2014; Bush and Martiniello 2016).

These groups battle to restore local, short-chain food systems that provide food security for local communities (Halweil 2004; Winne 2008), as well as maintain cultural traditions, local breeds of livestock and varieties of a diverse range of crops (agro-biodiversity), local knowledge and production methods - that are deemed as nature and ecosystem positive, as well as climate resilient (Altieri 1995; Holt-Giménez 2006; Gliessman 2007; Pereira et al. 2020; www.europeanfooddeclaration.org). Many of these groups argue the need for transparent food systems to be designed on ethical rather than market principles (Kirwan et al. 2017), claiming that food is a universal right, irrespective of whether the hungry have enough money in their pockets to purchase it. There are also calls for the adoption of food systems based on a degrowth agenda, one that is non-capitalist and where the market plays a more supportive rather than dominant role (Gibson-Graham 1996, 2006; Latouche 2016).

This contestation of the industrial global food system has resulted in a plethora of alternative approaches, which explicitly aim to provide scalable alternative food system approaches that are equitable, socially just (SAPEA 2020), and ecologically sustainable. Part 2 of this book explores alternative food system models/approaches that are being piloted and scaled across the globe (Fuchs and Glaab 2011; Constance et al. 2014). Chapter 9 takes a multi-level perspective (Geels 2005; Geels and Schot 2007), which examines these approaches (Constance et al. 2008; Rosin and Campbell 2009) and success factors associated with the so-called experimental niches of alternative forms of food systems. Central to the success of alternative approaches is their capacity to create institutions and practices that allow them to be scaled and stabilised within reconfigured power relations with the broader capitalist market economy and, still further, perhaps threaten the so-called wider frame of the capitalist and market-based economy (or landscape) (Geels 2010, 2011, 2014; Smith and Raven 2012; Ilbery and Kneafsey 1998; Ilbery and Maye 2005a; Goodman 2004).

3.4 Sustainable food systems

Whilst a broad consensus has evolved that the current trajectory of the industrial global food system, with its inherent food loss/food waste, reliance on fossil fuels, and generation of global environmental externalities, does not have the capacity to provide a meat-based western-styled diet to a rapidly expanding (dare I say unsustainable), global human population, there remains a ferocious debate as to what needs to be done to redress the situation (Benton and Bailey 2019; Benton et al. 2021). International, regional, national, and sub-national narratives on the subject range from often romanticised visions of what we want or would like, to more pragmatic narratives around the need to feed an unprecedented number of human mouths, to maintaining economic growth and human welfare, set against the absolute limits of planet earth (resource limitations, pollution, global warming, etc.) (Benton et al. 2021; Béné et al. 2019b; Caron et al. 2018; De Schutter 2017; IPES-Food 2016; SAM 2019; Steffen et al. 2015; Rees 2019; Marchetti et al. 2020; Herrero et al. 2020; Steiner et al. 2020).

Will the future see humanity papering over the ecological and biophysical cracks? Or will society look to reimagine humanity's social, economic, and political structures and our fundamental relationship with the planet that we live on (i.e. a move from exploitation to co-existence)? If we redesign, will this be incremental/minor changes to food systems, or will it be a more radical/disruptive change that reverberates much wider than food systems to a reconstruction of our socio-cultural, economic, and political structures?

Whilst there is no doubt that current global food will change, many different versions of what sustainable food systems should look like are being developed. Part 2 of this book explores measures being taken to adjust the current industrial global food system to meet growing demands for higher quantities of higher quality foods, whilst reducing fossil-fuel energy consumption and environmental externalities. This continued adherence to the growth paradigm envisages solutions found in increasing economic growth through productivity-enhancing financial and techno-fixes (such as biotechnologies - GMOs, gene editing, and nanotechnologies - vertical farming, conservation, precision, and climate-smart agriculture), and free markets that aim to maximise food outputs and minimise resource inputs required by food systems through the process of sustainable intensification (Economist 2010; Carlson 2016; Taylor and Uhlig 2016; Holden 2018; Rosin 2013; Foley et al. 2011; Tilman et al. 2011; West et al. 2014; SAM 2019; FAO et al. 2020; Herrero et al. 2020; OECD 2021; Steiner et al. 2020; Benton et al. 2021; Benton 2015; Baulcombe et al. 2009; Garnett et al. 2013; Bommarco et al. 2013; Doré et al. 2011; Tittonell et al. 2016; Marchetti et al. 2020; Pereira et al. 2018).

In parallel to ongoing adjustments to the industrial global food system, Part 2 of this book explores the emergence of sustainable niches/sites of innovation and experimentation (including organic, agroecological, regenerative, and permaculture-based agricultures), providing alternative food system models, at various geographical scales, which proposedly offer access to healthy and nutritious foods produced in a more environmentally friendly and resource-conserving manner (Benton et al. 2021; Benton and Bailey 2019; Béné et al. 2019b; SAM 2019; Steiner et al. 2020; Gomiero et al. 2011a, de Ponti et al. 2012; Seufert et al. 2012; Hansen 2019; Joseph et al. 2019; Wezel et al. 2009; Biel 2016; Gliessman 1989; Altieri 1995; Pretty 2008; Feenstra 1997; Pretty 1998; Rickerl and Francis 2004; IPES 2016; Bellamy and Ioris 2017; Frison and Clément 2020; Gliessman 2018a; Latouche 2012; Boraeve et al. 2020; Hirschfeld et al. 2020).

Chapter 7 also explores socio-cultural and political innovations, such as the Food Sovereignty, Slow Food Movement, Food Barter, Food Sharing, and Free Food movements, which aim to organise food systems in socially just ways. Chapter 7 also examines the relocalising (local, urban, regional) of food systems versus the continued expansion of the global food system (Lamine et al. 2019a,b,c; Ilbery and Kneafsey 2000; Murdoch et al. 2000; Renting et al. 2003; Barham 2003; Feagan 2007; Arfini et al. 2012; Rosset and Martínez-Torres 2012; Sherwood et al. 2017; Salbitano et al. 2019; Castro et al. 2018; Lohrberg et al. 2016; Altieri 2012; Camargo and Beduschi 2013; Greenpeace 2020b; Rees 2019; Rosin 2013).

Chapters 4–7 compare the productive, economic, environmental performance, and social and political acceptability of these different approaches. Lastly, Chapter 9 explores the likely spatially differentiated outcomes of food system change.

Part 2

Competing paradigms of food production

Chapter 4

Neo-productivist food systems

4.1 Introduction

The European Environment Agency defines the sustainability transition as 'a fundamental and wide-ranging transformation of a socio-technical system towards a more sustainable configuration that helps alleviate persistent problems such as climate change, pollution, biodiversity loss or resource scarcities' (EEA 2019). Whilst this definition is clear, in specifying transition outcomes, it is non-prescriptive with regards to the types/form of socio-technical transformations that need to achieve these outcomes. This ambiguity has led to a proliferation of food system innovations that claim to address the sustainability transition. Indeed, achieving the goal of global food security, together with environmental sustainability and social and economic justice, is one of the greatest challenges of the twenty-first century (Campos et al. 2019; Pereira et al. 2020). A wide range of actors are rapidly vying to provide a blueprint for a sustainable future global food system, one that both serves the needs of a growing population and the economies they depend on, whilst providing stewardship over the earth's limited biophysical resources. The following chapters focus on several key food system innovations, which claim to contribute to the sustainability transition.

The first chapter focusses on innovations that are Neo-Productivist in nature, such as the growing bioeconomy, synthetic foods (plant-based meat alternatives – PBMAS and cultured meat), biotechnology (GMO and gene editing), and vertical farming. The second chapter focusses on so-called Reformist approaches, such as precision agriculture, climate-smart agriculture, and Conservation Agriculture. The third chapter focusses on so-called progressive approaches such as regenerative agriculture, organic agriculture, and agroecological agriculture. The fourth chapter focusses on so-called radical approaches, such as permaculture, Slow Food, Food Sovereignty, free food, and degrowth.

Despite growing scepticism and condemnation, actors supporting the Neo-Productivist food system model (or Corporate Food Regime) remain convinced that they have the right techno-financial-spatial answers to both current and future challenges facing the global food system. This continued adherence to the growth paradigm envisages solutions found in increasing economic growth through productivity enhancement brought about by technical, financial, and spatial fixes, which aim to maximise food outputs and minimise resource inputs required by food systems through the process of sustainable and/or ecological intensification. These fixes include

1 growth of the bioeconomy - including the development of synthetic foods (plant-based meat alternatives (PBMAS) and cultured meat synthetic foods;
2 continued development of biotechnologies, including GMOs and gene editing; and
3 controlled environment agriculture - including vertical and container farming.

In turn, these technological and spatial fixes are underpinned by increasingly liberal national, regional, and international food markets and internationally mobile capital (Economist 2010; Carlson 2016; Taylor and Uhlig 2016; Holden 2018; Rosin 2013; Foley et al. 2011; Tilman et al. 2011; West et al. 2014; SAM 2019; FAO, IFAD, UNICEF, WFP and WHO 2020; Herrero et al. 2020; OECD 2021; Steiner et al. 2020; Benton 2015; Benton et al.2021; Baulcombe et al. 2009; Garnett et al. 2013; Bommarco et al. 2013; Doré et al. 2011; Tittonell et al. 2016; Marchetti et al. 2020; Pereira et al. 2018; Chiles 2013; Mouat et al. 2019; Kuch et al. 2020; Barrett and Rose 2020; Broad 2019, 2020a,b; Klerkx and Rose 2020; Clapp and Ruder 2020).

4.2 The bioeconomy: synthetic foods (petri-proteins and lab-grown meat), genetically modified organisms and gene editing

The potential of synthetic foods was first recognised in the late nineteenth century during the height of the Industrial Revolution and associated rapid advances in organic chemistry (Burton 2019). The idea that food and fibre would one day be chemically synthesised in industrial vats, like dyes, medicines, and perfumes, led to the 1882 journal article 'War against agriculture', as quoted by Burton (2019). Unfortunately, whilst the vision was evident, complexities associated with the biosynthesis of food and fibre meant that synthetic foods would remain in the realms of science fiction for the foreseeable future. In 1932, even Winston Churchill was convinced that microbes would be used to feed the world and that the advent of artificial meat would one day arrive (Webb and Hessel 2022). However, it was not until the 1960s and 1970s that NASA developed the first synthetic foods designed for astronauts bound for long space journeys.

Whilst the concept of the bioeconomy had been discussed since the early 1990s, it failed to be internalised as a subject in policy agendas until the early twenty-first century. Even then, a consensus had not yet been reached regarding its definition. According to Bugge et al. (2016), the bioeconomy remains an emerging concept. Bugge et al. (2016) attempted to categorise the bioeconomy in terms of vision. They defined three distinct visions:

1 Biotechnology (i.e. use of biologically based technologies such as fermentation to produce value-added products);
2 Bioresource (i.e. discovery and exploitation of biological resources such as biomass from plants); and
3 Bioecology (i.e. developing a self-sustaining, more eco-friendly economic model based on processing and recycling biological resources).

Across the globe, these visions are embraced to greater or lesser degrees. For example, in the case of the USA and OECD, the vision for a bioeconomy, which evolved during the 1990s, fundamentally equates to the increased application of biotechnology to produce high quantities of cheap bio-based products (Konstantinis et al. 2018; Maciejczak 2018). Here, the bioeconomy is predominantly linked to economic growth, with any social and environmental benefits considered as positive spillovers associated with the commercialisation of biotechnological innovations (Konstantinis et al. 2018).

Conversely, in the European Union, the bioeconomy concept, which evolved in the EU in the early twenty-first century, embraces all three visions. Developed by the European Commission (2012), the bioeconomy envisages the bioeconomy as the sum of all economic sectors involved in the production of renewable biological resources and their conversion into value-added products, such as food, drinks, animal feed, bio-textiles, wood products and furniture, paper, bio-based chemicals, pharmaceuticals, rubber and plastic, liquid biofuels and bioenergy, and other bio-based products (EC 2022).

In the EU the concept has evolved to incorporate the notion of circularity, which has recently grown in prominence (Konstantinis et al. 2018). The EC (2022) envisages that a gradual transition to a bioeconomy will 'modernise and strengthen the EU industrial base, creating new value chains and greener, more cost-effective industrial processes, whilst protecting biodiversity and the environment'. In addition to increasingly underpinning societies' material needs, which has attracted significant investment (Maciejczak 2018), the EU's vision of the bioeconomy explicitly embraces a quality of life/environment dimension. For example, within the EU, according to Konstantinis et al. (2018), France has been classified as having an explicit bio-ecology vision of the bioeconomy which, 'encompasses the whole range of activities linked to bioresource production, use, and processing. The purpose of bioresources is to provide a sustainable response to the need for food and to part of society's requirements for materials and energy, as well as providing society with ecosystem services'.

Whilst the transition to a bioeconomy encompasses aspects of both quantity (cheap and plentiful bio-based products) and quality (of life for people and protection of the planet), most investments made to date, by both public and private sector entities, have tended to follow a relatively biotech and bio-resources route, significantly deepening both the substitution and appropriation of biological resources witnessed during the First and Second Food Regimes (Goodman et al. 1987; Watson 2018; Burton 2019).

4.2.1 Synthetic foods: petri-proteins and lab-grown meat

Whilst fermentation has been used for thousands of years in the production of beer, wine, cheese, soy sauce, bread, and vinegar, recent advances in both *in vitro* (cell-free) and *in vivo* (using living cells) biosynthesis, especially genetically engineered living cells, have opened significant room for growth in this sector (Zhang 2010; Zhang et al. 2017; Forster and Church 2007; Ro et al. 2006). Indeed, according to Zhang et al. (2017), advances in 'biomanufacturing will become one of the most important cornerstones of the sustainability revolution happening in the twenty-first century'. Currently, there are two basic types of synthetic foods:

- petri-proteins or plant-based meat analogues (PBMAs); and
- lab-grown or cultured meat.

Other under-explored sources of cheap protein, such as insect protein (Chriki and Hocquette 2020; Bloomberg 2023a), or protein from seaweed and microalgae (Lei 2021), for both human and animal consumption, are not discussed here.

4.2.1.1 Petri-proteins/mycoproteins and PBMAs

Petri-proteins are proteins synthesised by closed fermentation by either naturally occurring or genetically engineered, yeasts, bacteria, fungi, or algae. The fermentation process converts sugar into proteins that look and taste like meat, eggs, or dairy products (ETC Group 2019a; Burton 2019; Credit Suisse 2021). Launched in the UK by Imperial Chemical Industries (ICI) in 1985, Quorn products (chunks, sausages, and burgers), which are produced from the fungus *Fusarium venenatum* in bioreactors/vats using the fermentation process, are probably the best-known synthetic meat substitutes (Wiebe 2002), yielding US$260 million a year (ETC Group 2019a). By 2021, over 600 start-up companies were developing protein food products made from synthetic proteins, especially meat substitutes (Credit Suisse 2021). Figure 52 illustrates the principal sources of proteins being used to produce PBMAs.

4.2.1.2 Lab-grown meat

There are many names attributed to lab-grown meat, such as healthy meat, slaughter-free meat, *in vitro* meat, vat-grown meat, lab-grown meat, cell-based meat, clean meat, cultivated or cultured meat, and synthetic meat (Wikipedia 2021b).

The first recorded biosynthesis of muscle fibres was achieved in 1971 (Ross 1971). However, it was not until 1991 that John Vien secured a US patent to produce artificial lab-grown meat that was deemed fit for human consumption (Wikipedia 2021b). This was followed in 2001 by a Dutch consortium, which secured a worldwide patent to produce artificial meat (Wikipedia 2021b). However, it was not until 2013 that the In-Vitro Meat Consortium produced the first beef burger made from artificial meat, at a jaw-dropping cost of US$300 000 (Ismail et al. 2020). Initial production of artificial meat was based on living stem cells extracted from umbilical cords and other animal-based sera (ETC Group 2019a; Wikipedia 2021b). However, more recently, companies in Japan, Canada, and the UK have focussed on the use of non-animal-based sera additives to produce artificial meats (Wikipedia 2021b).

In 2016, Impossible Burger released the first genetically modified alternative plant-based protein in the USA. Since its launch, fast food and restaurant chains, such as Burger King, KFC, and Dell Taco, have included products made

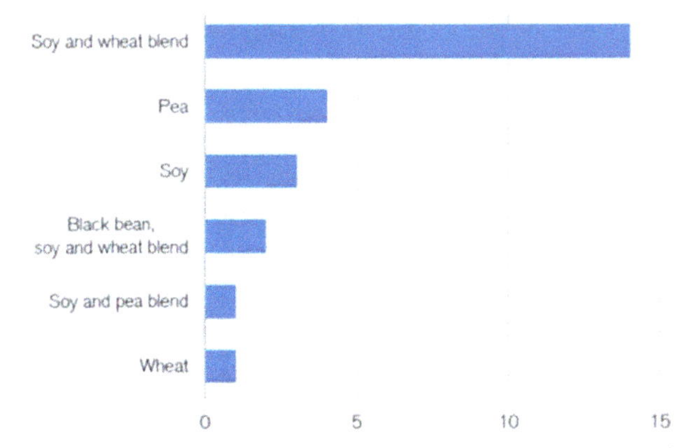

Figure 52 Plant protein bases of the top 25 plant-based meat products by US$ sales. Source: Adapted from: Credit Suisse (2021).

from 100% synthetic meat (Samuel 2019; Rogers 2022). Since 2016, Impossible Foods has released sausages, chicken nuggets, meatballs, and pork made from plant-based proteins (Impossible Foods 2022). The number of synthetic protein companies has risen significantly since the release of the first cultured meat burger in 2013 (Fig. 53).

Based on data from The Good Food Institute, most investments have been made in plant-based synthetic meat companies. By 2021, there were more than 600 companies either developing or selling alternative protein sources (Credit Suisse 2021), with at least US$1.4 billion invested in research and development (GLP 2022).

Figure 54 illustrates the significant investments made by pioneering governments in both research and infrastructure in alternative proteins. Denmark, Canada, and Australia have made significant investments in infrastructure, whilst the EU, Singapore, Israel, and the USA have invested heavily in research.

In addition to Impossible Foods, other companies commercially producing lab-grown meats and PBMAs include

- Beyond Meat, produces plant-based burgers (sold in McDonald's), ground meat, sausage, meatballs, and chicken (Khoury et al. 2014; Beyond Meat 2022);

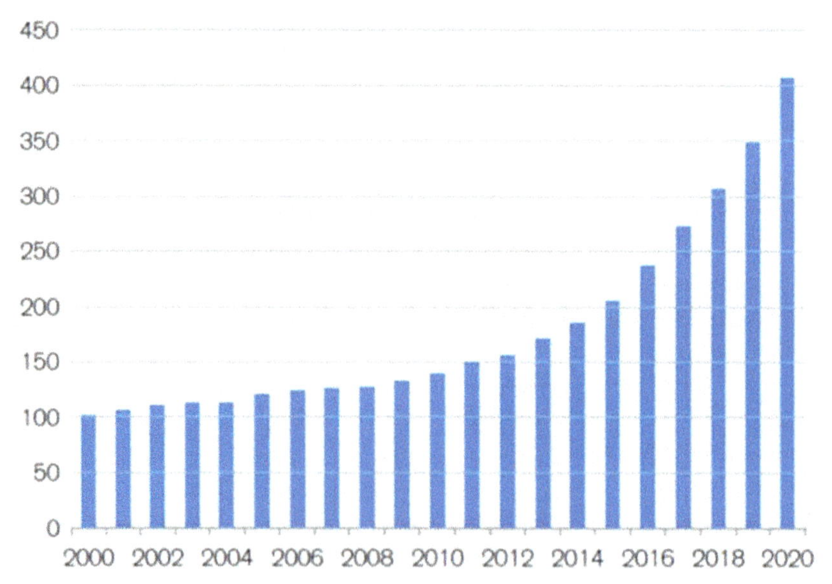

Figure 53 Number of alternative protein companies by year. Source: Adapted from: Credit Suisse (2021).

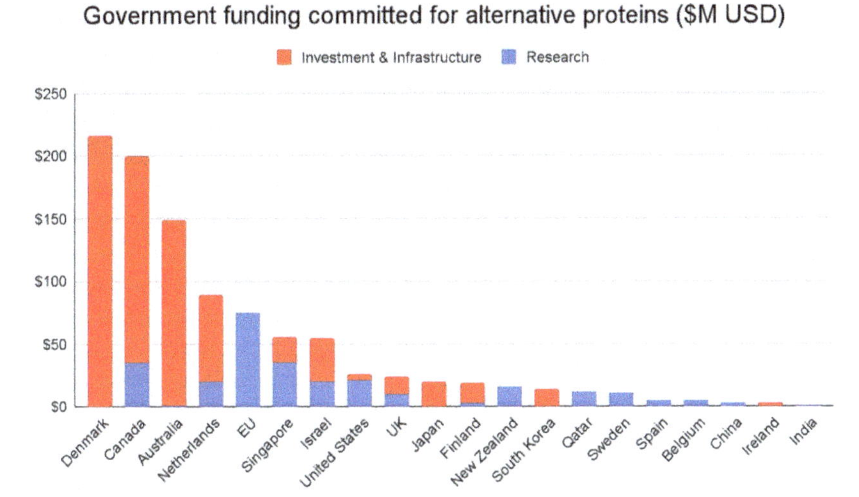

Figure 54 Government funding committed for alternative proteins in US$ million. Source: Adapted from: Green Queen (2023a).

- Upside Foods (originally Memphis Meats) produces chicken, fish, and beef products from plant-based proteins (Upside Foods 2022);
- SuperMeat produces chicken products from lab-grown meat (SuperMeat 2022);
- Mosa Meats produces textured steaks (Webb and Hessel 2022);
- Future Meat Technologies that produce cultured meat chicken, beef, and lamb (Future Meat 2022);
- Finless Foods produces both plant-based and lab-grown tuna (Finless Foods 2022); and
- BlueNalu and Wildtype produce cultured tuna and salmon respectively (Guardian 2022b).

The first cultured meat product (Chicken Bites - produced by Eat Just) hit the high street in Singapore in December 2020 (Carrington 2020). More recently, in 2023, both Upside Foods and Good Meat received final approval from the USDA to sell cultured meat (Al Jazeera 2023). In the EU, despite awarding European Commission grants to companies developing cultured meat, and continuous lobbying by producers, the European Food Safety Authority (EFSA) is currently unwilling to approve sales of cultured meats (GovGrant 2023). However, industry analysts suggest that, despite this, it will not be long before the UK and the EU approve the sale of cultured meat, and that Israel, Japan, China, and Qatar won't be far behind them (Chemistry World 2022).

Others are producing hybrid food products, such as KFC's hybrid chicken nuggets (20% cultured chicken cells and 80% plant-based meat substitute (Webb and Hessel 2022). Figure 55 illustrates patents taken out by the top ten developers of cultured meat. Of these, three are European (HigherSteaks, Mosa Meat, and Biotech Foods), Upside Foods, Tufts College, and Fork and Goode are from the USA, Integriculture is from Japan, Yonsei University is from South Korea, and both Aleph Farms and Yissum Research Development are from Israel (GovGrant 2023).

Given the market hype and associated investment bubble, many of the numerous start-up companies working on cultured meat and other plant-based meats have already been bought out by high-value individuals (ETC Group 2019a) or have received substantial capital influxes from venture capital firms. According to ETC Group (2019a), for example, Impossible Foods received at least US$372 million from Google Ventures, Bill Gates, Li Ka-shing (Hong Kong high-worth individual), and investment bank UBS. Future Meat Technologies has received capital from venture capital firms such as HB Ventures (operating in China and South-East Asia), as well as S2G Ventures, Emerald Technology Ventures, Manta Ray Ventures, Bits X Bites, and Agrinnovation (Food Processing-Technology 2022).

Mega-meat corporations, such as Tyson, ADM, Cargill, and Hormel, are also investing heavily in synthetic meat production (ETC Group 2019a). Indeed, both Tyson's and Cargill have invested directly in Upside Foods (Carrington 2020). In addition, established food manufacturers and distributors such as JBS, Müller Group, Rich's Products Corporation, Neto Group (Israel), and Monde Nissin, are also investing in the development of PBMAs and cultured meat products (PRNewswire 2021; Food Processing-Technology 2022). Figure 56 illustrates the amount of R&D investment in cultured meat. Here, as with other biotechnologies, the USA is a clear front-runner.

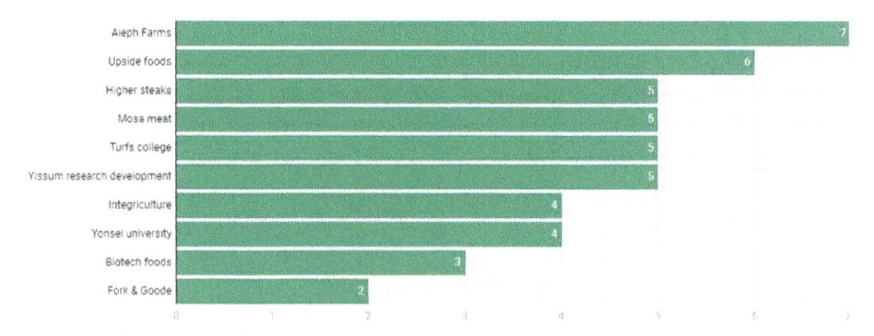

Figure 55 Top patent filers in the cultured meat industry (chart created using the patent database Questel Orbit). Source: Adapted from: GovGrant (2023); NB GovGrant now known as Source Advisors.

1	US	£1,360.24 million
2	Israel	£474.59 million
3	Netherlands	£123.92 million
4	Singapore	£100.67 million
5	UK	£28.55 million

Figure 56 Top five countries investing in the development of cultured meat (chart created using Pitchbook). Source: Adapted from: GovGrant (2023); NB GovGrant now known as Source Advisors.

More and more firms are getting involved in this fast-growing market, like Motif Foodworks (plant-based meat and dairy alternatives), Ginkgo Bioworks (custom-built microbes), BioMilq (lab-grown breast milk), Nature's Fynd (fungi-grown meat and dairy alternatives), Eat Just (egg substitutes made from plant proteins), Perfect Day Food (lab-grown dairy products), or NotCo (plant-based animal products made through AI), to name but a few (Shroff et al. 2021). Cargill invested in Cubiq to commercialise Go! Drop, an emulsion made from vegetable oils and water, gives PBMAs a more natural meaty taste (nxtaltfoods 2023; Financial Times 2022a). In 2022, more than US$600 million was invested in the development of cultured meats and PBMAs (OilPrice 2023).

4.2.1.3 Opportunities for cultivated meats and emergent PBMAs

One of the key selling points of PBMAs is that there is no need for animals to be slaughtered (ETC Group 2019a). Indeed, even in the case of cultured meats, only a limited number of animals need to die to harvest stem cells. Even here, several cultured meat companies have announced plant-based substitutes for foetal bovine serum (FBS), which are being developed for cultured meat production (Cosgrove 2017). Plant-based and cultured meat production also do not require the use of antibiotics (ETC Group 2019a; Carrington 2020; Credit Suisse 2021), have less unhealthy fat tissue and are free from animal faeces and zoonotic diseases, as well as bacteria such as *Salmonella* and *Escherichia coli* (Chemistry World 2022).

Another key selling point is that the production of both plant-based and cultured meats requires significantly fewer natural resources. For example, compared to farmed livestock, the production of plant-based meat is claimed to use 72–99% less water, and 47–99% less land (Burton 2019; Credit Suisse 2021; Food Processing-Technology 2022; Kozicka et al. 2023). Cultured meats are claimed to use 35–60% less energy and 98% less land (Webb and Hessel 2022). Indeed, according to recent food system modelling by Kozicka et al. (2023), 'if globally 50% of the main animal products (pork, chicken, beef, and milk) are substituted – a net reduction of forest and natural land is almost fully halted and agriculture and land use GHG emissions decline by 31% in 2050 compared to 2020. If spared agricultural land within forest ecosystems is restored to forest, climate benefits could double, reaching 92% of the previously estimated land sector mitigation potential. Furthermore, the restored area could contribute to 13–25% of the estimated global land restoration needs under target 2 from the Kunming Montreal Global Biodiversity Framework by 2030, and future declines in ecosystem integrity by 2050 would be more than halved. Other studies also suggest that GHGEs can be reduced by between 30% and 95% (ETC Group 2019a; Burton 2019; Credit Suisse 2021; Webb and Hessel 2022). For example, when compared to conventional production, if cultured meat is produced using renewable energy, GHGEs can be 90%, 52%, and 17% lower for beef, pork, and chicken, respectively (Sinke et al. 2023). However, some scientists are concerned that without improving the efficiency of cultured meat production and replacing fossil fuel energy with renewable sources of energy, GHGEs associated may exceed those of conventionally produced meat by up to 26% (Risner et al. 2023).

According to the IPCC, by 2030, cultured meat is expected to contribute to a 'substantial reduction in direct GHG emissions' (IPCC 2022). Both plant-based and cultured types of meat can also be produced adjacent to large population centres, significantly reducing food miles (Webb and Hessel 2022). Water pollution is also claimed to be substantially reduced (Credit Suisse 2021). Figure 57 illustrates the comparative energy requirements of cultured meat, both using conventional and sustainable energy sources. If sustainable energy sources are used for cultured meat production, the impact on the environment, especially global warming is highly significant. Currently, the small-scale/pilot production of cultured meat is primarily dependent on relatively high inputs of fossil fuels. Both scaling up production and moving towards more sustainable energy sources are likely to significantly reduce carbon emissions (Mattick et al. 2015; Smetana et al. 2015; Carrington 2020).

Founded in 2017, Solar Foods produces a natural protein Solein, which contains all nine essential amino acids needed in a healthy human diet. Solein is produced with 100% renewable energy. The company claims that Solein's GHGEs are only 1% of that produced by farmed animals and release 80% less GHGEs than plant-based meats made from soya beans and peas (Credit Suisse 2021). More broadly, the conversion of food and other biological wastes into edible carbohydrates and proteins is receiving significant attention. For example, Parchami et al. (2021) report on the use of spent brewers' grain as a substrate for fungi protein synthesis. Other opportunities include the harvesting of wild leguminous plants. For example, proteins have successfully been recovered from gorse and broom bushes, which grow wild in marginal farmland across Scotland. According to Carrington (2020), 'there's enough gorse protein to easily feed Scotland's population.' Danish scientists are also

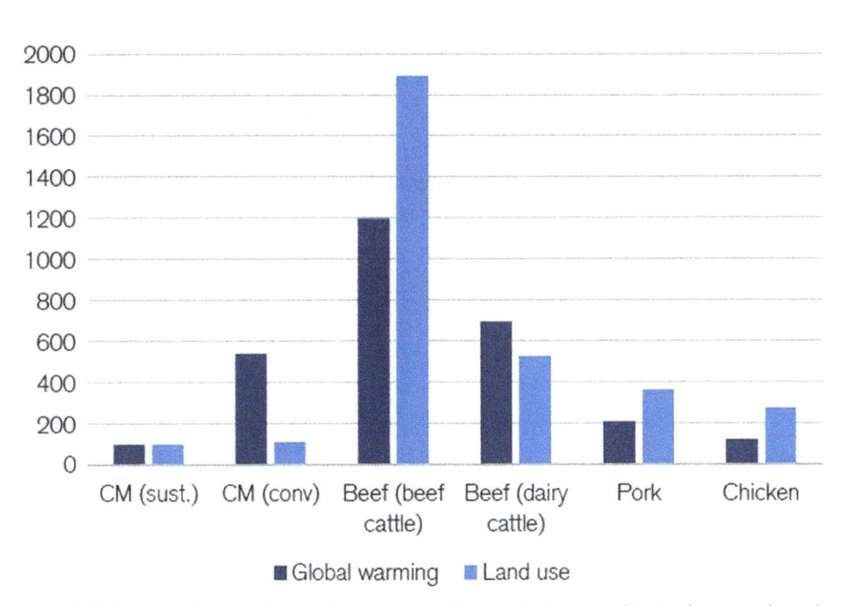

Figure 57 Impact of animal-based protein products relative to cultivated meat when the latter is produced using renewable energy. Source: Adapted from: Credit Suisse (2021).

working on extracting protein from grasses for human consumption (Cultivated Food 2023).

4.2.1.4 Challenges to the commercialisation of PBMAs and cultured meats

There are several challenges facing the commercialisation of PBMAs. Whilst the increasing convergence of the pharmaceutical, nutraceutical, and agricultural sectors has enhanced the transferability of new technological advances (Watson 2018; Burton 2019), the cost of production for PBMAs, especially for cultured meat, remains prohibitively high. However, since 2013, when the first cultured meat burger rolled off the production line at US$300 000, technological advances and economies of scale have slashed this down to US$200 per pound of beef by 2019 (Food Processing-Technology 2022), and US$139 per pound in 2022 (Garrison et al. 2022).

Even the cost of PBMAs needs to fall to increase sales (Bayford 2023). Currently, PBMAs are around twice the cost of conventional meat products. High levels of investment in both research and large-scale production facilities have led synthetic protein companies to speculate that, sometime between 2022 and 2026, the cost of a cultured beef burger could be less than US$10 (Future Meat 2022; Food Processing-Technology 2022). Indeed, optimistic speculation suggests that cultured meat may reach parity with conventionally farmed meat, and plant-based meats may reach parity with soy as early as 2030 (Credit

Suisse 2021). In addition to maintaining meat purity (against bacteria, etc.), the limited capacity of stem cells to divide is one of the key limitations associated with transitioning from the laboratory to industrial-scale production. However, recent advances in the generation of immortalised muscle stem cells may be able to address this barrier if they are able to shake off their resemblance to pre-cancerous cells (Pasitka et al. 2022; Stout et al. 2023; Bloomberg 2023b).

Other key constraints include regulatory hurdles, especially in the EU (Credit Suisse 2021; Food Navigator 2023a), pushback from the farming industry, especially livestock and dairy enterprises (Burton 2019; ETC Group 2019a), poor nutrient absorption (Chen et al. 2022; Mayer Labba et al. 2022; New Scientist 2022), heath issues (linked to excessive consumption of ultra-processed synthetic plant-based proteins (New Scientist 2021), and consumer acceptance (Food Navigator 2022a). In response to growing health concerns, Impossible Foods has recently launched a Beef Lite product, which contains zero cholesterol less than 75% saturated fat, 45% less total fat, and 33% less sodium compared to conventional minced beef (Green Queen 2023b). Recent studies have suggested that between 24% and 32% of consumers are willing to eat plant-based and cultured meats (Burton 2019; Credit Suisse 2021). However, similar studies have found that '35% of meat eaters and 55% of vegetarians would be too disgusted to try cultivated meat' (Reuters 2023).

4.2.1.5 Growth in the production and consumption of PBMAs and cultured meats

Unless plant-based and cultured meats experience the same level of public resistance as seen with the introduction of GMOs, market share for these food products seems set to soar, especially if price parity, or better, can be achieved. Whilst not yet considered mainstream, annual growth rates in several countries are close to 50% (Credit Suisse 2021). The USA, UK, Europe (Germany, Italy, France, the Netherlands, and Belgium), Sweden, Australia, and Brazil are key growth markets for plant-based meats, cultured meats, and PBMA/cultured meat hybrids (ETC Group 2019a; Credit Suisse 2021; Food Navigator 2022a; World Grain 2023a). However, there is also likely to be significant market potential for these products in both countries in transition and developing countries (FAO 2011d). Figure 58 illustrates market growth in synthetic plant-based protein sales in Europe between 2018 and 2020. Figure 59 illustrates the growth of different types of synthetic plant-based products versus animal-based products in the USA in 2020.

Figure 60 illustrates the value of synthetic plant-based product sales in Europe in 2020. In 2021, it was estimated that the global market size for synthetic plant-based food products was around US$14 billion. In 2020, the USA spent US$5 billion, closely followed by Europe, at around US$4 billion

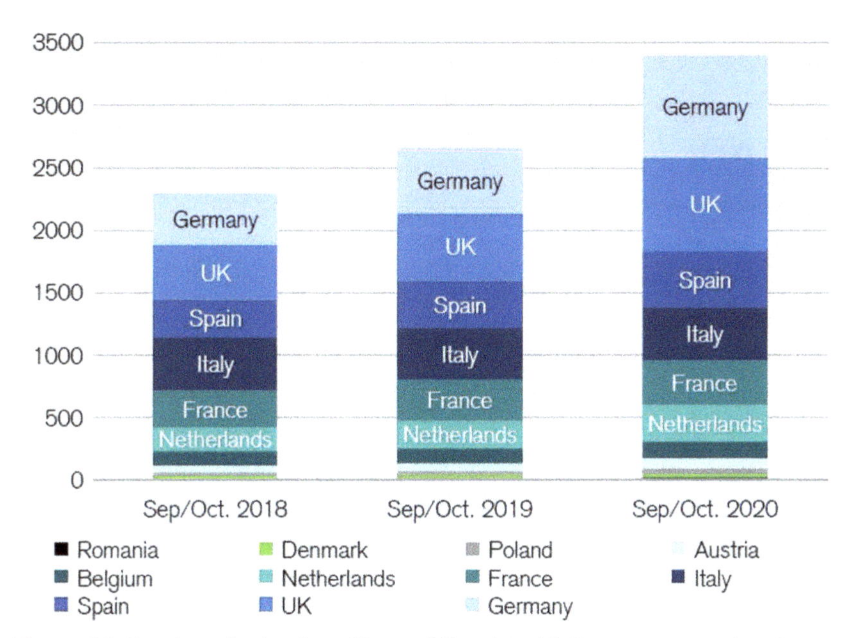

Figure 58 Plant-based sales (in millions of Euros) in 11 European countries. Source: Adapted from: Credit Suisse (2021).

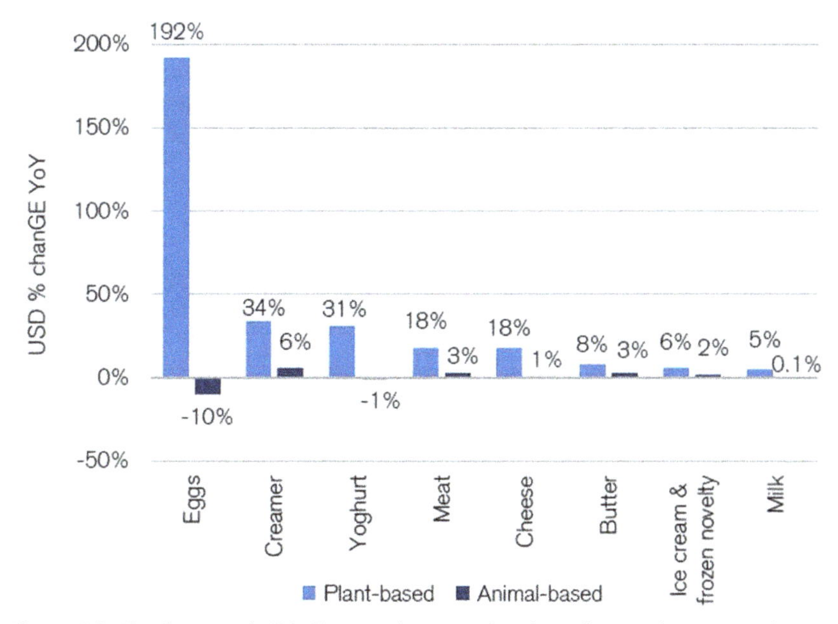

Figure 59 US sales growth (2019); animal versus plant-based meat alternatives. Source: Adapted from: Credit Suisse (2021).

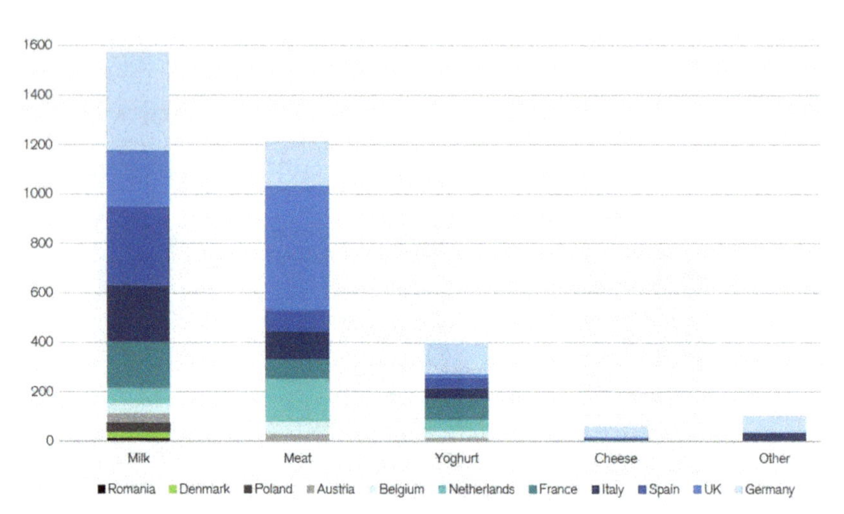

Figure 60 Synthetic plant-based product sales in Europe in 2020. Source: Adapted from: Credit Suisse (2021).

(Credit Suisse 2021). However, in 2022, venture capital investments in PBMAs and similar food tech companies declined by 44% between 2021 and 2022 (Financial Times 2023a).

Some food system analysts suggested that the sector had maxed out, citing poor sale figures, profitability, and market share of PBMAs, whilst others attributed the fall to the global rise in interest rates, 15–25% inflation, which increased labour, energy, transportation, and other input costs, staff lay-offs, and concerns about business models, and the process of market consolidation (Financial Times 2023c). Difficult market conditions in 2022, caused by the cost-of-living squeeze and consumers purchasing cheaper forms of protein, resulted in staff reductions of 19%, 20%, and 25% in Beyond Meat, Impossible Foods, and Greenleaf, respectively, and 33% and 8.4% reduction in sales revenue in Beyond Meat and Greenleaf, respectively (Bloomberg 2022; Reuters 2022a; Food Navigator 2022b; GBNews 2023). Some PBMA businesses have been permanently closed, such as US-based PlanterraFoods (Food Navigator 2022b).

Research from Bloomberg illustrates the dramatic fall in stock values for PBMA and cultivated meat companies from 2021 to 2023 (Financial Times 2023a). However, many analysts see this as nothing more than a blip. Ultimately, most market analysts expect that the market for PBMAs will continue to grow. Indeed, the market for PBMAs is expected to grow to around US$143 billion by 2030 (Credit Suisse 2021), US$450 billion by 2040 (Financial Times 2023a), and potentially US$1.4 trillion by 2050 (Credit Suisse 2021). Total market share of PBMAs could increase to 10% by 2027, and potentially 50% by 2050 (Credit Suisse 2021; World Grain 2023a).

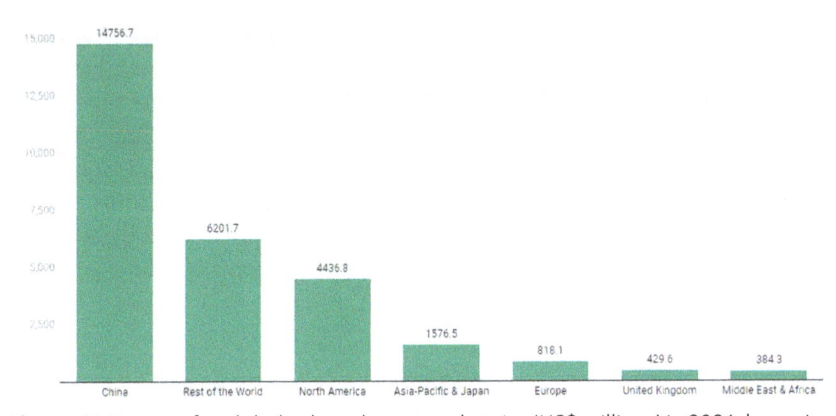

Figure 61 Forecast for global cultured meat market size (US$ millions) in 2026, by region. Source: Adapted from: GovGrant (2023). NB GovGrant now known as Source Advisors.

Figure 61 illustrates the projected market for cultured meat. Once regulatory approvals are received, the market for cultured meat is also expected to grow rapidly. Meticulous Research expects that this market will grow at a CAGR of 11.9% to US$74.2 billion by 2027 (nxtaltfoods 2023). Indeed, GovGrant (2023) predicts the European market to be worth $818 million by 2026. However, this is far behind China and North America, with a predicted market size of US$14.8 billion and US$4.4 billion respectively by 2026.

Similarly, some market analysts are predicting market shares for cultured meat ranging from 35% to 60% by 2040 (Credit Suisse 2021; Carrington 2020; GovGrant 2023). Figure 62 provides a projection of global conventional meat, PBMAs, and cultured meat consumption from 2025 to 2040.

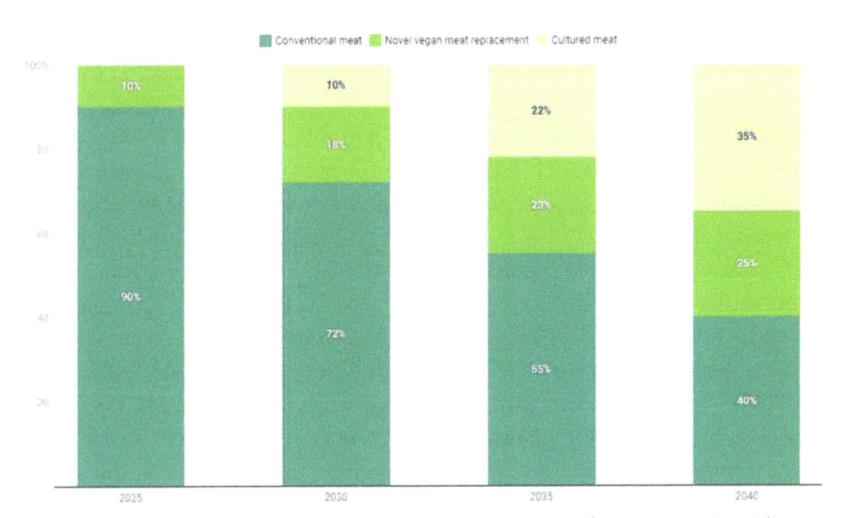

Figure 62 Forecasted breakdown of global meat consumption from 2025 to 2040, by type. Source: Adapted from: GovGrant (2023). NB GovGrant now known as Source Advisors.

These projections are largely based on encouraging customer surveys. In 2023, 52% of global consumers identified themselves as flexitarians, basically people consuming both animal protein and PBMAs. Most of the flexitarians belong to the Generation Z (<25 years old) age group (World Grain 2023a). In a poll undertaken by GlobeScan in 2022, 42% of the 30 000 respondents believed that PBMAs would replace traditional meat by 2032 (PBN 2022). Indeed, according to Webb and Hessel (2022), 'it's plausible that by the year 2040, many societies will think it's immoral to eat traditionally produced meat and dairy products'.

4.2.2 Genetically modified organisms and gene editing

Currently, there are three distinct types of genetically modified organisms (GMOs). These are

- first generation: input-traits;
- second generation: output-traits; and
- third generation: pharming.

The 1990s were undoubtedly the decade for input-trait biotechnologies (also called First-Generation biotechnologies), such as enhanced tolerance to abiotic stresses (drought, heat, salinity, etc.) and biotic stresses (weeds, pests, and diseases). Some of the most successful input-trait biotechnologies include Roundup-Ready maize, soya beans, canola, Bt maize and cotton, etc. (Carroll 2017). By the mid-1990s, input-trait biotechnologies were already grown on over 2 million ha of land globally, expanding rapidly to 63 million ha by 2001 (NGIN 2002). However, to date, much of this expansion has been achieved in countries that have wholeheartedly embraced GMOs, such as the USA, Canada, Australia, China, Brazil, Argentina, India, the Philippines, and South Africa (James 2015). By 2016, approximately 185 million ha were planted with GM seed (USA 72.9 million ha, Brazil 49.1 million ha, Argentina 23.8 million ha, Canada 11.6 million ha, and India 10.8 million ha) (ISAAA 2016). However, growth in the use of input-traits biotechnologies has tended to plateau. By 2020, 190 million ha of GM crops were planted globally, with soya bean (50%), maize (30%), cotton (13%), and canola (5%) making up the lion's share of input-trait GM crops (Turnbull et al. 2021). By 2022, there were 202.2 million ha sown with GM crops (AgbioInvestor 2023). Given the weakening of EU consumers' resistance to GM products, the need for more climate-tolerant crop varieties, and the growing move away from traditional synthetic pesticides, new input-trait biotechnologies, such as heat-tolerant and insect-resistant (HTIR) maize, heat-tolerant (HT) sugar beet, heat-tolerant

(HT) winter oilseed rape, heat-tolerant (HT) soya bean, drought-tolerant wheat, frost-tolerant barley, and Phytophthora-tolerant potato, are likely to break into the European market within the next decade (McFarlane et al. 2018; Cohen 2019).

Output-trait biotechnologies (also known as Second Generation biotechnologies) are those that alter the nature of the crop itself, either adding something desirable or valuable or reducing or removing something undesirable (Deutsche Bank 1999). Unlike successfully commercialised input-trait biotechnologies that now dominate production in GM-friendly countries, output-trait biotechnologies have struggled to find a niche. However, a growing number of consumer acceptance studies (Britwum et al. 2018; Swift 2020) suggest that, where consumers expect to benefit directly from output traits, attitudes may be changing, even in zones of staunch GM opposition such as the EU. New output-trait products include

- omega-3 producing oil seed rape/canola and maize (Watson 2018);
- low-fat soya bean (Groundup 2003a);
- low protein, and high dietary fibre wheat (linked to coeliac disease) (McFarlane et al. 2018); and
- vegetables containing enhanced vitamin C and antioxidant levels (Britwum et al. 2018).

An early example of an output-trait product was 'Golden Rice', a beta-carotene-enhanced rice genetically engineered to combat Vitamin A deficiency first developed in the late 1990s and early 2000s (Dubock 2017).

The so-called third generation biotechnologies refer to the production of a range of food and non-food pharmaceutical, nutraceutical, and other products from plant resources such as bioplastics, sometimes also known as 'pharming' in the case of pharmaceutical products (UNESCO 2003; Britwum et al. 2018). One range of products using these technologies, which overlaps with second-generation biotechnologies, is functional foods or nutraceuticals. Functional foods have been defined as foods providing a health benefits beyond simple nutrition. Health benefits are related to food components such as antioxidant compounds, stanols, prebiotics, and probiotics which are either present in foods or which can be enhanced or added to foods to provide a health benefit. Functional foods cover a wide range, from fresh fruits and vegetables (where research has identified nutraceutical benefits), to fortified food products and nutraceutical extracts provided as supplements.

In the field of biotechnology, the so-called new breeding techniques (NBTs) are creating quite a stir. A key innovation is a development of 'gene editing' (also known as genome editing or engineering), in which DNA in a plant or animal genome is 'edited', for example, by deleting or modifying part of the DNA sequence either to remove a harmful or enhance a desirable trait linked to that part of the sequence (Wilmann 2021). Gene editing has been distinguished from previous genetic modification techniques because it does not require the insertion of foreign DNA, a particularly controversial feature of earlier GM technology.

New and emerging approaches to genome-editing include clustered regularly interspaced short palindromic repeats (CRISPR), zinc finger nucleases (ZFNs), meganucleases (MNs), transcription activator-like effector nucleases (TALENs), and oligonucleotide-directed mutagenesis (ODM) (Sirinathsinghji 2020). These techniques are reportedly easier, cheaper, faster, and more accurate/precise than previous approaches in altering the characteristics of living organisms (Seyran and Craig 2018). Compared to GMOs, where commercial returns could only be assured from major agricultural crops, NBTs are able to cost-effectively edit a much wider range of crop species.

Just like first-generation biotechnologies, NBTs are being used to enhance crop plants' resistance to pests, improve water and nitrogen-use efficiency, and increase drought and heat tolerance (Lusser et al. 2011). They are also being used to modify the colour, palatability, flavour, shape, size, micro-nutrient, fatty acid, anti-inflammatory, anti-cancer, anti-oxidation, and anti-nutrient content, as well as the shelf life of a wide range of agricultural and horticultural crops (Liu et al. 2021). The first gene-edited (soya bean) product entered the US market in February 2019. Created using TALENs, the new soya bean oil is reported as having 0% trans fats, 80% oleic acid, and 'three times the fry life and extended shelf life' (Cohen 2019). In 2021, Corteva received clearance to introduce the first CRISPR gene-edited waxy corn/maize in the USA, Canada, Argentina, Brazil, and Chile (CBAN 2021). Ongoing CRISPR work aims to develop higher yielding wheat with up to 16% protein content (CropLife 2019). Table 1 highlights the crops that have already been changed using the CRISPR/Cas9 system, and Fig. 63 illustrates the number of CRISPR technology publications.

Given that the NBTs do not artificially transfer genes across species, many claim that they should not be lumped together with traditional GMOs, and therefore, should not be subject to the same level of regulation (Seyran and Craig 2018).

Countries that have already sanctioned the release of GMOs, such as the USA, Canada, Australia, Japan, Argentina, Brazil, Columbia, and Chile, have classified NBTs alongside traditional breeding approaches, which require less stringent testing before they are released (ABC News 2019; Liu et al. 2021). In the UK the Genetic Technology (Precision Breeding) Act became law in March 2023, allowing farmers to grow gene edited crops (https://www.gov.uk/government/news/genetic-technology-act-key-tool-for-uk-food-security). Despite the European Court of Justice ruling of 2018 treating gene-edited the same way

Table 1 Gene-edited crops (Liu et al. 2021)

Categories	Species
Feed crops	Alfalfa
Fiber crops	Cotton
Food crops	Apple, banana, barley, basil, blueberry, cabbage, carrot, cassava, chickpea, chill, citrus, coconut, cowpea, cucumber, date palm, grapefruit, grapes, kale, kiwifruit, *Lactuca sativa*, lemon, lettuce, lychee, maize, melon, oats, orange, papaya, pear, pepper, potato, pumpkin, rice, saffron, strawberry, sugar beet, sweet potato, tomato, watermelon, wheat, yam
Crops for industrial use	*Cichorium intybus*, coffee, dandelion, *Hevea brasiliensis*, *Jatropha curcas*, millet, papaver, Parasponia, *Salvia miltiorrhiza*, Sorghum, sugarcane, switchgrass, Tragopogon, *Tripterygium wilfordii*
Oil crops	Canola, flax, oil palm, oilseed rape, soya bean, sunflower
Ornamental crops	Lily, lotus, petunia, poplar, rose, Sedum, snapdragon, *Torenia fournieri*

as GMOs, recent evidence suggests that the EU is moving closer to approving the first tranche of gene-edited crops (Financial Times 2022b, 2023b). Biotech companies hope that WTO support for gene-editing technologies will help to reduce regulatory obstacles to the release of gene-edited crops. However, whilst unintended effects of gene-editing processes occur infrequently, this still raises concern (Blythman 2020; Liu et al. 2021). Figure 64 visualises the current, mostly positive, regulatory status of gene-edited and GMOs across the globe.

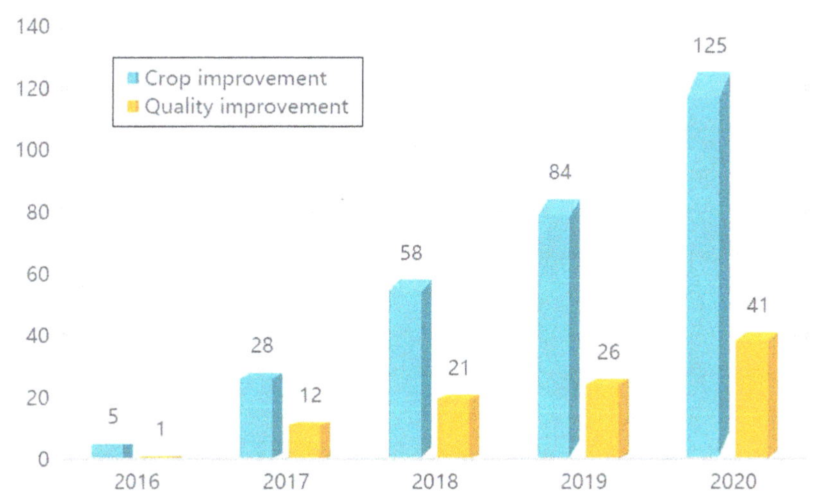

Figure 63 Number of publications mentioning the CRISPR technology using a Web Of Science search. Source: Adapted from: Liu et al. (2021).

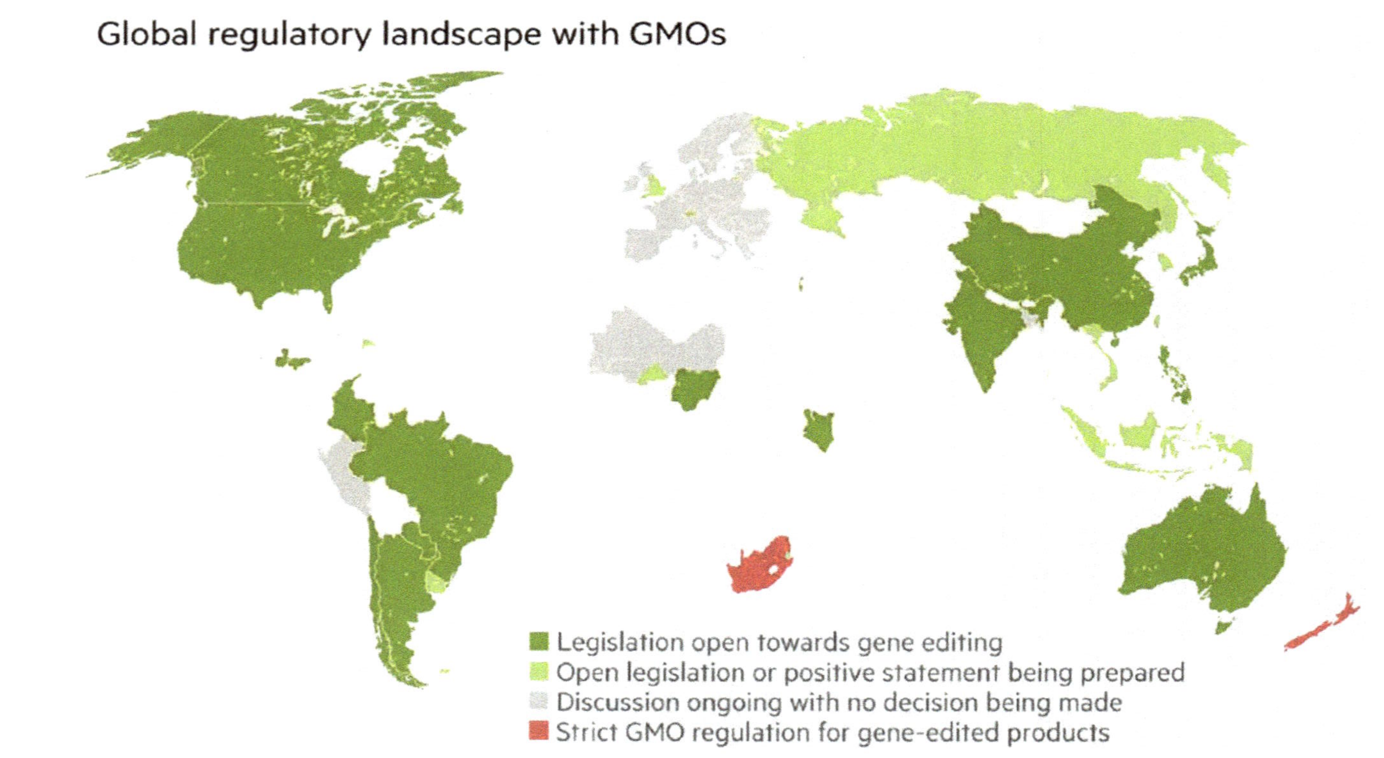

Figure 64 Global regulatory landscape with GMOs. Source: Adapted from: Sprink et al. (2022).

4.3 Controlled environment agriculture

Controlled environment agriculture (CEA) combines the use of both crop husbandry and structural engineering approaches in pursuit of the efficient production of high-quality and high-yielding horticultural crops. Earliest records of CEA date back to AD 14–37 (Higgins 2022), but their use for commercial food and flower production (in glasshouses) did not become widespread until the 1950s and 1960s, especially in Belgium, Holland, and other parts of northern Europe (FAO 2021).

The commercial use of polythene tunnels, a cheaper version of glasshouses, evolved during the 1960s, 1970s, and 1980s, firstly in southern France, Italy, and then Spain, spreading to the Middle East and North Africa in the 1990s, and then across mainly temperate areas of the Far East, especially China (FAO 2021). Today, CEA (including plastic mulches) covers approximately 5 000 000 ha (FAO 2021). Figure 65 illustrates the geographic distribution of CEA.

During the past 20 years or so, the design of CEA structures has changed significantly, tending to grow vertically rather than horizontally, following high population densities rather than geographies that receive high levels of sunshine all-year-round. A wide variety of CEA structures have come to be known as vertical farms, and they are the primary focus of this section. Inspired by Dickson Despommier and Toyoki Kozai (Despommier 2010; Kozai et al. 2016), the first vertical farms were highly experimental in nature, focussing on proof of concept rather than profitability. Energy costs associated with the provision of artificial lighting scuppered any thoughts of producing crops commercially.

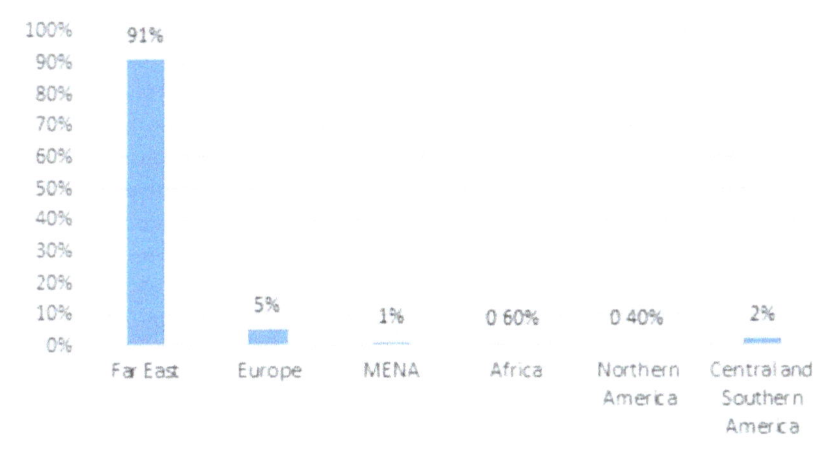

Figure 65 Protected agriculture distribution in the world. Source: Adapted from: FAO (2021).

However, as technologies rapidly improved, especially with advances in light-emitting diodes (LED) as a source of light, some of the earliest commercial vertical farms sprung up around the end of the twentieth and beginning of the twenty-first century (Mitchell and Sheibani 2020; Munns 2014). They first developed in the USA, Japan, and Holland and then spread rapidly across much of the developed world (Canada, UK, etc.) and parts of the developing world (Arabia, Malaysia, etc.) (Fi Global Insights 2021; Hofman 2021; Agrizon 2021).

4.3.1 Key components of vertical farms

Vertical farming involves growing crops in vertical layers in precisely controlled environments, where key production variables such as light, temperature, water, carbon dioxide, oxygen, humidity, nutrition, and growth media are managed in such a way as to create a bespoke growing environment for each crop (Agrizon 2021; Little 2021; van Delden et al. 2021). Figure 66 illustrates the key differences between open-field farming and vertical farming.

Vertical farms can be above ground or below ground and can be based on either natural lighting (such as Shockingly Fresh) or artificial lighting (such as Growing Underground) (Guardian 2021). They can be soil-based or soil-less (Little 2021)

- aeroponic (with plant roots suspended in the air and fed by a mist of nutrient solution); and
- hydroponic (with plant roots growing in a nutrient solution).

They can involve two or more layers of plants and are designed to produce yields all year round (Credit Suisse 2021; The Market Herald 2021).

To capture economies of scale and to enhance cost-competitiveness, vertical farms have tended to grow bigger and bigger. For example, Europe's (then) largest vertical farm (5000 m^2) was constructed by Jones Food Company, in Scunthorpe, UK, in 2019, supplies between 15% and 30% of the basil market in the UK (The Telegraph 2023). In January 2021, Affinor Growers in the UK leased a 15 000 sq. ft vertical farm facility (The Market Herald 2021). In the same year, an American firm acquired a 77 000 sq. ft warehouse on the outskirts of Atlanta, USA (Popular Science 2021). This facility plans to produce 10 000 000 lettuces a year (Popular Science 2021). US-based AeroFarms is currently building a 90 000 sq. ft vertical farm in Abu Dhabi (Fletcher 2020). Believer Meats (formerly Future Meats) is constructing a 200 000-square-foot facility in the USA, whilst Upward Farms is currently building a mega vertical farm of 250 000 square feet (The Optimist Daily 2022), and Group Amana is constructing a 330 000 sq. ft vertical farm in Dubai South (Arabian Business 2022). Sky Green in Malaysia currently

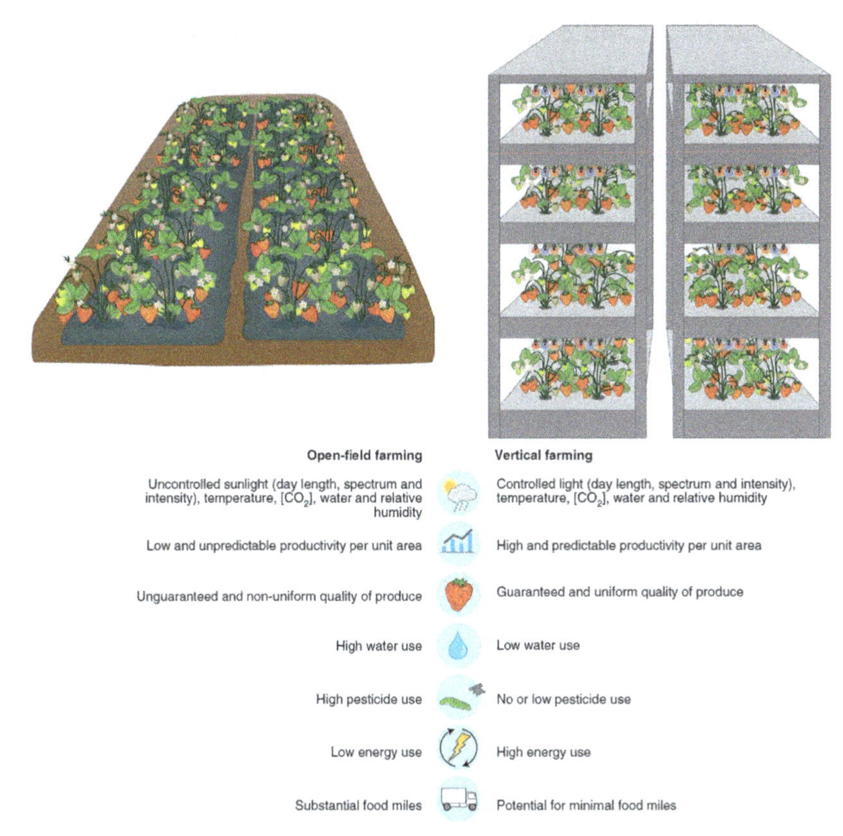

Figure 66 Key differences between open field and vertical farming. Source: Adapted from: van Delden et al. (2021).

has almost 2000 vertical farming towers, which produce between 5 and 10 tonnes of vegetables each day (Fletcher 2020).

Vertical farms concentrate on the production of a wide range of leafy vegetables (lettuce, spinach, kale, cannabis, pak choi, pea shoots, watercress), fruits, berries (strawberries), herbs (basil), nutraceuticals, medicinal, and plant-based cosmetics, and other high-value plants, primarily destined for a range of restaurants, high-end grocery stores, supermarkets chains (such as Walmart), and other industrial users (Fletcher 2020; The Market Herald 2021; Popular Science 2021; Guardian 2021; Little 2021; van Delden et al. 2021). In addition to the hundreds of leafy vegetables and fruits, entrepreneurs are now experimenting with the production of cocoa, pineapple, avocados, and even rice and wheat in vertical farms (Independent 2017; Asseng et al. 2020; Little 2021). However, it is likely to take several decades before these types of crops can be grown commercially, if at all (The Telegraph 2023).

4.3.2 Advantages and disadvantages of vertical farms

One of the major advantages of vertical farming is the generation of extremely high crop yields per area of land. As a result of precisely controlled growing environments, crops grow 30–40% faster than those conventionally farmed (Little 2021). Due to all year round production in multi-layered structures, estimates suggest that vertical farms can produce between 300 and 350 times more than conventional farming per unit of land (Credit Suisse 2021; Little 2021). Quality is often cited as another key advantage of vertical farms. It is claimed that crops are fresher, tastier, more nutritious, and safer, with little to zero pesticide and fertiliser residues and other contaminants (Marcelis 2019; Business Insider 2021; Credit Suisse 2021; The Market Herald 2021; Little 2021).

Given their location in, or extremely close to, large urban centres, several existing vertical farms have the capacity to competitively meet 80% or more of local demand for fresh leafy vegetables (Credit Suisse 2021), thus extending shelf life, reducing food waste, and potentially minimising transport costs, food miles, and greenhouse gas emissions (GHGEs) (The Market Herald 2021; Popular Science 2021; van Delden et al. 2021; Eastern Daily Press 2021). Given the extremely precise use of land, water, nutrients, and pesticides, vertical farming is also credited with being more resource-use efficient than conventional farming (van Delden et al. 2021; Guardian 2021). For example, vertical farms are estimated to use up to 95% less water than conventional farms (Independent 2017; Marcelis 2019; Credit Suisse 2021; Little 2021; The Optimist Daily 2022). Additionally, due to the precise production methods used, vertical farming releases few, if any, pollutants into the environment (van Delden et al. 2021; Popular Science 2021).

Whilst burgeoning vertical farms have significant advantages, they also have several disadvantages. One of the major challenges facing vertical farms is building costs. The size, design, siting (on expensive urban real estate), and high-tech nature of their operation mean that significant investment, often in the form of loans or venture capital, is required to construct large-scale vertical farms (Hofman 2021; Agrizon 2021; Benis and Ferrão 2018; Eigenbrod and Gruda 2015; Benke and Tomkins 2017). Once constructed, running costs for a vertical farm can be substantial.

Firstly, the energy bill, especially for climate control (light, heat, humidity, irrigation, fertilisation, managing CO_2 and O_2, etc.) is significant (Marcelis 2019; Popular Science 2021). Vertical farms use around 3500 kWh/m^2 each year of electricity (Independent 2018), which comprises approximately 25% of production costs (Sifted 2022). The industry's reliance on cheap energy exposed cracks in the business models of many vertical farms when energy costs increased (The Telegraph 2023). A rapid increase in energy prices, caused by the food, fuel, and fertiliser crisis (Watson 2022), combined with the increased cost of living, which tightened consumers' belts, resulted in the closure of several vertical farms. This included the Dutch company Glowfarms, and Future Crops, Eider Vertical Farming in the UK, Agricool in France, AeroFarms, AppHarvest, and Fifth Season in the USA (Food Navigator 2023b; Sifted 2022; Bloomberg 2023c; Fresh Plaza 2023). Given that vertical farms require skilled labour (engineers, plant scientists, and computer programmers), labour costs for vertical farming are high, usually 25–30% of total production costs (Kozai et al. 2020; Little 2021). Vertical farms, such as the UK's Infarm and the USA's Iron Ox, were obliged to lay off 500 staff (over half the workforce) in 2022 (The Telegraph 2023) and 2023, respectively (BBC 2023).

Another disadvantage is that many of the early vertical farms were highly experimental/exploratory in nature. Nearly half of the operatives had no, or little, experience in applied agriculture or horticulture before establishing vertical farms (Agrizon 2021). Often, due to the experimental/trial and error approaches taken, the choice of crop, production regime, and technologies turned out to be sub-optimal, both from a crop growth and quality and profitability perspective. A survey conducted by Agrizon (2021) highlighted that, with hindsight, 73% of vertical farm operators would choose different crops, equipment, and technology. Additionally, unlike the early days of vertical farming, where operators shared lessons, both good and bad, with their peers, recent years have witnessed a much more secretive, competitive, and proprietary environment (Agrizon 2021). The highly technical, complex, and exploratory nature of crop production in vertical farms, and the general lack of sharing of experiences, means that production lines often do not perform as well as expected, increasing the cost of the final product, which is usually already at the high end of what consumers are willing to pay (Credit Suisse 2021; Little 2021).

The high-profile nature of the vertical farming sector has attracted interest from several high-worth individuals and venture capital firms, which invest around a US$1 000 000 000 per year in the sector (Popular Science 2021). The hype around the sector has led to many vertical farms being left well-funded but struggling to pay back loans or make overly ambitious returns on capital promised to investors, leading to growing disillusionment in the sector (Agrizon 2021). In Agrizon's 2021 CEA census, '70% of farming operators agreed that CEA is susceptible to excessive greenwashing' (Agrizon 2021).

4.3.3 Prospects

Whilst the future of vertical farms may appear to hang in the balance, advances in technology and production platforms, along with the experimental learning acquired during the past 20 years, lead most sector analysts to predict a rosy future for vertical farming. The market size for vertically farmed produce was around US$4.4 billion in 2019 and is predicted to be around US$6.4 billion by 2023, US$15.7 billion by 2025 (Little 2021), and possibly US$24.11 billion by 2030 (The Market Herald 2021). More recent figures suggest that the global market size for vertical farming in 2022 was around US$4.16 billion and is expected to be US$5.05 billion in 2023. According to Fortune Business Insights (2023a), the market size is expected to grow to US$27.42 billion by 2030. Indeed, the Ellen MacArthur Foundation believes that by 2050, it is possible that vertical farms could be producing up to 80% of food consumed in urban areas (Credit Suisse 2021; Fi Global Insights 2021). Figure 67 illustrates Statista's projections for the global market size of vertical farming between 2022 and 2030.

Technological advances offer the most promising route to ensure that vertical farms become both profitable and environmentally sustainable. These advances are emerging in different components of vertical farming systems. One of the most promising is developments in more efficient and effective low-heat LED lighting systems, leading to improvements in light use efficiency, better placement of lights close to the crop canopy, and reduced energy costs for lighting and cooling (Eastern Daily Press 2021; Pattison et al. 2018; Popular Science 2021). According to Little (2021), LED efficiency increased by more than 59% in the past 6 years. Additionally, continuing improvements in the design of vertical farms, such as the use of vertical racks and rotating towers under natural sunlight, and other developments such as wavelength selective films, offer significant opportunities for increasing plant performance at the same time as reducing costs of production (Graamans et al. 2020; SkyGreens 2010; Ravishankar et al. 2021; The Market Herald 2021; Agrizon 2021). Some

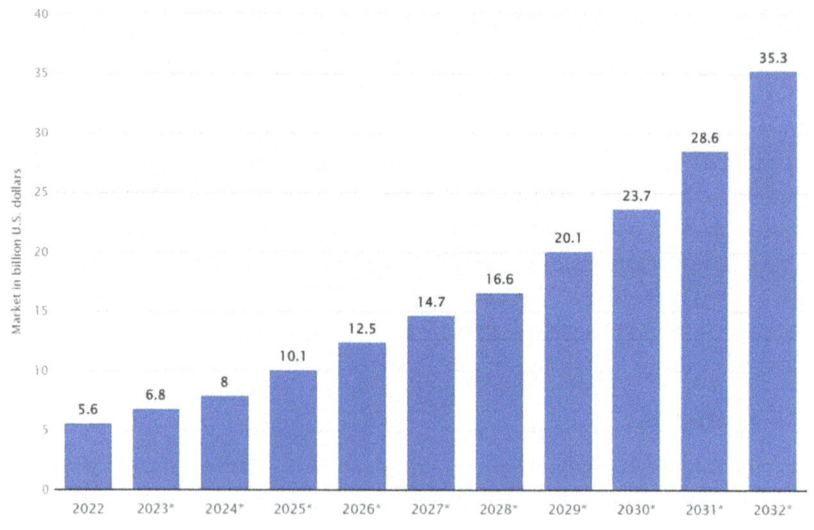

Figure 67 Projected vertical farming market worldwide from 2022 to 2032. Source: Adapted from: Statista (2023b).

companies, such as Vertical Future, are developing low-cost equipment in-house to reduce initial CAPEX requirements and boost profitability (FreshPlaza 2022).

The use of artificial intelligence, especially machine learning to infer patterns in the data, and real-time monitoring using sensors, offers opportunities to create ideal conditions for crop growth on a per-species basis, including bespoke temperature, light quantity, and quality, both day and night, fertigation (water and nutrient management), etc. (van Delden et al. 2021; Eastern Daily Press 2021). For example, Little (2021) reports that the use of sensors has been able to increase commercial crop yields by 23% and reduce the period from planting to harvest of baby leafy greens from 20 to 14 days. The benefits of machine learning and real-time monitoring of crop growth are expected to be amplified when combined with the use of robotic systems for most, if not all, operations currently undertaken by expert but expensive staff (Jahnke et al. 2016; Arad et al. 2020; Lehnert et al. 2020; Ling et al. 2019; Xiong et al. 2020; Van Henten et al. 2006; Fi Global Insights 2021). However, it must be noted that this is contingent on the actualisation of significant improvements in the field of robotics (Bac et al. 2014; Kootstra et al. 2021; Blok et al. 2021). Intelligent Growth Solution's vertical farm has already reduced energy costs by 50% and labour costs by 80% through the application of artificial intelligence and bespoke power and communication technologies (Independent 2018). Many new vertical farms, such as Dream Harvest, are selecting renewable energy sources to both cut their energy bills in the medium term and to mitigate their GHGEs (Popular Science 2021; Guardian 2022b). Indeed, some vertical farms aspire to become carbon-negative (Agrizon 2021; Eastern Daily Press 2021).

Ultimately, the developments outlined above are only possible by building on significant advances in the science of vertical farming (FreshPlaza 2022). For example, an improved understanding of the use of far-red light supplementation has resulted in increased crop yields of between 33% and 40% (Ji 2020). Beta carotenoids and lutein content can also be increased using blue and red light (Rouphael et al. 2018; Taulavuori et al. 2016; Lefsrud et al. 2006, 2008). New science has also evolved around the rhizospheres (root zone) of crop plants in vertical farms. Specifically, more precise control, including timing of nutrients (both day and night), pH, water, temperature, light (especially blue and red wavelengths), and oxygen has been demonstrated to increase crop growth rates, nutrient content, and increase the shelf life of produce (Marschner 2012). For example, adjustment of the rhizosphere can result in biofortification of crops through the uptake of potassium, iron, zinc, selenium, iodine, etc. (Voogt et al. 2010; Eldridge et al. 2020). Advances in understanding the role of elevated CO_2 levels have resulted in yield increases of 40% (Poorter et al. 1997). Precise control of the cropping environment can also improve crop quality by reducing the production of phytochemicals, such as phenolics, which are oxidised in the presence of oxygen and turn bruised tissue brown. Precise control can also reduce crops' production of anti-nutrients/toxins, such as nitrite, lectin, phytate, and glucosinolates, which bind to nutrients making them biounavailable (Rouphael et al. 2018).

Emergent vertical farms are also experimenting with new, more circular, and ecologically sensitive, production systems (Fi Global Insights 2021; van Delden et al. 2021). For example, the company Aquaponics has developed a hybrid system that combines hydroponics and aquaculture (fish farming). In this system, fish are produced and sold onto the local market, and their wastes are incorporated into the hydroponic production of crop plants as fertiliser (Credit Suisse 2021; Fi Global Insights 2021; The Optimist Daily 2022). Other vertical farms are using bees to pollinate strawberry crops (Business Insider 2021). As the efficiency of vertical farms improves, it is possible that, one day soon, water and nutrients will only be lost from vertical farms in the produce sold to the local market, approaching 100% resource use efficiency (Graamans et al. 2018). There is also a great opportunity here for the safe recycling of urban wastes, such as organic refuse and treated sewage water, to act as essential nutrients for vertical farms (van Delden et al. 2021).

Ultimately, given the recent disruption of supply chains caused by the COVID-19 pandemic and the 2022 global food crisis, along with growing concerns around urban food security, together with the ability to locate vertical farms in, or adjacent to, areas of high urban population, vertical farms are seen as an attractive technology by both customers and policymakers (Eastern Daily Press 2021; Agrizon 2021). Whilst much work is still required to optimise technological production platforms, reduce costs, improve profitability, and secure social and political acceptance, vertical farms are increasingly being seen as delivering safe, nutritious, climate-resilient, and environmentally friendly alternatives for the local production of many popular fruits and vegetables (Hofman 2021; Little 2021; Agrizon 2021; Eigenbrod and Gruda 2015; Benke and Tomkins 2017). Many in the sector see the current situation as just the beginning. They see the potential for vertical farms to expand into a much wider range of crops, including orphan and genetically engineered crops (Fi Global Insights 2021; SharathKumar et al. 2020; Brynjolfsson et al. 2010; Anderson 2004). Indeed, in places such as the Gulf, vertical farming is envisaged as a route to ensure national food security (Fi Global Insights 2021).

Chapter 5

Reformist food systems

5.1 Introduction

Measures are currently being taken to adjust the industrial global food system to meet growing demands for higher quantities of higher-quality foods, whilst reducing fossil-fuel energy consumption and environmental externalities. This chapter explores some of the key technologies and strategies employed by food system actors, both public and private sector, which underpin minor reforms to the existing Neo-Productivist model of capitalist food systems.

The Reformist approach also continues adherence to the growth paradigm, which argues that the sustainability transition needs to be built on increased economic growth brought about by technical, financial, and spatial fixes. These fixes include

1. the development of high output/lower input production systems - including precision agriculture, climate-smart agriculture, and Conservation Agriculture, which aim to maximise food outputs and minimise resource inputs required by food systems through the process of sustainable intensification; and
2. the development of resource efficient and environmentally friendly technologies, including nanotechnologies and biopesticides.

In turn, in the same vein as the Neo-Productivist paradigm, these technological and spatial fixes (also known as Agriculture 4.0 technologies) (Rose and Chilvers 2018) are underpinned by increasingly liberal national, regional, and international food markets and internationally mobile capital (Economist 2010; Carlson 2016; Taylor and Uhlig 2016; Holden et al. 2018; Rosin 2013; Foley et al. 2011; Tilman et al. 2011; West et al. 2014; SAM 2019; FAO 2020; Herrero et al. 2020; OECD 2021; Steiner et al. 2020; Benton 2015; Benton et al. 2021; Baulcombe et al. 2009; Garnett et al. 2013; Bommarco et al. 2013; Doré et al. 2011; Tittonell et al. 2016; Marchetti et al. 2020; Pereira et al. 2018; Chiles 2013; Mouat et al. 2019; Kuch et al. 2020; Barrett and Rose 2020; Broad 2019, 2020a,b; Klerkx and Rose 2020; Clapp and Ruder 2020).

5.2 High output/lower input production systems: Precision agriculture, Climate Smart Agriculture and Conservation Agriculture

5.2.1 Precision agriculture

Precision agriculture (PA), also known as smart farming, is a farm management approach that uses site-specific information to target the biologically, economically, and environmentally efficient application of farm inputs to optimise crop and livestock performance (Sishodia et al. 2020; EU 2017; Poppe et al. 2015; Sonka 2015; Lajoie-O'Malley et al. 2020; Cisternas et al. 2020; Balafoutis et al. 2020). Site-specific information is central to PA because it acknowledges the heterogeneity of field conditions. PA recognises that key variables associated with crop or livestock production, such as water content, soil fertility, pH, pests, disease, and weed pressure, vary both spatially and temporally across farmers' fields. As such, instead of applying equal amounts of irrigation water, fertiliser, soil amendments (such as lime), and pesticides based on standard guidelines and practices, PA aims to tailor the application of inputs to crop and livestock needs at a much more granular, almost bespoke, scale (Cisternas et al. 2020). PA applications include controlled traffic farming (restricting tractors and other machines to pre-selected 'tram lines' in a field to optimise fuel use and minimise soil compaction whilst still ensuring targeted delivery of inputs across a field), precision tillage, variable-rate seeding, variable-rate, site-specific application of fertilisers and pesticides, and precision weed control (Stafford 2018). PA technologies have also been applied in livestock farming, e.g. in the use of sensors to monitor animal behaviour and health as well as in grazing management (Berckmans 2022).

To enable the targeted application of inputs, a farmer needs to be able to accurately quantify the status of key variables and accurately calibrate and apply variable rates of farm inputs to optimise farm productivity and reduce the likelihood of excessive input applications, especially of fertilisers and pesticides, which can cause environmental pollution. Whilst testing for the spatial and temporal variability in fertility and pest and disease pressure has been possible since the industrial revolution, the cost-effectiveness of monitoring in-field variability, especially during the decades of cheap farm inputs, combined with the inability of farm equipment to easily vary input application across individual fields, meant that PA could not come of age until appropriate supportive technologies were in place.

PA did not really start to emerge until the 1980s (Finger et al. 2019). It did so on large commercial farms across the USA, Canada, Australia, and Western Europe. Some of the early PA pioneers began utilising data emerging from satellite remote sensing, especially Landsat 1, which enabled more usable land cover information (Sishodia et al. 2020). However, it was not until the late 1990s and early 2000s that satellites, such as IKONOS, could produce images at a resolution of 4 m, in both visible and near infrared bandwidths, at regular time intervals, providing information that could start to be used to map soil composition and health, crop nutrient and water status and patterns of disease and pest damage across individual fields (Finger et al. 2019; Seelan et al. 2003; Enclona et al. 2004; Sullivan et al. 2005; Yang et al. 2014; Dhoubhadel 2020). However, it was not possible to produce multispectral images at a resolution of 2 m on a daily or more frequent basis until the launch of the Pleiades-1A and Worldview-3 Satellites in 2011, allowing precise, localised detection of key variables such as disease pressure, nutrient needs, and water stress (Navrozidis et al. 2018; Bannari et al. 2017; Salgadoe et al. 2018).

Early examples of the commercial application of PA in the 1990s included some of the world's largest farms across the Midwestern Plains and the Northwestern corn belt of the USA. PA offered an opportunity for these 'corn barons' to better target inputs, especially fertilisers (Humberto et al. 2022), spreading the cost of doing so across thousands of hectares of maize, wheat, and soybeans (Humberto et al. 2022). However, it was not until the turn of the twenty-first century that the technologies to allow cost-effective variable rate application (VRA) of fertilisers and seed, and variable tillage, emerged (Finger et al. 2019). Indeed, automatic section control, which eliminated double planting or fertilisation of headlands (field edges), did not emerge until around 2005 (Nowak 2021). Precise fertiliser applications help to minimise leaching of nutrients to deeper soil layers, which are inaccessible to plant roots, as well as minimising surface runoff and contamination of ground water (Mikula et al. 2020). The use of VRA has been demonstrated to reduce field-level pesticide applications between 11% and 90% (Finger et al. 2019), and nitrogen application by more than 13% (Finger et al. 2019). The use of VRA for irrigation management has been shown to reduce water use by 20–25% (Finger et al. 2019).

From 2010, PA has become mainstream in many areas of intensive agriculture (Zhang et al. 2002). Currently, remote sensing, via satellites, remains the most important source of information for PA applications. This is probably due to the versatility of modern satellites, which can provide high-quality spatial, spectral, and temporal information on temperature, humidity, pH, soil organic matter content, soil nutrients, and water/irrigation management, etc. (Jha et al. 2019; Elijah et al. 2018; Blasch et al. 2015; Kalambukattu et al. 2018; Castaldi et al. 2019; Khanal et al. 2018). In the last

20 years, significant advances have been made in commercial application of ground-based platforms/proximal remote-sensing precision technologies, namely handheld, free-standing, and tractor/machinery mounted units (Pallottino et al. 2019). Some of the main technologies currently in commercial use include: CropCircle®, OptRX®, GreenSeeker®, Yara N-Sensor/Field Scan®, Crop-Spec®, WeedSeeker®, WEEDit®, ASD FieldSpec®, and CropScan® (Pallottino et al. 2019).

It must be noted that, due to their relative cheapness and capacity for real-time analysis and generation of high-quality digital data, the use of unmanned aerial vehicles (UAVs) and unmanned aerial systems (UAS) (also known as drones) has grown recently. Drones, tractor/machinery mounted sensors, and handheld sensors provide higher-resolution images than publicly available satellite images and can more accurately measure real-time variables, such as crop biomass, growth stage, photosynthetic efficiency, nitrogen content, and a range of soil properties (Finger et al. 2019). UAVs and mechatronics (agro-robots) are also being used to apply both pesticides and fertilisers, saving labour, time, fuel, inputs, and reducing environmental pollution (Finger et al. 2019; Koutsos and Menexes 2019; Yatribi 2020; Rovira-Más et al. 2020). For example, Song et al. (2021) developed a UAV granular fertiliser spreader that, when linked to GIS and GPS systems, can apply variable rates of fertiliser, increasing farmer profitability at the same time as reducing the environmental impacts of fertiliser over-application. In South Korea and Japan, the use of UAVs for pesticide application in arable crops is increasing rapidly (Finger et al. 2019).

Other key technologies include the use of Global Positioning System (GPS), via smart phones, wireless sensor networks, and other kinds of sensors, such as emergent intelligent nano-sensors, to accurately fix points across open fields (Sekhon 2014; Fraceto et al. 2016; Usman et al. 2020). GPS guidance systems can reduce driver fatigue, operating costs (due to more precise control of tractors and equipment), environmental impact, greenhouse gas emissions (GHGEs), and increase productivity (faster operations), yields (inputs attuned to plant needs), quality, and human safety (Koutsos and Menexes 2019). Nano-sensors are increasingly being developed to detect real-time, field-wide changes in soil pH, water, nutrient status, pesticide residues, and humidity, as well as detection of pests and diseases (Fraceto et al. 2016). Ultimately, using yield monitors, these technological advances have allowed farmers to generate digitalised Geographic Information System (GIS) maps, which accurately display the spatial and temporal variability of soil water, nutrition, seed rate, pests, diseases, and weeds, as well as crop yields (Cisternas et al. 2020). For example, compared to conventional drilling (sowing of seeds), precision drilling can increase crop yields between 10% and 30% (Zhai et al. 2014).

Significant investments have also been made by the private sector in both the collection and analysis of 'big data.' Global revenues for big data analytics grew from US$130 billion in 2016 (Carolan 2018) to US$231.43 billion in 2021 and are predicted to grow to US$549.73 billion by 2028 (Global News Wire 2021). Ultimately, with the increased use of artificial intelligence/machine learning, internet of things (IoT), and cloud computing, private companies are increasingly able to make sense of the mountains of farm-level data to better inform farm management decision-making in field crops, pasture, and livestock, as well as high-value horticultural and viticultural crops (Finger et al. 2019; Pallottino et al. 2019; Benyezza et al. 2021; Liao et al. 2021; Mikula et al. 2020; Song et al. 2021; Hadi Ishak et al. 2021; Ghafar et al. 2021; Lin et al. 2020; Bai and Gao 2021; Beukes et al. 2019). These approaches are already being used to estimate evapotranspiration, soil moisture content, and crop yields to tailor irrigation schedules, and fertiliser, herbicide, and insecticide applications (Boursianis et al. 2020; Weiss et al. 2020; Huang et al. 2020; Partel et al. 2019; De Castro et al. 2017; Monteiro et al. 2021). Indeed, UAVs have been used in China to map weeds in rice fields with up to 90% accuracy (Huang et al. 2020).

Ultimately, the opportunities being offered by PA are increasingly being capitalised by global agricultural machinery producers, often in concert with global agricultural input suppliers, especially seed, fertiliser, and pesticide suppliers. This increased vertical integration of input supplies and agronomic advice, based on big data, artificial learning, and real-time crop monitoring, offers a wide range of services to large commercial farmers to maximise input productivity and maintain high yields, which are destined for often predetermined markets (IPES-Food 2017b).

Of the global agricultural machinery corporations, John Deere, who has been investing in this space for over 20 years, leads in the development of PA solutions (Carolan 2018). Indeed, John Deere has been fitting a GPS link in its farm machinery as a standard feature since 2001 (Watson 2018). John Deere was quick to establish a dedicated unit called the Intelligent Solutions Group and, in 2017, acquired Blue River Technology, to develop smart robotics and telematics (GPS tracking of machinery), and machine learning for PA.

As with biotechnology, life science corporations quickly realised the utility of intelligent machinery to enhance delivery of their own products, namely seeds, pesticides, and technical agronomic knowledge (Watson 2018). As early as 2007, Syngenta and John Deere had teamed up to launch an integrated Force CS insecticide delivery system. In 2013, John Deere teamed up with DuPont Pioneer to link DuPont's Pioneer Field 360, a suite of precision agronomy software, with 'John Deere Wireless Data Transfer, JD Link and MyJohn Deere' (Watson 2018). 'See and Spray', developed by Blue River Technology, can detect weeds in a standing crop, and target individual weeds with the application of herbicide, reducing herbicide use by up to 90% (GAP 2020).

Whilst John Deere continued to dominate the space, by 2013, other major machinery manufacturers began to invest heavily in PA. Currently, the Allis-Gleaner Corporation, better known by its acronym AGCO, is John Deere's principal competitor. AGCO is known for its notable range of global tractor brands such as Massey Ferguson, Deutz-Fahr, Fendt, and Valtra (Watson 2018). AGCO has already collaborated with life science corporations such as DuPont on a global wireless data transfer system, with Monsanto to integrate Monsanto's Precision Planting System in AGCO's planters, and with BASF on its 'Farm Management Information Systems' (Watson 2018). Monsanto possesses more crop performance data than any other private sector company (Carolan 2018). AGCO also collaborated with Bayer to integrate Bayer's Climate Corp Field View system into AGCO machinery. Indeed, Bayer's head of digital farming stated that the company will transition from selling seeds and pesticides to 'selling positive social and environmental outcomes through the sale of big data generated information' (Maity 2018). Just like John Deere, AGCO also has a dedicated unit called Fuse, which focusses on the development of PA technologies such as sensors, machine control, and actionable insights, which are claimed to increase farmer profitability by 20%, and, in 2017, acquired the specialist company Precision Planting (farmindustrynews.com 2017).

5.2.1.1 Advantages of precision agriculture

Whilst the adoption of PA technologies globally has been relatively slow, some specific technologies have really taken off, especially for certain crops in certain production geographies. For example, the number of farmers adopting GIS and Global Navigation Satellite System and GPS guidance systems with automatic section control and yield monitoring/mapping is growing by approximately 4% per year (Nowak 2021). By 2016, it was claimed that between 60% and 80% of arable farmers use these technologies (Nowak 2021). Indeed, in some USA states, such as Kansas, adoption rates for GIS and GPS systems have been reported at more than 80% (Finger et al. 2019). In Europe, >65% of arable farmers in the Netherlands, the most advanced country for PA, use GPS technologies (Michalopoulos 2015). However, in other countries, such as Germany, PA adoption rates are much lower at around 10–30% (Finger et al. 2019). In 2015, it was estimated that there were 30 million devices gathering agricultural information, a figure that was predicted to increase to 75 million by 2020 (Finger et al. 2019).

The adoption of other technologies, such as those for soil mapping and VRA of seeds and fertilisers, has been much slower at around 2% per year (Koutsos and Menexes 2019; John Deere 2020; Loures et al. 2020; Stefanini et al. 2019; Larson et al. 2016; Torbett et al. 2007; Nowak 2021). By 2016, between 20% and 33% of arable farmers had adopted soil mapping and VRA technologies

(Nowak 2021). To a lesser extent, VRA has also been adopted for pesticides and lime applications (Pallottino et al. 2019; Lajoie-O'Malley et al. 2020; John Deere 2020; Young et al. 2003; Agrimonti et al. 2021). Over 50% of maize, wheat, soybean, and cotton farmers in the USA use GIS, GPS, and VRA technologies for managing inputs (Schimmelpfennig 2016). However, even in Kansas, adoption of VRA for fertiliser application is only just above 25% and VRA seeding is only at 20%, with average adoption rates for VRA technologies across the USA at around 19% (Finger et al. 2019). By 2030, John Deere predicts that its VRA technologies will have improved farmers' nitrogen use efficiency and crop protection efficiency by >20% and will have reduced farmer CO_2 emissions by 15%. (John Deere 2021). Adoption has also been spurred on by reduced fuel consumption, reduced labour time required per task, and reduced CO_2, fertiliser, and pesticide pollution (John Deere 2020, 2021). Closely tied to the VRA technologies are the yield monitor technologies, which, in 2016, due often to the sale of combines with factory-fitted yield monitors, had an adoption rate of between 60% and 80% (Nowak 2021). Ultimately, the key drivers of PA adoption are the perceived production of higher or more stable yields through the application of more cost-efficient, and environmentally sensitive, crop inputs, and overall increased profits (Liu et al. 2017; Nawar et al. 2017; Calegari et al. 2013; Jayakumar et al. 2017; West and Kovacs 2017; Yatribi 2020; Ullah et al. 2021; ISPA 2019; John Deere 2020, 2021).

Whilst not widely documented, some studies have attempted to measure PA's positive impact on the environment (Yatribi 2020). GPS has been shown to reduce fuel consumption by up to 6% and fuel costs by up to 25% (Finger et al. 2019). In Germany, VRA of nitrogen fertiliser in maize crops reduced nitrous oxide emissions by 34% (Finger et al. 2019).

John Deere's global footprint of PA technologies, which incorporates the technologies outlined above, grew from 100 million engaged acres in 2018 to 165 million acres in 2020 and 329 million acres in 2022, with aspirations to reach 500 million engaged acres by 2026. Currently, John Deere has more than 500 000 connected machines globally and aspires to reach 1.5 billion by 2026 (John Deere 2021, 2022).

Utility, or at least perceived utility, especially if it is witnessed first-hand, has been identified as the major determinant of adoption (Barnes et al. 2019; Brown and Roper 2017; D'Antoni et al. 2012; Dela Rue and Eastwood 2017; Griffin et al. 2017; Mengistu and Assefa 2019; Ng'ang'a et al. 2019). This is followed by age and wealth/access to finance. Basically, younger, and more educated, innovative, and tech-savvy farmers are likely to be wooed by the promise and accessibility of PA (Paustian and Theuvsen 2017; Brown et al. 2019; Kaliba et al. 2020). Ultimately, given that acquiring large-scale PA technology is not cheap, it is generally only the larger farms and richer farmers that have profits to invest, or those that can easily access bank loans, who can afford this type of technology (Reichardt and Jürgens 2009; Griffin et al. 2017; Miller et al. 2019; Ng'ang'a et al. 2019; Barnes et al. 2019; Yatribi 2020).

As mentioned above, farmer age, farm size, and wealth have a significant bearing regarding the adoption of PA technologies. The cost of high-tech PA technologies has limited their adoption to some of the biggest or most profitable farming operations across the globe. The cost of PA technologies has excluded access to many farmers across the world. For example, across the whole of SSA, only large-scale farmers in South Africa have been able to make use of satellite imagery, GPS, and GIS technologies (Onyango et al. 2021). The relatively exclusive nature of this technology has spurred the parallel development of smaller-scale, cheaper, and more accessible technologies, such as hand-held sensors (for nutrient needs), drones (for crop monitoring as well as fertiliser and pesticide applications), smart mobile phones with intelligent applications/algorithms, and capacity to take high-quality images (for irrigation management, disease detection, soil mapping, crop yield estimation, extension services, access to credit and market information, etc.), which are only currently limited by poor network connectivity (Nielson et al. 2018; FAO 2018g; Mendes et al. 2020).

5.2.1.2 Disadvantages of precision agriculture

Whilst there is growing interest in the utility of PA, there are several perceived disadvantages that are holding up adoption of PA technologies. Of these, the cost of PA technologies is probably the most important obstacle to adoption. The initial investment needed to benefit from PA is significant (Cisternas et al. 2020; Cosby et al. 2016; Rogovska et al. 2019). Consequently, aside from some capital-intensive businesses, such as intensive livestock and viticulture, adoption of PA has primarily been restricted to large and highly mechanised arable farms, in countries such as the USA, Canada, and Brazil, where capital is available to fund the initial investment, where significant intra-field variability exists, and where costs can be spread over large areas of land (Yatribi 2020; Nowak 2021).

The complexity and multidimensionality of PA technologies have been highlighted as another important obstacle to adoption (Pallottino et al. 2019; Pathak et al. 2019; Nowak 2021; Humberto et al. 2022). Complexity and multidimensionality challenges are faced at several levels. First, farmers must choose which technology packages to adopt. This involves determining which technologies are needed to address needs in both the present and future. The expensive sunk-cost nature of technologies lock farmers into arrangements for hardware, software, and intellectual property, (Higgins et al. 2017), all of which must be compatible/interoperable with existing farm machinery and operations and involve a mix of open and closed data (Holden et al. 2018; Yatribi 2020; Robertson et al. 2012).

A second barrier is that farmers need enhanced technical skills to manage the inputs going into the GIS system and outputs coming out of the GIS systems (Cisternas et al. 2020; Yatribi 2020; Sarri et al. 2020). Even in low-cost approaches to PA in places like SSA, there continues to be a mismatch between the actual skills of users and extension agents and the actual skills required (Onyango et al. 2021). This links to a third potential barrier in that farmers need to factor in the need for continuous support (agronomic, software, and hardware) (Jellason et al. 2021; Humberto et al. 2022). A fourth issue is that farmers are concerned about corporate and government access to, and ownership of, their farming information (Carolan 2018; Eastwood et al. 2019; Jakku et al. 2019; Bronson 2018, Bronson 2019; Carolan 2017; Rotz et al. 2019 and Yattibi (2020)).

Fifth, several PA technologies still require further development to be able to sense and model the exceedingly dynamic environments for weeds, pathogens, and even nitrogen (Pallottino et al. 2019). For example, in the case of nitrogen, VRA is hailed as having significant potential, but both the current hardware and software configurations struggle to accurately monitor and model the impact of temperature, rainfall, soil type, management approaches, and high spatial variability on the cycling of this nutrient (Rogovska et al. 2019). Ultimately, the costs associated with PA, combined with the challenges of complexity, compatibility, and requirement of enhanced user and support personnel skills, lead several to question the economic returns of PA for many farmers (Castle et al. 2016; Pallottino et al. 2019; Yatribi 2020; Nowak 2021; Pathak et al. 2019; Kernecker et al. 2020; Dhoubhadel 2020).

5.2.1.3 The future of precision agriculture

Despite current challenges to adoption, PA technologies are continually advancing, and the cost of access to these technologies is continually decreasing (Finger et al. 2019; Honrado et al. 2017; Abdullahi et al. 2015). Many agricultural analysts are convinced that PA has significant room for growth (Sishodia et al. 2020). Indeed, it is estimated that, by 2030, PA could add up to US$500 billion to the global GDP, and by 2050, increase farm productivity by up to 70% (Credit Suisse 2021). Smart crop monitoring (especially by drones), linked by inter-operable networks of sensors (especially intelligent nano-sensors), machine learning (utilising uber amounts of data), drone farming (for seed, fertiliser, and pesticide application - Jacobs 2020; TropoGo 2022; DJI Enterprise 2022), smart livestock monitoring, smart irrigation, and autonomous farming are envisaged to be the most significant PA technologies (Credit Suisse 2021; Nowak 2021). Indeed, the global agricultural drone market in 2023 was worth US$4.98 billion and could reach US$5.7 billion by 2025 (TropoGo 2022), and US$18.22 billion by 2030 (Fortune Business Insights 2023b). Given the increasing use of sensors along the whole food chain, from production (Wang et al. 2006), to post-harvest management (Pérez-Marín et al. 2010), to transportation logistics, to processing (Woodcock et al. 2007), wholesale and retail storage and purchases (Abad et al. 2009), and consumer storage, nutrition, and health (Chi et al. 2008; Holden et al. 2018), it is likely that the 'Internet of Food', will soon become a reality (Internet Society 2020).

5.2.2 Climate-smart agriculture

The term climate-smart agriculture (CSA) was coined by the World Bank in 2009 (World Bank 2011). However, it was not until The Hague Conference on Agriculture, Food Security and Climate Change the following year that the FAO proposed CSA as a key component of climate resilient food systems. According to Lipper et al. (2014), the concept of CSA is defined by three objectives/pillars:

1. Sustainably increasing agricultural productivity to support equitable increases in incomes, food security and development;
2. Adapting and building resilience to climate change from the farm to national levels; and
3. Developing opportunities to reduce GHG emissions from agriculture compared with past trends.

CSA attempts to meet the objectives of, and manage the trade-offs between, food security, climate adaptation, and climate mitigation (Lipper et al. 2014; Vermeulen et al. 2012; Thornton et al. 2018). CSA is an approach that attempts to coordinate the actions of key actors, namely farmers, researchers, private sector, civil society, and policymakers towards evidence-based and socially embedded, climate resilience.

On the policy front, agriculture and climate change were first discussed together in Article 2 of the Earth Summit in 1992, but agriculture did not become a mainstream topic of climate change debate until the United Nations Framework Convention on Climate Change (UNFCCC) 17th Conference of the Parties (known as COP17) held in Durban (South Africa) 2011, when UNFCCC requested that agriculture be considered by the Subsidiary Body for Scientific and Technological Advice (SBSTA) (Muldowney et al. 2013). The creation of the Global Alliance for Climate Smart Agriculture (GACSA) during the 2014 UN Summit on Climate Change was an important step forward. However, it was not until COP22 in Marrakech (Morocco) that CSA programmes and practices became mainstream (FAO 2016a; CTA 2016). In 2017, COP23 identified a spectrum of agricultural challenges including

1 methods and approaches for assessing adaptation, adaptation co-benefits, and resilience;
2 improved soil carbon, soil health, and soil fertility under grassland and cropland as well as integrated systems, including water management (Loboguerrero et al. 2019);
3 improved nutrient use and manure management towards sustainable and resilient agricultural systems;
4 improved livestock management systems; and
5 socioeconomic and food security dimensions of climate change in the agricultural sector (Loboguerrero et al. 2019).

Ultimately, whilst CSA has a lot in common with PA, its principal difference relates to the roles of the private versus public sectors. In the case of PA, the private sector is firmly in the driving seat. PA technologies and approaches are designed by the private sector for other actors in the private sector. Any lobbying that needs to be done to create an enabling environment for PA, such as legislation allowing the use of drones for PA tasks, is undertaken by the private sector. Conversely, CSA is primarily public sector and NGO led, targeting regions where there is an acute need for CSA, but where financial, technical, infrastructural, and institutional constraints exist that limit the development and application of CSA technologies and approaches (Campbell et al. 2018).

On the technical front, the gauntlet of CSA was taken up by key research institutions across the world, including the University of Wageningen (The Netherlands), University of California-Davis (USA), CIRAD (France), and CGIAR (a global alliance of leading agricultural research centres) (Saj et al. 2017). Ultimately, CSA aims to facilitate the context-specific and socially acceptable transition to climate-resilient food systems to increase food security and reduce GHGEs (Lipper et al. 2014). To date, much of the focus of CSA has been in developing countries, especially in sub-Saharan Africa (SSA), where climate change challenges have been most acute and where smallholder farmers have limited capacity to adjust to climate change.

5.2.2.1 What technologies/approaches does CSA entail?

CSA encapsulates a wide range of technologies and approaches. As alluded to above, many of the technological aspects of CSA are borrowed from PA. These so-called smart farming technologies include GIS, GPS, ICT, big data, AI, IoT, UAVs (drones), robotics, and VRA (Bacco et al. 2018, 2019; Nawar et al. 2017; Alameen et al. 2019; Stamatiadis et al. 2020; Boursianis et al. 2020; Muangprathu et al. 2019; Falco et al. 2019; Stombaugh 2018; Kamilaris et al. 2017; Bronson and Knezevic 2016; Wolfert et al. 2017; Tsouros et al. 2019; Hajjaj and Sahari 2016; Balafoutis et al. 2020; Edan 1999; Brewster et al.

2017; Gebbers and Adamchuk 2010). In a similar vein, these technologies are leveraged to improve fertiliser, pesticide, and water-use efficiencies (Lipper et al. 2014; Makate 2019; Murray et al. 2016). As with PA, the success of CSA relies on access to quality extension advice. This includes support to access CSA farming inputs (including finance), timely advice on input use (reducing input costs), field management operations, weather advise (Williams et al. 2015; Hansen et al. 2011; Tall et al. 2014; Abegunde et al. 2019), access to weather-indexed insurance, harvesting, post-harvest processing, and storage management advice, research and innovation systems, and access to markets (Cooper et al. 2008; Hansen et al. 2011). Recent efforts have been made to digitalise most types of information, accompanied by significant investments in ICT (Adamides 2020). The combination of big data, AI, and ICT, operationalised via smart phone technologies, is seen as a major step in the scaling up and scaling out of CSA (Campbell et al. 2018). Whilst both PA and CSA focus on optimising the synergies between crop genetics and agronomy, given the acute climatic challenges CSA aims to address, the selection of climate-resilient (drought and heat-tolerant and nutrient-efficient) germplasm and livestock breeds takes higher precedence (Singh et al. 2016; Wissuwa et al. 2002; Chin et al. 2011; Adebayo and Menkir 2014; Adebayo et al. 2015; Fisher et al. 2015; Makate 2019; Ramirez-Villegas et al. 2015; Challinor et al. 2018).

Unlike PA, CSA stresses the need for integrated and landscape-wide approaches to agricultural production. These include integrated crop, grassland, livestock, aquaculture, forestry, and agroforestry production (Lipper et al. 2014; Murray et al. 2016); crop diversification (Makate 2019); inter-cropping (Murray et al. 2016); crop rotation, especially with legumes; improved fallow; residue valorisation and recycling; and organic farming techniques (Agrimonti et al. 2021), to enhance soil carbon storage, biomass production, and agro-ecosystem resilience, and reduce GHGEs - usually via the sustainable intensification of food production, reduced soil disturbance, and reduced deforestation/re-afforestation (Field 2014; IPCC 2014; Branca et al. 2021).

Given that most areas targeted by CSA are marginal, subject to unpredictable rainfall, and containing nutrient-mined soils, soil fertility management is usually central to CSA interventions. Where soil organic matter and nutrients are limiting factors to agricultural production, integrated soil fertility management practices, such as mulching, rotations, Conservation Agriculture (comprising minimal soil disturbance, crop residue management, and cover crops and increased soil water storage), agroforestry (for nitrogen fixation and soil rehabilitation - usually focussed on increasing soil organic matter and fertility), are often deployed to restore degraded soils (Lipper et al. 2014; Makate 2019; Branca et al. 2021; Murray et al. 2016).

Given the private sector's limited role in most CSA target areas, socio-economic and institutional systems, such as public-private partnerships and local institutions and innovation networks, also need to be put in place to ensure success of CSA interventions (Wise et al. 2014; Dowd et al. 2014). For example, due to the need for capital to invest in CSA technologies and approaches, one of the central tenets of CSA is ensuring access to credit and crop/livestock insurance (Barnett et al. 2008; Makate 2019; Murray et al. 2016; Hansen et al. 2018; Carter et al. 2014; Linnerooth-Bayer and Hochrainer-Stigler 2015; Jensen and Christopher 2017). The multidimensionality of CSA also calls for a holistic approach. In some cases, this has taken the approach of Climate Smart Villages (CSVs) (Lipper et al. 2014), which takes an Agricultural Innovation Systems approach to test the proof of concept for both technical and institutional approaches to CSA in discrete locations (Aggarwal et al. 2018). The Climate Change, Agriculture and Food Security Program of the CGIAR has widely utilised the CSV approach, as have many NGOs (Totin et al. 2018).

Ultimately, in areas that face (sometimes) intractable challenges, productive social safety nets need to be put in place, including cash and in-kind transfers (Johnson et al. 2013; Steinbach et al. 2016). Adaptive social protection can provide credit, agricultural inputs, extension, and support, and reduce finance risks (Arnall et al. 2010; Davies et al. 2008, 2009, 2013; Gilligan et al. 2009; Hoddinott et al. 2012; Soares et al. 2016). These types of safety nets can benefit marginal smallholder farmers in three ways (1) by providing vital support to maintain farming operations where attaining a harvest is still possible; (2) providing credit to pay for household essentials in times where, temporarily, farming does not offer a livelihood; and (3) providing enough capital assets for families to leave agriculture and find alternative livelihood opportunities elsewhere, perhaps in nearby cities (Loboguerrero et al. 2019).

5.2.2.2 Adoption of CSA technologies

The adoption of CSA technologies/approaches is highly heterogeneous, depending on the specific agro-ecological and climatic characteristics, market linkages (both input and output), extension services, the institutional and policy environment, etc. (Mizik 2021). Of the technologies and approaches highlighted above, climate-resilient cultivars (Loboguerrero et al. 2019; Mizik 2021; Makate et al. 2018; Mutenje et al. 2019), the use of organic manure (Mizik 2021; Abegunde et al. 2020a,b), planting of cover crops, mulching (Abegunde et al. 2019), crop rotation (Abegunde et al. 2019, 2020a,b), and soil conservation (Abegunde et al. 2019; Branca et al. 2021; Mizik 2021) are usually those CSA technologies/approaches more frequently adopted by smallholder farmers Mizik (2021). The adoption of minimum tillage, crop residues, cover crops, and integration of legumes is generally associated with higher economic returns (Branca et al. 2021). Higher economic returns have been found to be higher in semi-dry compared to sub-humid environments (Branca et al. 2021). Of the factors affecting the adoption of CSA, farm size and income/wealth/access to credit, and farmers' level of education/skills/awareness of what is available are often the most instrumental (Onyeneke et al. 2018; Knowler and Bradshaw 2007; Makate et al. 2018).

Conversely, factors affecting the lack of adoption include lack of education, skills, knowledge, information, high cost of initial investments (including often high demand for labour), insecure land tenure, delayed benefits associated with these investments (such as CA and agroforestry), and uncertainties about perceived benefits (Abegunde et al. 2019; Mizik 2021; Setshedi and Modirwa 2020; Branca et al. 2021). An enabling institutional and policy environment, which supports farmers in accessing CSA technologies, agricultural inputs (including climate resilient cultivars), extension services, credit, and markets, is also important for scaling out CSA (Dwivedi et al. 2017; Makate 2019; Lipper et al. 2014).

5.2.3 Conservation Agriculture

5.2.3.1 History of Conservation Agriculture

The concept of zero tillage (non-inversion of the topsoil) gained traction in the USA after the Dust Bowl of the 1930s and the publication of 'Plowman's Folly' by Edward H Faulkner in 1943 (Faulkner 1943), which proclaimed an urgent need to protect the topsoil. This spurred the development of direct drills that could sow seeds in the ground without the need for prior cultivation. The adoption of Conservation Agriculture was given further impetus during the energy crisis of 1974/75 when farmers sought to reduce crop establishment costs (Haggblade and Tembo 2003). During the 1970s, 1980s, and 1990s, Conservation Agriculture spread to both South America and West Africa, and became mainstreamed across many areas of Argentina, Brazil, and Paraguay (Farooq and Siddique 2015).

Conservation Agriculture involves the application of three basic agronomic principles:

1. Minimum or at least reduced soil disturbance;
2. Retention of crop residues; and
3. Crop rotations, especially with legumes.

These principles are shown in Fig. 68.

The application of no-till (NT) increases soil water infiltration and conservation, yield stability (especially during drought), soil organic matter (at least in the top 5-10 cm) (Shumba et al. 2023), stable soil aggregate structure, soil fertility, and soil biodiversity (microbial biomass - especially fungal and both soil micro- and macro-fauna). It also reduces crop production costs, soil

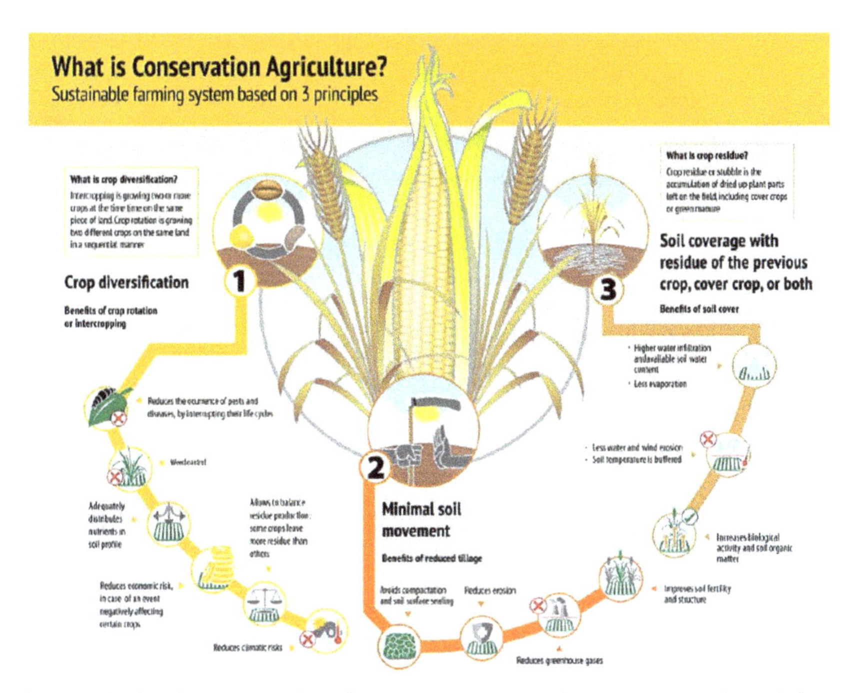

Figure 68 The three principles of Conservation Agriculture. Source: Adapted from: Donovan (2020).

erosion (wind and water), soil degradation, water runoff (including sediments, nutrients, and pesticides - even at wider watershed level), and evaporation. Additionally, the application of CA reduces daytime soil temperatures (Verhulst et al. 2010; Palm et al. 2014; Gupta et al. 2019; Page et al. 2020; Sun et al. 2020).

There is mixed evidence regarding the performance of Conservation Agriculture regarding crop yield, profitability, biological control of pest species, carbon sequestration, and greenhouse gas emissions (Palm et al. 2014; Giller et al. 2015; Jat et al. 2020; Hunt et al. 2020). The performance of Conservation Agriculture is highly dependent on both scale and climatic conditions (Giller et al. 2015; Sun et al. 2020).

For Conservation Agriculture, including NT on its own, to be successful, there must be a significant production of biomass to return organic matter and nutrients back to the soil. This is especially important concerning the use of Conservation Agriculture to increase yields and restore degraded soils (Palm et al. 2014). Conservation Agriculture struggles in low (synthetic and organic) fertiliser input systems. However, in low nutrient input systems, there is some evidence that undisturbed growth of arbuscular mycorrhizal fungi (AMF) plays an important role in nutrient acquisition and drought resistance (Palm et al. 2014).

With the exception of South Asia, Conservation Agriculture works well at scale, in the USA, Canada, Brazil, Argentina, Paraguay, Western Australia, and Asia Minor, where either timeliness or water use efficiency of farming operations is essential, and large cost savings can be made during the establishment of crops (Verhulst et al. 2010; Jat et al. 2020). Nearly 50% of land under Conservation Agriculture management is currently in South America, with 32% in the USA and Canada, and in Australia and New Zealand (Farooq and Siddique 2015). The land area managed under Conservation Agriculture continues to expand from 2.8 million hectares back in 1973/74 to 6.2 million hectares in 1996/97, and 148 million hectares in 2023 (Project Drawdown 2023).

There is mixed evidence, leaning towards the negative, regarding the feasibility of Conservation Agriculture in smallholder systems, especially in SSA (Giller et al. 2015). Conservation Agriculture also works well in drier and warmer regions but does not perform well in cold humid and tropical humid climates (Sun et al. 2020). The adoption of Conservation Agriculture in China, India, and sub-Sahara leads to an increase soil organic matter but has no effect on yields.

The evidence suggests that yield increases under Conservation Agriculture are possible but uncommon, and often only observed after several years and in certain contexts. Benefits also vary across the three practices and combination of the three practices. Zero tillage tends to reduce yields when compared to conventional tillage (Brouder and Gomez-Macpherson 2014; Rosenstock et al. 2014; Pittelkow et al. 2015), but has resulted in higher yields in places such as the Indo-Gangetic wheat-rice system (Gathala et al. 2014). However, even in these systems, low or zero-till has led to yield losses for many in the short term. Adoption is often constrained by low profits in the short term, although there is also some evidence of higher profits in spite of yield losses in certain contexts (Ahmad et al. 2014; Stevenson et al. 2014). Trade-offs regarding the use of crop residues for fodder or fuel also disincentivises the adoption of zero-till (Corbeels et al. 2014a).

5.3 Nanotechnologies

Nanotechnologies refer to the development/use of nanomaterials (measuring between 0.2 and 100 nm) in physical, chemical, biological, electronic, and engineering processes (Chaudhry 2020). However, larger materials are being developed up to 400 nm or more (Sekhon 2014). Nanomaterials can occur naturally, such as nano-clays and viruses, or be manufactured. Nanomaterials can be either two-dimensional (e.g. nanowire or nanotube) or three-dimensional (e.g. nanoparticle) (Sekhon 2014), and possess intrinsic properties linked to their size (Kah and Hofmann 2014; Mukhopadhyay 2014). Research into the development and use of nanotechnologies began in earnest in the 1990s and has increased rapidly since then, especially during the past 15 years (Kah and Hofmann 2014; Mukhopadhyay 2014; Servin and White 2016; Jain et al. 2018).

In a similar vein to biotechnologies, many scientists and industrialists are convinced that nanotechnologies are essential for achieving sustainable food systems (Sekhon 2014; Kumar et al. 2017; Prasad et al. 2017; Campos et al. 2019). Nanotechnologies are being developed for a wide range of food system applications, including

- nano-fortification and nano-nutraceuticals;
- nano-pesticides;
- nano-fertilisers;
- bio-nanotechnologies for crop and livestock breeding;
- nano-food packaging and preservation;
- nano-removal of contaminants from soil and water;
- nano-soil erosion and water management; and
- nano-sensors.

These have been explored in a wide range of studies (Fraceto et al. 2016; Mukhopadhyay 2014; Sekhon 2014; Campos et al. 2019; Calo et al. 2015; Chaudhry 2020; Lu and Ozcan 2015; Jain et al. 2018; de Oliveira et al. 2014; Chen and Yada 2011).

5.3.1 Nano-fortification and nano-nutraceuticals

In a similar vein to biotechnologies, due to their size and specific characteristics, nanomaterials can be used to carry both fortified foods and nutraceuticals, such as colour or flavour additives, dietary supplements, antioxidants, anti-cancer, and anti-diabetic medicines (Fraceto et al. 2016; Prasad et al. 2017; Jain et al. 2018; He et al 2019; Chaudhry 2020), in both humans and farmed animals (Sekhon 2014). Several bio-fortified nano-foods have been on the market for more than a decade (McClements et al. 2009; Dasgupta et al. 2016). These include Canola Active Oil, which is claimed to reduce cholesterol uptake in the human body (Fast Company 2022), and Nanoceuticals™ Slim Shake Chocolate, which claims to increase the chocolate taste whilst reducing the sugar content (Nanotechproject 2022).

Given the health issues associated with ultra-processed food, nanomaterials are being used to significantly reduce sugar, salt, and fat content (Jain et al. 2018). For example, it is claimed that the use of nano-salt can maintain the same flavour whilst reducing the salt content of crisps by 90% (The Conversation 2022). Similar claims are made about nano-sugars, which maintain the same sweetness whilst reducing sugar content by 80% (DagangHalal 2022; Diabetes UK 2022). Fortified bread is now being produced that contains nano-capsules of omega-3 fatty acids (Jain et al. 2018).

Nanomaterials are also being used as efficacious carriers for a range of medicines. For example, Kole et al. (2013) reported that the use of nano-carriers for the anticancer phytomedicines cucurbitacin-B and lycopene increased their efficacy by 74% and 82%, respectively, and increased the efficacy of Charantin and Insulin (antidiabetic phytomedicines) by 20% and 91%, respectively. Further research determined that the use of a nano-carrier significantly prolonged the release of Charantin (Nirupama et al. 2019).

5.3.2 Nano-pesticides

Pesticides delivered by both nano- and micro-emulsions are becoming increasingly popular in industrial crop protection (Hazra et al. 2017; Shao et al. 2018; He et al. 2019). Nano- and micro-pesticides, including biopesticides, have been developed for pest, disease, and weed control, some of which are already in commercial use (Ulrichs et al. 2005; Cioffi et al. 2004; Park et al. 2006; Jo et al. 2009; Esteban-Tejeda et al. 2009; Guan et al. 2008; Mukhopadhyay 2014; de Oliveira et al. 2014; Kah and Hofmann 2014; Bhattacharyya et al. 2017; Prasad et al. 2017; Campos et al. 2019). Figure 69 illustrates the dramatic reduction in applied active ingredients (AIs) since the 1950s. In the 1950s, pesticide applications rates were between 1200 and 2400 g of AI per hectare. By the 2000s, the application of AIs had fallen by 95% to between 40 and 100 g per hectare (Phillips McDougall 2019).

Nano-pesticides include a range of surfactants, organic polymers, and inorganic metal nanoparticles, such as iron-oxide, titanium-oxide, aluminium oxide, silicon oxide, zirconium dioxide, copper, nickel, gold, and silver (Sekhon 2014; Usman et al. 2020; Singh et al. 2021). Silver and copper nanoparticles

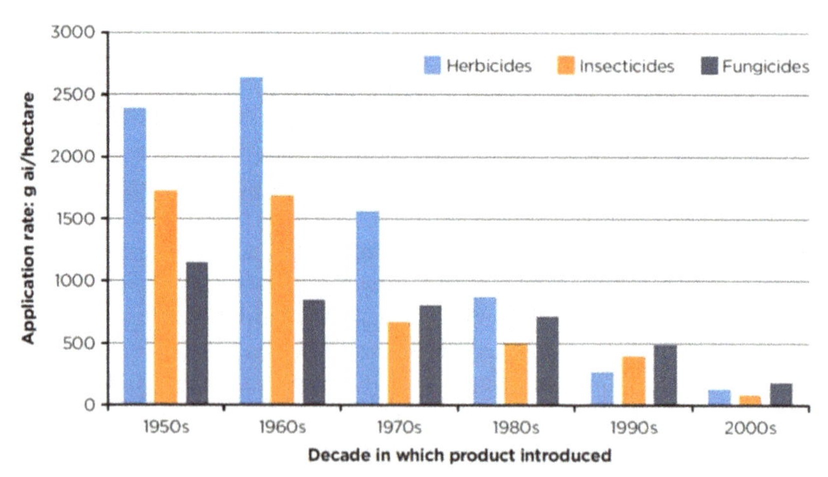

Figure 69 Average active ingredient application rates over time. Source: Adapted from: Phillips McDougall 2019).

are effective against rice blast disease and *Botrytis cinerea*, respectively (Singh et al. 2021). There are several key benefits associated with the use of nano- and micro-pesticide formulations. One of the most important is that the use of nano- and micro-formulations, of both new and existing pesticides, can dramatically increase pesticide efficiency and reduce the physical amount of pesticide applied (lower doses), which has a positive impact on the environment (Singh et al. 2021). As little as between 1% and 10% of conventional pesticides hit the target organism. The rest misses the target and potentially compounds the problem of pesticide pollution (Usman et al. 2020; Singh et al. 2021). Nano-pesticide formulations also enhance both the targeting and penetration of the target pest, disease, or weed, leading to greater efficacy of the pesticidal AI (Usman et al. 2020).

One of the most widely used nanotechnologies is encapsulation, which is basically a nano-coating protecting the AI from degradation and promoting controlled release over time (Usman et al. 2020). For example, encapsulated thymol in nanoparticles of Zein stabilised with sodium caseinate and chitosan chlorohydrate is more than 50% more effective than conventional formulations of thymol (Singh et al. 2021). Similarly, the efficiency of nano-encapsulated curcumin increases by up to 67% (Singh et al. 2021).

Another key benefit of nano-pesticides is the controlled release of the AI, as well as longer-lasting effectiveness, through nano-encapsulation, leading to reduced frequency of pesticide applications (Sekhon 2014; Campos et al. 2019; Chhipa 2017; Chaudhry 2020). Commercial applications of nano-encapsulation include one of the most popular herbicides, 2,4-dichlorophenoxy acetic acid, and the fungicide carbendazim (Kumar et al. 2017; Cao et al. 2018).

Many existing AIs are unstable in direct sunlight and high temperatures. Nano-encapsulation offers a cheap and effective way to protect AIs, allowing them to do their intended job (Kah and Hofmann 2014). Nano-emulsions also increases AI solubility, application, and uptake (Kah and Hofmann 2014). The use of nano-pesticides also reduces the amount of water needed when farmers spray their fields, as well as reducing the amount of energy and labour required to apply the AIs (Usman et al. 2020). Nanogels have been proposed for use in plant protection products as a possible way to meet organic farming standards, with pheromones, essential oils, or copper as the AIs (Kah and Hofmann 2014; Singh et al. 2021).

5.3.3 Nano-fertilisers

Nano-fertilisers are nano-sized materials that either act as carriers or additives for nutrients (sometimes encapsulated) or are nutrients themselves (Usman et al. 2020). The use of nano-fertilisers has a range of advantages over conventional fertiliser formulations. One of the most promising attributes of this emerging technology is the capacity for controlled and precise release of essential plant nutrients (Naderi and Abedi 2012; Moaveni and Kheiri 2011; Prasad et al. 2017; Campos et al. 2019; Usman et al. 2020; Verma et al. 2022). For example, nano-formulations of nitrogen-based fertiliser released nitrates for up to 60 days, compared to conventional fertilisers (urea and ammonium nitrate), which released nitrates for just 30 days (Kottegoda et al. 2011). The use of nano-fertilisers is also more efficient than conventional formulations (Lu et al. 2002; Mukhopadhyay 2014; Liu and Lal 2014).

Whilst estimates vary, applied nutrients are often either chemically bound to soil particles or leached away into water courses, thus both failing to benefit a crop and creating a potential pollution risk. For example, only between 30% and 60% of nitrogen-based fertilisers are taken up by crop plants. Potassium use efficiency is around 35–40%. In the case of phosphorous, due to chemical bonding in the soil, 10% or less is available for plant uptake (Giroto et al. 2017; Iqbal et al. 2020; Verma et al. 2022). Conversely, due in part to their controlled release, encapsulation, and physical properties, nano-fertilisers experience ultrahigh absorption (INIC 2014; Usman et al. 2020). This is especially important for micro-nutrients, which, like phosphates, often become tightly bound in soils, especially clay and organic soils. Compared to conventional fertilisers, nano-fertilisers enhance uptake by up to 29% (Kah et al. 2018), and increase crop growth rates by up to 32%, and seed yield by up to 20% (Liu and Lal 2014). Similarly, nano-growth stimulants have been shown to enhance both seed germination (Nadiminti et al. 2013) and post-emergence growth (Aslani et al. 2014). For example, experimental yields have recorded a 44% increased growth in the dry weight of spinach (Zheng et al. 2005; Yang et al. 2006). The use of Fullerol, a carbon-based nanoparticle, increased the yield of bitter melon by 128% (Sekhon 2014). Carbon-based nanoparticles and nanotubes of gold, silver, silicon, zinc, and titanium have been shown to increase both the uptake and utilisation of nutrients (Fraceto et al. 2016). In addition to nano-fertilisers reducing the absolute volume of fertilisers applied, they also cut the cost and time required to apply nutrients (Naderi and Danesh-Shahraki 2013; Verma et al. 2022). From an environmental point of view, the use of nano-fertilisers significantly reduces the loss of mobile nutrients into water course, significantly reducing nitrate pollution and eutrophication (Liu and Lal 2015; Kah et al. 2018). Whilst many nano-fertiliser formulations remain at the advanced experimental stage (Liu and Lal 2015; Dimkpa and Bindraban 2018), it is likely that more will move into commercial application.

5.3.4 Bio-nanotechnologies for crop and livestock breeding

Due to their specific properties (size, shape, surface charge, and physical/chemical properties), nanoparticles, nanofibers, nanotubes, and nanocapsules, especially gold, silver, silicon, zinc, and titanium nanomaterials, are increasingly being used for gene transfer in crop and livestock breeding (Torney et al. 2007; Kuzma 2007; Maysinger 2007; Lu and Ozcan 2015). Nanoparticles have already been used in breeding for pest and disease resistance, abiotic stresses, and crop growth (Forsberg and de Lauwere 2013; Joseph and Morrison 2006; Owolade et al. 2008; Knauer and Bucheli 2009; Mukhopadhyay 2014; Prasad et al. 2017; Campos et al. 2019; Ahmar et al. 2021). For example, carbon nanofibers have been used to genetically modify golden rice (AZoNano.com 2013). Nanobiotechnology provides yet another tool to genetically modify genes and even create new organisms (Singh et al. 2021).

5.3.5 Food packaging and preservation

Nanotechnologies have been developed to monitor the freshness and safety of food products and to enhance the shelf life of foods (Sekhon 2014; Mukhopadhyay 2014; Calo et al. 2015; Fraceto et al. 2016; Prasad et al. 2017; Jain et al. 2018; Campos et al. 2019; Chaudhry 2020; Singh et al. 2021).

Often called smart, intelligent, or active packaging, nanotechnologies contained in the packaging of both lightly and ultra-processed foods has been designed to monitor the deterioration of food. Nano-biosensors monitor the minute changes in the quality of packaged foodstuffs (humidity, pH, temperature, light exposure, etc.) (Fraceto et al. 2016). If, for example, the quality of stored food deteriorates by a certain amount, biological change (i.e. changes in oxygen or carbon dioxide levels, odour, or other indicator chemicals) is transformed into electrical signals, which can be easily detected (Singh et al. 2021). Many of these biosensors are in commercial use, such as rapid detection biosensors, enzymatic biosensors, E-nose (Electronic nose), and gold nano-biosensors (Singh et al. 2021). Using a bioswitch in response to these changes can prompt the automatic release of preservatives (anti-bacterial or anti-fungal agents), also contained in the packaging, to prevent further deterioration (McClements et al. 2009; Huang et al. 2010; Yu and Huang 2013; Chaudhry 2020). This type of technology can be used along the whole food chain from post-harvest processing and storage, transportation, food processing, storage, and display (Vanderroost et al. 2014).

Nanotechnologies have also been developed to enhance the packaging of foods. Nanomaterials, such as, nano-clays, nano-silica, nano-calcium carbonate, carbon nanotube, and nano-cellulose, have been used due to their inherent flexibility, durability, and stability under a wide range of temperature and moisture conditions. Nanomaterials also have specific barrier properties, which regulate the movement of water and gases and prevent tainting (Jain et al. 2018). Nano-films containing titanium dioxide nanoparticles have also been developed that scavenge oxygen, which is primarily responsible for the oxidation (browning) of food (Singh et al. 2021). Similarly, GuardIN Fresh slows the ripening of fruits by scavenging the ethylene gas responsible for ripening (Calo et al. 2015). Active packaging containing cinnamon oil controls browning and reduces weight loss in packaged mushrooms better than conventional paraffin-based packaging (Echegoyen and Nerin 2015).

Nano-preservatives are also being added to fresh fruit, vegetables, and meats, which are often not stored or displayed in air-tight packaging. Many naturally occurring organic nanomaterials (essential oils) such as eugenol, menthol, thymol, lemongrass, limonene, and other naturally occurring compounds, such as pectin and chitosan, have been successfully used as edible coating which preserve foods (Calo et al. 2015). For example, the use of eugenol, menthol, and thymol coatings on strawberries have been shown to slow postharvest decay (Wang et al 2007). Essential oils extracted from bay leaf were able to extend the shelf life of fresh Tuscan sausages by 2 days (da Silveira et al. 2014).

5.3.6 Remediation of contaminated water and soil

Nanotechnologies have been deployed during the past 20 years as a fast, cheap, and easy solution to remove contaminants from both soil and water (Araújo et al. 2015; Duhan et al. 2017; Fraceto et al. 2016; Lavicoli et al. 2017; Mukhopadhyay 2014; Nair et al. 2010; Rai and Ingle 2012; Sekhon 2014; Torney et al. 2007; Verma 2017). Due to their small size, large surface area, and physical/chemical properties, nanoparticles are increasingly being used to reduce, oxidise, or adsorb contaminants in both soil and water (Guerra et al. 2018). Nanoparticles can be both organic, such as polymers and carbon nanotubes, and inorganic, such as nanometals and metal oxides (Khin et al. 2012). Some nanoparticles are magnetic (Usman et al. 2018). Nanoparticles can be used for the bioremediation of inorganic pollutants, including heavy metals and radioactive elements such as cadmium, lead, chromium, salt, hydrocarbons, and synthetic chemicals and pesticides in contaminated soils (Mukhopadhyay 2014; Zheng et al. 2020; Singh et al. 2021). For example, Zheng et al. (2020) reported at least 80% immobilisation of lead arsenate and hexavalent chromium in contaminated soils. Nanoparticles and membranes are also effective in removing a range of contaminants from water, including inorganic and organic pollutants, such as salt, heavy metals, and pesticides (Singh et al. 2021).

5.3.7 Improved soil erosion management

Nanotechnologies are also being used to reduce soil erosion and to improve soil water management (Mukhopadhyay 2014; Chen and Yada 2011; Salamanca-Buentello et al. 2005). For example, the use of polyacrylamide modified magnetite nanoparticles reduced soil erosion by 90% (Usman et al. 2020), and titanium dioxide reduced organic matter breakdown, and subsequent release of CO_2 by creating stronger bonds between humic molecules, which are a key component of soil organic matter (Nuzzo et al. 2016; Simonin et al. 2015). Likewise, the application of zinc oxide and ferric oxide nanoparticles reduced organic carbon decomposition by up to 130% (Rashid et al. 2017b), and CO_2 emissions by up to 30% (Rashid et al. 2017a; Shi et al. 2018).

Nanomaterials, such as multiwall carbon nanotubes, carbon nanofibers, laponite, hydrogels, nano-clays, nano-zeolites, and nano-saturated colloidal silica, can also be used to improve soil texture and water-holding capacity and improve soil fertility (Sekhon 2014; Fraceto et al. 2016; Singh et al. 2021; Mukhopadhyay 2014). These nanoparticles are particularly useful for the remediation of degraded lands for agricultural or reafforestation purposes (Mukhopadhyay 2014; Fraceto et al. 2016).

5.3.8 Nano-sensors

The development and use of nano-sensors has been one of the most active areas of nano-research. Nano-sensors include nanotubes, nanowires, nanoparticles, and nanocrystals. These nanomaterials have useful physical, chemical, thermal, electrical, and optical properties (Fraceto et al. 2016), which allow for analyte detection at ultra-low concentrations (Scognamiglio 2013). Nanomaterials have been used to develop biosensors for a wide range of purposes, including the detection and quantification of vitamin content, pests, diseases, and viruses in fields and storage (Jones 2006; Brock et al. 2011; Otles and Yalcin 2010; Farrell et al. 2013; Mukhopadhyay 2014; Campos et al. 2019; Hu et al. 2019; Euronews 2022); nutrient and water deficiency (Sekhon 2014; Campos et al. 2019); acidification of irrigated lands (Mukhopadhyay 2014); spoiled food (Chaudhry 2020), toxins, and pesticide residues, soil pH, and pollutants (Rai and Ingle 2012; Jones 2006; Brock et al. 2011).

5.3.9 The need for life-cycle analysis

Nanotechnologies in the food system seem to provide a multitude of solutions for increasing agricultural productivity and increasing resource-use efficiency, reducing costs and waste, whilst at the same time, significantly reducing negative environmental externalities (Scott and Chen 2013; Kah 2015; Servin and White 2016; Verma et al. 2022).

However, much uncertainty remains concerning the fate of nanomaterials, whether in humans, animals, plants, or the wider environment (Ragaei and Sabry 2014). Much of this uncertainty is for the same reasons nanomaterials have been found to be so useful: their incredibly small size and unique physical/chemical properties (Jain et al. 2018). For example, their size means that nano-pesticides can pass through human cell membranes and into the blood stream after inhalation, or into the gut after ingestion (Rico et al. 2011; Usman et al. 2020). It is also known that nanomaterials applied at high concentrations can affect the cardiovascular system and be mutagenic and carcinogenic in animals (Jain et al. 2018). In plants, nanomaterials have been shown to be phytotoxic, damaging DNA and disrupting cellular processes, which, in turn, have led to poor germination, reduced production of biomass, especially leaf numbers and root elongation, and reduced capacity to photosynthesise (Lin and Xing 2007; Racuciu and Creanga 2007; Doshi et al. 2008; Lee et al. 2010, 2012; Barchowsky and O'Hara 2003; Pulido and Parrish 2003; Valko et al. 2005; Singh et al. 2021). Little is known about the complexity of eco-interactions and the behaviour of nanomaterials in soils (Simone et al. 2009; Strand and Kjølberg 2011; Tavares et al. 2015; Usman et al. 2020). Of the studies conducted to date, which have tended to focus on high dose/short-term effects and not long-term effects of lower doses, several indicate the potential for eco-toxicity if nanomaterials are improperly managed. In one study, high levels of Ag, Cu, and TiO_2 nanoparticles were found to be toxic for earthworms (Heckmann et al. 2011; Servin and White 2016). Another study demonstrated that the nanoparticle ceric oxide significantly reduced the rates of nitrogen fixation in the root nodules of soybean (Priester et al. 2012).

Whilst most studies suggest that the use of nanomaterials at low concentrations comes with minimal to moderate risk (Servin and White 2016), both the global scientific community and public regulatory authorities have embraced the need for more research on the fate of nanomaterials through holistic life-cycle research (Kah and Hofmann 2014; Servin and White 2016; Fraceto et al. 2016; Jain et al. 2018; Usman et al. 2020; dos Santos and Gottschalk Nolasco 2017; Retzbach and Maier 2015; Stokes 2013; Strand and Kjølberg 2011). Early analyses of the effects of nanomaterials have led market-based organisations, such as the International Federation on Organic Agriculture Movements, and the Canadian Government to reject the use of nanotechnologies in organic agriculture (Sekhon 2014).

5.4 Biopesticides

Naturally occurring biopesticides have been used for millennia in countries such as India, China, and Egypt to control pests (de Oliveira et al. 2014). The Romans and Greeks used hellebore (a naturally occurring biopesticide) to control rats, mice, and insects. Likewise, rotenone, made from roots of the *Derris* plant, was used in South America to paralyse fish from the seventeenth century, later spreading to Asia to control insect pests in the 1840s (Watson 2018). Other naturally occurring biopesticides include nicotine and pyrethrum, introduced into Europe during the seventeenth century to control crop pests such as plum curculio and lace bugs (Boardman 1986). Unfortunately, their tendency to rapidly breakdown in the field limited the use of nicotine and pyrethrum in agriculture (Watson 2018). Fungal spores were also used during the late 1800s to control insect pests (Abdollahdokh et al. 2022).

However, particularly with the growth of organic agriculture, the number of biopesticides has expanded rapidly since the 1960s (Fig. 70) (Fortune Business Insights 2020). Indeed, since the turn of the twenty-first century, some years have witnessed more biopesticides released than synthetic pesticides.

5.4.1 What are biopesticides?

Biopesticides are based on naturally-occurring substances (both toxic and non-toxic - and pheromonal) that can be extracted from plants, animals, bacteria, fungi, viruses, and protozoa (Raja 2013; Sekhon 2014; EPA 2022). Whilst many of the toxic biopesticides are highly selective, several exhibit non-selective and broad-spectrum antimicrobial activity (Bhat et al. 2012; Conti et al. 2014; George et al. 2010; Spyrou et al. 2009; Mossa 2016). Non-toxic biopesticides act either by disrupting feeding or mating, or via disturbance, desiccation, and suffocation (Abdollahdokh et al. 2022). The US Government also includes the introgression of Bt genes into crop plants as a biopesticide (EPA 2022).

Biopesticides to control insect pests include entomopathogenic nematodes, mites, fungi, and bacteria which predate insects, semiochemicals and allelochemicals that disrupt mating and foraging activity, insertion of insecticidal *Bacillus thuringiensis* (Bt) genes into crops, and peptide-based insecticides based on natural venoms (Birch and Glare 2020). Biocontrol agents also include use of bacteria such as *Bacillus* spp. and *Pseudomonas* spp. as microbial bioprotectants or biocontrol agents (BCAs) to boost crop defences against disease (Köhl and Ravensberg 2021). The category has also been

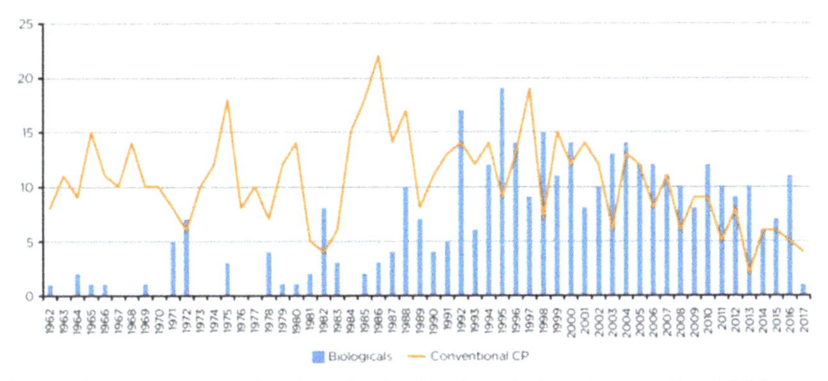

Figure 70 Annual new product introduction for biologicals and conventional CP. Source: Adapted from: Phillips McDougall (2019).

extended to incorporate biostimulants such as humic substances, seaweed extracts, protein hydrolysates, silicon, plant growth-promoting rhizobacteria, and AMF. These substances enhance plant nutrient use efficiency and other functions, improving crop resilience to disease (Rouphael et al. 2020)

Two popular biopesticides are Bt, which is derived from soil-living bacteria, and pyrethrin, an essential oil found in some *Chrysanthemum* species (Chakraborty and Deshmukh 2022). Essential oils (EOs), also known as volatile or etheric oils, essences, or aetheroleum, are secondary metabolites produced in aromatic plants. These secondary metabolites, predominantly mono-terpenes, act as the plant's natural defence against attack from pests and diseases (Mossa 2016). There are approximately 17 500 aromatic plants, containing 3000 EOs. Over 300 of these essential oils are currently being utilised as biopesticides, pharmaceutical drugs, cosmetics, perfumes, etc. (Mossa 2016). Examples of essential oils include lemongrass, eucalyptus, rosemary, vetiver, clove, thyme, and citronella (Said-Al Ahl et al. 2017).

There is currently an interesting mix of start-up companies, SMEs, off-patent pesticide manufacturers, as well as all the major pesticide corporations that are both producing and supplying biopesticides. Some of them are listed here: AgBioChem, Inc., AgBiTech Pty Ltd., Ajay Bio-Tech Ltd., Amit Biotech Pvt. Ltd., Andermatt Biocontrol AG, Arizona Biological Control, Inc., BioWorks, Inc., BASF SE, Bayer AG, Dow-Dupont, Syngenta AG (ChemChina), Marrone Bio Innovations, Inc., Certis USA L.L.C., Kemin Industries, Andermatt Biocontol AG, Som Phytopharma India Limited, W. Neudorff GmbH KG, International Panaacea Ltd, Valent USA LLC (part of Sumitomo Chemical Co., Ltd.), The Stockton Group, Koppert B.V., InVivo Group, FMC Corporation, Gowan Company, and Novozymes A/S (Chakraborty and Deshmukh 2022; Meticulous Research 2021).

5.4.2 Future of Biopesticides

Sales of biopesticides have grown steadily. In 1993, the global biopesticide market was worth US$100 million, growing to US$1 billion by 2009, US$3 billion by 2016 (Phillips McDougall 2019), US$7 billion by 2020, and is predicted to reach to US$8.5 billion by 2025, potentially reaching US$33.6 billion by 2031 (Chakraborty and Deshmukh 2022; Allied Analytics LLP 2022). The market for microbial biopesticides is growing at an impressive rate of 13.8% per annum, driven by their cost competitiveness and efficacy (Chakraborty and Deshmukh 2022). However, despite rapid growth, biopesticides still only account for around 10% of the pesticide market (Phillips McDougall 2019). Figure 71 illustrates the expected growth in biopesticides by product type, and Fig. 72 illustrates growth in biopesticides by crop type. Geographically, in order of prominence, the principal markets for biopesticides are North America, Europe, Asia-Pacific, Latin America, and the Middle East and Africa.

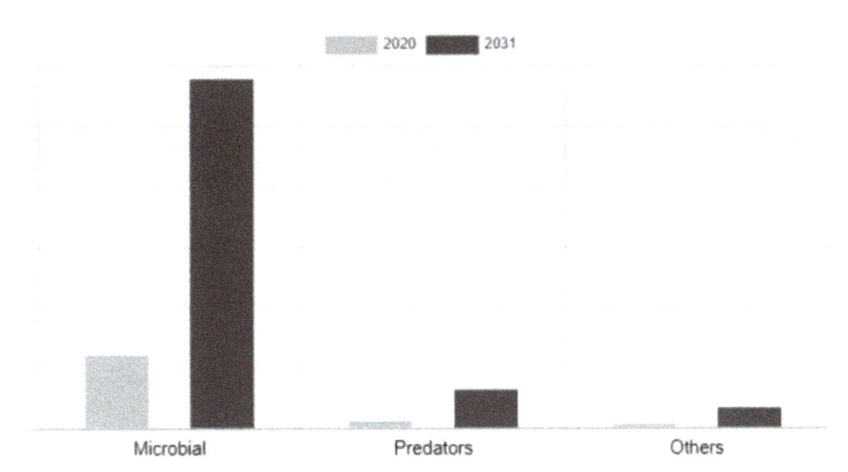

Figure 71 Global biopesticide market by product type. Source: Adapted from: Chakraborty and Deshmukh (2022).

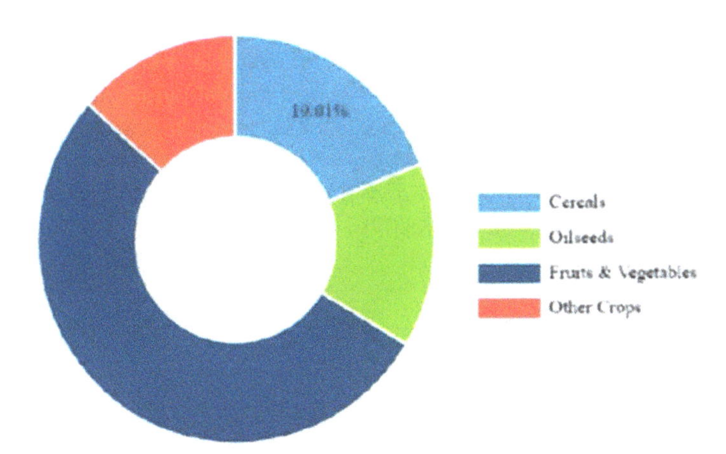

Figure 72 Global biopesticide market share by crop. Source: Adapted from: Fortune Business Insights (2020).

Whilst growth in the biopesticides market has been driven by the easier, faster, and cheaper regulatory processes, increased use of biopesticides as part of integrated pest management (IPM) programs, and increased demand from farmers (Phillips McDougall 2019), the principal driver of the biopesticide market is the concomitant growth in organic agriculture. Biopesticides are seen to offer cost-effective, eco/farmer/consumer-friendly pest and disease control whilst being complaint with organic farming regulations (Raja 2013; Mossa 2016; Fortune Business Insights 2020). As consumer groups continue to lobby governments for low/no pesticide residues (Jacobs 2020), eco-friendly biopesticides that are rapidly catabolised in the environment, and offer low/no toxicity to fishes, mammals, and birds, are increasingly being seen as offering win-win crop management solutions (Said-Al Ahl et al. 2017).

Whilst there is no doubt that the use of biopesticides will almost certainly continue to grow, challenges remain that manufacturers will need to address. One challenge is that the 'average annual sales at maturity for a biological product is of the order of $10m, compared with an average of $75m for a conventional crop protection product' (Phillips McDougall 2019). If companies wish to make a profit from their investments, biopesticides will increasingly need to remain efficacious over the long-term and achieve noticeable market share to ensure sufficient sales (Phillips McDougall 2019). A key issue is that, as biological products, biopesticides are more sensitive to environmental conditions, have a limited shelf life, often require direct contact with the target to be effective, and may be less efficacious than more tailored synthetic pesticides. They typically require combination within an IPM programme combining cultural, mechanical, and biological methods of control. This situation is not helped by the fact that industry analysts remain concerned about the development of resistance to biopesticides (Dara 2017). Another issue, which needs to be considered, is the toxicity of some biopesticides. Nicotine, for example, is classified as highly hazardous (WHO Class Ib). Both rotenone and natural pyrethrum are highly toxic (Isman 2013). Several biopesticides, such as citronella, eucalyptus, garlic, pyrethrum, and neem products are also toxic to, or have non-lethal effects on, non-target/beneficial insects (Maia and Moore 2011; Ndakidemi et al. 2016; Rousidou et al. 2013; Spyrou et al. 2009).

Chapter 6

Progressive food systems

6.1 Introduction

Progressive food systems differ from neoliberal and reformist food systems in several ways. A key theme that unites progressive approaches (such as regenerative agriculture, organic agriculture, and agroecological agriculture) is a focus on reducing reliance on synthetic chemical inputs (whether fertilisers or pesticides) in favour of leveraging natural resources to deliver equivalent benefits in a way that is considered more sustainable. Examples include adding organic matter to soil to boost its nutrient content (organic agriculture), understanding agriculture as a modified ecosystem to manage material and energy flows to optimise ecosystem function and benefits (agroecology), or using reduced/no-till techniques to preserve and enhance the physical, chemical, and biological properties/health of soil (regenerative agriculture).

All these approaches are explicitly or implicitly based on agroecology, which has been defined as 'the application of ecological concepts and principles to the design and management of sustainable agroecosystems' (Gliessman 2018b). Agroecology views agriculture as an adapted form of an ecosystem in which an understanding of fundamental biogeochemical flows/cycles and functional interactions between organisms within this type of ecosystem can be used to optimise its ability to deliver a range of ecosystem services, including food production. Examples of practices seen as promoting agroecosystem function include optimising the physical, chemical, and biological health of soils to promote crop growth and health (optimising nutrient and other material flows), promoting biodiversity (e.g. to encourage beneficial species such as pollinators or predators of crop pests), and ensuring diversity in agricultural landscapes (to optimise functional interactions between species) (Wezel et al. 2020).

This focus on understanding, conserving, and regenerating natural, organic processes also means that progressive food systems tend to favour less intensive, smaller-scale, and localised food production. Although it may not be a formal requirement of organic certification, progressive systems often promote local foodsheds (food from somewhere) in which urban consumers have direct linkages with rural producers via mechanisms such as farmers' markets (Kloppenburg et al. 1996; Holt-Giménez and Shattuck 2011). This is contrasted with international foodsheds (food from anywhere) where urban consumers have little idea where their food originates, a common characteristic of globalised neoliberal and reformist food systems. Although again not a formal requirement, progressive food systems are often based on food produced on smaller family farms, compared to large-scale and corporate farms typical of neoliberal and reformist food systems. Progressive food systems may also give greater emphasis to quality compared to yield, favouring 'good, clean, fair', natural (relatively unprocessed), and authentic foods (Petrini 2005; Patel et al. 2009), compared to high-volume, standardised, low-cost, and often highly processed foods produced within neoliberal and reformist food systems. However, in a similar vein to neoliberal and reformist food systems, progressive food systems primarily operate within the economic and political structures of free market capitalism (Patel et al. 2009).

As noted, progressive food systems are those that seek a balance with nature rather than a relationship of domination, which is inherent throughout much of the neoliberal and reformist food systems. The main progressive food system approaches are

1. regenerative agriculture;
2. organic agriculture; and
3. agroecological agriculture.

Progressive food systems are coordinated by a multitudinous network of highly heterogeneous producer and consumer groups. For example, in the UK, organic farmers are represented by several organisations, such as the Soil Association and the Organic Growers Alliance. Internationally, organic associations tend to be federated into umbrella organisations, such as the International Federation of Organic Agriculture Movements (IFOAM), which represents around 800 country-level organic groups in 117 countries.

6.2 Regenerative agriculture

6.2.1 History of regenerative agriculture

The term 'regenerative agriculture' was first coined in the late 1970s by Robert Rodale of the Rodale Institute, USA. Since the early 1970s, the Rodale Institute has been conducting comparative experiments on regenerative organic agriculture versus conventional agriculture (Gabel 1979; Civil Eats 2020) and spearheading organic agriculture through its magazine *Organic Gardening and Farming* (Giller et al. 2021). Many struggle to tease out the fundamental differences between regenerative agriculture, organic agriculture and agroecological agriculture (Whyte 1987). Indeed, even the Rodale Institute refers to it as regenerative organic agriculture (Electris et al. 2019; Rodale Institute 2020).

Whilst regenerative agriculture stimulated significant interest during the 1980s and 1990s, it dropped off the radar until around 2015, when it began rekindling the interest of researchers, non-governmental organisations (NGOs) (such as The Nature Conservancy, World Wildlife Fund, Greenpeace, and Friends of the Earth), corporate foundations (such as the IKEA Foundation), mainstream media, politicians, private investors, and corporate businesses (especially food aggregators, manufacturers and retailers such as General Mills, Cargills, Unilever, Danone, Nestlé, PepsiCo, Kellogg's, McCain, Hormel Foods, Stonyfield, Patagonia, Target, Walmart, Whole Foods Market, McDonald's, Land O' Lakes, and the World Council for Sustainable Business Development), (Electris et al. 2019; McCarthy et al. 2020; EIT Food 2020; Newton et al. 2020; Giller et al. 2021; Wilcox 2021; Uldrich 2021; Marquis 2021; Fisher 2022). In 2017, the Rodale Institute launched its own Regenerative Organic certification and label, which provides strict rules on soil health (no synthetic pesticides or fertilisers), animal welfare, and social fairness (Beyond Pesticides 2019; Civil Eats 2020).

In September 2019, during the UN Climate Action Summit, Danone, the Balbo Group, Barry Callebaut, DSM, Firmenich, Google, Jacobs Douwe Egberts, the Kellogg Company, Kering, Livelihoods Funds, L'Oreal, Loblaw Companies Limited, Mars, Migros Ticaret, McCain Foods, Nestlé, Symrise, Unilever, and Yara formed the coalition called One Planet Business for Biodiversity, with the aim to promote alternative farming practices (Beste 2021). Also, in 2019, the Intergovernmental Panel on Climate Change listed regenerative agriculture as a sustainable approach to land management capable of building agroecosystem resilience (Newton et al. 2020).

To date, there is no legal or regulatory definition of regenerative agriculture. Indeed, regenerative agriculture practitioners define their approaches in a variety of ways (Newton et al. 2020). According to Rodale Institute (1983), regenerative agriculture is an approach that leads to 'increasing levels of productivity, and increases our land and soil biological production base. It has a high level of built-in economic and biological stability. It has minimal to no impact on the environment beyond the farm or field boundaries. It produces foodstuffs free from biocides. It provides for the productive contribution of increasingly large numbers of people during a transition to minimal reliance on non-renewable resources'. In addition to the many other definitions of regenerative agriculture, Rodale's definition appears to be both a desired outcome and an approved list of farming systems approaches (Merfield 2019; Soloviev and Landua 2016; Beyond Pesticides 2019). Based on the work of Francis et al. (1986), Fig. 73 illustrates the basic components of regenerative agriculture.

When examined from a desired outcomes perspective, regenerative agriculture aims to achieve several important end-states, which resonate with much of the contemporary food systems thinking (IDH 2020). These desired outcomes include (Rodale 1983; Beyond Pesticides 2019; McCarthy et al. 2020; IDH 2020; Schreefel et al. 2020; Newton et al. 2020; Rhodes 2017; Sherwood and Uphoff 2000)

1 restored ecosystem services at the landscape level (improved biodiversity - especially soil macro- and micro-organisms - and water quality, etc.);
2 improved soil health (enhanced fertility, organic matter - especially via carbon sequestration to mitigate climate change - and biological activity);
3 improved water cycle (hydrology, storage, and reduce pollution);
4 increased carbon sequestration and reduced greenhouse gas emissions (GHGEs);
5 improved animal welfare;
6 increased or maintained farm productivity (crop and livestock yields);
7 diverse, nutritious, productive, and resilient food systems, which are both equitable and profitable for farmers; and
8 food systems in which farmers and consumers have a direct and close relationship, built on shared values and trust.

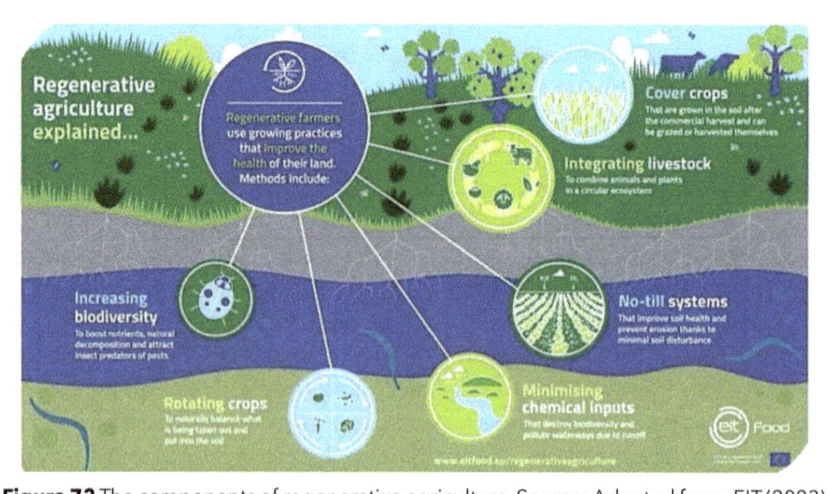

Figure 73 The components of regenerative agriculture. Source: Adapted from: EIT (2023).

Regenerative agriculture can be seen as a low-input, moderate/high-output agriculture. In many respects, it represents a return to the type of semi-closed/circular/mixed farming/holistic, integrated, and even indigenous agricultural systems that existed before the agricultural industrialisation of the 1950s, 1960s, and 1970s, where the animals feed the plants and the plants feed the animals (EIT Food 2020; Pearson 2007; Fisher 2022). Instead of purchasing expensive synthetic off-farm inputs to achieve high output, farmers rely on inputs sourced primarily from the farm itself and focus on farm profitability, rather than focussing on the attainment of high yields. Ultimately, the regenerative agriculture philosophy is that high-input/high-output agriculture may not be the most profitable, climate-smart, resilient, or environmentally sensitive use of resources (EIT Food 2020).

When examined from an approved list of farming systems approaches, regenerative agriculture borrows technologies and approaches from both organic, agroecology, and sustainable intensification (Giller et al. 2021). The following principles of regenerative agriculture are based on LaCanne and Lundgren (2018), Francis et al. (1986), Rhodes (2012), Merfield (2019), Moyer et al. (2020), IDH (2020), Roulac (2021), Reeder and Schuman (2002), Newton et al. (2020), Giller et al. (2021), and Fisher (2022):

1 Increased soil productivity by increasing the depth, fertility, and physical characteristics of the upper soil layers. This often involves the adoption of Conservation Agriculture, zero-till, minimum-tillage, or controlled traffic approaches (or the proactive rebuilding of soil communities following cultivation), and permanent soil cover, using cover crops, crop residue retention, composts and mulches, green manures, and animal manure, and the use of soil conditioners such as biochar (charcoal added to composts);
2 Integrated nutrient-flow systems based on the efficient use of soil flora and fauna, leading to increased crop nutrition and reduced environmental damage;
3 Low or no use of external inputs. Crop production based on biological interactions and either zero or judicious use of synthetic biocides;
4 Substances which disrupt biological structuring of the farming system (such as present-day synthetic fertilisers) should not be used;
5 Restore natural habitats and use ecological or natural principles or systems;
6 Integrated and highly diversified livestock, crop, and tree systems, often embracing permaculture, holistic grazing, agroforestry, silvo-pastural systems, which are largely self-reliant in nitrogen through biological nitrogen fixation, nutrient cycling, integrated nutrient management, lengthened crop rotation, integrated pest management, diversifying

field margins using conservation mixes, inter-cropping, promotion of non-crop plants between crop rows and 'weed cycling';

7 Animals in agriculture should be fed and housed in such a manner as to preclude the use of hormones and the prophylactic use of antibiotics;
8 Agricultural production should generate increased levels of employment; and
9 A regenerative agriculture requires national-level planning but a high degree of local and regional self-reliance to close nutrient-flow loops.

6.2.2 What are the key strengths of regenerative agriculture?

According to staunch advocates, the key strengths of regenerative agriculture include increased soil health, biodiversity (especially soil-biodiversity), carbon sequestration, yields, profitability, system resilience, and social acceptability.

Several would argue that the principal focus of regenerative agriculture is the improvement of soil health. Indeed, many use the organic adage of 'feed the soil and not the crop' (Giller et al. 2021). Soil health is generally equated to the biomass and diversity of macro- and micro-organisms, especially those responsible for the biological cycling of nutrients (de Vries et al. 2013; Tsiafouli et al. 2014). Most, if not all, Regenerative Agriculture practices, such as those causing minimal soil disturbance, cover crops, crop residue retention, composts, biochar, green manures, and animal manure, plus reduced application of pesticides and synthetic fertilisers, generally lead to increased quantities and diversity of soil macro- and micro-flora and fauna.

The above practices are also inextricably linked to the build-up of soil organic matter (SOM), which is an important indicator of soil fertility (Reeves 1997; Xia et al. 2023). Organic matter eventually leads to the creation of humic acid, which is a relatively stable store of carbon. Thus, regenerative agriculture is lauded as being able to remove carbon dioxide from the atmosphere and sequester it in both the SOM, perennial crops (such as trees), and wetland management (EIT Food 2020). For example, Shroff et al. (2021) claim that beef produced using Regenerative Grazing can have a lower carbon footprint than an equivalent quantity of 'beef' contained in plant-based impossible burgers. Indeed, World Food Price Winner Dr Rattan Lal claims that 'increasing the carbon content of the world's soil by just 2% would entirely return greenhouse gases in the atmosphere to safe levels' (EIT Food 2020). Claims have been made that the application of regenerative agriculture grazing approaches could sequester more than 100% of current global CO_2 emissions (Rodale Institute 2014). In the USA, efforts are currently being made to make carbon credits available to farmers applying soil regenerative practices (Civil Eats 2020).

In addition to increased micro- and macro-organisms and carbon storage, higher quantities of organic matter also improve soil structure and increase infiltration of water into the soil, as well as water holding capacity, supply of nutrients, and reduce soil erosion and run-off of nutrients and any pesticides used (Johnston et al. 2009; Watts and Dexter 1997). Increasing SOM through the application of regenerative agricultural practices has been demonstrated to maintain high yields, whilst allowing a reduction in nitrogen application of 60 kg/ha (Impey 2021b). Indeed, it is claimed that, due to greater water infiltration, the application of regenerative agriculture soil management has even rekindled water flows from once dried up springs (EIT Food 2020). Organic matter also increases soil fertility by increasing both the nutrient holding capacity of soils and nutrient cycling.

Under certain conditions, regenerative agriculture is also claimed to lead to higher yields of major crops, such as maize, wheat, rice, soybean, and sunflower, compared to the same crops farmed conventionally (EIT Food 2020). For example, studies demonstrated that silvo-pastural practices, the traditional practice of combining tree and livestock production, could improve productivity by 55% compared to livestock and tree production practiced separately (Arguelles Gonzalez 2022). Even in situations where regenerative agriculture results in lower yields compared to conventional agriculture, price premiums (either for Regenerative Organic or more liberal interpretations of regenerative agriculture) and much lower input costs (also linked to reduced pest and disease pressures due to increased biodiversity) often make crop production more profitable. For example, whilst regenerative agriculture maize production in the Northern Plains of the USA resulted in a 29% yield reduction, the price premium attracted increased profits by 78% (Beyond Pesticides 2019; EIT Food 2020).

In the UK, commercial trials with regenerative agriculture resulted in higher net margins compared to conventional agriculture. Whilst average output per hectare and gross margins were lower (21% and 24% lower), net margins were either similar or even higher than those gained from conventional agriculture (Impey 2021a). In 2020, net margins where £27/ha for regenerative agriculture and £7/ha for conventional agriculture. Improvements in net margins were gained from an 18% reduction in variable costs and a 32% reduction in labour and machinery costs. The cost of machinery, especially on conventional arable farms, is the most significant cost, especially when factoring in depreciation (Impey 2021a). Regarding reduced input costs, Hutchison et al. (2010) and Lundgren et al. (2015) suggest that due to biological control (higher number of natural predator species), increased crop diversity, and increased crop rotations, insect pest populations on regenerative agriculture fields can be as little as just 10% of those found in conventionally managed fields.

Regenerative agriculture also claims to increase yield and food system resilience, especially against extreme temperatures, droughts, floods, etc. This is primarily achieved through increased soil health, crop diversity, biodiversity, and farmer profitability (Civil Eats 2020; IDH 2020). For example, during drought conditions, fields using regenerative agriculture approaches absorb and store increased amounts of water due to the higher biomass/organic matter content and improved soil structures (EIT Food 2020).

Regenerative agriculture has attracted significant attention from corporate food system actors. For example, General Mills Restorative Farming Project covers 650 000 ha across Kansas. Through saving water, increasing soil health, and the sequestration of carbon, General Mills and Danone aim to make their farmer suppliers and surrounding communities more profitable and more resilient (EIT Food 2020).

6.2.3 Key weaknesses of regenerative agriculture

Regenerative agriculture has also met with significant criticism linked to the exaggeration of benefits - holistic grazing and carbon sequestration, high yields, and increased biodiversity.

Whilst holistic grazing and other regenerative practices have been demonstrated to sequester carbon, many suggest that the extent of soil carbon accumulation and reduction of greenhouse gases is greatly exaggerated (Briske et al. 2014; Garnett et al. 2017; IDH 2020; McGuire 2018; Ranganathan et al. 2020). For example, studies have shown that, if widely adopted, regenerative agriculture could at best sequester 30% of annual anthropomorphic CO_2, compared to the 100% which is claimed. Indeed, these studies suggest that the average figure could be much lower, around 10–15% (Gao et al. 2018).

Some of the reasons for this exaggeration link to the varying capacities of soils to assimilate higher amounts of SOM, which is comprised of approximately 60% soil organic carbon (SOC) (Impey 2021b). There are basically three ways to increase SOC:

1 Stimulate plant root exudates;
2 Increase application of biomass (crop residues, animal manures, etc.); and
3 Reduce cultivation.

However, the lion's share of carbon building is from the exudation of carbon-based molecules from plant roots. Root exudates build carbon between 5 and 30 times faster than biomass applied to the soil surface (Impey 2021b). Whilst highly cultivated and degraded clay soils offer significant opportunities to increase SOM and carbon, increasing SOM in light/sandy soils is extremely challenging (Chivenge et al. 2007). For example, in the case of Australia's soils, deep bodied clay soils generally contain around 1-2% SOC. With enough rainfall, conducive temperature, and minimal cultivation, it may be possible to increase the SOC content up to 4%. Permanent pasture, under ideal conditions, has SOC of up to 14%. In the case of sandy soils, which contain as little as 0.5% SOC, increasing SOC past 2% will present a significant challenge (Burke 2021).

In addition, well-managed and/or highly fertile soils may already be close to carbon or C saturation potential and offer little potential for increasing the carbon content. Even management practices designed to increase the return of biomass to the soil, such as the use of animal manures, green manures, and crop residues, generally only lead to small increases in SOM (Poulton et al. 2018; van der Esch et al. 2017). The integration of agroforestry offers much greater potential to sequester carbon by assimilating carbon both below and above ground (Feliciano et al. 2018; Rosenstock et al. 2019).

Whilst proponents of regenerative agriculture report moderate to high yields, many studies have demonstrated a clear reduction in yields when applying regenerative practices (EIT Food 2020; IDH 2020; Benton et al. 2021). Studies of regenerative agriculture in the USA demonstrated a 29% reduction in maize production compared to conventionally produced maize (EIT Food 2020). Other studies have demonstrated that when significant amounts of nutrients (in the form of grains, pulses, vegetables, etc.) are transported and consumed off-farm, off-farm nutrients (either organic or synthetic) are required to replace them. Without the import of plant nutrients, regenerative practices would eventually deplete fertile soils of their nutrients and yields would plummet (Giller et al. 2021). Indeed, without the use of significant quantities of off-farm nutrients, the application of regenerative agriculture on highly weathered and nutrient-mined soils of Africa would tend worsen agricultural performance rather than improve it (Giller et al. 2021).

Lastly, whilst regenerative agriculture may well lead to increased biodiversity on the farm, associated lower yields would require an expansion of agricultural production into previously non-agricultural/natural and semi-natural habitats, potentially leading to a net loss of biodiversity (Giller et al. 2021).

6.2.4 Future of regenerative agriculture

In many respects, the lack of a clear definition of regenerative agriculture will either be its saviour or downfall. Ultimately, if detractors unpick the threads of regenerative agriculture, the whole movement could unravel. However, if its advocates can maintain momentum in its spread, building greater political and social buy-in, the prospects for regenerative agriculture look promising. Corporate actors, especially powerful food processors and retailers, are key to the scaling of this approach. Several companies have significant plans to scale regenerative agriculture. Their strategies are designed to be flexible, focussing primarily on easily saleable key outcomes, such as regenerated soils, improved water quality, carbon sequestration, improved ecosystems, and care for farmers and their communities, and less on the practices that deliver them (Klein 2021; Fastler 2021; Nunes 2021).

This flexibility allows for the use of many means to these politically and socially acceptable ends. Such flexibility provides space for the use of GM crops and pesticides. With the aim of reducing its carbon footprint, General Mills has set a goal of 1 million acres of land under regenerative agriculture practices by 2030. It also aspires to become net zero in carbon emissions by 2050 (Wilcox 2021). Similarly, Archer-Daniels-Midland Company (ADM) managed to enrol 1 million acres of farmland into its regenerative agriculture programme in 2022 and aimed to increase this to 2 million acres in 2023, and 4 million acres by 2025 (World Grain 2023b). In turn, PepsiCo has announced that it intends to promote regenerative agriculture on 7 million acres of its suppliers' farmland. Likewise, Cargill has also committed to promote regenerative agriculture on 10 million acres of its suppliers' farmland, and Walmart has done the same for 50 million acres of its suppliers' farmland (Uldrich 2021; Klein 2021). As part of its Project Gigaton, Walmart is committed to reducing CO_2 emissions from its farmer suppliers by 7 million metric tons by 2030 (Klein 2021). Both Danone and McCain aim to source a 100% of ingredients from suppliers practicing regenerative agriculture (Marquis 2021). In Brazil, Bunge now promotes regenerative agriculture across 250 000 ha in the states of Bahia, Maranhão, Mato Grosso, Paraná, Piauí, and Tocantins (Bunge 2023).

In addition to setting such lofty targets, these food corporations seem willing to put their money where their mouth is by providing technical support and financial incentives for the rapid scaling of regenerative agriculture. For example, Danone and General Mills have made significant investments in assisting their farmers and suppliers' transition to regenerative agriculture (EIT Food 2020). ADM provides premium payments of up to US$25/acre/year (World Grain 2023b).

Nestle has committed US$1.3 billion over the next 5 years to assist their farmers and suppliers' transition to regenerative agriculture (Nunes 2021). Nestlé has committed to deploy its researchers and agricultural extensionists, co-invest with its 500 000 farmers and 150 000 suppliers, and assist them to access loans for necessary capital investments. Nestle is also committed to providing price premiums for agricultural commodities produced using regenerative agricultural practices (Nunes 2021; Holmes 2021). Nestlé hopes that transitioning to regenerative agriculture will assist in reducing its GHGEs by 50% by 2030 and achieving net-zero emissions by 2050 (Nunes 2021; Holmes 2021). Whilst originating in the USA, regenerative agriculture continues to take root across the globe, from Europe, to India, to Brazil, to Australia and New Zealand (EIT Foof 2020; Marquis 2021).

6.3 Organic agriculture

The definition of organic has been shaped by organic producers, processors, and retailers of organic foods, consumers of organic foods, and regulators of organic foods (Seufert et al. 2017). Whilst not as opaque as the definitions of regenerative agriculture, organic agriculture has often meant different things to different food system actors (Rigby and Cáceres 2001; Guthman 2004; Hughner et al. 2007).

For early organic practitioners, such as Sir Albert Howard, organic agriculture is encapsulated by a series of principles. One such principle, captured in Howard's 1943 book entitled *An Agricultural Testament*, and shared by another pioneer Rudolph Steiner, is 'the Law of Return' (Gomiero et al. 2011). This law focusses on the recycling of all organic wastes, including human sewage, back into the land to maintain SOM/humus, and fertility. This principle is echoed in Lord Northbourne's book *Look to the Land*, where organic agriculture is seen as the holistic management of a farming system which encourages the diversity of farmed and non-farmed organisms and recycles nutrients through relatively closed natural biological processes (Gomiero et al. 2011). Implicit or explicit in these early definitions was avoiding any use of artificial inputs, notably synthetic fertilisers, and pesticides. For many, this prohibition of synthetic chemical use (with very limited exceptions) has become the key defining characteristic of organic agriculture.

According to organisations like IFOAM, organic agriculture is based on four basic principles (Krstić et al. 2017):

1 Health;
2 Ecology;
3 Equality; and
4 Sustainability.

Organic agriculture aims to sustain and enhance ecosystem health, balancing the needs of soil, plants, animals, and humans for current and future generations. Organic agriculture works with and emulates local ecological systems and the cycles contained therein. Organic agriculture combines traditional knowledge with science and innovation and promotes quality of life and fairness (Nowak et al. 2015; Darnhofer et al. 2019; Gomiero et al. 2011).

Unlike other alternative food system approaches, organic agriculture sets itself apart by its adherence to a range of sector-specific regulations, which vary slightly from one country to the next. However, there are several rules that fundamentally frame organic agriculture. As far as possible, organic farmers rely on the recycling of nutrients within relatively closed ecosystems using integrated crop, livestock, pasture, and agroforestry, or through intercropping of nitrogen-hungry cereals and vegetables with nitrogen-fixing trees or legumes, polyculture, cover crops, and mulching, as well as diverse crop rotations (Gomiero et al. 2011). Any additional nutrients required can only be in the form of organic (farmyard manure, food waste, etc.) materials that are free from synthetic chemical contaminants, or naturally occurring inorganic elements such as rock phosphate. No synthetic fertilisers are permitted.

Since synthetic pesticides are also not permitted, weeds in organic systems are managed through a mix of cultural methods of control such as complex crop and livestock rotations, land cultivation and false seedbeds, mulching, timing of sowing and transplanting, as well as physical/mechanical methods of control such as flaming, flailing, etc. (Howard 1943; Lampkin 2002; Lotter 2003; Altieri and Nichols 2004; Koepf 2006; Kristiansen et al. 2006; Gliessman 2007). Pest and disease control is primarily managed through diverse crop rotations, management of pest/prey relationships, intercropping and mixed cropping, and the use of locally adapted pest and disease-resistant/tolerant crops and breeds. If economically damaging pest and disease outbreaks occur, biological pest control can be used, as well as some natural biological pesticides (primarily extracted from plants), as well as a limited range of simple inorganic compounds with pesticidal activity, such as copper sulphate.

Organic agriculture precludes the use of GMOs (currently including gene-edited crops), ionising irradiation for the treatment of food, and synthetic pesticides. In addition, the use of animal hormones is prohibited, as is the use of antibiotics. However, in cases where animals are suffering unduly, organic regulations allow the use of antibiotics as a last resort (Adamtey et al. 2016; Krstić et al. 2017; Seufert et al. 2017). The ban on synthetic chemical use, and reliance instead on supporting and optimising natural agroecosystem processes, means that many consumers of organic produce across the globe strongly believe that organic foods are more natural, healthier, more nutritious, tastier, and safer than industrially produced foods (Davies et al. 1995; Chang and Zepeda 2005; Dahm et al. 2009; Sirieix et al. 2011).

6.3.1 The development of organic agriculture

Some of the early pioneers of organic agriculture include agronomist Franklin Hiram King in the USA, who, in 1911, published a book entitled *Farmers of Forty Centuries*, which described traditional fertilisation, tillage, and farming practices present in China, Korea, and Japan just after the turn of the twentieth century. Likewise, in Germany, during the early 1920s, Rudolf Steiner developed the concept of biodynamic farming, which was developed as an alternative approach to the increasingly industrial nature of modern agriculture (Darnhofer et al. 2019). By 1927, a cooperative (Association Demeter) had been established to market biodynamic products and, by 1928, quality control standards had been developed for biodynamic products (Gomiero et al. 2011).

Whilst primarily based on the work of British organic agriculture pioneers, Sir Albert Howard and Lady Eve Balfour, and the early work on biodynamic agriculture, the term 'organic farming' was coined by Walter James (Lord Northbourne), in his book *Look to the Land* published in 1940 (Paull 2014). Lord Northbourne had been a student of biodynamic agriculture. In 1939, Alice Debenham and Lady Eve Balfour established the Haughley Farm Research Project in the UK, which aimed to scientifically prove the benefits of what was to become known as organic agriculture compared to the conventional agriculture of the period (Clunies-Ross 1990).

In 1943, Lady Eve Balfour's book *The Living Soil* described the interrelationship between farming practice and plant, animal, human, and environmental health (Gomiero et al. 2011). Public interest whipped up by Balfour's book quickly led to the establishment of the UK 'Soil Association' in 1946 with its own certification body as well as associated organic standards (Gomiero et al. 2011). Driven by founders Hans Müller, Maria Biegler, and Hans Peter Rusch, the concept of organic agriculture also emerged in Switzerland during the 1940s. Other terms, such as 'humus farming', 'biologic' (used in France, Italy, Portugal, and Holland), and 'ecologic' (used in Denmark, Germany, and Spanish-speaking countries) were developed almost in parallel. However, in the post-war period, the UK, and other governments in Europe prioritised the expansion of food production, effectively putting their weight behind more intensive production methods through policy and legislation (e.g. the UK 1947 Agriculture Act), effectively marginalising the early organic movement (Tomlinson 2010).

From the 1960s onwards, concerns about the environmental and health impacts of modern, intensive farming (e.g. in the use of pesticides) stimulated greater consumer interest in alternative approaches such as organic agriculture. The growing number of national organic organisations led to the establishment of the IFOAM in 1972. In recognition of the growing popularity of organic produce, state-level regulation of organic agriculture (e.g. defining what could be labelled as organic) also appeared in the USA during the 1970s (Seufert et al. 2017). However, it was not until the 1980s, when the environmental externalities of modern agriculture began catching more widespread attention amongst the public and politicians alike, that organic agriculture began to gain traction (Fromartz 2007; Lockeretz 2007). For example, in the UK, Government attitudes towards organic agriculture changed from that of scepticism to an acknowledgement that organic agriculture had a serious role to play (Tomlinson 2010). Much of this change in attitude was driven by a new generation of organic pioneers, such as Peter Segger, Patrick Holden, and Lawrence Woodward. These energetic and articulate individuals revitalised the UK organic movement by catalysing the establishment of the Elm Farm Research Centre (EFRC) in 1980, the Organic Growers Association (OGA) in 1981, and British Organic Farmers (BOF) in 1982 (Tomlinson 2010).

The establishment of the 'British Organic Standards Committee' (BOSC) in 1981, which included Organic Farmers and Growers (OF&G), the Henry Doubleday Research Association (HDRA - an organic gardening research association), and the IFOAM, enabled the organic agriculture movement to lobby government more effectively for policy changes (Tomlinson 2010). Indeed, I remember being a fresh-faced lecturer in sustainable agriculture and the environment at this time, sitting alongside a member of the Soil Association in Ministry of Agriculture Fisheries and Food (MAFF) steering committee meetings on the development of the first agri-environmental programmes in the UK.

Support for organic agriculture also began to sprout in the UK Government's Agricultural Development and Advisory Service, especially after the arrival of Roger Unwin in 1982. Roger was highly supportive of organic farming research (Tomlinson 2010). After 1984, the MAFF began to take organic agriculture seriously, commissioning research on organic farming techniques and visiting organic trials (Tomlinson 2010). The media also rediscovered organic agriculture. In 1983, the BBC Radio 4 programme 'On the Farm', which usually focussed on conventional agriculture, dedicated a full-length programme to organic farming fronted by Peter Segger and Patrick Holden (Tomlinson 2010). The early 1980s was also the time when mainstream supermarkets became interested in organic foods. In the UK, the supermarket Safeway introduced a range of organic product lines in its London and Bristol stores (Tomlinson 2010).

Organic agriculture also began taking root in several developing countries during the late 1980s and 1990s. For example, growing international demand for organic products, the existence of production systems using few, if any, synthetic inputs (fertilisers and pesticides), plus an enabling policy and economic environment, led to the establishment of organic production in Uganda during the late 1980s. By 2016, Uganda had the largest area of certified organic production and the highest number of organic farmers in Africa (Hauser and Lindtner 2017). In India, organic farming was initially promoted by the government's National Programme for Organic Production (NPOP) in 2001, with supporting third-party organic certification established through the Participatory Guarantee System in 2005 (Khurana and Kumar 2020). In 2015, Sikkim became India's first 100% organic state, with others, including Andhra Pradesh and Himachal Pradesh, also aiming to become 100% organic (Khurana and Kumar 2020). Despite this achievement, only 2% of India's cultivated land is organic and only 1.3% of farmers are registered as organic (Khurana and Kumar 2020). Figure 74 illustrates the growth in land farmed according to

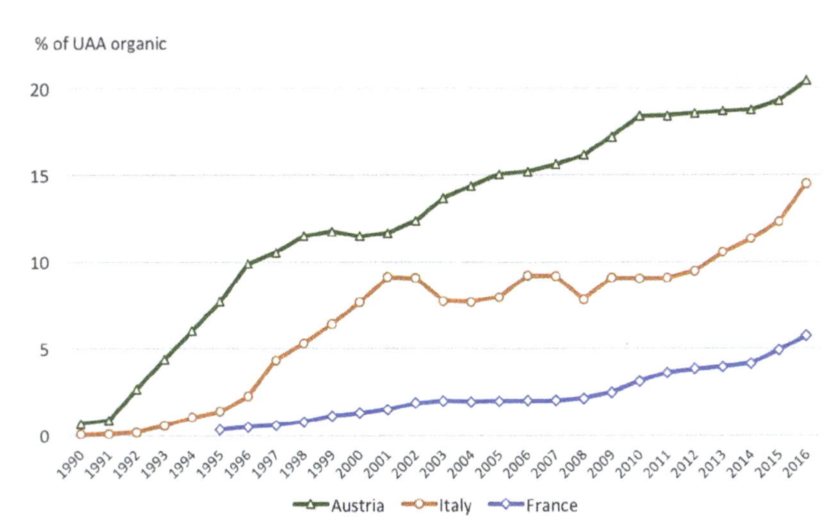

Figure 74 Share of total organic area in the total utilised agricultural area for Austria, Italy, and France. Source: Adapted from: Darnhofer et al. (2019).

organic agriculture standards in Austria, Italy, and France between 1990 and 2016 (Darnhofer et al. 2019).

Organic agriculture also began receiving policy support in both the EU (Darnhofer et al. 2019) and USA (Seufert et al. 2017) during the 1990s. In the EU, organic agriculture has become established in public policy as the principal alternative approach to conventional agriculture and is perceived as addressing public concerns such as protecting the environment, maintaining rural areas, and improving animal welfare (EC 2004; Häring et al. 2004; Nieberg et al. 2007). Indeed, the European Commission (EC), the main policy-development body in the EU, has orchestrated a number of official regulations, with dedicated action plans developed and implemented by member states, to support organic farming, especially through country-level agri-environmental programmes (Darnhofer et al. 2019). Tailor-made payments to assist farmers in their transition to organic farming have been instrumental in expanding the area of land under organic production (Sanders et al. 2011). By 2021, 76 countries across the globe had fully implemented regulations governing standards in organic agriculture.

Figures 75 and 76 illustrate the variability of adoption country by country. For example, Fig. 75 illustrates the adoption of organic agriculture in the EU 27, as of 2019.

It can be seen that, whilst only 8.7% of agricultural land in the EU as a whole is dedicated to organic agriculture, countries such as Austria (25.3%), Estonia (22.3%), and Sweden (20.4%), have a significantly higher proportion of agricultural land under organic agriculture. Other countries, such as Ireland (1.6%), the UK

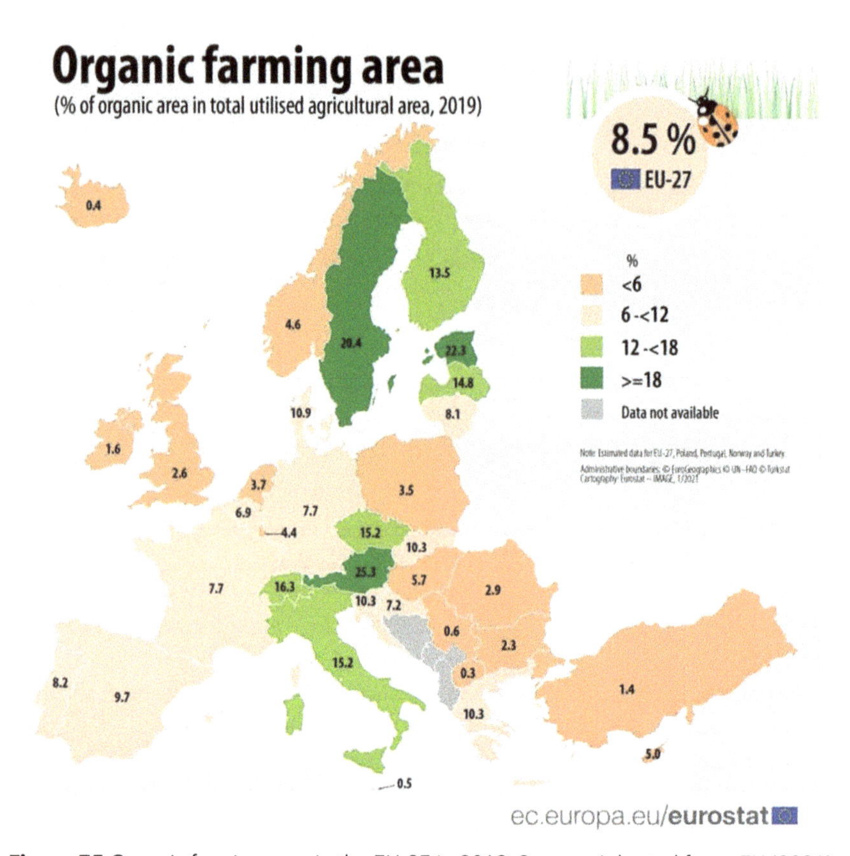

Figure 75 Organic farming area in the EU-27 in 2019. Source: Adapted from: EU (2021).

(2.6%), and the Netherlands (3.7%), have significantly less land dedicated to organic production (EU 2021). There is also significant concentration of organic agriculture at sub-national levels. In the Salzburg region of Austria, for example, over half of the farmed land is organic (52 %). Calabria and Sicilia in Italy have 29% and 26% of the farmed area classified as organic production (EU 2021).

A similar situation exists in the USA, where organic agriculture represents 1–2% of US farming as a whole. Figure 76 shows that organic production is primarily concentrated along western coastal states (California, Washington, and Oregon) and northeast states (New England, Maine, down to Pennsylvania), and the Northern Great Lakes area (Minnesota, Wisconsin, and Michigan), serving predominantly large urban areas (Kuo and Peters 2017).

Rates of adoption of organic agriculture can also be quite dynamic, with countries often having periods of both growth and decline. For example, in 2003, 4% of agricultural land in the UK was farmed organically, 1.4% higher than the area of organic agriculture reported in 2019. Between 2009 and

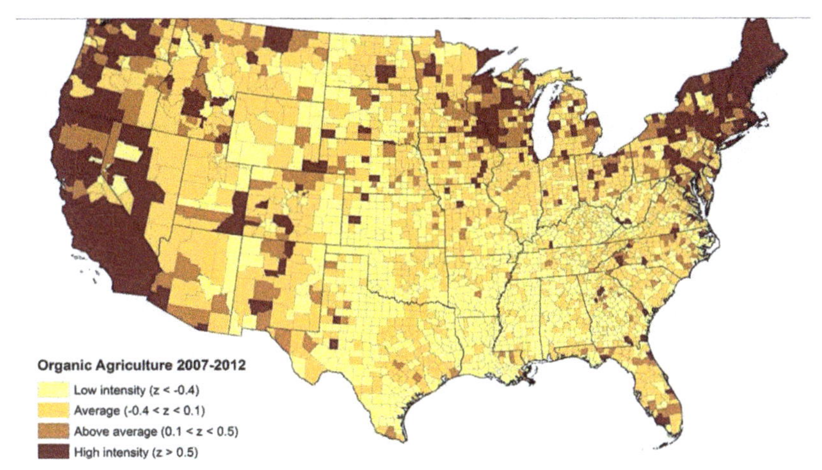

Figure 76 Intensity of organic agriculture across the USA. Source: Adapted from: Kuo and Peters (2017).

2011, the area organically farmed noticeably declined for cereals, other arable crops, and fresh vegetables. This was primarily due to a 5% reduction in retail sales, a reduction of choice in organic produce, weak advertising, and reduced investment in own-label organic ranges by supermarket chains.

Despite some fluctuations, Fig. 77 illustrates that land dedicated to organic agriculture globally has expanded steadily from 15 million ha in 2000 to 76.4 million ha in 2021 (Willer et al. 2023). This equates to approximately 1.6% of the global farmed area (Willer et al. 2022). Of this, 36 million ha are in Oceania, which is basically low productivity grassland in Australia (Gomiero et al. 2011), 17.8 million ha in Europe, 9.9 million ha in Latin America, 6.5 million ha in Asia, 3.5 million ha in Northern America, and 2.7 million ha in Africa (Willer 2023).

Figure 78 illustrates the steady growth in global organic agriculture between 2000 and 2021, whilst Fig. 79 provides a breakdown of land use types and crops under organic agriculture. At 65%, permanent grassland dominates land use.

Of the arable crops grown, Fig. 80 illustrates that cereals, green fodder, and oilseeds dominate production.

By 2021, there were 3 669 201 certified organic farmers worldwide (Willer et al. 2023). Figure 81 lists the top ten countries with the highest number of organic producers. It can be seen that India contains the largest number of organic farmers with nearly 1.6 million.

Whilst India has the largest number of organic farmers, most organic farms tend to be small. Figure 82 illustrates the countries with the greatest area of land dedicated to organic cereal production. Here, China dominates production with over 1.4 million ha, followed by Germany, France, Italy, and

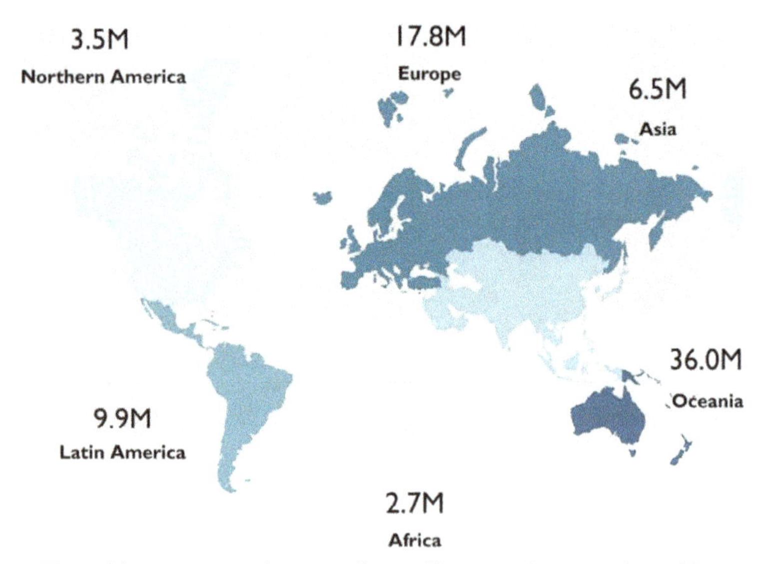

Figure 77 World organic agriculture in millions of hectares. Source: Adapted from: Willer et al. (2023).

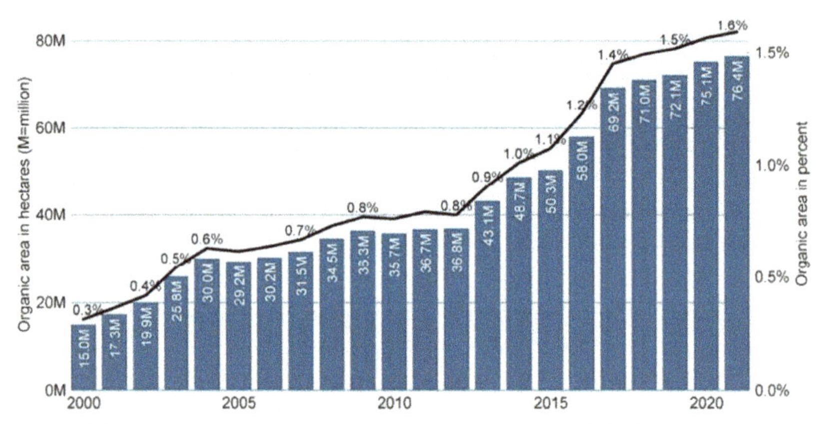

Figure 78 World growth of organic land and organic share 2000–20. Source: Adapted from: Willer et al. (2023).

Canada. Austria, Estonia, and Sweden have the highest proportion of organic cereals compared to conventionally produced cereals.

Figure 83 illustrates growth in the global market for organic produce between US$21 billion in 2001 to US$135.5 billion in 2021. Figure 84 illustrates the key markets for organic produce. It can be seen that the USA dominates the demand for organic produce with nearly €50 billion. Figure 85 illustrates the increasingly global nature of organic production. It can be seen that the

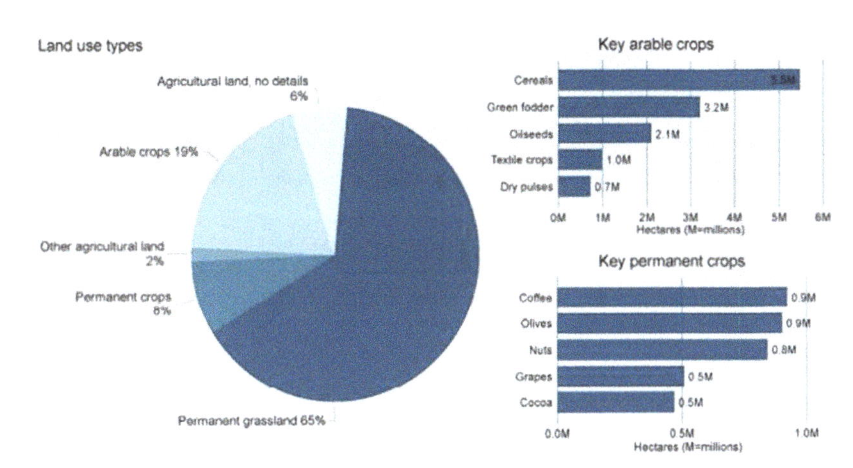

Figure 79 World distribution of main land use and key crop categories 2021. Source: Adapted from: Willer et al. (2023).

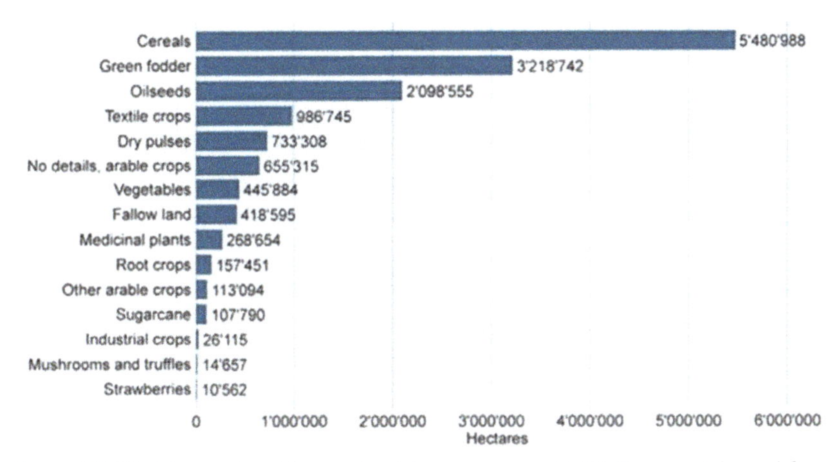

Figure 80 World use of arable cropland by crop group 2021. Source: Adapted from: Willer et al. (2023).

EU relies heavily on other countries for the supply of tropical or out-of-season organic produce.

6.3.2 Key strengths of organic agriculture

According to its advocates, organic farming has a number of key benefits. These include enhanced produce safety, sensory and nutritional quality (thus improving consumer health); better protection of the environment and agroecosystem health; more resilient farming systems; and increased profitability. Whilst organic production usually leads to reduced yields of

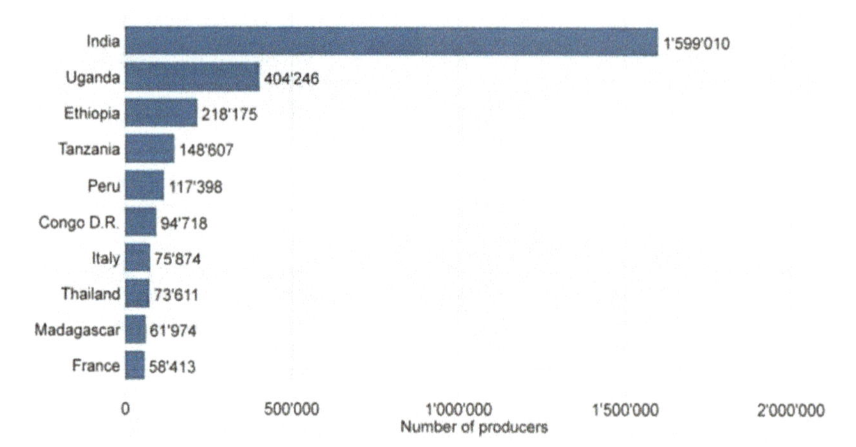

Figure 81 World organic producers - top ten countries. Source: Adapted from: Willer et al. (2023).

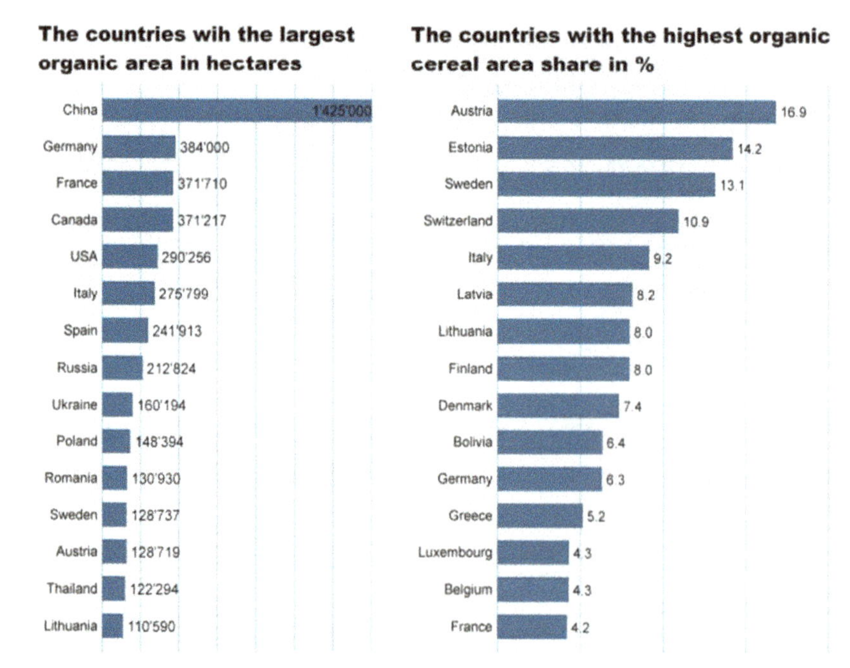

Figure 82 Organic area of cereals (hectares) and percentage share of organic cereals to conventionally produced cereals in 2021. Source: Adapted from: Willer et al. (2023).

between 20% and 25%, lower input costs and higher premiums for organic produce tend to lead to equal or even greater profitability when compared to conventional agriculture (Ramankutty et al. 2019; Cisilino et al. 2019).

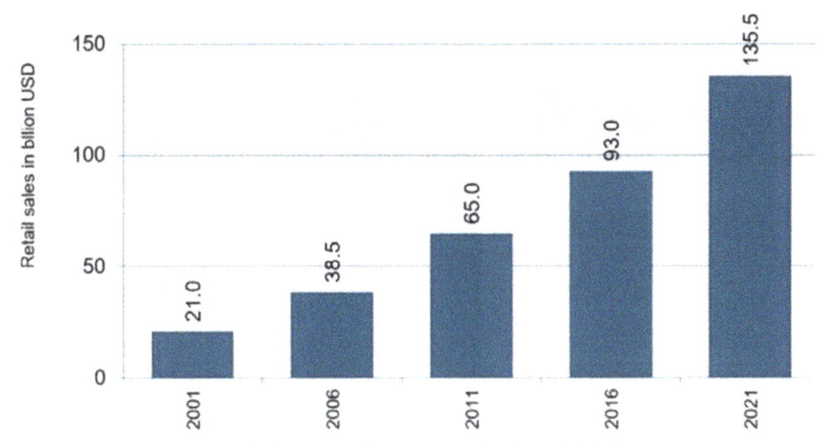

Figure 83 Growth in global market for organic food and drink 2001-2021. Source: Ecovia Intelligence.

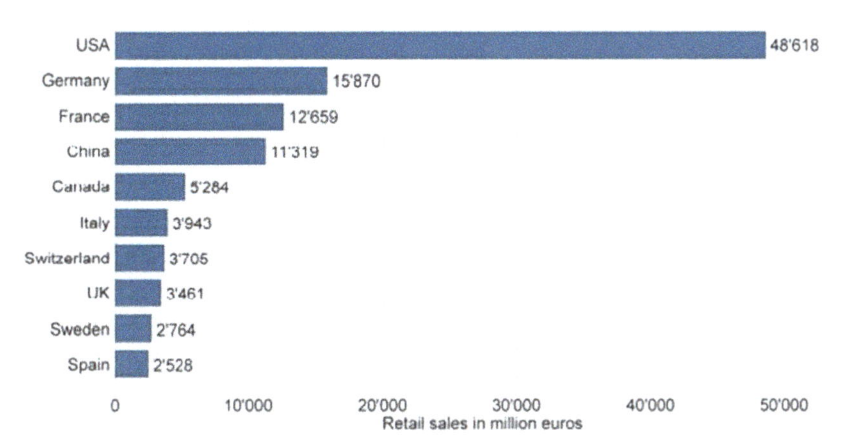

Figure 84 The countries with the largest markets for organic food. Source: Adapted from: Willer et al. (2023).

Organic agriculture claims to enhance human health through reduced exposure to synthetic pesticides, GMOs, synthetic fertilisers, and other artificial chemicals used in conventional agriculture (Reganold and Wachter 2016). Studies have suggested a number of direct health benefits associated with organic foods, such as higher levels of beneficial phytochemicals (antioxidants such as polyphenols) (Baranski et al. 2014; Zamora-Ros et al. 2013). Other studies suggest that, compared to conventional production, organic milk and meat contains up to 60% more healthy fatty acids (including omega-3 fatty acids) (Srednicka-Tober et al. 2016a,b). Organic crops have been shown to

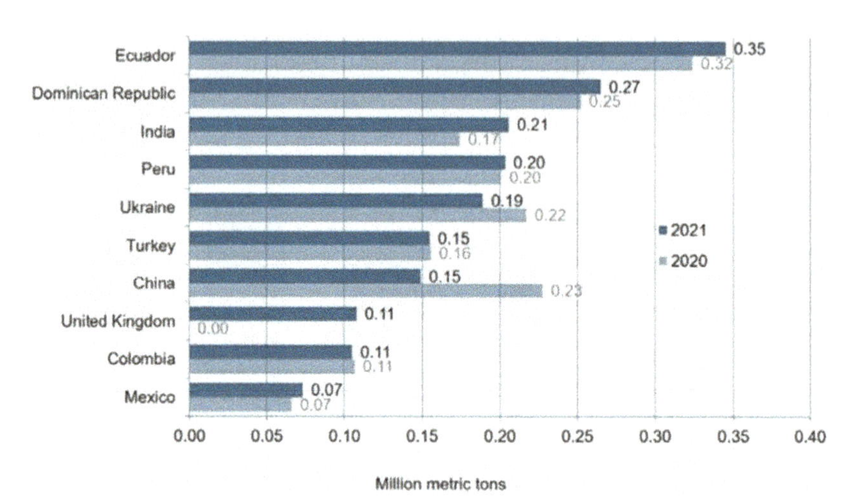

Figure 85 EU organic import volumes: top ten countries exporting to the European Union. Source: Adapted from: Willer et al. (2023).

contain up to 90% more vitamin C, increased micro-nutrients, and 50% more secondary metabolites than conventionally produced crops (Hunter et al. 2011; Hattab et al. 2019; El-Hage Scialabba 2013).

Studies suggest that ecosystem health and biodiversity are enhanced under organic systems (Letourneau and Bothwell 2008; Norton et al. 2009; Bengtsson et al. 2005; Mäder et al. 2002a; Pimentel 2005). When the management of farmland is better integrated with the surrounding ecosystem, especially in developing a more heterogeneous landscape over a large area, both are expected to flourish (Gomiero et al. 2011). Proponents of organic agriculture stress the importance of balancing food production, biodiversity, and ecosystem services. Under organic production, biodiversity is increased, and habitats are conserved. In return, crop and livestock production benefits from weed, disease, and pest control, improved water cycling and nutrient management, and enhanced pollination services (Kremen and Miles 2012; Lin 2011; Mijatović et al. 2013).

Figure 86 illustrates the impact on biodiversity in organic systems compared to conventional systems. Under organic production, biodiversity can increase by up to 30%, and the abundance of organisms can increase by up to 50% compared to conventional farming (IPES-Food 2016).

Organic agriculture also increases SOM, which, in turn, increases soil biodiversity, soil fertility (nutrient holding and cycling), and water cycling (holding water during dry spells and allowing water to drain after heavy rains) (Mäder et al. 2002; Fließbach et al. 2007; Gattinger et al. 2012; Gomiero et al. 2011; Reganold and Wachter et al. 2016).

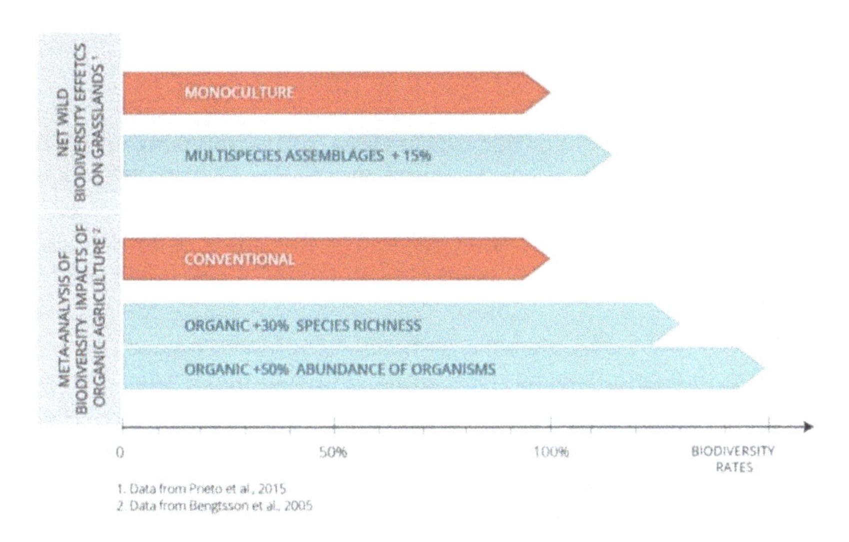

Figure 86 Species abundance in organic versus conventional agriculture. Source: Adapted from: IPES-Food (2016).

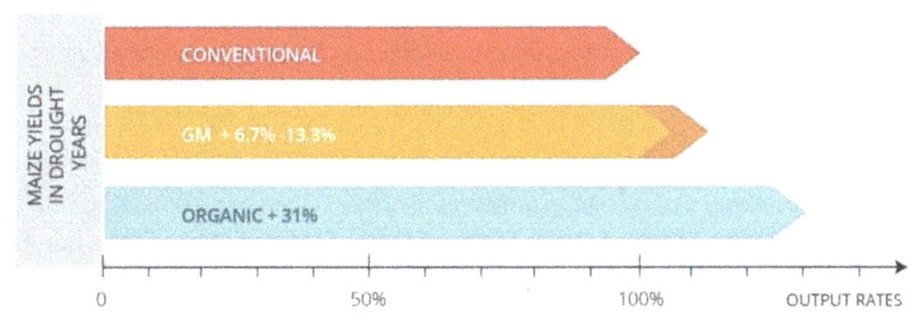

Figure 87 Organic maize yields during drought versus conventional and GM maize. Source: Adapted from: IPES-Food (2016).

Evidence suggests that organic farming builds farming system resilience to biophysical shocks, such as droughts, pests, and diseases. For example, due to higher SOM content, soils on organic farms have higher water-holding capacity and therefore greater resilience to spells of dry weather and even drought (Lotter 2003; Mijatović et al. 2013). Indeed, Gomiero (2018) suggests that, under severe drought stress, organic farms can produce up to 70–90% more yield than conventionally managed farms due to the better water storage capacity of organically managed soils. Figure 87 illustrates a 31% maize yield advantage under drought conditions for organic management compared to conventional agriculture.

6.3.3 Key weaknesses of organic agriculture

6.3.3.1 Low crop yields

The performance of organic agriculture varies from crop to crop, farming system to farming system, and geography to geography (Trewavas 2001b; Rigby and Cáceres 2001; Degré et al. 2007; Letourneau and Bothwell 2008). However, a number of studies suggest that, in general, crop yields tend to be lower under organic management compared to conventional management (Mäder et al. 2002a; Kaut et al. 2008). In one of the most comprehensive comparative studies, Seufert et al. (2012) found that (depending on factors such production system, crop, and geography) organic yields tended to be between 5% and 35% lower than conventionally produced crops. Other reviews suggest that, on average, organic agriculture produces around 20–25% lower yields than conventional agriculture (Benton et al. 2021; Joseph et al. 2019). Kirchmann (2019) has estimated that yields from organic farming are around 35% lower than from conventional farming. As he points out, an organic farm generally relies on integrating green manures into the cropping system. These are crops that are grown and then ploughed back into the land to build organic matter and fertility, without being harvested. Organic farms also rely on the frequent use of low-yielding legumes to build fertility, thus lowering the overall yield of the farm (Kirchmann 2019).

This lower yield performance raises the concern that organic agriculture would struggle to meet current and expected future demand for food (Badgley et al. 2007; de Ponti et al. 2012; Seufert et al. 2012; Gomiero 2018; Kirchmann et al. 2008). Lower yields mean that an expansion of agriculture into natural and semi-natural habitats by up to 50% may be necessary to meet future food demand (Agrimonti 2021; Gomiero et al. 2011; Seufert et al. 2012; Reganold and Wachter 2016; Gomiero 2018; Kirchmann 2019).

6.3.3.2 Environmental externalities

Whilst organic agriculture professes to reduce environmental externalities, a number of studies suggest that, due to factors such as lower yields and increased area of farmland required to produce a given volume of produce, fossil-fuel energy use and GHGEs per unit of produce may actually be higher than conventional agriculture (Gomiero et al. 2011; de Ponti et al. 2012). It has been argued that avoidance of synthetic inputs does not automatically lead to positive environmental outcomes (Seufert et al. 2017). For example, poorly managed organic cow slurry can just as easily lead to nutrient run-off and the pollution of watercourses as conventionally managed cow slurry (Kirchmann and Bergström 2001). Indeed, according to Clark and Tilman (2017), on a per unit of output basis, organic systems lead to higher levels of eutrophication potential.

Whilst purported to be central to organic farming, explicit rules regarding the management of soil, water, and biodiversity are primarily absent from organic regulations (Seufert et al. 2017). For example, in most countries, there are no absolute rules governing crop and soil management approaches such as the use of cover and catch crops (Altieri and Rosset 1996; Tonitto et al. 2006), or crop mixtures and crop associations (Altieri and Rosset 1996; Zhu et al. 2000). Few regulations include water-use efficiency (Rosegrant et al. 2009). Indeed, some organic practices can be even more environmentally polluting and degrading than conventional practices. For example, mechanical weeding on sloping land can lead to increased soil erosion and silting of watercourses compared to the use of chemical weed control (Khurana and Kumar 2020).

Organic farms often also rely on the importation of nutrients from conventional farms. Studying French organic farms, Nowak et al. (2013a,b) found that organic farms relied heavily on nutrients derived from conventional farmers. Indeed, 23% of the nitrogen, 53% of phosphorous, and 73% of potassium used on organic farms was transferred from conventional farms in the form of straw, animal manures, organic fertilisers, and feed materials. This dependence on external inputs needs to be factored into an assessment of environmental impact.

6.3.3.3 Conventionalisation of organic production

Lastly, and probably most importantly, the idyllic perception of organic agriculture as small-scale and local labour of love is not always matched by the realities on the ground. Indeed, many organic farms are large-scale, intensive, highly commercial, and firmly integrated into global supply chains (Guthman 2004; Seufert et al. 2017; Gliessman et al. 2019). Current codification of organic agriculture has not prevented the growth of corporate farms which, aside from not using synthetic inputs (outlined above), in all other ways resemble large-scale conventional farms (Allen and Kovach 2000; Goodman 2000; Buck et al. 1997; Guthman 2004). This mainstreaming of organic farming has become known as the conventionalisation of organic production (Darnhofer et al. 2019).

Buck et al. (1997) define conventionalisation as 'agribusiness finding ways to industrialise organic production by reconfiguring farm processes as inputs so they can be produced in the factory and by adding value and asserting control in the processing, distribution, and retailing links of the commodity chain'. The control of organic sales by large-scale global and regional retailers has especially raised concerns (Laville and Vidal 2006; Baqué 2012; Dion 2013; Mercury News 2014; Hielscher 2017). Here, retailers have continually driven down the premiums offered to organic farmers in exchange for the access they provide to large numbers of consumers of organic produce (Desquilbet et al. 2018; Hansen 2019).

Far from the idyllic image of the smallholder farmer turning his/her hand to just about any farming task, the reality of many large-scale organic farms is reliance on cheap migrant labour, working long hours for low wages (Sacchi et al. 2018; Desquilbet et al. 2018).

6.3.4 Future of organic agriculture

The future of organic farming is somewhat obscure. It is possible that its popularity with farmers, processors, retailers, and consumers will continue to grow. However, the premiums currently paid for organic produce may limit the size of the consumer base. In Europe, plans to increase the area under organic production to at least 25% by 2030 have come under fire with new concerns around food insecurity, exacerbated by Russia's invasion of Ukraine in 2022. There may be more scope for the adoption of organic production in developing countries, following India's example. However, the linking of the announcement by the Sri Lankan government of a nationwide transition to 100% organic agriculture in April 2021, including a ban on agrochemical imports, to the subsequent collapse of the economy in the following year, may make other countries more cautious in their approach. Whilst the link has been contested, fears about global food shortages and a poorer global economic outlook may stifle government ambitions for a larger-scale transition to organic farming (Malkanthi 2021).

Given the growing corporate backing for regenerative agriculture, and the fact that many regenerative agriculture practitioners are organic, there may be an opportunity for organic agriculture to ride on the coat tails of regenerative agriculture and expand well beyond the current 1–2% of agricultural land under organic cultivation. In many respects, the future of organic agriculture will rest on its adaptability. Organic may increasingly conventionalise, perhaps allowing gene editing, or even GM, to build resistance to pests and diseases, and other yield-enhancing characteristics. Conversely, organic agriculture may break from its current trend towards production intensification and corporate market integration, and revert back to its original roots, serving local consumers with seasonal produce.

6.4 Agroecological agriculture

6.4.1 History of agroecological agriculture

The term agroecology was first coined by the Russian agronomist Bensin in 1928 (Wezel et al. 2009). In a similar vein to organic agriculture, agroecology emerged as an alternative to industrial agriculture (Rosset and Altieri 1997; Vandermeer 2010). Drawing on both biological sciences and agronomy, agroecologists aimed to develop sustainable agroecosystems based on sound scientific ecological concepts and principles (Gliessman 1998; Wezel et al. 2009). Much of the initial work of agroecological pioneers, such as Bensin, Klages, and Hanson, utilised zoology and ecology to better understand plot and field-scale management of pests by naturally occurring predators (Wezel et al. 2009). During the 1950s, Tischler, a German ecologist/zoologist, published several articles on the agroecology (farm ecology) of insect and pest ecology and soil biology (Wezel et al. 2009).

During the 1960s, the study of agroecology helped to formally integrate the scientific disciplines of agronomy and ecology (Hecht 1995). In 1965, Tischler wrote the first book specifically on agroecology in which he explained the various interactions of humans, plants, animals, soils, and climate within discrete agroecosystems (Tischler 1965). In 1967, the French agronomist Hénin conceptualised the science of agroecology as being 'an applied ecology to plant production and agricultural land management' (Hénin 1967). The term 'agroecosystem' was first coined by Eugene Odum in 1969 (Odum 1969), where he described an agroecosystem as a 'domesticated, human-managed ecosystem'. Work on the ecology of agriculture at landscape scale was given further impetus during the 1960s by Rachel Carson's book *Silent Spring*, which provided a damming indictment of industrial agriculture and spurred emergence of the environmental movement in the USA (Carson 1964; Altieri 1989, 1995; Hecht 1995).

The 1970s witnessed a significant burst of interest in agroecology (Wezel et al. 2009). It was during this time that agroecology evolved from science to practice (based on solid agroecological science). Agroecological agriculture began to be packaged as a set of sustainable agricultural practices providing an alternative model to industrial agriculture (Wezel et al. 2009). Support for agroecological agriculture grew in the USA throughout the 1960s and 1970s. At the same time, agroecological farming emerged both as a practice and social movement in Mexico and other Latin American countries, providing overt resistance to the spread of intensive agricultural practices associated with the Green Revolution (Jonas 2021).

Set against the industrialisation of food systems in the developed world, it was during this time that agroecology began to draw on principles and lessons from farming systems research focussed on more traditional tropical and subtropical farming systems in developing countries, especially in Mexico and other Latin American countries (Janzen 1973; Gliessman et al. 1981; Altieri 1989; Hecht 1995; Bentley et al. 1994; Toledo 1990). Researchers were particularly interested in farming systems that contained a balance between crop and livestock production and the sustainable management of natural resources. The key scientific principles and practices drawn from these studies formed the basis of what was described as 'sustainable agriculture' (Douglass 1984; Altieri 1989, 1995; Gliessman 1990, 1997).

During the 1980s and early 1990s, the interest in agroecology in developed countries continued to intensify. In the USA, several universities began to teach and research agroecology and sustainable agriculture (Altieri 1993; Wezel et al. 2009; Fernandez et al. 2013). The first was the University of California-Santa Cruz in 1981 (Francis et al. 2003), followed by the University of California-Davis in 1986, the University of Maine in 1986, Iowa State University in 1987, the University of Illinois 1988, the University of Wisconsin Madison in 1989, the University of Minnesota in 1991, and Washington State University in 1991. In 1993, the University of California-Santa Cruz established the Center for Agroecology and Sustainable Food Systems (Fernandez et al. 2013). By 2013, there were more than 55 land grant and private colleges and universities in the USA teaching sustainable agriculture and food systems, with approximately 25% providing modules and degree pathways in agroecology (Fernandez et al. 2013). In parallel, the USDA launched the Low Input Sustainable Agriculture Programme, now called the Sustainable Agriculture Research and Education (SARE) Programme, which promotes agroecological agriculture (Fernandez et al. 2013).

It was at this juncture that agroecological agriculture took on a more explicit transformational role, expanding from its roots as a science and farm-level practice into a sub-discipline that analysed the ecological dynamics of whole food systems, from food production to food processing and marketing, political economy, and food consumption (Wezel et al. 2009; Altieri 1989; Francis et al. 2003; Gliessman 2007). Whilst agroecology remained as a science and practice in most developed countries, it morphed into both social and political movements in several developing countries, especially across Latin America (Dalgaard et al. 2003; Wezel et al. 2009). Since the early 1980s, hundreds of NGOs have promoted agroecology projects across Latin America, Asia, and Africa. These projects were based on practices/approaches that incorporated both traditional knowledge and agroecological science designed to support local communities (Pretty 1995; Altieri et al. 1998; Uphoff 2002). Figure 88 provides an overview of the evolution of agroecology.

In the case of Brazil, agroecology as a movement was built on the promotion of traditional farm practices, autonomy, and sovereignty of small family farms and resistance to the spread of industrial agriculture and its associated environmental externalities. The movement helped to counteract the pressure on small-holder farmers to consolidate farmland and adopt more intensive practices to remain competitive (Norgaard 1984). In 1980, the Federation of Brazilian Agronomists convened the first 'National Meeting of Alternative Agriculture' and lobbied for the establishment of the 'Advisory Body and Services for Projects in Alternative Agriculture' (AS-PTA) (Wezel et al. 2009). In 2003, agroecology was given formal recognition under Brazilian law, enshrining participatory guarantee systems and political support for agroecology as a

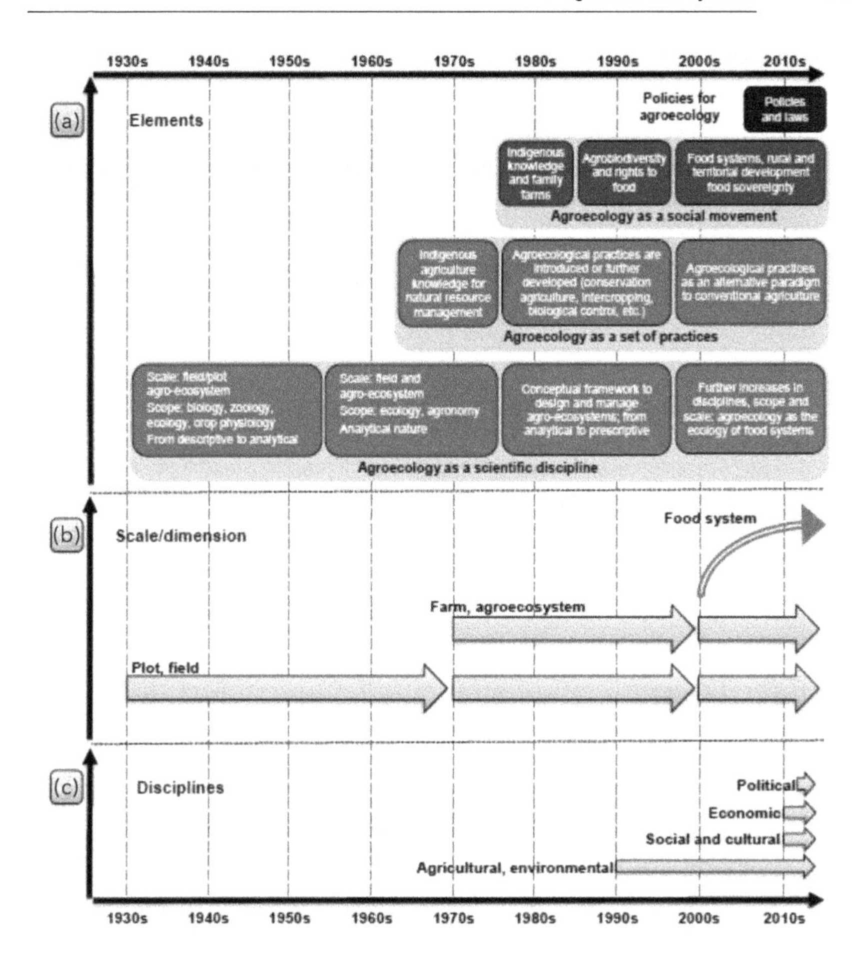

Sources: (A) adapted from Silici (2014), based on Wezel *et al.* (2009) and Wezel and Soldat (2009); (B) adapted from Wezel *et al.* (2009).

Figure 88 The evolution of agroecology 1930s–2010s. Source: Adapted from: HLPE (2019).

mechanism to support small farmers and rural communities (Byé et al. 2002; Bellon and Abreu 2006). Government technical support and extension services began promoting the implementation of agroecological agriculture (MDA 2004), and public food procurement programmes provided price premiums for agroecological produce from family farms (Grisa et al. 2011).

Commissioned by the World Bank and the FAO to chart the future of agricultural science and technology, the 2009 International Assessment of Agricultural Knowledge, Science and Technology (IAASTD) Report gave renewed impetus to agroecology and, within the next few years, the term, especially related to food security and climate adaptation, was being mentioned in 5% of agriculture-related peer-reviewed research articles (Saj et al. 2017). Agroecological agriculture was given a further boost when the FAO held the first symposium on Agroecology for Food Security and Nutrition in 2014 (FAO 2018e). In 2016, Agroecology Europe (AEEU) was founded to share knowledge between scientists working on agroecological research as a foundation of sustainable development of farming and food systems (Migliorini et al. 2020). By 2018, the FAO was prioritising agroecological agriculture as an approach that could help transform food systems and contribute to the attainment of UN Sustainable Development Goals (SDG) (FAO 2018g). Agroecology has also become part of university curriculums across most European countries, and is well established in Norway, Sweden, Germany, France, Belgium, the UK, the Netherlands, and Spain (Wezel et al. 2018). By this time, agroecological agriculture was seen as a central tenet of both re-localised/re-territorialised food systems and the global food sovereignty movement.

Figure 89 illustrates the multidimensional nature of agroecology, which brings together science, practice, as well as economic, social, and political dimensions (IPES-Food 2020).

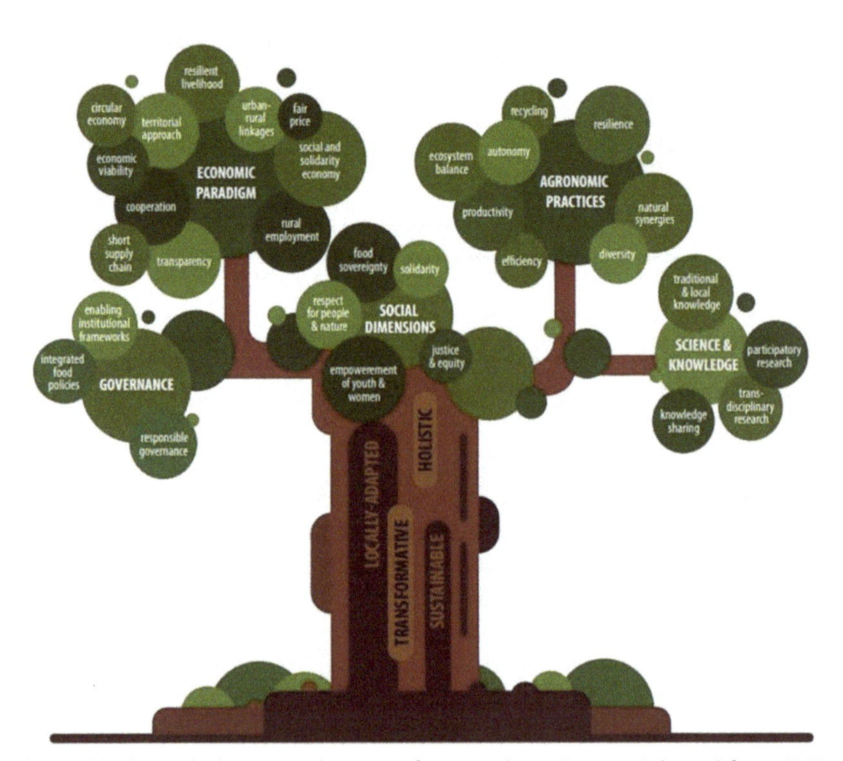

Figure 89 The multidimensional nature of agroecology. Source: Adapted from: IPES-Food (2020).

Gliessman (2007) has identified five different types/levels of agroecology, which can be differentiated by the degree to which each adapts or transforms conventional farming practices and current food systems. These levels are shown in Fig. 90.

In ascending order of transformational capacity, these levels are:

1 increase efficiency of existing input use;
2 replace conventional (synthetic) inputs with organic alternatives;
3 redesign agroecosystems to be more self-sufficient (e.g. by recombining crop and livestock production or introducing more complex rotations to optimise soil health and keep nutrients in the system);
4 develop local food supply chains that directly link farmers and consumers; and
5 redesign national and global food systems.

This section will focus on both the science and application of agroecological agriculture (primarily levels 1–3). The economic, social, and political dimensions of agroecological agriculture are dealt with in the section on food sovereignty.

Figure 91 summarises the five levels required for the transformation of industrial food systems to agroecological food systems and compares this to the FAO's ten elements of agroecological agriculture (FAO 18b), and the 13 principles of agroecological agriculture of the High-Level Panel of Experts on Food Security and Nutrition (Ewert et al. 2023).

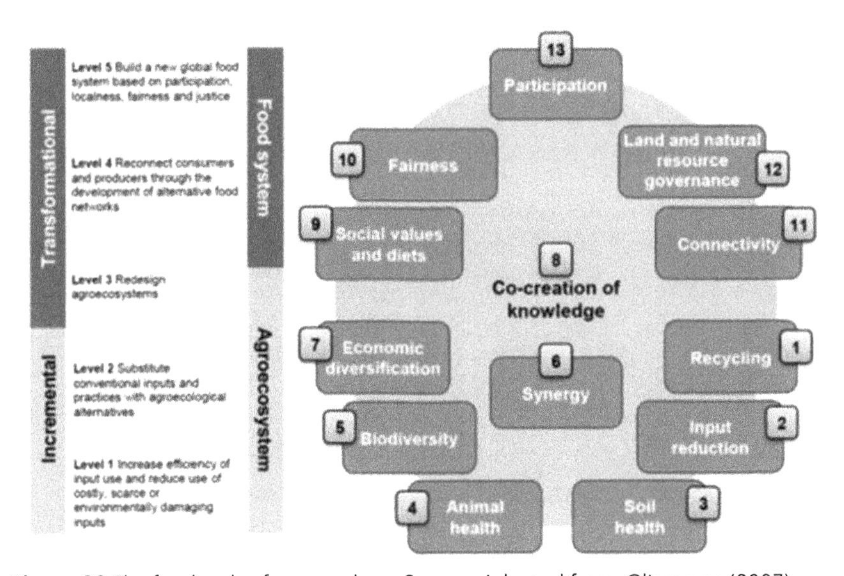

Figure 90 The five levels of agroecology. Source: Adapted from: Gliessman (2007).

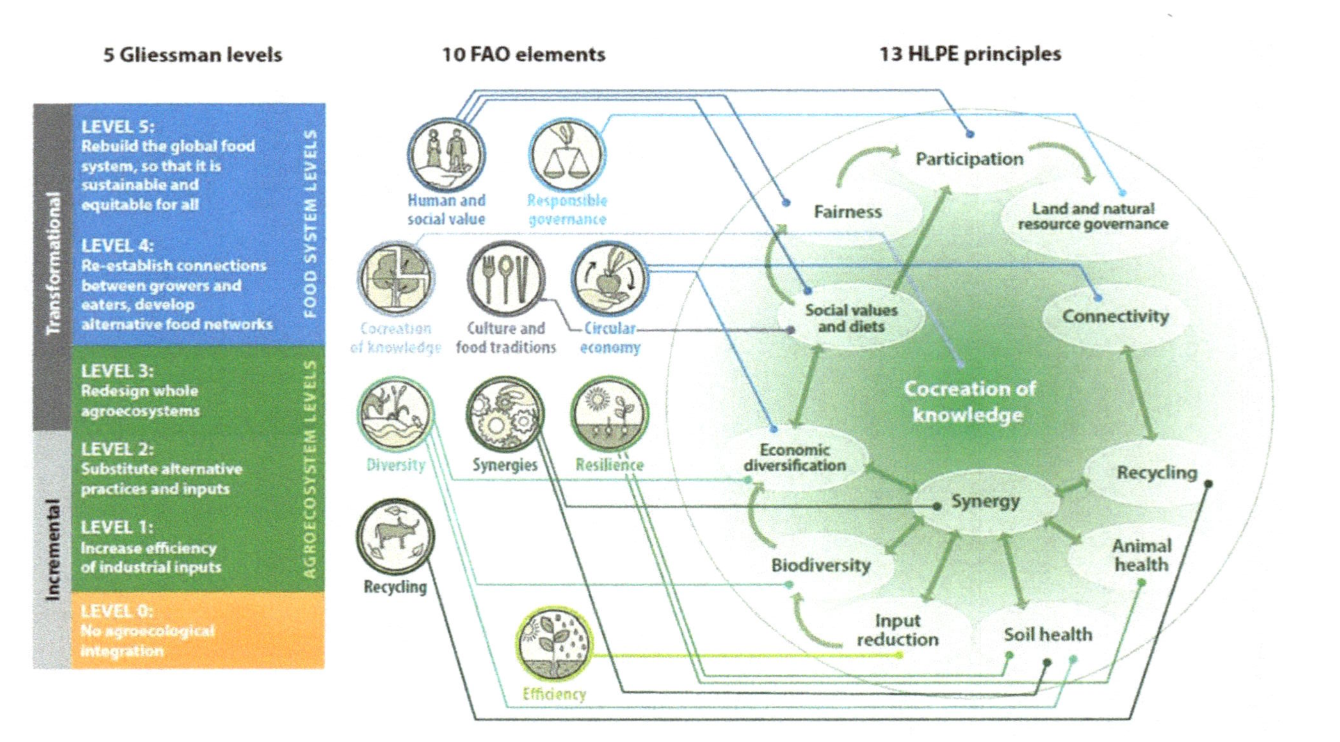

Figure 91 Levels, elements, and principles of agroecology. Source: Adapted from: Ewert et al. (2023).

Based on Vaarst et al. (2018), agroecological science and practice is based on eight principles:

- Resource recycling and minimising losses;
- Minimal external inputs;
- Contextuality;
- Resilience;
- Multifunctionality;
- Complexity and integration;
- Equity; and
- Nourishment.

These are discussed here.

- **Resource recycling and minimising losses**

Agroecological agriculture is based on the principle of circularity, which recycles organic matter and nutrients within the farm, and between farms and landscapes, and even rural/urban complexes, within relatively closed systems (Vaarst et al. 2018; Wezel et al. 2018b). The application of agroecological science in nutrient cycling includes the establishment of integrated farming systems, where, for example, the integration of dairy cows recycles nutrients and organic matter in the form of manure and urine, as well as producing valuable products such as meat and milk, and important co-by-products, such as leather. Nitrogen-fixing crops and trees are integrated into agroecological farms to increase soil fertility. Integrated aquaculture systems are another example of the recycling of nutrients. Here, farm products and wastes are repurposed as fish feed, and, in turn, pond sediments are used as crop fertilisers (Soussana et al. 2015). In integrated rural/urban systems, farmers produce food, fuel, and fibre, and, in addition to providing a ready market for farmers produce, urban centres return nutrients and organic matter in the form of food, other organic wastes, and treated human sewage (Vaarst et al. 2018).

- **Minimal external inputs**

Agroecological agriculture focusses on the use of local and renewable resources (Vaarst et al. 2018; Luna et al. 2019). The recycling of nutrients, and efforts to minimise nutrient and carbon loss, also minimise farmers' reliance on external inputs, such as synthetic fertilisers, pesticides, and seeds (Wezel et al. 2018b). Farmers rely on solar energy, biomass, and hydropower, rather than fossil fuels (Altieri and Nicholls 2005; Sarandón and Flores 2014). Indigenous knowledge is preferred over propriety knowledge used by corporations to produce synthetic fertilisers, fuel, pesticides, antibiotics, seeds (GMO and gene edited), high-tech mechanisation, and agronomic advice (Migliorini et al. 2020). Agroecological agriculture attempts to mimic and work with functional natural ecosystems rather than to dominate or even replace them (Jackson et al. 2020). It aims to promote ecological processes and services and enhance genetic diversity, which aids in nutrient recycling and biological pest control (Altieri 1995). Complex crop rotations, multi-cropping, and mixed cropping (growing several crops in the same field), and the use of biological pesticides, etc., reduce both the occurrence and severity of crop diseases (Luna et al. 2019; Kerr et al. 2021). Agroecological agriculture also focusses on the conservation and optimal use of soil, water, and genetic resources (Altieri and Nicholls 2005; Sarandón and Flores 2014). Because of these factors, other than access to and control over land, agroecological agriculture presents few obstacles to individuals or groups establishing successful smallholder farming (Levidow 2015; AFSA 2019; Isgren 2018).

However, unlike organic agriculture, agroecological agriculture does not have hard and fast rules. This means that, whilst modern technologies such as GMOs and gene-edited crops, smart phones, drones, etc., are not generally envisaged to be part of agroecological agriculture's arsenal, they are not explicitly banned, which leads to constant debates about the scope and goals of agroecology (Migliorini et al. 2020).

- **Contextuality**

Unlike industrial agriculture, agroecological agriculture is explicitly guided by local biophysical, socio-cultural, and economic conditions (FAO 2018a; Vaarst et al. 2018). Agroecological food systems are designed/built from the bottom-up to address local needs (FAO 2018a). Using primarily local knowledge and resources, farming communities create diverse and resilient food systems that aim to provide local food, fibre, and fuel needs, and maintain economic viability, whilst promoting social justice and protecting local identity and culture (Wezel et al. 2018b; Isgren and Ness 2017).

- **Resilience**

Agroecological agriculture lays claim to multifaceted resilience. Increased ecological diversity, of farmed and unfarmed areas, is claimed to lead to greater food systems resilience (Perfecto and Vandermeer 2010; Elmqvist et al. 2003; Altieri et al. 2015; Kerr et al. 2016).

- **Multifunctionality**

Multifunctionality refers to the multiple functional outcomes produced by agroecological farming. In addition to the production of food, fibre, and fuel, farmers applying agroecological principles generate a range of public goods such as carbon sequestration and biodiversity conservation (Altieri and Nicholls, 2004; Wezel et al. 2018a,b; Tittonell 2020; Vaarst et al. 2018).

- **Complexity and integration**

It is argued that the success of agroecological approaches relies on the integration of a diverse and complex array of biophysical, socio-cultural, and economic factors (Vaarst et al. 2018), which create synergistic ecological processes (Fig. 92). Examples include processes such as nutrient recycling and nitrogen fixation, biological control of pests, diseases, and livestock parasites, weed control (through crop rotations and biocontrol), biomass production, and organic matter accumulation within relatively closed agroecological systems (Bàrberi 2019).

- **Equity**

Equity and fairness, especially as they relate to knowledge and natural resource management, wealth, and value, are seen as cornerstones of agroecological food systems. Part of the impetus behind the development of agroecology has been to protect the rights and livelihoods of local communities and smallholder farmers (Conway 1987; Peeters et al. 2013; Jansen 2015; Le Mire et al. 2016;

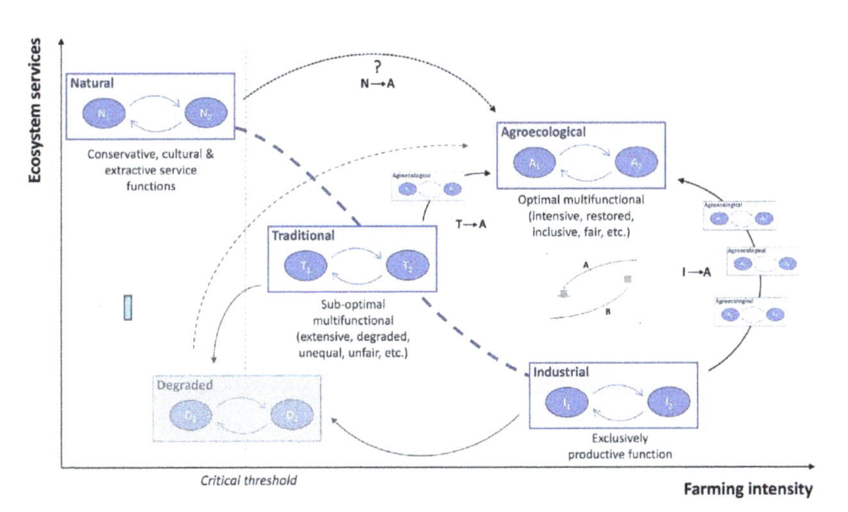

Figure 92 States and transitions in agroecosystems. Source: Adapted from: Tittonell (2020).

Wezel et al. 2016, 2018b; Astier et al. 2017; Beudou et al. 2017; Khadse et al. 2018; FAO 2014, 2018d; HLPE 2019; Nair 2014).

- **Nourishing**

Agroecological agriculture is encapsulated as nourishing both people, environment and, especially, the soil (Wezel et al. 2018; Wezel et al. 2018b; Vaarst et al. 2018).

6.4.2 Key Strengths of agroecological agriculture

Strengths of agricultural agriculture include

- stewardship of natural resources;
- climate change mitigation;
- resilience; and
- food and nutrition security.

These are discussed below.

- **Stewardship of natural resources**

The stewardship of natural resources is probably agroecological agriculture's primary strength. The diversity of farming systems, created through crop diversification, inter-cropping, mixed cropping, and increased SOM, results in significantly enhanced agrobiodiversity (IPES-Food 2018; Brandmeier et al. 2023). Broader landscape heterogeneity, containing uncropped areas and diverse farmland, further contributes to ecosystem diversity and integrity (Wanger et al. 2020; Kremen and Merenlender 2018) and to the sustainable provision of ecosystem services (such as reduced soil erosion, increased soil fertility, pest and disease management, and improved watershed management) (IPES-Food 2018; Santos et al. 2019; Shroff et al. 2021).

- **Climate change mitigation**

On a cultivated area basis, agroecological food systems are claimed to mitigate climate change (Rodale Institute 2020; Owen 2020). This is achieved in part through reduced reliance on applying synthetic inputs as well as encouraging renewable solar and hydro-energy compared to fossil fuels, significantly reducing carbon dioxide emissions. Shorter food chains also reduce food waste (Grain 2011), and the energy (generally from fossil fuels) needed to transport food, fibre, and fuel to market (Altieri et al. 2020). Due to the focus on increasing SOM, agroecological food systems also sequester larger amounts of CO_2 compared to industrial food systems (Oakland Institute 2018).

- **Resilience**

Agroecological food systems possess greater resilience to both biophysical and economic shocks. Diversified agricultural landscapes that integrate highly diverse tree-crop-livestock systems reduce vulnerability to climate shocks by spreading risk (Snapp et al. 2021). Different sowing and harvesting dates for diverse crops ensures that if one crop fails due to abiotic stresses such as lack of rainfall, others may succeed. Livestock may survive drought by browsing on the leaves of deep-rooted trees/shrubs or migrating to areas that have received rainfall. Livestock are also able to move to higher ground and other safe areas during floods. Increased SOM content due to agroecological practices creates higher soil moisture-holding capacity that buffer against erratic rainfall and allows easier adsorption of water during heavy rains, thus reducing soil erosion (ACF 2017; FAO 2015c; IPES-Food 2020; Wezel et al. 2018; Snapp et al. 2021).

In addition, the relatively closed nature of agroecological food systems, which reduces reliance on often inaccessible and expensive off-farm inputs (such as improved seed, synthetic fertilisers, pesticides, and fossil fuels), and the generally short distance to sites of food consumption, enhances the economic resilience of farmers (Gliessman et al. 2019; Migliorini et al. 2020; Mbow et al. 2019). This resilience was admirably demonstrated during the 2019-2022 COVID Pandemic and 2022 Food, Fuel, and Fertiliser Crisis (Watson 2022), where links to both input supply chains and output markets were severely disrupted (Shroff et al. 2021). Reliable local production at affordable and consistent prices provides economic and social resilience for both producers and consumers in local communities (Van der Ploeg et al. 2019). A third aspect of resilience is that communal ownership and stewardship of natural resources promoted by agroecology enhances resilience to ecosystem shocks (Holling et al. 1998; Holt-Giménez 2001; Gonzales De Molina 2012). A final factor is that, by promoting conservation and biodiversity, food and other farming outputs can be supplemented by harvesting wild produce, for example, edible forest plants (Kerr et al. 2021).

- **Food and nutrition security**

Most scientific research (~80%) supports a clear link between the adoption of diverse agroecological practices (such as integrated crop, livestock, agroforestry-based food systems, and improved soil management) and increased food and nutrition security amongst smallholder farming communities (Luna-Gonzalez and Sorensen 2018; Azupogo et al. 2019; Bisht et al. 2018; Fernandez and Méndez 2019; Jones et al. 2018; Walingo and

Ekesa 2013; Whitney et al. 2018). Improvements in the nutrition of both adults and children are generally linked to increased diversity in home gardens, as well as field crops, and increased access to meat and milk from integrated crop-livestock farming (IPES-Food 2016; Bellon et al. 2016; Hernández et al. 2017; Azupogo et al. 2019; Bisht et al. 2018; Fernandez and Méndez 2019; Jones et al. 2018; Walingo and Ekesa 2013; Whitney et al. 2018). Reduced expenditure on off-farm inputs also increased the amount of income that could be used to buy a wider range of food types from local markets (IPES-Food 2016; Frison and Clément 2020; Kerr et al. 2021). Broader health benefits are also suggested to occur from the elimination of pesticides and other hazardous synthetic chemicals, and increased consumption of local food produce rather than ultra-processed foods (Reganold and Wachter 2016).

Ultimately, the transition to agroecological food systems is suggested to improve farmer livelihoods and foster more resilient and vibrant local communities (Shroff et al. 2021). A reduced dependence on off-farm inputs generally leads to a reduced unit cost of production and higher farmer profitability (D'Annolfo et al. 2017). In an analysis of 74 case studies, Kerr et al. (2021) found that adoption of agroecological practices increased yields by an average of 61%. Farm profitability also increased in 66% of the case studies. Several case studies from both tropical and sub-tropical countries have demonstrated yield increases of nearly 50% when compared to conventional agriculture (Reganold and Wachter 2016). Highly diversified crop and livestock production systems generally spread risks and stabilise incomes (Gliessman 2007a; Johnston et al. 1995).

Close integration with local output markets, facilitated through farmers' organisations, is often linked with improved food and nutrition security. These linkages deliver predictable markets for farm outputs and often higher prices (Warner 2006; Kerr et al. 2021). According to Wezel et al. (2016), 'the emergence of new norms rooted in direct exchange, proximity, transparency, and ethical production and consumption - a shift from a global "food from nowhere regime" to a "food from somewhere regime" - has been emphasized as central to agroecological transition'. The 'solidarity economy', involving the sharing of risks and benefits between producers and consumers, lies at the core of many community-supported agriculture (CSA) schemes and other transition initiatives (Anderson et al. 2015). For some, agroecological transitions are characterised by the involvement and shared ownership of 'a great diversity of stakeholders beyond farmers and consumers, such as actors in food chains (including food processing industries and marketing operators), actors from the voluntary sector (environmental or social organisations at the community or the national level), and policymakers, funders, and implementers' (European CSA Research Group 2016).

6.4.3 Key Weaknesses of agroecological agriculture

Weaknesses include

- yield variability;
- complexity and context specificity; and
- lack of an enabling environment.

These are discussed here.

- **Yield variability**

Yield variability is one of the most important factors limiting the spread of agroecological agriculture (Bellamy and Ioris 2017; De Ponti et al. 2012; Muller et al. 2017). Yields can vary significantly depending on factors such as crop, geography, and management practices (Ponisio et al. 2015). As noted earlier, some studies have suggested adoption of agroecological practices can increase yields by 50–60% (Reganold and Wachter 2016; Kerr et al. 2021). However, other research suggests that, when compared like for like with industrial systems, yields from agroecological farms are, on average, approximately 20–25% lower (Seufert et al. 2012). In the case of the European Union, it has been estimated that adoption of agroecological agriculture could reduce food production by up to 35%.

However, in other contexts, multi-cropping, combined with well-planned crop rotations, can significantly reduce any yield gap between agroecological and industrial yields (Ponisio et al. 2015). One example of this is the historic Milpa System of Mexico and Central America. Milpa's success is based on multi-cropping of a range of complimentary food crops such as maize, legumes, and squash. Milpa fields are established on freshly cleared areas of forest or on existing arable fields. Farmers plant numerous crops in the same field, such as maize, squash, beans, melon, tomatoes, chilis, sweet potato, jícama, amaranth, and even tree crops such as avocados (Hernández et al. 2017).

Another good example is the rice-duck-fish system found in the Hani terraces of Yunnan Province, China. Here, the fish and ducks eat both weeds and pests, and improve rice growing by disturbing the soil and providing essential nutrients contained in their faeces, whilst being sheltered from the sun and predators by the rice. The farmer receives a good crop of rice as well as protein from consuming some of the fish and ducks (HLPE 2019). This mixed farming approach is estimated to be up to 7.8 times more profitable than farming hybrid rice alone (Zhang et al. 2017).

- **Complexity and context specificity**

Whilst diversity is obviously a key strength of agroecological agriculture, it can also be a significant weakness. Managing complex rotations and mixed cropping of a diverse range of crop species, livestock, and tree crops is both knowledge and labour-intensive (Frison and Clément 2020). Some fast-growing crops may, for example, smother other slower-growing crops. Failure to effectively manage crops that need, or provide, different quantities of key nutrients at different times of the year can result in potential nutrient deficiencies. If local knowledge is not readily available, farmers may require training and extension support to be successful (IPES-Food 2020). Factors such as recycling adequate amounts of biomass/compost and the management of mixed cropping, mixed varieties, complex rotations, cover crops, etc., tend to limit the physical size of farms able to transition to agroecological agriculture, as does the proximity of wild pollinators or natural predators (Garibaldi et al. 2016).

A companion of diversity is context. What works well in one location may not work in another location (Horlings and Marsden 2011). This may be due to microclimate (temperature, rainfall, and wind), soils (structure, texture, and inherent nutrient content), or pest and disease pressures, or access to markets, etc.

This complex relationship between the success of agroecological agriculture and the local biophysical, socio-cultural, and political economy of a particular locale has thwarted many attempts to scale-up successful approaches. This has been complicated further by missed opportunities to properly document key success factors (Horlings and Marsden 2011). It has been suggested that emergent work on political agroecology research uncritically idealises farmers and their communities, which tends to overstate the sustainability of the agroecological approach (Bellamy and Ioris 2017). Lastly, whilst agroecological farming is deemed highly accessible to the poor, due to reduced need for off-farm inputs, establishing a highly diverse agroecological farm/community from scratch can require significant upfront capital. For example, farmers may need to invest in fencing for livestock, dig wells for irrigation, source and purchase numerous varieties of crop seeds, and invest in storage and perhaps even primary processing equipment.

- **Lack of enabling environment**

Lack of an enabling environment has been identified as the key institutional barriers to the scaling-up and scaling-out of agroecological agriculture (Pretty and Hine 2001; FAO 2018g; Andrée et al. 2019; Bellamy and Ioris 2017). Like organic agriculture, support is required for agroecological agriculture to grow (Frison and Clément 2020). This requires policies that promote the transition to agroecological farming and revision of existing policies that may discriminate against agroecological agriculture (IPES-Food 2018).

Some of the key enabling policies include securing 'access to land, water, forests, common property resources, and seeds; policies providing access to credit; supporting urban and peri-urban agroecological production, particularly of small- and medium-sized enterprises; reorienting national and international trade policies to reverse the incentives for export-oriented monocultures; agreeing on the valuation and incorporation of externalities in national and international markets; or providing incentives for multifunctional agriculture and the provision of ecosystem services (ActionAid 2012; Anderson et al. 2015; ARC2020 2015; Ecumenical Advocacy Alliance 2012; Fitzpatrick 2015; IATP 2013; Silici 2014; Vaarst et al. 2017; van Walsum et al. 2014; Watts and Williamson 2015; Wezel et al. 2016).

The promotion of agroecological agriculture requires financial and technical support at the farm level to transition from industrial agriculture, as well as investments in state-funded research and market infrastructure (Lieblein et al. 2000, 2007a,b; IAASTD 2009a,b; Bellamy and Ioris 2017; Herrero et al. 2017; HLPE 2019; Delonge et al. 2016; Pimbert and Moeller 2018). For example, in 2013, both the USA and UK Governments allocated less than 2% of their research budgets to agroecological research (Carlisle and Miles 2013; Pimbert and Moeller 2018; Niggli and Riedel 2020).

6.4.4 The future of agroecological agriculture

There is growing global support for agroecological agriculture. Intergovernmental organisations, such as the FAO, for example, have proclaimed 'agroecology as a model for the future of agriculture' (Migliorini et al. 2020). The FAO's First and Second International Symposia on Agroecology, in 2014 and 2018, respectively, and regional events such as the 2016 FAO Regional Symposium Conference for Europe and Central Asia on agroecology for sustainable agriculture and food systems, clearly demonstrates increasing support for agroecological agriculture (Frison and Clément 2020; Migliorini et al. 2020). Agroecological agriculture is now part of the EU biodiversity strategy and the EU Farm to Fork strategy (European Commission 2020).

Many national governments, such as France, Germany, and Switzerland, have also embraced agroecology and the process of ecological intensification as their favoured food system model (de Sartre et al. 2019). Ecological intensification aspires to increase efficiency in the use of natural biophysical resources in the quest for multifunctional landscapes (Niggli et al. 2008; Bellamy and Ioris 2017).

Agroecological food and farming systems are increasingly acknowledged for their explicit contribution to achieving objectives set out in international initiatives such as the 'Paris Climate Agreement, the Strategic Plan for Biodiversity of the Convention on Biological Diversity for 2011–2020, the 2014 Rome Declaration on Nutrition, and the Decade of Action on Nutrition' (FAO 2018c,f). Groundbreaking publications, such as the IAASTD Report

(2009), the UN High Level Panel of Experts on Food Security and Nutrition - Agroecological and other innovative approaches for sustainable agriculture Report (2019), and IAASTD + 10 report (2020), continue to champion the role of agroecological agriculture in sustainable food systems (HLPE 2019; Herren Haerlin & IAASTD+10 Advisory Group 2020).

Many acknowledge the need for new technologies, such as digitalisation (Asseng and Asche 2019; Basso and Antle 2020) and crop and livestock breeding (Qaim 2020), for more productive agroecological food systems (Bellon-Maurel and Huyghe 2017; Gascuel-Odoux et al. 2022). Recent work on ecological intensification is harnessing artificial intelligence in a process known as 'computational agroecology'. This approach enables the modelling of complex systems (crops, livestock, soil types, local climate, etc.) to design diverse farming ecosystems, which increase crop and livestock yields, soil health, biodiversity, and resilience (Tittonell 2015; Gonzalez de Molina and Guzman Casodo 2017; USC 2023; Runck et al. 2023).

Chapter 7

Radical Food Systems

7.1 Introduction

This chapter focusses on the discussion of so-called Radical Food Systems, most of which have been identified as alternative food provisioning systems, which are broadly aligned with the degrowth agenda. These include movements such as food sovereignty (based on agroecological principles), slow food, permaculture, food-sharing and free food (edible landscapes and urban gardening and community-based agriculture – CBA), and urban food systems.

7.2 The degrowth agenda

7.2.1 History of degrowth

The term degrowth, or decroissance, was first coined in 1972 by the Franco-Austrian leftist intellectual and philosopher André Gorz (Perryer 2019). Gorz wrote extensively about the theory of society, political ecology, and alternatives to capitalism (Gomiero 2017). Gorz was concerned by the inherent need for growth under capitalism and its impact on the biophysical world, especially climate and natural resources (Demaria et al. 2019). His contributions were built on the ongoing work in the 1960s and early 1970s on the over-dependence of capitalist growth on fossil fuels and its negative environmental externalities and structural inequities (Markantonatou 2016).

Two parallel schools of thought emerged. In a similar vein to Gorz, one set of academics primarily focussed on the social, political, and cultural dimensions of the growth versus degrowth debate. They were driven by the desire to create development paradigms based on just and equitable socio-ecological systems (Gorz 1972; Schumacher 1973; Illich 1975; Smith et al. 2021). Given extra zeal after the 1974 food and energy crisis, Gorz's degrowth quickly became a rallying cry for left-wing French intellectuals to stimulate thought and action for democracy, justice, and equitable development in a world consumed by the unstoppable spread of capitalism (Latouche 2009; Cosme et al. 2017; Demaria 2019; Dunlap et al. 2021). In 1980, Gorz wrote 'even at zero growth, the continued consumption of scarce resources will inevitably result in exhausting them completely. The point is not to refrain from consuming more and more, but to consume less and less – there is no other way of conserving the available reserves for future generations' (Paulson 2017).

Other academics focussed on the biophysical dynamics (especially thermodynamics) of the emergent growth versus degrowth debate. Key literary contributions came from Ezra J. Mishan of the London School of Economics (LSE) in his 1967 publication entitled 'The cost of economic growth', and the landmark Club of Rome publication the 'Limits to Growth', also known as the Meadows Report (Meadows et al. 1972), which argued that current levels of population growth and mass production/consumption were not possible within the constraints of the earth's finite resources (Paulson 2017; Robra and Heikkurinen 2019; Belmonte-Ureña et al. 2021).

Another key figure was the Romanian mathematician and economist, Nicholas Georgescu-Roegen. As an advocate of organic agriculture, Georgescu-Roegen was a firm believer that 'mankind should gradually lower its population to a level that could be adequately fed only by organic agriculture' (Georgescu-Roegen 1975). Georgescu-Roegen's analysis of economic growth and biophysics also hypothesised that 'all natural resources are irreversibly degraded when put to use in economic activity' (Bonaiuti 2011; D'Alisa et al. 2015). Together, Georgescu-Roegen's 1971 book entitled *The entropy law and the economic process*, which stressed the need for new systems of accounting that merged both the biophysical and socioeconomic dimensions of development, and H. T. Odum's 1971 book entitled *Environment, power and society* initiated the new sub-discipline of ecological economics (Paulson 2017). In 1979, Jacques Grinevald and Ivo Rens published a collection of Georgescu-Roegen's essays, entitled 'Demain la decroissance: entropie-écologie-économie' or 'Tomorrow's Degrowth: Entropy-Ecology-Economy' (English translation) (Demaria et al. 2013). Both degrowth supporters and emergent ecological economists agreed on the urgent need to re-size global economic activity to fit within planetary limitations and to acknowledge that economic activity is fundamentally embedded in socio-ecological systems.

In 1976 and 1979, Leach and Pimentel, and Pimentel, respectively, published the first analyses of the energy balances of industrial agricultural systems. These analyses suggested that, instead of being more energy-efficient, industrial agriculture's reliance on fossil fuels for synthetic fertiliser, pesticide production, and mechanical traction made it increasingly less efficient (Amate and de Molina 2013). Similar studies conducted in Spain concluded that energy efficiency in agriculture reduced from 6.1 J J-1 in 1950/1951 to 1.2 J J-1 in 1977/1978 (Amate and de Molina 2013).

Some academics, such as the ecologist Paul Ehrlich (1968), Hardin (1968), and Commoner (1971, 1975), combined both the biophysical and socio-political dimensions of degrowth, linking excessive population growth to environmental destruction. Ehrlich's most important contribution was articulated through his landmark book *The population bomb* (Ehrlich 1968). Here, in neo-Malthusian form, Ehrlich postulated that 'there was not enough food to feed the world population and forecasted a 'massive famine' to occur in the 1970s' (Gomiero 2017). In a similar vein, Garrett Hardin's article 'The Tragedy of the Commons' raised concerns over the uncontrolled negative impacts of selfish individuals on the environment. Both believed that birth control or similar restrictions should be imposed by the state (Dunlap et al. 2021). On the other hand, Commoner (1971, 1975) suggested that poverty leads to population growth, and not the other way around (Gomiero 2017).

However, after memories of the 1974 energy crisis had waned, and the neoliberal intensification of capitalism had firmly taken root, the once-vibrant degrowth movement became increasingly quiet (Paulson 2017). Indeed, it was not until 2001 that French environmental activists re-invented degrowth as a slogan to launch environmentalism back onto the global political agenda (D'Alisa et al. 2015; Kallis et al. 2012; Martínez-Alier et al. 2010; Demaria 2019). By the mid-2000s, the conversation around degrowth had spread throughout much of Western Europe (as 'decrescita' in Italy and 'decrescimiento' in Spain) (Kallis et al. 2015; Latouche 2016). The topic of degrowth appeared in the influential newspaper Le Monde Diplomatique in 2003 (Latouche 2003). However, it took the Food, Fuel, and Financial Crisis of 2008 before the degrowth movement was able to make political headway, organising their first International Degrowth Conference for Ecological Sustainability and Social Equity, which took place in Paris in April 2008 (Degrowth 2022). Subsequent international degrowth conferences took place in Barcelona in 2010, Venice in 2012, Leipzig in 2014, Budapest in 2016, Malmö 2018, Manchester 2021, and The Hague 2021 (Kallis et al. 2012; Degrowth 2022; Demaria et al. 2013; Kallis et al. 2018; Sekulova et al. 2013). According to Perryer (2019), it was the first two International Conferences (in Paris 2008, and Barcelona 2010) that 'cast this radical movement onto the global stage'.

7.2.2 Principles of degrowth

Whilst the degrowth literature is exceedingly broad-church, it can be simply conceptualised as 'the voluntary and fair transition from an unsustainable growth path to a stationary and sustainable state of the economy' (O'Neill 2012). Unsustainable growth is defined as growth pathways that exceed the biophysical limits of the planet as well as those that are socially inequitable and exploitative (Germain 2017). Discussions around degrowth focus on managing people, the economy, and society in a way that is in balance with nature, protect the biosphere and operates within the planet's human carrying capacity (including humanity's generation of environmental externalities) (Fournier 2008; Jackson 2009; Martinez-Alier et al. 2010; Dunlap et al. 2021).

According to Belmonte-Ureña et al. (2021), paraphrasing Vandeventer et al. (2019), degrowth should be conceptualised 'as a radical niche innovation to the capitalist-growth regime, proposing that degrowth fundamentally seeks to undermine the hegemonic idea of equating a single measure – in the capitalist-growth regime this is growth – with well-being and pleasurable existence. Degrowth rejects the notion that there is a singular, external measure of the good life'. Kallis et al. (2015) frame degrowth as a conscious transition to a world governed by conviviality, simplicity, sharing, caring, and the commons. Political economist Serge Latouche (2009) suggests that degrowth is based on eight principles: re-evaluate, reconceptualise, restructure, redistribute, relocalise, reduce, re-use, and recycle (Paulson 2017). Whilst discussion of degrowth in food systems has been predominantly lacking (Gomiero 2017), Amate and Gonzalez de Molina (2013) suggest four additional food system degrowth strategies of reterritorialisation of production, relocalisation of markets, revegetarianisation of diet, and reseasonalisation of food. Degrowth principles overlap with concepts such as urban agriculture with its focus on local food hubs linking producers and consumers as well as ideas about building circular bio economies in which organic raw materials are used to produce food and other materials, with co-products recycled back into the system.

7.3 Food sovereignty

7.3.1 History of the food sovereignty movement

The food sovereignty movement has its roots back in the late 1950s when a growing number of indigenous landed and landless farmers and farm workers, especially across Latin America, struggled to feed their families. The pre-food sovereignty movement manifested itself in the struggles of indigenous peoples to control enough productive farmland to ensure food security (Boyer 2010). Honduras was at the centre of these struggles during the 1960s and 1970s.

Against a backdrop of economic and political instability, during the 1960s, the Honduran government attempted to ramp up export-oriented agriculture by adopting the US Government's 'economic diversification and expansion' model (Boyer 2010). As such, large tracts of primarily public lands across Honduras were enclosed to facilitate the expansion of banana and sugar cane plantations and cattle ranches, which caused uproar amongst local indigenous communities (Boyer 2010).

This public outcry quickly led to the establishment of numerous peasant-led movements campaigning for national agrarian reforms, with the Aguan Valley peasant associations at the centre. With a united voice, during the 1970s, these peasant associations, or campesinos, managed to secure reforms, which led to the establishment of, and granting of, land to cooperatives and new settlements for over 20% of the nation's landless and near-landless campesinos. These kinds of peasant associations, such as Campesino à Campesino (Farmer to Farmer) (Gliessman 2016), emerged in other neighbouring countries, such as Mexico, Guatemala, Nicaragua, and Cost Rica, and began forming strategic alliances (Boyer 2010; Edelman 2014). Aided by NGOs, such as World Neighbours, peasant farmers were taught the principles of successful agroecological farming techniques and provided with basic agricultural tools (Boyer 2010).

Whilst the initial struggles of the Campesinos may have focussed inwardly on local land and food security, during the 1980s, attention began being focussed outwardly on the global spread of industrial agriculture, and its supporting institutional infrastructure, such as the World Bank, International Monetary Fund (IMF) and their 'structural adjustment policies', the free-trade agenda of the World Trade Organisation (WTO), and farm subsidies in developed countries, as part of the Second Food Regime (Edelman 2014; McMichael 2005, 2016). Driven by increased global financialisaton in search of low-hanging profits, foreign direct investment (FDI) in large-scale land purchases as an investment asset (a practice known as 'land-grabbing'), and increasing international pressures on farmers to produce 'cheap food' using the industrial agriculture model, whilst embracing the principle of 'free trade', Campesino groups found themselves battling with their national governments to guarantee a fair living, land tenure rights, and the traditional way of life for small-holder farmers (Boyer 2010; Bernstein 2014; Edelman 2014; McMichael 2016).

At this time, land use across many Central American countries was becoming increasingly bifurcated. On the one hand, there were large areas of often foreign-owned or controlled land devoted to the industrial production of bananas, palm oil, sugar, cotton, coffee, melon, broccoli, and cashews, produced for global markets. On the other was the primarily subsistence agroecological production of traditional foods (Boyer 2010). Because of this bifurcation, peasant associations also campaigned to ensure that the profits of large-scale export-led industrial production remained within the respective countries and were not used to line the pockets of already rich corporations from the USA and Europe (Edelman 2014). Critics argued that adherence to the principles of free trade privileged 'corporate rights over state and citizen rights' (McMichael 2016). In many respects, the emerging food sovereignty movement promoted 'cooperation, diversity, and independence over the standardisation, competition, and dependence contingent in the free market' (Bernstein 2014).

It was on the back of these struggles that, in 1983, the Mexican Government proclaimed the objective of the National Food Programme (Programa Nacional de Alimentación, PRONAL) 'to achieve food sovereignty' or 'soberanía alimentaria' (Heath 1985). Here, food sovereignty encompassed not only the goal of food self-sufficiency but also that of regaining national control over the whole food system and the profits ensuing from it (Edelman 2014; McMichael 2016). It is still debated whether Mexico was the origin of the food sovereignty movement or whether, due to similar drivers, the movement spontaneously arose across several Central American countries during the same period (Esteva 1984; Heath 1985; Sanderson 1986).

La Via Campesina was founded in 1993 during an international conference in Belgium convened in response to the Uruguay Round of the General Agreement on Tariffs and Trade (GATT), which included agriculture in its discussions for the first time and which was seen as promoting a neoliberal agenda of market-driven, globalised food production (Alonso-Fradejas et al. 2015). However, it was not until 1996, during a second international conference organised by Vía Campesina in Mexico, that food sovereignty was first openly discussed by the organisation. Over 69 organisations from 37 countries attended this conference (Edelman 2014). La Via Campesina developed into an international organisation representing 148 separate organisations from 69 countries, with membership comprised of farmers, peasants, small producers, and farm workers (Huambachano 2018). Its raison d'être is to promote an alternative agroecological food model in the face of the neoliberal model of industrial agriculture and to advocate for the right of both people and nations to control their own food systems (Gliessman 2016; Holt-Giménez and Altieri 2013; Fernandez et al. 2013; Altieri and Nicholls 2012; Anderson et al. 2015; Perfecto et al. 2009; Huambachano 2018; Martinez-Torres and Rosset 2010; Altieri and Toledo 2011).

Food sovereignty received a boost of credibility when it promoted its alternative food system, based on local food production, during the World Food Summit in Rome in 1996 (Boyer 2010; Lamine 2014; Edelman 2014). La Vía Campesina issued a statement that 'Food security cannot be achieved without taking full account of those who produce food. Any discussion that ignores our contribution will fail to eradicate poverty and hunger. Food is a basic human right. This right can only be realised in a system where food sovereignty is guaranteed. Food sovereignty is the right of each nation to maintain and develop its own capacity to produce its basic foods respecting cultural and productive diversity. We have the right to produce our own food in our own territory. Food sovereignty is a precondition to genuine food security' (NGLS Roundup 1997). Indeed, it was this forum that was credited with persuading the FAO to include local food self-sufficiency in its definition of food security (Boyer 2010). Organisations such as the FAO began to champion the concept of 'food from somewhere' in contrast to the standard narrative of 'food from anywhere' as the foundation of food security.

In addition, the NGO Forum Statement to the Summit, entitled 'Profit for few or food for all' and subtitled 'Food sovereignty and security to eliminate the globalisation of hunger' (Edelman 2014), contained six key elements (NGO Forum 1996; Shaw 2007):

- Strengthening family farmers, along with local and regional food systems;
- Reversing the concentration of wealth and power through agrarian reform and establishing farmers' rights to genetic resources;
- Reorienting agricultural research, education, and extension towards an agroecological paradigm;
- Strengthening states' capacity for ensuring food security through a suspension of structural adjustment programmes, guarantees of economic and political rights, and policies to 'improve the access of poor and vulnerable people to food products and to resources for agriculture';
- Deepening the 'participation of peoples' organisations and NGOs at all levels; and
- Assuring that international law guarantees the right to food and that food sovereignty takes precedence over macroeconomic policies and trade liberalisation.

La Via Campesina has subsequently made a formal declaration that agroecological agriculture is the only solution to the food crisis (Martinez-Torres and Rosset 2010; La Via Campesina 2010; Soper 2020; Jonas 2021).

By the turn of the twenty-first century, the cause of food sovereignty advanced even further with the establishment of the International Planning Committee (IPC) for Food Sovereignty in 2002 (Edelman 2014). In 2007, in Nyéléni, Mali, La Vía Campesina hosted the first International Forum for Food Sovereignty, which attracted 500 delegates from more than 80 countries (Huambachano 2018). One of its major outputs was the Declaration of Nyéléni 2007, which asserted, 'Food sovereignty is the right of peoples to healthy and culturally appropriate food produced through ecologically sound and sustainable methods, and their right to define their own food and agriculture systems' (Huambachano 2018; Gliessman et al. 2019). Nyéléni 2007 established seven principles of food sovereignty. These are

- food as a basic human right;
- the need for agrarian reform;
- protection of natural resources;
- reorganisation of food trade to support local food production;
- reduction of multinational concentration of power;
- fostering of peace; and
- increasing democratic control of the food system.

The global food crisis of 2008 provided a further impetus since it highlighted the impact of food export bans imposed by major food-producing countries on food availability, accessibility, and affordability amongst poorer importing countries (McMichael 2013). This persuaded the FAO in 2009 to reform its Committee on World Food Security (CWFS) to include input from civil society (McKeon 2015).

By the early 2000s, grass-roots food sovereignty groups were springing up across large swathes of Latin America (including Ecuador and Brazil), sub-Saharan Africa (SSA) (Wilson 2011), and Asia (People's Coalition on Food Sovereignty) (PCFS 2007). By the 2010s, the Food Sovereignty movement had established itself in many parts of the developed world. For example, Sedgwick Town, in the USA, passed a Food Sovereignty ordinance in 2011, with towns across a further seven US states (Vermont, Massachusetts, Georgia, North Carolina, Utah, Wyoming, and Montana) passing similar food sovereignty ordinances (Fernandez et al. 2013). Food sovereignty groups are increasingly active across most EU countries. Food sovereignty has been integrated into district and even national constitutions in countries such as Venezuela, Senegal, Mali, Nicaragua, Ecuador, Nepal, and Bolivia (Beauregard 2009; Gascón 2010; Muñoz 2010; Beuchelt and Virchow 2012; Field and Bell 2013).

In 2015, intended as a follow-up to the 2007 International Forum for Food Sovereignty, Nyéléni, in Mali, hosted the first International Forum on Agroecology (IFA 2015). The Forum's final declaration embraced agroecology as a social and political movement, one that was deeply intertwined with the goal of food sovereignty (Wezel et al. 2009; Lambek et al. 2014; Gliessman et al. 2019).

7.3.2 Principles of the food sovereignty movement

Food sovereignty is guided by 15 principles as follows:

- Food sovereignty is based on peasant agriculture and predominantly small family farms. Peasant agriculture is diverse, dignified, and self-sustaining, and drawing upon the traditional cultural practices of both men and women farmers, pastoralists, and fisherfolk. It is claimed that the traditional peasant food system model can meet the food requirements of the growing global population whilst providing social, economic, and political sustainability (Bernstein 2014; Edelman 2014; McMichael 2014; Robbins 2015; Alonso-Fradejas et al. 2015; Peano et al. 2020).

- Food sovereignty is built on autonomy, namely, it relies on relatively closed ecological systems that sustainably conserve and manage land, seeds, soil, biodiversity, water, local knowledge, etc. Food sovereign food systems are predominantly independent of external inputs such as synthetic fertilisers and pesticides, and modern hybrid seeds (especially GMO and gene-edited) (Bernstein 2014; Alonso-Fradejas et al. 2015).
- Food sovereignty ensures access and control over the use of the means of production. In other words, unlike the industrial food system, where corporations control the price and availability of inputs, and determine prices for output, in food sovereign food systems, small-holder farmers, and their communities maintain control over their lands (secure tenure), the inputs (predominantly natural) required to produce food, any value addition that takes place, and marketing of final or intermediate products (Edelman 2014; Bernstein 2014; Alonso-Fradejas et al. 2015; Peano et al. 2020).
- Food sovereignty promotes highly diversified small-holder food systems, which may include up to 12 crops grown on the same farm, at the same time, in addition to livestock and trees (Alonso-Fradejas et al. 2015; Peano et al. 2020).
- Food sovereignty is based on the provisioning of food, fibre, and fuel outputs to local and domestic markets for fair (negotiated) prices before any surplus outputs are sold onto regional or international markets (Pimbert 2008; Boyer 2010; Edelman 2014; Robbins 2015; Peano et al. 2020; Soper 2020). Food sovereignty maintains the rights of sovereign governments to protect farmers from the dumping of cheap food imports and the effects imports can have on the food security, economic stability/volatility, and dependency of local food systems created by global commodity markets (Edelman 2005; van der Ploeg 2014; Alonso-Fradejas et al. 2015). Food sovereignty reserves the right of any sovereign government to support small-holder production 'via subsidised farm inputs, technical assistance, marketing assistance, public procurement, and price floors for basic commodities such as corn, rice, bananas, and milk to protect producers and consumers from volatile price fluctuations' (El Diario 2008; El Universo 2008).
- Food sovereignty secures the provision of local high-quality, nutritious, diverse, and culturally appropriate food, and avoids both the economic and environmental costs associated with long (international) supply chains

(Altieri and Toledo 2011; Bernstein 2014; Alonso-Fradejas et al. 2015; Peano et al. 2020).

- Food sovereignty ensures the use of appropriate technologies that are environmentally sustainable and climate-sensitive, controlled by peasant men and women, and which integrate both traditional and modern knowledge. Technologies such as hybrid and GMO crop varieties, synthetic fertilisers, and pesticides are excluded (Edelman 2014; Alonso-Fradejas et al. 2015).
- Food sovereignty promotes the use of local indigenous knowledge, which helps to maintain traditional production methods, conserves local livestock breeds, crop types, and cultivars, and local biodiversity (Edelman 2014; Alonso-Fradejas et al. 2015, Antonelli 2023).
- Food sovereignty promotes agroecological agriculture, which is deemed the most sustainable form of agricultural production for local food systems. Agroecology is seen as both economically sustainable and socially acceptable and protects and conserves biodiversity and the natural environment (Pimbert 2008; Boyer 2010; Edelman 2014; van der Ploeg 2014; Alonso-Fradejas et al. 2015; Robbins 2015; Soper 2020; Peano et al. 2020).
- Food sovereignty is distinctly embedded in a specific territory or geographic space, where farmers and their communities have existed in harmony with nature for generations (Alonso-Fradejas et al. 2015).
- Food sovereignty acknowledges peasants and small-holder farmers as the pillar of the local economy, providing employment for rural labourers and providing the economic lifeblood and socio-political stability for local communities, which, in turn, strengthens national economic development (Pimbert 2008; Alonso-Fradejas et al. 2015).
- Food sovereignty is based on cooperation and not competition, in local arrangements where all people (men, women, old and young, rich, or poor, and all ethnicities) are treated equally and are free from oppression (Pimbert 2008; Boyer 2010; Edelman 2014; Bernstein 2014; van der Ploeg 2014; Alonso-Fradejas et al. 2015; Peano et al. 2020).
- Food sovereignty is built on direct linkages between both the producers and consumers of food (Edelman 2014; van der Ploeg 2014; Alonso-Fradejas et al. 2015).
- Food sovereignty promotes health care, education, adequate incomes, and dignified life for all (van der Ploeg 2014; Alonso-Fradejas et al. 2015).
- Food sovereignty promotes the coordination of autonomous peasant organisations able to benefit directly, and fully, from the fruits of their labour and control their own destinies.

7.3.3 Success of the food sovereignty movement

Whilst the IAASTD Report of 2009 and the FAO's International Symposium on Agroecology for Food Security and Nutrition in 2014 launched the science of agroecology onto the global stage, the food sovereignty movement's International Forum on Agroecology in 2015 lent significant credibility to the grassroots application of agroecological agriculture by millions of smallholder and peasant farmers. Several studies have demonstrated that agroecological intercropping and multi-cropping can lead to higher yields per hectare, higher profitability, and greater ecological sustainability when compared to mono-cropping industrial agricultural production (Altieri 2009; Chappell and LaValle 2011; Perfecto et al. 2009). Additionally, the application of agroecological agriculture has been linked to a so-called ecological revolution, bringing the relationship between society and nature into balance (Martinez-Alier 2011). The practice of agroecological agriculture has also been promoted by the food sovereignty movement as a proven way to reverse climate change, claiming that peasants can cool down the planet whilst providing balanced diets for a growing global population (Martinez-Alier 2011; Van der Ploeg 2014).

In countries such as Ecuador, Bolivia, and Brazil, the growth of agroecological agriculture, and the socio-political, economic, and ecological, arrangements, at both state and national levels, which evolved along with it, have created relatively stable alternative food system models for other countries to potentially emulate (Holt Giménez and Altieri 2013). These models gained greater legitimacy during the recent food crises of 2008/2009, 2011/2012 (McMichael 2016), and 2022. In these geographies, governments have enacted both protectionist and redistributive policies to protect peasant livelihoods, enhance national economic growth, and protect/conserve habitats and biodiversity (McMichael 2014; Hoey and Sponseller 2018; Gliessman et al. 2019; Soper 2020). For example, food sovereign governments have enacted laws to prevent the dumping of cheap, often highly subsidised, food commodities in their home markets and resisted market-based land grabs and land reforms (Rosset et al. 2006; Borras and Franco 2012).

7.3.4 Challenges faced by food sovereignty movement

There are several weak, if not contradictory, propositions contained within the doctrine of the food sovereignty movement, such as

- the oversimplification of binaries between industrial and peasant agriculture;
- its claims to be able to feed the world;

- congruence between rural and urban needs; and
- both the scope and form of food sovereignty.

On the first proposition, peasant farmers are unrealistically, almost romantically, portrayed as the bastions of sustainable food production, fighting for survival against the spread of industrial agriculture, with its ensuing environmental destruction and capitalist tyranny (Soper 2020; Van der Ploeg 2014). However, the truth is much more complex and fluid. For example, across the developed world, many farmers have turned their backs on industrial agriculture and established organic, agroecological, and permaculture-based food systems. It can also be argued that agroecology as both a science and practice was developed in the West and exported to developing countries (Clark 2017)

Conversely, in many developing countries, peasant and small-scale agriculture is often highly heterogeneous (Altieri and Toledo 2011). Indeed, many peasant farmers are commercially focussed, willingly use synthetic agrochemicals such as pesticides, have scant regard for their ecological surroundings, and often export food and fibre products in search of profit (Bernstein 2014). This heterogeneity calls into question the wholesale acceptance and adoption of agroecological farming by peasants, which is one of the essentialisms of food sovereignty. In many cases, commercially focussed peasants and small-holder farmers have sought to resist the adoption of agroecological agriculture (Soper 2020).

This heterogeneity also calls into question the exclusive focus of food sovereignty on domestic markets. Indeed, the whole concept of ecologically, socially, and economically sustainable peasant production for local markets can often feel somewhat contrived. Several commentators have highlighted the potentially protectionist nature of food sovereignty. The original goals of preserving a traditional agrarian way of life and protecting livelihoods, often by enacting protectionist trade policies, have morphed into contemporarily appealing narratives, such as the defence of Mother Earth, cooling of the planet, eco-friendly agroecology, local production, and short supply chains (Field and Bell 2013; Martínez-Torres and Rosset 2014; Brass 2015; Robbins 2015). However, whilst indigenous communities and some peasant and smallholder farmers produce exclusively for their households and the local community, millions of commercially focussed peasant and smallholder farmers produce almost exclusively for international markets (Burnett and Murphy 2014; Edelman 2014). In the words of Edelman (2014), 'What will become of the millions of smallholders who depend for their livelihoods on export production and whose incomes would plummet if they were required to switch, say, from cacao or African palm production to cassava and maize?'.

The reliance on low input and predominantly self-provisioning peasant and small-holder farmers to feed the growing world population has also been called into question (Bernstein 2014). An analysis of farm production across 55 countries and 154 crops demonstrated that, whilst farms under 2 ha occupy only 24% of farmed land, they only produce between 30% and 34% of the world's food (Ricciardi et al. 2018). In other words, whilst acknowledging that diverse, mixed, and multi-cropping systems, dependent on high labour intensities per hectare, produce more than industrial mono-cropping, smallholder farmers may in practice still only be able to feed a fraction of the 70–80% of humanity it has been claimed they could supply (FAO 2014).

Whilst food sovereignty advocates talk of equitable arrangements between rural and urban communities (Altieri 2010), concerns abound about how these arrangements work in practice (Bernstein 2014). In addition to the need for financial support for peasant and small-holder agriculture, if diets were to be restricted to locally available seasonal foods, urban populations would lose access to many imported exotic and processed foods, potentially impacting food choice, nutrition, and price (Edelman 2014).

On an even more fundamental basis, food sovereignty advocates are remarkably vague regarding 'who or what is "the sovereign" in "food sovereignty"'. Namely, does it refer to the nation-state, a district or region, a specific community, or to individuals themselves (Edelman 2014)? Most, but not all, analysts focus on the nation-state, i.e. the sovereign state, as the epicentre of food sovereignty. For some, this is problematic, as peasant and small-holder farmer-based food sovereignty movements generally mistrust their own governments and are often at loggerheads with them (Edelman 2014). Ceding enough power to the government to control food production transportation, processing, retail, and consumption would also most likely result in unwieldy and authoritarian governments, rather than institutions that would promote equity, democracy, and freedom of choice (Edelman 2014).

7.3.5 Future of food sovereignty movement

Whilst the food sovereignty movement has significant appeal as a counterweight to the neoliberal paradigm of free-market and globalised food production, it lacks conceptual clarity and successful working models remain elusive. Where proponents point to practical exemplars of food sovereignty, they often use countries that are poor and sparsely populated, or, in the case of Brazil, a country where food sovereignty is only practiced in discrete areas of the country and juxtaposed to some of the most industrial forms of agricultural production on the planet. These hybrid models generally fall short of the 'complete overhaul of food and farming - along with associated changes in values' called for by Hassanein (2003).

Many commentators are convinced that food sovereignty needs to grow from discrete farming communities practicing agroecological agriculture into coherent territories, which may be defined by discrete biophysical or political boundaries. As they grow, they develop the food sovereign enabling socio-cultural, economic, and political infrastructures required for sustainability (IPES-Food and Frison 2016). Conversely, others are convinced that industrial farmers can break away from the seed, chemical, and debt dependencies spurred by capitalism to re-establish functional and vibrant rural communities based on the principles of food sovereignty (Blesh and Wolf 2014). This concept is not too dissimilar from the funding of structural transformation in the EU's less-favoured areas and areas with environmental restrictions (Bryden and Hart 2001). Here, rural communities receive basic services and support for the conservation of traditional heritage, support for adding value and marketing of quality local farm produce, and income diversification, including conservation of biodiversity, landscapes, and watersheds, as well as the development of tourism and crafts (Bryden and Hart 2001).

7.4 Slow food movement

7.4.1 History

Led by Carlo Petrini, the Slow Food International movement was initiated by a group of activists in Rome in 1986. Slow Food's aim was to 'defend regional traditions, good food, gastronomic pleasure and a slow pace of life' (Slow Food 2022). Slow Food International aimed to prevent the loss of traditional cuisine and food diversity, and was designed as a rejection of industrial agriculture, fast food, and homogenised 'food from nowhere' (Irving and Ceriani 2013; Slow Food International 2016). Three years later, in 1989, the international slow food movement was officially founded in Paris, with an accompanying manifesto.

During the 1990s, initial national chapters of the slow food movement were established (Germany in 1992; Switzerland in 1993), as well as the 'Ark of Taste' in 1996, a reference work that aimed to 'rediscover, catalogue, describe and publicize forgotten foods' (Slow Food 2022). A year later, in 1997, the slow food movement established the Slow Food Foundation for Biodiversity, which aims to 'defend local food traditions, protect food communities, preserve food biodiversity and promote quality artisanal products, with an increasing focus on the global south' (Slow Food 2022).

In 2004 Slow Food launched its Terra Madre network, which was designed to enhance the voice (agency) for small-holder farming communities, especially in the developing world, that produce food, protect the environment, and sustain communities and traditions (Slow Food 2022). However, it was not until 2012, during the UN Permanent Forum for Indigenous Issues, that the Slow

Food president embraced more explicit support for food sovereignty. During this pivotal meeting, Carlo Petrini stated:

> *'For twenty-five years now, Slow Food has sought to preserve agricultural and food biodiversity as a tool for ensuring a future for our planet and humanity as a whole,' ...... 'It is necessary to point out, though, that it would be senseless to defend biodiversity without also defending the cultural diversity of peoples and their right to govern their own territories. The right of peoples to have control over their land, to grow food, to hunt, fish, and gather according to their own needs and decisions, is inalienable. This diversity is the greatest creative force on earth, the only condition possible for the maintenance and transmission of an outstanding heritage of knowledge to future generations.' (Slow Food 2022)*

Since its initiation, Slow Food evolved into a global movement championing a broad food systems and food sovereignty approach (Duncan and Pascucci 2017).

7.4.2 Principles of the slow food movement

Slow Food's manifesto reads thus:,

> *'Born and nurtured under the sign of Industrialization, this century first invented the machine and then modelled its lifestyle after it. Speed became our shackles. We fell prey to the same virus: 'the fast life' that fractures our customs and assails us even in our own homes, forcing us to ingest 'fast-food'. Homo sapiens must regain wisdom and liberate itself from the 'velocity' that is propelling it on the road to extinction. Let us defend ourselves against the universal madness of 'the fast life' with tranquil material pleasure. Against those - or, rather, the vast majority - who confuse efficiency with frenzy, we propose the vaccine of an adequate portion of sensual gourmandise pleasures, to be taken with slow and prolonged enjoyment. Appropriately, we will start in the kitchen, with Slow Food. To escape the tediousness of 'fast-food', let us rediscover the rich varieties and aromas of local cuisines. In the name of productivity, the 'fast life' has changed our lifestyle and now threatens our environment and our land (and city) scapes. Slow Food is the alternative, the avant-garde's riposte. Real culture is here to be found. First of all, we can begin by cultivating taste, rather than impoverishing it, by stimulating progress, by encouraging international exchange programs, by endorsing worthwhile projects, by advocating historical food culture, and by defending old-fashioned food traditions. Slow Food assures us of a better-quality lifestyle. With a snail purposely chosen as its patron and symbol, it is an idea and a way of life that needs much sure but steady support.' (Slow Food 1989)*

The slow food movement is based on three core principles:

- GOOD: a fresh and flavoursome seasonal diet that satisfies the senses and is part of our local culture;

- CLEAN: food production and consumption that does not harm the environment, animal welfare, or our health; and
- FAIR: accessible prices for consumers and fair conditions and pay for small-scale producers (Slow Food 2022).

Guided by the three priorities outlined above, the slow food movement is built around three priorities (see Fig. 93):

- Defence of biological and cultural diversity;
- Influencing policies, both public and private; and
- Education, inspiration, and mobilisation of citizens.

As it has developed, the slow food movement has championed several key social and environmental causes, such as family farming, agroecological farming, conservation of biodiversity, anti-land grabbing, and reducing food waste. Proponents of the slow food movement argue that family farms, especially smallholder and peasant farmers, need to play a central role in counteracting the social and environmental problems brought about by industrial agriculture (Peano et al. 2020; Slow Food 2022). They are concerned about the economic, social, and environmental challenges faced by these farmers and, just like the food sovereignty movement, see them as a better investment in a more sustainable future. Specifically, they see these farmers as the most efficient and effective way of 'preserving biodiversity, eradicating hunger, improving health, ensuring food security, maintaining rural (and urban) livelihoods, managing natural resources, and protecting the environment' (Slow Food 2022). According to Slow Food, in addition to producing a range of diverse and quality foods and being the social and economic engines of the rural economy, these farmers possess indigenous knowledge that endows them with the skills and tools to sustainably manage soil, water, biodiversity, and environmental conservation. However, slow food proponents acknowledge the vulnerability of these farmers and stress the acute need to support them (Slow Food 2022).

Agroecological farming and the conservation of agro-biodiversity and ecosystem services are pivotal to Slow Food's support for family farming (Peano et al. 2020). Even more central is Slow Food's aversion to GMOs, especially the control over seeds exerted by global corporations. GMOs are seen as 'unreliable from a scientific point of view, inefficient in economic terms, and environmentally unsustainable. Little is known about them from a health perspective, and from a technical standpoint, they are obsolete. They have a severe social impact, threatening traditional food cultures and the livelihoods of small-scale farmers' (Slow Food 2022). In addition to its stance on GMOs, the slow food movement actively supports the use of Participatory Guarantee

Systems (PGS), which assures the quality of local produce without the need for full-blown third-party organic certification. Additionally, the slow food movement adds a social dimension to organic agriculture, aiming to ensure fair incomes and working conditions for farmers and farmworkers, whilst also promoting crop and livestock diversity and local markets (Slow Food 2022).

The slow food movement has also made a determined stand against 'land grabbing' by large (often foreign) corporations, buying up large areas of land either as an investment or for productive use, often using intensive production methods at scale to produce major agricultural commodities for export. Smallholders, especially peasants and indigenous peoples, particularly those without secure tenure on public lands, are highly vulnerable to land grabs. In 2010, Slow Food launched a global campaign to try to prevent further land grabbing, which involved direct support of affected communities as well as a global awareness campaign (Slow Food 2022).

As part of its direct support, Slow Food launched two key initiatives (Slow Food 2022):

- The Presidia producers project; and
- The Gardens in Africa communities project.

The Presidia producers project is essentially twofold. First, it focusses on supporting small-holder farmers who are committed to preserving indigenous breeds of livestock and crops, as well as the farming systems of which they are a fundamental part, whilst also maintaining traditional husbandry techniques and keeping local knowledge alive (Slow Food 2021). Secondly, the project assists in developing sustainable and profitable demand-driven markets for the high-quality outputs that these farmers produce (Peano et al. 2014).

Gardens in Africa are schools and communities where gardens have been established to grow good, clean, affordable food. The objective of these gardens is to raise awareness about the importance of fresh, healthy, and diverse local foods, especially amongst youth. Training is provided to garden leaders to advocate for the conservation of these foods and the farmers who provide them (Slow Food 2022). The slow food movement is also dedicated to reducing food waste, especially post-harvest (initial processing and storage) in developing countries, as well as at the retail and consumption stages in developed countries (Slow Food 2022) .

7.4.3 Success of the slow food movement

During the past 30 years, the slow food movement has gradually expanded from its humble beginnings to over 100 000 members and 1 000 000 supporters in 160 countries (see Fig. 94 below). Slow Food's Gardens in Africa initiative has

Figure 93 The three priorities of Slow Food. Source: Adapted from: Slow Food (2021).

Figure 94 Slow Food Worldwide. Source: Adapted from: Slow Food (2022).

established more than 3000 gardens (with a target to establish 10 000), and the Ark of Taste initiative has prevented the loss of nearly 5000 culinary dishes, crop varieties, and livestock breeds (Slow Food 2022).

In a gesture of defiance, the slow food movement, along with other like-minded organisations, including most food sovereignty organisations, boycotted the UN Food Systems Summit in 2021. They opted instead to host a counter-summit given the perceived railroading of the UN Summit by global agribusinesses and others. The UN Special Envoy for the Food Systems Summit, Agnes Kalibata, met with significant criticism because of the perceived close ties between agribusiness and AGRA, the institution of which she was President (AGRA is an NGO set up to promote agricultural development in Africa). The UN secretary-general received a letter signed by 176 organisations from 83 countries, which suggested that Dr Kalibata's appointment was 'a deliberate attempt to silence the farmers of the world" and signalled the "direction the summit would take' (Guardian 2020b). It was claimed that up to 9000 people joined the counter-summit (Slow Food 2021).

Slow Food also made its presence at the UN international summit on climate change, COP26, through its Slow Food Climate Action campaign. Slow Food's dogged determination led to the signing of a Slow Food Climate Declaration by more than 2000 representatives of organisations from 103 countries (Slow Food 2021). Other policy successes include the collection of over 1 200 000 signatures on a petition calling for an end to pesticides in Europe (Slow Food 2021).

7.5 Permaculture

7.5.1 History of permaculture

The term permaculture, as opposed to shifting cultivation, was officially coined by Bill Mollison and David Holmgren in 1978 (Ferguson and Taylor Lovell 2014). However, several pioneers have been exploring experimental approaches to develop sustainable permanent agriculture or permanent culture since the early 1900s (King 1911; Smith 1929; Howard 1940; Ulbrich and Pahl-Wostl 2019). For example, as early as 1929, Joseph Russell Smith published one of the first treatises on integrated agricultural systems entitled *Tree Crops: A Permanent Agriculture*. Others, such as Brand and Fukuoka, pioneered no-till gardening and no-till orchards (Hart 1996). During the mid-1900s, Yeomans developed integrated approaches for holistic water, soil, and tree management (Yeomans 1954, 1958, 1971, 1981).

However, permaculture's holistic systems approach to land management was most strongly influenced by H. T. Odum, the younger brother of Eugene Odum, who helped develop the agroecosystem concept (Holmgren 1992). H. T. Odum was credited with developing early theories on energy flows within ecosystems, especially aspects around ecological engineering, functional relationships, self-organising systems, and species selection, which shaped the aspirations of permaculturists (Odum 1994; Mollison and Holmgren 1978; Mollison 1988; Holmgren 2004; Hemenway 2009). During the 1980s and 1990s, permaculture spread throughout Asia and Central America.

7.5.2 Definition and principles of permaculture

Whilst there is no universally agreed definition, Mollison and Holmgren originally defined permaculture as 'an integrated, evolving system of perennial or self-perpetuating plant and animal species useful to man ... in essence, a complete agricultural ecosystem, modelled on existing but simpler examples' (Mollison and Holmgren 1978). This later morphed into 'Permaculture ... is the conscious design and maintenance of agriculturally productive ecosystems which have the diversity, stability, and resilience of natural ecosystems' (Ferguson and Taylor Lovell 2014).

The central tenets of permaculture are (EEA 2018):

- people care (equity and equality for all);
- fair shares (limiting population growth and living within planetary boundaries); and
- earth care (environmental stewardship).

Permaculture is also a spiritual and self-healing journey, which facilitates a reconnection between people and the Earth (Ulbrich and Pahl-Wostl 2019). Recognising the intrinsic connection between food, water, and energy, permaculture is based on a set of core principles that lead to the purposeful design and optimisation of socio-ecological systems, which mimic natural processes to deliver food, fibre, and fuel (Holmgren 2002; Ferguson and Taylor Lovell 2014; Jackson et al. 2020). Figure 95 illustrates the 12 principles of permaculture as laid down by Holmgren (2002).

Observe and interact - According to Holmgren, before acting, practitioners should allocate sufficient time to observe nature, such as micro-climate, soils, flora and fauna, ecosystem relationships, as well as surrounding human and ecological context, etc., before attempting to design permaculture solutions (Holmgren 2002; Ferguson and Taylor Lovell 2014).

Catch and store energy - The design of permaculture solutions should catch and store energy when it is abundant so that it can be used when needed. This primarily refers to the capture of sunlight, during summer months or times of good rainfall, and storing the energy in biomass, or designing dwellings that capture and store the heat of the sun (Holmgren 2002).

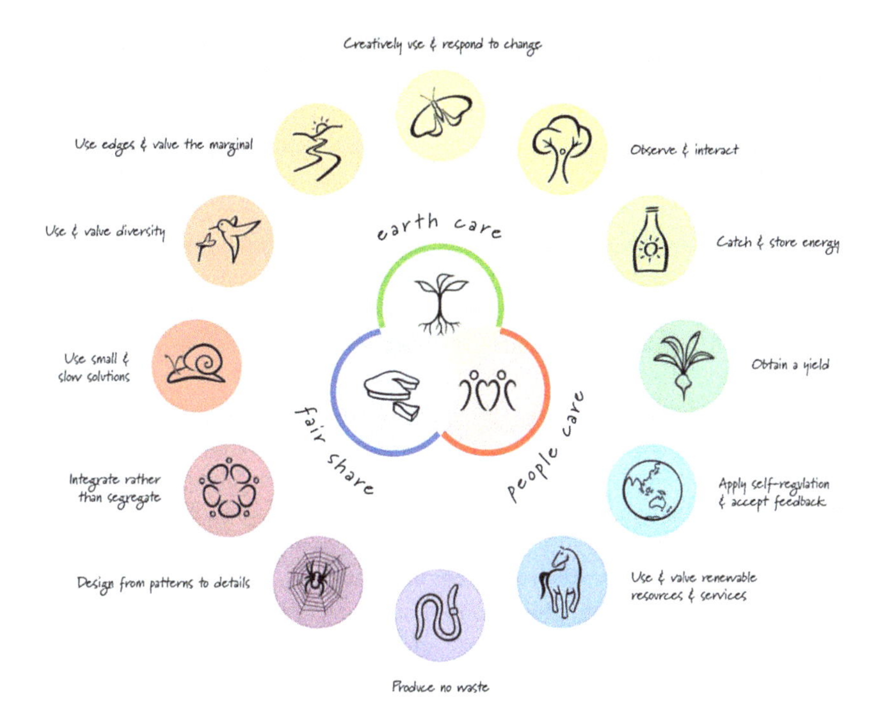

Figure 95 The 12 principles of permaculture. Source: Adapted from: Holmgren (2002).

Obtain a yield - The practice of permaculture should always generate a yield of some kind, which is basically a reward for the energy that people have invested (Holmgren 2002). When thinking about yield, permaculturists work in three dimensions or seven layers. In classic permaculture designs, layer 1 the canopy, includes trees to be used for timber and fuel, or they might be leguminous and aid in the fixing of atmospheric nitrogen. In smaller-scale designs, nut trees, or tall fruit trees such as cherry, may form the highest canopy (Long 2017). In all cases, timber trees are thinned to allow light to penetrate. Trees also provide leaf litter for composting and mulching. Beneath the high canopy, the understorey trees, layer 2, make use of the light that penetrates the high canopy. This layer is dominated by fruit trees such as apple, apricot, plum, peach, pear, etc. (Long 2017). Beneath the understorey layer, lies the shrub layer, layer 3. This layer generally consists of currant and berry bushes, such as red and blackcurrants, buckthorn, gooseberry, blackberry, and raspberry (Long 2017). Next is the herbaceous layer, layer 4. This layer generally consists of culinary and medicinal herbs, 'artichokes, asparagus, beans, beets, sunflowers, and brassicas as well as plants that provide habitats and food for beneficial insects' (Long 2017). Then comes layer 5, the soil surface and ground cover layer, which often overlaps with layer 4. Crops in this layer either contribute directly to reducing soil erosion or increase soil nutrients, especially nitrogen, and soil organic matter. Plants in this layer might include mint, lemon balm, strawberries, valerian, borage, yarrow, absinthium, and oregano (Long 2017). Layer 6 is the rhizosphere, or root zone. In addition to crops that produce yield underground, or partially in the ground, such as garlic, onions, potatoes, and sweet potatoes (Long 2017), this layer includes plant roots, fungi (especially mycorrhiza), insects, nematodes, worms, etc. The final layer is the vertical layer 7, which integrates many, and sometimes all layers. In this layer, climbing plants, such as runner beans, hops, grapes, kiwi, and vines, begin in the rhizosphere and grow up through the various layers towards the light (Long 2017).

Apply self-regulation and accept feedback - This principle works on the premise that natural systems constantly adjust due to negative and positive feedback loops, linked to changes in day length and light quality, temperature, rainfall, nutrients, pH, pests, diseases, etc. A permaculturist should observe and then change activities based on this feedback. Feedback is also taken to mean verbal feedback/advice from fellow permaculturists (Holmgren 2002; Ferguson and Taylor Lovell 2014).

Use and value renewable resources and services - This principle focusses on making the best use of nature's abundance, especially increasing reliance on natural and renewable resources, and reducing dependence on non-renewable resources (Holmgren 2002). Rainwater harvesting from buildings and storm drains, and in-field channelling, through the Keyline design, is a key practice of permaculture. Nutrient recycling is another key example of valuing natural resources. In permaculture, nutrients are recycled through the production of composts made from animal manure and treated human sewage, vegetative biomass, food, and other organic wastes. Compost is then applied to the ground to provide nutrients to growing crops (both annual, biennial, and perennial), as well as providing a range of other functions such as smoothing weeds, increasing soil organic matter and humic acid, improving soil structure, fertility, and water retention. Permaculturists also strive to moderate household consumption to reduce the pressure on both renewable and non-renewable resources.

Produce no waste - Permaculturists endeavour to minimise waste, including from human endeavours and energy use. For example, wastewater from bathing, laundry, or dishwashing is recycled for irrigation. Human sewage and animal manure are also often used to create biogas, which can be used for cooking and heating (Holmgren 2002)

Design from patterns to details - Permaculturists aim to design interventions based on observed patterns in nature and society (Holmgren 2002). With the underlying intention of achieving maximum gain for minimal energy expenditure, permaculturists aim to enhance the performance of ecosystems to leverage increased yields for human use (Lefroy 2009; Hatton and Nulsen 1999). Permaculturists employ the 'Guild' concept, which aims to maximise useful yield by combining polycultures of plants, bacteria, fungi, etc. For example, mycorrhizal fungi are valued for their symbiotic relationship with crop plants, helping to provide essential nutrients such as phosphate (Mollison and Slay 1997).

Integrate rather than segregate - Considered designs promote synergistic relationships, which facilitate relationships between biophysical components of the landscape (Holmgren 2002). This way, the whole becomes greater than the sum of its parts.

Use small and slow solutions - Permaculturists believe that small and slow systems generally result in more sustainable outcomes. Permaculturists tend to promote small-scale designs and perceive large-scale designs as riskier, i.e. much more likely to fail/collapse (Holmgren 2002).

Use and value diversity - Permaculturists strongly believe that diversity enhances resilience against both biophysical and economic shocks and allows humans to make full use of the environment (Holmgren 2002). The inclusion of diverse enterprises is also associated with increased labour efficiency (Dey et al. 2010) and generates multifunctional benefits, including food production, conservation of the environment, as well as economic and social benefits (Tipraqsa et al. 2007).

Use edges and value the marginal - Permaculturists prioritise the management of boundaries (edges) between discrete zones, referred to as 'ecotone' in ecology. These interfaces are often deemed as the most dynamic and productive spaces (Mollison and Holmgren 1978; Mollison 1988; Holmgren 2004; Jacke and Toensmeier 2005). Because of this, when designing interventions, permaculturists endeavour to increase the area of edges. Maximising the edge effect is a central tenet when designing alley cropping, shelterbelts, and water features (Holmgren 2002). Permaculturists exploit the edge effect by designing their interventions into ergonomically efficient zones (Mollison and Holmgren 1978; Mollison 1988; Mars 2005; Holmgren 2004; Bell 2005; Hemenway 2009; Bane 2012). Figure 96 illustrates the zone concept.

Zone 0 is the home or centre of the design. Zone 1 is close to the house and consists of plants and animals that require the greatest level of attention. Plants in this zone are things such as raised beds salad crops, herbs, vegetables, and soft fruit. Animals in this zone may include caged rabbits. Other structures, such as greenhouses, cold frames, propagation areas, compost bins, rainwater barrels, and fuel storage, are also found in this zone. This zone generally receives 100% sheet mulching and is often serviced with full drip irrigation (Deep Green 2022). Zone 2 consists of perennial plants that need less management. This often includes soft fruit, fruit trees and orchards, perennials, long-duration vegetables, beehives, ponds, chickens, ducks, geese, etc., and housing for

Figure 96 The permaculture zone concept. Source: Adapted from: Deep Green (2022).

larger animals. Irrigation is usually present, and sheet mulching is generally practiced in this zone (Deep Green 2022). Zone 3 is dedicated to field crop production and grazing for large livestock such as cows and sheep. Dams for water storage may be in this zone. Orchards and trees may also be planted in this zone. Green mulching with ground cover plants may be used in this zone. Produce generated in this zone is for both domestic use and for sale to local markets (Deep Green 2022). Zone 4 is partially wild and used for harvesting timber/firewood, animal forage, and rough pasture for grazing. Where trees grow, they are actively thinned to allow adequate sunlight to reach the layers below (Deep Green 2022). Lastly, zone 5 is a conservation area, a refuge for wild plants, animals, fungi, bacteria, etc. In some designs, a wildlife corridor is created from zone 5 through to zone 0.

Creatively use and respond to change - Lastly, permaculturists embrace change and constantly look for opportunities to positively build on any changes that occur (Holmgren 2002).

Ultimately, aside from the use of detailed designs and a clear focus on labour and energy-use efficiency, permaculture is fundamentally based on solid agroecological approaches (described in Chapter 6). Table 2 illustrates the similarities and differences between permaculture and agroecology principles.

Table 2 Comparing permaculture and agroecology

Permaculture principles	Agroecology and related principles
Diversity	
Diversity, plant stacking, and time stacking (PDM, IPM), use and value diversity (PPBS)	- Species and genetic diversification of the agroecosystem in time and space (Reijntjes et al. 1992). - Contain pests through complex trophic levels (Malézieux 2011). - Maintain landscape heterogeneity and capture environmental gradients (Fischer et al. 2008)
Interaction	
Edge effects (PDM), use edges and value the marginal (PPBS) Everything Gardens (PDM) Relative location (IPM) Each important function is supported by many elements (PDM); each element performs many functions (PDM)	- Optimise available resources through synergies between 'plants animals, soil, water, climate, and people' (Pretty 1994, Vandermeer 1995). - Use complementary functional traits to ensure production and resilience (Malézieux 2011). - Enhance beneficial biological interactions and synergisms (Reijntjes et al. 1992). - Enhance recycling of biomass and optimising nutrient availability and balancing nutrient flow. (Reijnties et al. 1992).
Creativity and innovation	
The problem is the solution (PDM) The yield of a system is theoretically unlimited (or only limited by the imagination and information of the designer) (PDM) Make the least change for the greatest possible effect (PDM)	*No corollary agroecological principles.*
Adaptive management	
Observe and interact (PPBS)	Management by experiment (Nudds 1999)
Apply self-regulation and accept feedback (PPBS)	Mobilise capacity for inquiry (Blann et al. 2003)
Creatively use and respond to change (PPBS)	Detect and foster novelty (Blann et al. 2003)
Accelerating succession and evolution (PPBS)	Come opportunities for self-organisation (Folke et al. 2003)

7.5.3 Strengths of permaculture

Self-governance and self-organisation are promoted by permaculturists as core strength. The connection to nature, the rewards of one's labour, and the feeling of place and belonging create high levels of motivation amongst permaculturists and cannot be overestimated (Scharf et al. 2019).

In a similar vein to agroecology, principles of diversity and multi-functionality, such as promoting soil improvement to deliver a range of ecosystem service benefits, are widely recognised as key strengths of permaculture (Ulbrich and Pahl-Wostl 2019). The presence of diverse polycultures provides significant biophysical resilience to pests and diseases and to periods of drought and floods. Greater self-sufficiency and self-reliance also enhance resilience during times of economic shocks (Mollison and Holmgren 1978; Hemenway 2009; Shepard 2013). The focus of permaculture on both understanding and managing key relationships between plants, animals, and soil flora and fauna is a great strength and generally missing in other food system configurations (Cavazza 1996; Veldkamp et al. 2001; Hatfield 2007; Osty 2008; Benoit et al. 2012). Permaculturists also promote the conservation of agrobiodiversity through the selection of both new and underutilised crops and animal species/cultivars and breeds (Mollison and Holmgren 1978; Jacke and Toensmeier 2005; Shepard 2013).

7.5.4 Weaknesses of permaculture

It is argued that the lack of a clear definition of permaculture causes confusion and limits detailed deliberations of its value and application (Ferguson and Taylor Lovell 2014; Ulbrich and Pahl-Wostl 2019). Permaculturists are also charged with the oversimplification of what are exceedingly complex socio-ecological systems and processes (Ferguson and Taylor Lovell 2014; Ulbrich and Pahl-Wostl 2019). Permaculture's populist and grassroots stance leads to claims that solutions to environmental and social challenges are both simple and known (Ferguson and Taylor Lovell 2014). Indeed, Mollison and Holmgren claim that 'Though the problems of the world are increasingly complex, the solutions remain embarrassingly simple' (Permaculture Institute 2013). They infer that no new knowledge is required but simply an innovative recombination of existing knowledge (Mollison and Holmgren 1978). Indeed, most permaculture textbooks were written between 40 and 60 years ago (Mollison 2003).

Permaculturists are also charged with making unsubstantiated, over-reaching, and un-replicable claims about the impacts of their work (Ulbrich and Pahl-Wostl 2019; Ferguson and Taylor Lovell 2014). In part, this is due to the permaculturists' limited engagement with mainstream science, lack of peer-reviewed publications, and limited reference to peer-reviewed science in permaculture publications, especially to respective developments in agroecology and agroforestry (Scott 2010; Chalker-Scott 2010; Ferguson and Taylor Lovell 2014).

Permaculture has therefore often been accused of being pseudo-scientific (Chalker-Scott 2010). Indeed, some of the terminology and assumptions used by permaculturists have been judged as misleading or down-right inaccurate. For example, the term 'guild' in permaculture literature refers to a group of plants living in synergy. In mainstream ecology, a 'guild' is a group of plants that compete for the same resources within a given space and which do not work well together (Simberloff and Dayan 1991). The term 'dynamic accumulators' in permaculture literature refers to deep-rooted plants that draw nutrients up from the lower layers of soil and make them available (through grazing, harvesting, or leaf-fall) to the soil surface and upper soil horizons (Whitefield 2004; Jacke and Toensmeier 2005; Bell 2005; Jacke and Toensmeier 2005; Hemenway 2009). Whilst the distribution of nutrients through the soil profile is a scientifically recognised function of trees and other deep-rooted plants, the term dynamic accumulators are only used in permaculture literature (Chalker-Scott 2010; Jobbágy and Jackson 2004; Callaway 1995; Porder and Chadwick 2009). Whilst acknowledging the nature of ecotones, some ecologists have questioned permaculturists' claims that creating complex shapes, with increased edges, leads to increased productivity (Mars 2005; Hemenway 2009; Bell 2005). Whilst met with counterclaims, the use of exotic/non-native species in permaculturists' novel ecosystem designs is also criticised by ecologists (Ferguson and Taylor Lovell 2014). Natural scientists also express concerns over permaculturists' modification of landscapes, via impoundments, etc., and other water drainage interventions that may cause increased erosion of topsoil and even flooding downstream (Rockström et al. 2010).

Compared to other alternative food networks, the permaculture movement is depicted as disorganised and poorly coordinated, which tends to both limit the movement's capacity to coordinate action beyond the local and to mobilise political support (Borras et al. 2008; Rosset et al. 2011; Martínez-Torres and Rosset 2010; Grayson 2010; de Molina 2012). Concerns also abound regarding labour productivity and overall profitability in highly diverse and complex multi-functional designs (Reich 2010; Tipraqsa et al. 2007; Amekawa et al. 2010; Kremen et al. 2012).

7.5.5 Future of permaculture

Whilst permaculture is practiced by up to 3 million practitioners in over 120 countries (Ulbrich and Pahl-Wostl 2019), it has failed to prevent the spread and intensity of industrial agriculture. Recently, however, permaculturists have made increasing efforts to publish literature in peer-reviewed journals, opening the approach up to greater scientific rigour as well as new ideas emerging from the fields of agroecology and ecological engineering (Ulbrich and Pahl-Wostl 2019). Indeed, permaculturists' work on the 'productivity of multi-strata silvopasture, integrated with multi-species rotational grazing' has prompted interest from agroecological scientists. It is likely that the integration of permaculture and agroecology theory could create productive synergies (Shepard 2013).

7.6 Food sharing

7.6.1 History of food sharing

The act of sharing food with family members or as part of wider cultural and religious institutions, rules, and norms, focussing on the act of giving, particularly to those less fortunate, has occurred since the beginning of humankind. Food sharing has often been associated with religious institutions such as churches, synagogues, temples, and mosques.

However, the significance and incidence of food sharing have increased in recent times due in part to a period of economic instability (e.g. the financial crisis of 2007-08 and its aftermath) and long-term declines in standards of living for poorer sections of the population in both developed and developing countries. Lack of access to affordable, safe, and nutritious food has also been juxtaposed with the high levels of food losses and waste associated with modern, globalised agri-food supply chains (Michelinia et al. 2018). Indeed, it is suggested that total annual global food waste is sufficient to feed approximately four times the current number of food-insecure families across the globe (Michelinia et al. 2018).

In the developed world, food sharing has manifested itself in the establishment of local food policy councils, food banks, social supermarkets, and food-sharing restaurants and cafes (Michelinia et al. 2018). Food policy councils were established as a mechanism to enable local people (farmers, agri-businesses, processors, retailers, civil society, and public officials) to shape local food policy to deal with the problem of food poverty and to ensure the supply of healthy and competitively priced foods for all sections of the community. The first food policy council was established in 1982 in Knoxville (Tennessee) in the USA to address food poverty during a period of economic recession.

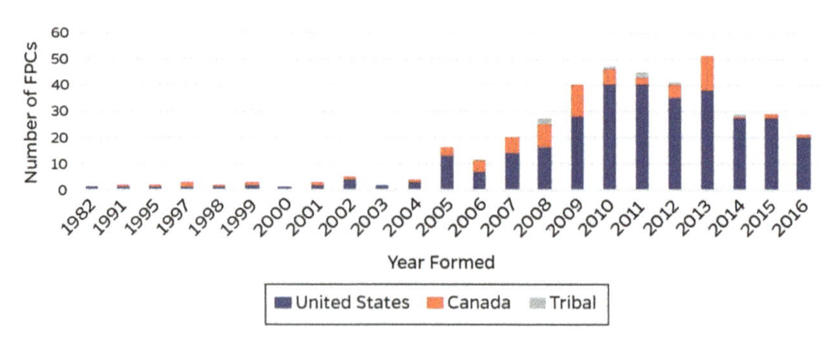

Figure 97 The growth of Food Policy Councils 1982-2016. Source: Adapted from: Sussman and Bassarab (2017).

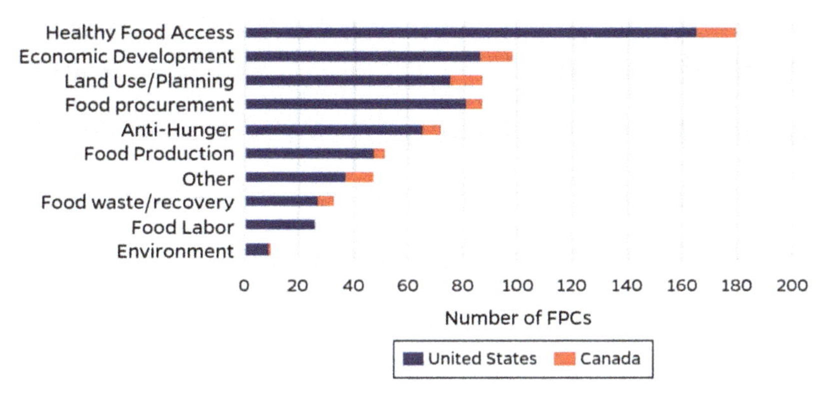

Figure 98 Food Policy Council priorities. Source: Adapted from: Sussman and Bassarab (2017).

Since first emerging in Knoxville, food policy councils have been established across the USA, Canada, and the UK to promote community-based development of just, sustainable, and local food systems (Fig. 97) (IPES-Food 2016; MWA 2015).

Key food policy council priorities generally include access to healthy foods, local economic development, local land-use planning (often urban farms and gardens), and food procurement by local institutions (schools, hospitals, prisons, community-based groups, etc.) (Fig. 98).

Food banks are community-based-organisations that collect surplus food from retailers and redistribute it to the poor free of charge (Teron and Tarasuk 1999). The first Food Bank was established by John Arnold Van Hengel in Phoenix (Arizona) in the USA in 1967, shortly after he witnessed a mother and her ten children foraging for food in a nearby waste bin (Osborne 2014). John was unable to accept the fact that, at the same time the poor were struggling to find enough food to eat, restaurants and supermarkets were throwing away significant volumes of perfectly good food (Schneider and Scherhaufer 2015). Using donations of food from food retailers and the public, food banks sprang up quickly across many large conurbations across the USA and Canada, providing a 3-day emergency food pack to meet the energy and nutritional needs of each qualifying family.

It was not until 1984 that Cécile Bigot borrowed the USA and Canadian model and established the first European food bank in Paris, France (Osborne 2014). This quickly led to the establishment of the European Federation of Food Banks in 1986, and, within the next 6 years, the rapid development of food banks, along with supportive logistics, across Spain, Italy, Ireland, and Portugal, and to Poland, Greece, and Luxembourg by 2001 (Osborne 2014; Michelinia et al. 2018). The demand for food banks rapidly grew during the 2008/2009, 2011/2012 (Osborne 2014), and 2022 food and economic crises.

By 2015, it was estimated that 4.8 million people in France, more than 15% of the population, relied on food banks and other supportive mechanisms for access to sufficient food (Perrott-White 2019). Food banks in the UK expanded from just 2 in 2004 to 423 in 2014, feeding nearly 1 000 000 people (Osborne 2014), and to 2547 food banks serving 3 000 000 people by 2022/23 (Pratt 2023). In the USA, a national network of food banks 'Feeding America' distributes more than 3 billion meals each year (Michelinia et al. 2018).

Other free food initiatives have been established to ensure that safe and edible food, which is unsuitable for Food Banks, can be made available to the poor. One such initiative is Foodsharing.de, which was established in Germany in 2012 and which latterly spread to German-speaking areas of Austria. Foodsharing.de takes delivery of food much faster than Food Banks, can take delivery of much smaller volumes of food (e.g. a dozen carrots, or two crates of tomatoes, or a dozen bread rolls) than Food Banks can, and can take food which is past its statutory sell-by date. The online and in-store platform is comprised of Food Savers (often individuals or charities in need of food) and Food Sharers (individuals, retailers, and farmers), who make food available free of charge (Collective Green 2022).

Social supermarkets (SSMs) are small, not-for-profit, volunteer-run, retail outlets that receive free food from farmers, food manufacturers, and fellow retailers, which is either blemished or misshapen in some way, close to its sell-by date, or highly perishable and destined for landfill or composting. This food is then sold at highly discounted prices (30–70% lower) to poor consumers, with any profits being invested back into local social projects (Holweg et al. 2010; Holweg and Lienbacher 2011; Schneider and Scherhaufer 2015; Michelini et al. 2018). According to Knežević et al. (2021), the principal aim of social supermarkets is the reduction of both institutionalised food waste and systemic poverty. The first social supermarket was established in Austria in 1990 in response to growing financial hardship sparked initially by the economic crisis and government austerity (Michelinia et al. 2018). Social supermarkets spread rapidly, and by 2013, there were an estimated 1000 social supermarkets scattered across Austria, Belgium, France, Germany, Luxembourg, Romania, Spain, Switzerland, and the UK (Cocozza 2013).

More recently, through the advent of online platforms and mobile digital apps, outlets such as pubs, restaurants, hotels, and cafes now have the capacity to retail surplus meals and reduce waste. Leloca is a for-profit mobile application that was founded in New York City in 2011 and has since spread across cities in both the USA and Canada. The company provides a platform for restaurants to sell surplus inventories of meals, which would otherwise have been disposed of as waste. Restaurants provide discounts of between 30% and 50% to consumers who can provide a Leloca code for a specific discounted meal within 45 min of an issue (Michelinia et al. 2018). Too Good to Go is a similar model which was created in 2015 in Denmark. The online platform allows customers to purchase food at bargain prices from pubs, hotels, restaurants, cafes, bakeries, and supermarkets. Food, that would otherwise have been thrown away, is collected within an hour of closing time and consumed off the premises (Michelinia et al. 2018).

7.6.2 Free food establishments

Established in 2012, in Berlin, the Baumhaus Community Kitchen Project relies on volunteer staff to prepare dinners made from food that would otherwise have gone to waste. For a small donation, these dinners are made available to the community. The Baumhaus Community Kitchen also provides storage for a community-supported agriculture (CSA) initiative and food cooperative (Scharf et al. 2019). The kitchen acts firstly to reduce food waste and secondly to foster local social and cultural capital, with people from all walks of life, sitting, talking, eating, and learning about the impacts of food waste (Scharf et al. 2019).

Established in 2014 by two local art students, Iris and Rebecca, in Groningen, Netherlands, the Free Café is a place where every Wednesday and Sunday between 40 and 80 people are treated to free dinners. The Free Café's objective is to reduce food waste, distribute healthy free meals, and provide a safe space without boundaries for people to interact, a space that is free from everyday market forces. The Free Café exists totally outside the market economy. No one gets paid or receives any form of monetary reward. The premises are rent-free, and even the materials used to construct the Café were donated. Dishes are prepared by volunteers from surplus/waste food collected from local open markets, supermarkets, and even food banks. Over a quarter of Groningen's population are students, who often make up a substantive proportion of the clientele. Recently, Free Cafés have been established in other Dutch cities, such as Amsterdam, Dordrecht, and Utrecht (TEDx University of Groningen 2015; Meesterburrie and Dupuy 2018).

7.6.3 Non-market food sharing

Over the past two decades, there has also been a resurgence in informal and non-market provisioning (gift-giving, sharing, or exchanging) of food, which has been generated in home gardens (Kneafsey et al. 2008; McEntee 2010; Schupp and Sharp 2012; Schupp et al. 2016; Goodman et al. 2012; McClintock 2014). This form of food provisioning remains popular in Eastern Europe. For example, whilst only 2% of the Czech workforce is directly engaged in commercial agriculture, at least 38% of households across the country grow fruit and vegetables, etc., in their gardens or allotments to reduce their weekly food expenditure (Jehlička and Daněk 2017). Of these households, 64% share a percentage of their production, and 37% gift food (Jehlička and Daněk 2017). The act of sharing is primarily motivated by the sense of satisfaction in helping others, feelings of social justice, and the importance of strengthening the cohesion of local communities, rather than the principle of reciprocity or barter (Jehlička and Daněk 2017).

7.6.4 Edible landscapes and gleaning

Lastly, it would be amiss not to include a discussion on the gleaning (gathering at no cost) of wild foods (mushrooms, fruits, berries, etc.) from public spaces, and of leftover crops (often unpicked fruit or vegetables) from commercial farmers' fields. For example, in Germany, Mundraub (English translation: mouth robbery) provides an online platform that maps edible landscapes/food commons (Scharf et al. 2019). In the UK, the tradition of gleaning has been resurrected, with at least 25 committed groups of volunteers that have formed up and down the country (Positive News 2022). These groups collaborate with local farmers to collect fruit and vegetables from the fields, that are too small, too large, or too misshapen to conform to supermarket standards or surplus to contract requirements, or cannot be harvested due to labour shortages or bad weather. It is estimated that approximately 16% of fruit and vegetables in the UK are never harvested, which is equivalent to 1.6 million MT, worth approximately £500 million (Positive News 2022). After gleaning, the harvested crops are distributed to food banks, community kitchens, and food projects (Guardian 2022). Additionally, some local governments are even purposefully planting fruit trees in public spaces to provide free fruits to passers-by (Trainer 2020).

7.6.5 Benefits of food sharing

There are basically three types of benefits associated with food sharing (Schneider and Scherhaufer 2015):

- Social;
- Environmental; and
- Economic.

Social benefits are accrued through strengthened social capital (social inclusion, self-confidence, dignity, and self-worth), enhanced social cohesion (an enhanced feeling of belonging), and increased social responsibility in all food system actors involved in the food sharing (Knežević et al. 2021). Some stress that this represents a sharing economy, existing alongside the market

Figure 99 The benefits of FoodCloud's distribution of surplus food. Source: Adapted from: FoodCloud (2021).

economy, where people share food or make good food available at low cost because of the joy of giving (Maye and Duncan 2017; Jehlička and Daněk 2017).

Environmental benefits are accrued through the redistribution of existing surplus food and reduction in overall food waste (Mair and Reischauer 2017; Scharf et al. 2019; Knežević et al. 2021; Collective Green 2022). As suggested by Bootsman and Rogers (2010), 'a redistribution market encourages reusing and reselling old items rather than throwing them out, and also significantly reduces waste and resources that go along with new production'. In the case of home gardens, a significant number of food producers use little or no synthetic inputs and, depending on their model of food sharing, can drastically reduce food miles (Jehlička and Daněk 2017).

Economic benefits are accrued via the more effective allocation of scant household budgets, enhanced food security, and quality of life (Knežević et al. 2021). Figure 99 summarises some of the key outcomes of FoodCloud's activities. Similarly, Foodsharing Germany has redistributed more than 32 000 tonnes of food to over 73 000 people (or Foodsavers), through 7000 outlets. As of 2022, the Too Good To Go App has redistributed 173 million meals to 66.9 million users through 174 027 outlets (Toogoodtogo 2022).

7.7 Urban agriculture: localisation and regionalisation of supply chains

7.7.1 History of urban food systems

Significant levels of urbanisation occurred across Western Europe and North America between the late nineteenth and mid-twentieth centuries, with developing countries following suit during the late twentieth and early twenty-first centuries. It was not until the emergence of the First Food Regime in 1870 and the growth of the global economy and international trade in food that rapidly growing urban populations began to rely on global value-chains and 'food from nowhere', rather than their traditional rural foodsheds (i.e. a network of primarily local farms supplying a neighbouring town or city) (see Chapter 1 for more details).

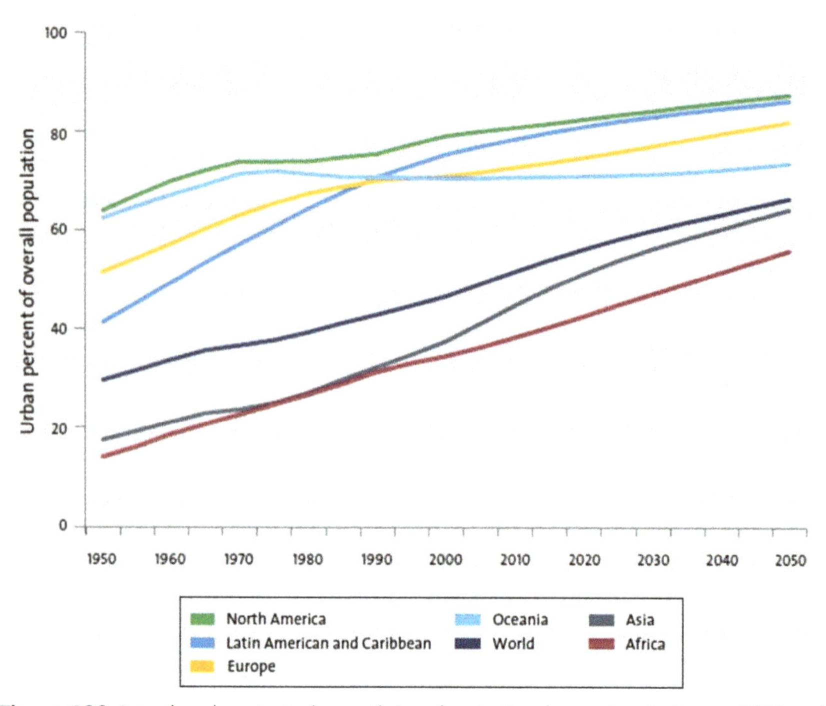

Figure 100 Actual and projected growth in urbanisation by region, between 1950 and 2050. Source: Adapted from: Bereuter et al. (2016).

Figure 100 illustrates the rapid pace of urbanisation across all continents since the 1950s through to predicted levels of urbanisation by 2050. According to Tefft et al. (2017), approximately 26% of food consumed globally is imported, with many of the large global economies, such as the USA, Western Europe, Japan, and China, importing the largest volumes of food. Some city-states, such as Singapore, import 90% or more of their food needs. Whilst international value chains are seen as providing an incredible variety of foods at competitive prices, as well as buffering nations against localised crop failure, many urban markets have become increasingly detached from local food production (Vaarst et al. 2018).

However, whilst the internationalisation of supply chains has become commonplace, much of the food eaten across the globe is still produced close to where it is consumed. Urban and peri-urban agriculture and horticulture (i.e. produced in zones where urban and rural land uses overlap) still provide a significant proportion of the food that urban populations consume. Indeed, according to Tefft et al. (2017), 15% of the world's agricultural land is in urban areas, and up to 40% of agricultural land is located within a 20 km radius of urban centres (i.e. peri-urban). In many Asian cities, up to 90% of vegetables (especially leafy vegetables) are sourced from within this 20 km radius. In SSA, many towns and cities are dependent on local production for most of their food needs.

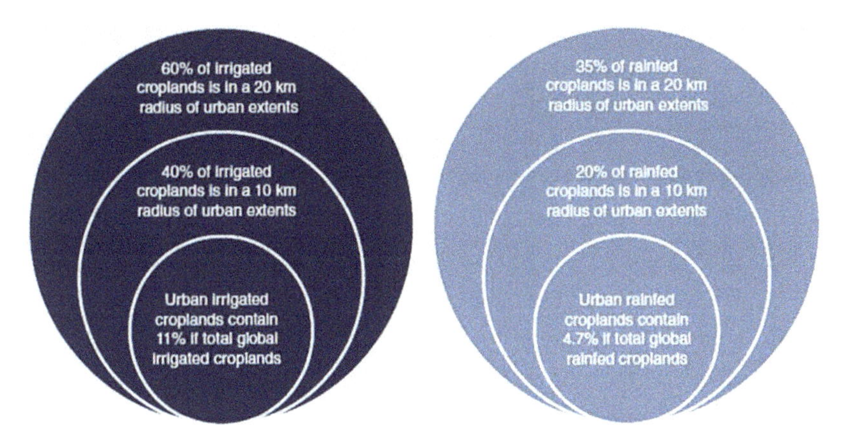

Figure 101 Share of urban and peri-urban land in total global agricultural land. Source: Adapted from: Tefft et al. (2017).

Figure 101 shows that 60% of irrigated land and 35% of rainfed land globally is located within a 20 km (12.5 miles) radius of urban centres (Tefft et al. 2017).

Urban and peri-urban agriculture has been defined as the production and distribution of food, fibre, and fuel in and around cities, which is fully integrated into urban, ecological, and social systems (Tefft et al. 2017; Morel et al. 2017; Daniel 2017; Bryant et al. 2016). The Global Partnership on Sustainable Agriculture and Food Systems (RUAF) defines urban agriculture as (Wiskerke 2020)

> Agricultural production (of crops, livestock, fish, and trees) in urban and peri-urban areas for food (e.g. vegetables, milk, eggs, poultry, and pig meat) and other uses (e.g. herbs, flowers, and fodder), the related input supply, transport, processing, and marketing of the agricultural produce and the provision of non-agricultural services (such as agritourism, urban greening, and water storage).

A defining characteristic of urban and peri-urban agriculture, and a key difference with rural agriculture, is that it is an agricultural production system that is closely intertwined with the urban socioeconomic, ecological, and legal system. This means that urban and peri-urban agriculture make use of urban resources, deliver produce and services to urban consumers, and are influenced by urban laws and market forces.

Types of peri-urban and urban agricultural production include

- medium-sized farms (close to urban settlements);
- small or micro-farms (<5 acres) within urban areas;
- allotments (<1 acre) and gardens;
- vacant or unused spaces;

Table 3 Types of urban and peri-urban agriculture (Tefft et al. 2017)

Type	Location	Growing container	Growing medium	Technology	Labor	Products
Small-medium scale gardening, primarily for self-consumption						
Open held garden family, allotment, squatter, community	Peri-urban vacant plots	Ground, box, shacks	Soil		Family	Vegetables, orchards, flowers and herbs, seedlings, livestock
constrained space	Rooftop, balcony, camps, indoor	Bags, bottles, containers trays	Soil, pellets, rocks	Hydroponics Aero-ponics Vertical	Family	Vegetables, some fruits, some grains and legumes
Institutional garden	SchoolsChurches Hospitals Prisons	Ground, box	Soil		Student Military CM Servants Prisoners	Fruits, vegetables., flowers, herbs, seedlings, restock
Medium-large scale fanning, primarily for commercial operation						
Multi-functional farms agro-tourism aquaculture tree	Peri-urban	Ground, ponds, greenhouse	Ground soil, water, ponds	Aqua-pomes, irrigation	Salary	Trees, tourism. horticulture, social livestock, fish, education, research
Open-field commercial farms	Urban peri-urban	Ground	Ground soil, water	Irrigation	Salary family	Crops, livestock, nuts, nursery beds, composting, seeds, composting
Greenhouse	Industrial buildings	Ground	Perlite Vermiculite Soil moist	Climate-controlled hydroponics Aero-ponics	Salary	Horticulture, hops
Rooftop	Urban	Light-weight container, greenhouse	Soil nutrients		Salary	Horticulture shallow root crops
Indoor farms	Buildings Vertical underground	Trays, box, container	Relets, Sol. rocks	LED light hydroponics Aero-pomes vertical	Salary	Vegetables, fruits, herbs

- rooftops and balconies; and
- protected agriculture and horticulture (greenhouses, polythene tunnels, shipping containers, vertical farms, etc.).

These types of urban and peri-urban agriculture are extremely diverse and constantly evolving (Table 3). A related concept is that of a 'foodshed'. First coined by Hedden in 1929 (Zazo-Moratalla et al. 2019), urban foodsheds are defined as both the geographical area of rural land adjacent to urban centres that are needed for the sustainable production of most of the urban centre's food, and the networks of producers and consumers involved in these foodsheds (Zazo-Moratalla et al. 2019).

A key characteristic of urban and peri-urban agriculture is its reliance on short food supply chains (SFSCs). This is especially the case across much of the developing world, particularly across Africa and Asia, where, despite the growing presence of Long Food Supply Chains (LFSCs), traditional urban food systems predominate in many cities. For example, a long-standing SFSC continues to supply the lion's share of food needs to the city of Dar es Salaam in Tanzania, which has a population of more than four million people (Wegerif and Hebinck 2016).

According to Chiffoleau et al. (2017), short food chains can be characterised as either

- involving direct sales between farmer/collective producer and urban consumer; and
- sales between the farmer/collective producer and urban consumer involving a maximum of one intermediary (e.g. farm shop).

Figure 102 illustrates the organisational diversity of short food chains in France.

Even in developed countries, where supermarket-dominated LFSCs are most established, SFSCs have managed to survive and even thrive. For example, in the case of France, and much of Southern and Eastern Europe, whilst experiencing periods of expansion and contraction, some SFSCs based on traditional local markets have survived reasonably intact since Roman times (Holleran 2012). Governments have sometimes played a role in the survival of SFSCs. For example, as early as 1905, the French Government sought to protect small, non-industrial farmers via geographically delineated 'produce of origin' regulations, which meant, for example, that champagne could only be marketed as such if grown in the Champaign region. Similarly, cheese, such as Camembert, could be marketed (both domestically and internationally) as Camembert if it was produced in the Camembert region (Colonna et al. 2013). Whilst this also helped promote exports of such regional products, it also helped to maintain local producers serving local markets. This provides a foundation for self-sufficient urban farming systems.

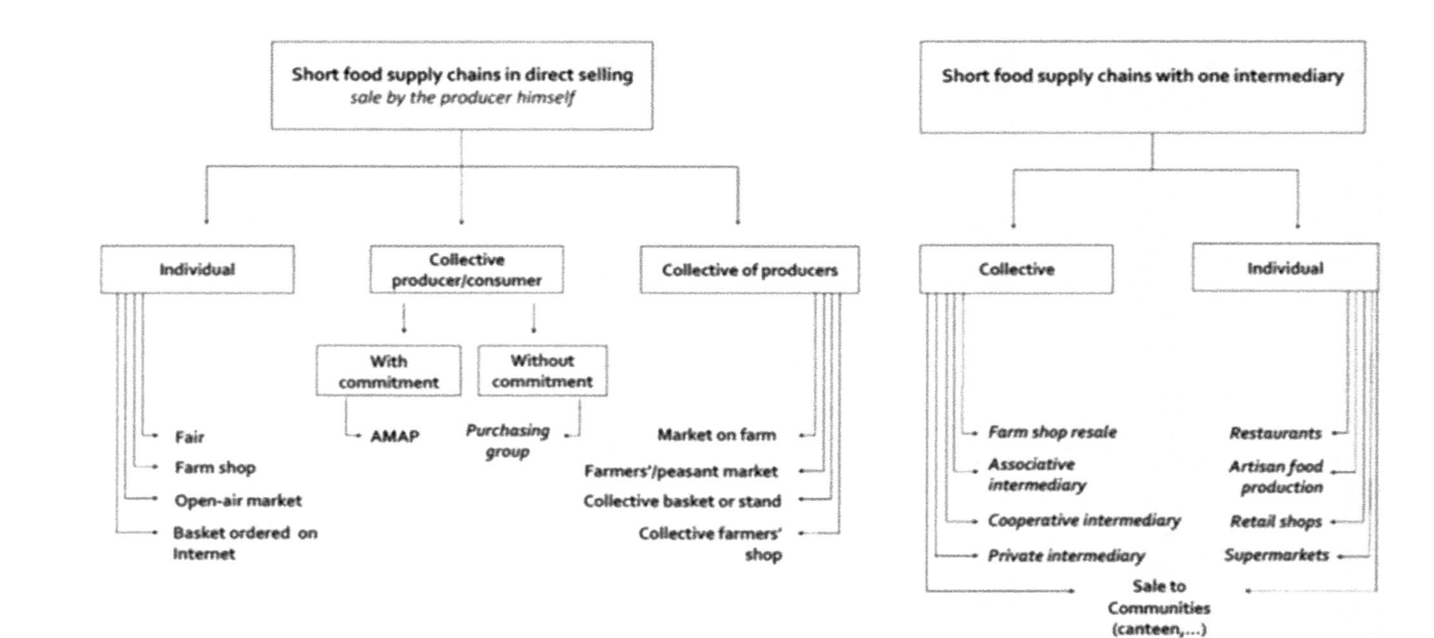

Figure 102 Types of short food supply chain in France. Source: Adapted from: Chiffoleau et al. (2017).

The development of modern urban agriculture and urban food systems

According to Fernandez et al. (2013), modern urban agriculture dates back to the 1890s, which coincides with the rapid expansion of industrial towns and cities in Western Europe and the USA. Whilst urban food security has always been on the agenda of national, regional, and local governments (Ligutti and Rawe 1940; Morris 1987), the official policy generally favoured the intensification of agriculture, promotion of trade, and globalisation of food chains.

Much of the recent renewed interest in urban food systems has been driven by issues of food security and sustainability (FAO 2010). Rapidly growing urban areas, especially the increasing number of mega-cities (>10 million inhabitants), began to present significant logistical challenges in ensuring the availability, accessibility, and affordability of adequate quantities of nutritious food (Dubbeling et al. 2009; Grewal and Grewal 2012; Pothukuchi and Kaufman 1999). There was also growing concern over the sustainability and ecological footprint/impact of LFSC (e.g. the resources required, environmental impacts such as pollution and biodiversity loss from intensive methods of production, the costs and impacts of globalised food distribution, as well as high levels of food losses and waste) (Haysom et al. 2019; Meerow et al. 2016; Wackernagel et al. 2006; Kirwan and Maye 2013).

The recent resurgence of interest has been underpinned by three key factors:

- local and regional government support;
- innovative small-scale farmers; and
- the concerted efforts of dedicated food activist groups and concerned individuals.

These are discussed here.

As noted earlier, most local and regional governments across the developed world, and much of the developing world, initially took a back seat when planning for food security. During the immediate post-war period, with the huge expansion and intensification of agriculture, food was seen as an easily available commodity traded along global value chains and regulated by the market mechanism of supply and demand (Friedmann and McMichael 1989). In this time of self-reliance, rather than self-sufficiency, if consumers had the financial resources to pay, food from across the world (food from nowhere) would always be available.

For example, historically, food production in the UK was the sole remit of the Ministry of Agriculture. Whilst local governments had detailed planning maps covering the land for water, energy, minerals, etc., use, the Ministry of Agriculture could demarcate 'white land' dedicated to food and fibre production, over which local and regional government had little or no say (Haysom et al. 2019). This white land was predominantly detached from the urban centres it surrounded. Indeed, food produced in these agricultural lands was as likely to end up on French, German, or Malaysian tables as it was on local tables. For example, lamb produced on the hills surrounding Sheffield in North England was just as likely to end up in French markets as it was in supermarket shelves in Sheffield, which were often stacked with New Zealand lamb.

However, during the past 20 years or so, an increasing number of both local and regional governments across the globe began to question the prudence of leaving such a critical public good to the market (May 2017; Dubbeling et al. 2009; Grewal and Grewal 2012; Pothukuchi and Kaufman 1999). Indeed, concerns over the reliability and resilience of food supplies, and the growing desire for urban/regional self-sufficiency, and even food sovereignty, began to take centre stage, especially during the global financial crises of 2008/2009, 2011/2012, and 2022, and the period during and immediately after the COVID-19 pandemic. These periods saw significant spikes in food prices and, in some cases, disruption of food supplies.

Many local and regional governments, including the notable examples such as London and Toronto, have attempted to take back local control of foodsheds by developing urban food strategies and plans (Reynolds 2009; Blay-Palmer 2009; Jarosz 2008, Pothukuchi and Kaufman 1999; Blay-Palmer et al. 2018; Carey 2011; Andrees et al. 2020; Filippini et al. 2016; Gray and Nuri 2020). They sought to ensure that

- plentiful supplies of quality, nutritious, and healthy food are available to, and affordable by, all;
- the ecological and climate footprint of the local food system is minimised; and
- food producers and other food system actors are able to earn a fair living from their endeavours.

In the UK, both Greater Manchester and Sheffield, e.g. have developed urban food strategies to ensure the resilience of food supplies, address malnutrition (especially of poor citizens), and reduce both food waste and the ecological and environmental impacts associated with intensive LFSCs.

According to Haysom et al. (2019), in a similar vein to the evolution of the food sovereignty movement, local and regional governments have particularly promoted SFSCs to counteract the dominance of transnational corporations over food production, flows, and prices. SFSCs are increasingly seen as underpinning Urban Food System strategies (Doernberg et al. 2016; Morgan and Sonnino 2010; Wiskerke 2009). An increasing number of municipalities and regional planning authorities are turning to SFSCs to enhance the sustainability of food supplies. Momentum in this area increased after the launch of the Milan Urban Food Policy Pact in October 2015, which was signed by 117 mayors from across the globe (Forster et al. 2015; Blay-Palmer et al. 2018). Here, municipalities aim to re-embed local food production networks into urban ecosystems (Chiffoleau et al. 2017)

The role of small-scale farmers

As noted, local SFSCs are seen as a key component of urban food strategies. Despite the rapid growth of supermarket-driven LFSCs as a way to feed urban populations, SFSCs have been preserved, often through the efforts of small-scale farmers. These farmers can be broken down into two distinct types (21 in Chiffoleau et al. 2017):

- Traditional small-scale or peasant farmers who resisted the move to industrial agriculture; and
- Innovative new entrants to farming, often driven by a need to 'return to nature'.

Many of these farmers grow organic foods sold directly to local consumers via farm shops or nearby urban markets (Deléage 2011). For example, in France, by the late 1980s, 27% of farmers were involved in the direct selling of food products to local consumers (Capt et al. 2014).

The role of buyer activism

Purchaser groups and individual consumers have also been instrumental in both the promotion and sustainability of local food systems (Haysom et al. 2019; Provė et al. 2019). Solidarity purchase groups, such as Teikei (提携), first arose in Japan during the mid-1960s. The English translation of Teikei means cooperation and is one of the first examples of community-supported agriculture (CSA). The first CSA farms were small-scale, situated close to urban centres, generally organic, often non-profit, and reliant on volunteer labour. Initially coined by two European farmers, Jan Vander Tuin and Trauger Groh, the CSA concept, which was built on the principles of biodynamic agriculture, spread to the USA during the mid-1980s (McFadden 2004). By 1986, two CSA systems had been established, one at Great Barrington in Massachusetts, and the other at Temple-Wilton in New Hampshire (Lamine 2005). The UK saw the emergence of the 'Sustainable Food Cities' movement (Sheffield City Council 2014; Moragues-Faus and Sonnino 2018), again predominantly based on organic production.

Many of the CSA-styled approaches were strongly rooted in the Food Justice Movement (Alkon and Norgaard 2009; Mares and Alkon 2011), which arose during the 1960s in the USA. This movement sought to gain universal access to healthy, nutritious, and affordable food (Fernandez et al. 2013; Gray and Nuri 2020). In addition to Food Justice, other allied social movements sought to both conserve natural resources and support the economic viability of farming communities (Holt Giménez and Shattuck 2011; Fernandez et al. 2013; Feldmann and Hamm 2015).

7.7.2 Perceived strengths of urban food systems

Improved food security, through the integration of urban food consumption and rural food production, is generally seen as a key benefit of SFSCs (Piorr et al. 2011; Opitz et al. 2015; Zasada 2012; Galli and Brunori 2013).

Enhanced environmental sustainability is also often quoted as a principal benefit of SFSCs (Sonnino 2014, Smith 2008; Renting et al. 2003; Sundkvist et al. 2005; Dubbeling and Zeeuw 2011; Lee-Smith 2010; Maxwell et al. 1998; Hedberg and Zimmerer 2020; De Schutter et al. 2020). SFSCs are attributed with reducing food miles, greenhouse gas (GHG) emissions, packaging, plastics, etc., which are traditionally associated with LFSCs (NRDC 2007; Blay-Palmer et al. 2018; Mundler and Rumpus 2012; Penker 2006; Bródy and de Wilde 2020; Garnett 2010).

SFSCs are also attributed with creating opportunities for integrated land-use planning. Here, local foodsheds can be more easily managed by local or regional planning authorities, promoting both food production as well as the conservation of soil, water, local habitats, biodiversity, etc. (Sonnino 2014; Vaarst et al. 2018). This is juxtaposed against food-from-nowhere LFSCs, where ensuring good environmental management of production platforms becomes extremely challenging, if not impossible (Haysom et al. 2019; Bródy and de Wilde 2020).

SFSCs also provide opportunities for developing more circular food systems, where energy, water, and nutrients can be managed more efficiently (Berdegué et al. 2014; Vaarst et al. 2018; Joseph et al. 2019). For example, SFSCs offer the opportunity for the recycling of energy and nutrients contained in human urine and faeces as fertilisers for local agricultural and horticultural production – and the production of animal feed, composts, or even biofuels (waste cooking oil, etc.) from organic wastes generated in the urbanisations (Atkins 2007; Barles 2007; Billen 2011; Billen et al. 2009; Blay-Palmer et al. 2018). Indeed, according to Vaarst et al. (2018), this approach to recycling energy and nutrients has been practiced in Paris's local foodshed for over 1000 years.

Ultimately, whilst ensuring ample supplies of local food to feed growing urban populations and the pursuit of environmental sustainability may be the principal drivers of SFSCs, there are other, more value-laden drivers that have underpinned the increasing embeddedness of SFSCs. Here, embeddedness transcends economic rationality by embracing trust, regard, moral, ethical, ecological, and environmental concerns (Penker 2006; Joseph et al. 2019; Migliore et al. 2015). Industrial LFSCs are increasingly seen as uniform, impersonal, and unequal Chiffoleau et al. (2017), whilst SFSCs are seen as the re-embedding of food production within local social, cultural, economic, and biophysical ecosystems (Goodman and Goodman 2009; Hinrichs 2000; Winter 2003; Migliore et al. 2014), where quality, taste, trust, traceably, provenance, ethics, and safety, etc., take prominence over cheap food from anywhere (Blay-Palmer et al. 2018; Kneafsey et al. 2014; Hinrichs 2000; Kirwan 2006; Sage 2003; Tregear 2011; Mount 2012; Feldmann and Hamm 2015).

SFSCs are seen to facilitate the social reconnection between producers and consumers (Holloway et al. 2007; Hvitsand 2016; Ilbery and Maye 2006; Marsden et al. 2000). This reconnection helps to rebuild social capital, promote social inclusion (Holland 2004), and civic participation (Baker 2004; Dimitri et al. 2016). Some aspects of SFSCs, such as community gardens, are also redistributive, ensuring that the urban poor have access to adequate amounts of quality nutritious foods (Schmelzkopf 1995). Many of the SFSCs are also attributed to improving mental and physical health. For example, community gardens, which have flourished over the last quarter-century are seen as a social activity, promoting urban health (Armstrong 2000; Wakefield et al. 2007) and social cohesion (Firth et al. 2011; Veen et al. 2015).

Indeed, many of these small-scale forms of urban and peri-urban agriculture and horticulture, such as private vegetable gardens, roof-top gardens, allotments, gorilla gardening, and community gardens, have cyclically waxed and waned in tune with broader political and economic drivers. For example, community gardens (or '*Jardin d'ouvrier*/workers' garden) were first documented during the industrial revolution and the process of rapid urbanisation across Europe and North America when some citizens turned their backs on emergent forms of industrial agriculture. The number of community gardens, at the time known as 'Liberty Gardens' and 'Victory Gardens', jumped during the First and Second World Wars in Europe when the local production of food became a necessity and economic opportunity. Again, community gardens have flourished during the food and energy crises of the mid-1970s (Bhatt and Farah 2016), 2000s, and 2010s, and during both the COVID-19 pandemic and the 2022 food crisis when many communities struggled to ensure affordable access to nutritious foods (WWF 2020).

In addition to urban food production, peri-urban agriculture and horticulture developed with parallel goals and values. For example, unlike industrial agriculture, where traditional farming families may have been managing the same business on the same land for hundreds of years, many local market gardens and micro-farms have been established within the past 20–30 years by new entrants to the industry. Indeed, some new entrants have no agricultural background whatsoever. Many of these new, often young, farmers embrace organic farming and are motivated by the vision of a better, more sustainable world and quality of life, as opposed to the more profit-motivated values associated with industrial farming (Ferris et al. 2001; Holland 2004). Indeed, in 2013, 33% of new farms in France were started by young people with no agricultural background (Morel et al. 2017). Of these new entrants, 63% adopted an organic farming approach and 58% produced food for SFSCs (Jeunes Agriculteurs 2013).

The reconnection between urban centres and their hinterlands has injected renewed enthusiasm into the farming industry (Vittersø et al. 2019). For example, direct sales between local producers and local consumers have generally resulted in more employment opportunities for young farmers and higher product prices for farmers (Bereuter et al. 2016; Aubry and Kebir 2013; Duram and Oberholtzer 2010). In turn, through regular face-to-face interactions, local producers are much quicker to adapt to the changing demands of local consumers. (Le Grand and Van Meekeren 2008). In the USA, the number of farmers' markets increased by 76% between 2008 and 2014 (USDA 2014a), with the direct sales of food doubling between 1997 and 2007 (USDA 2014b). Across the EU, more than 15% of farms are involved in direct sales to consumers, although this varies considerably from country to country: <10% in Malta, Austria, and Spain; 18% in Hungary, Romania, and Estonia; and 25% in Greece (Augére-Granier 2016).

According to Gray and Nuri (2020), SFSCs supply food for over 700 million urban consumers, approximately a quarter of the world's urban population. Approximately 15% of all food consumed in urban areas is supplied through SFSC. This percentage is much higher for some products, such as meat (34%) and eggs (70%) (ETC Group 2017). In the Global South, as many as 2.5 billion urban consumers source their foods, either directly, or indirectly (through kiosks and street vendors), from SFSCs (ETC Group 2017). In France, approximately 20% of farmers (circa 100 000) sell at least part of their food products through SFSCs, with over 40% of farms directly selling 75% of their produce to their urban neighbours (Chiffoleau et al. 2017). Most of these farms are small (around 39 ha) and produce high-value foods such as fruits, vegetables, and honey (Chiffoleau et al. 2017). More rural farm businesses tended to produce cereals, legumes, roots, and animal products (Vaarst et al. 2018). Conversely, 42% of urban consumers purchase food from SFSCs (Loisel et al. 2013). The role of SFSCs in supplying food to urban consumers is expected to double between 2017 and 2037 (ETC Group 2017).

7.7.3 Perceived weaknesses of urban food systems

7.7.3.1 Definitional boundaries of SFSCs

Before delving into the more practical weaknesses of SFSCs, there is a conceptual weakness that first needs to be dealt with. That is, the current definition of SFSCs is incredibly broad, covering virtually any form of production that is not obviously captured in the conventional/industrial food systems model (Holloway et al. 2007; Venn et al. 2006). This broad-brush definition fails to capture the different motivations and practices of producers, the type and quality of relationships with consumers, and the focus on, and the level of embeddedness in, the broader territory.

For example, whilst many farmers involved in SFSCs are motivated by the quality of life and their relationships with consumers, many continue to be motivated by profit (Venn et al. 2006; Tregear 2011). Not all farmers producing for SFSCs are organic or use agroecological farming approaches, and many continue using unsustainable practices (Watts et al. 2005). Indeed, some organic farmers are caught up in the quandary around the conventionalisation of organic agriculture by corporate agribusinesses (Forney and Häberli 2015). At some stage, questions may be asked about the motivations and technologies employed by vertical farming businesses that are currently categorised as part of urban agriculture alongside very different practices such as growing vegetables on allotments. Additionally, evidence abounds that engagement in SFSCs does not always produce more sustainable social and economic outcomes (Hinrichs 2003; Stephenson et al. 2008; Alkon and McCullen 2011; OECD 2021).

7.7.3.2 Economic viability

Whilst it is possible for small-scale urban and peri-urban agricultural and horticultural businesses to make a profit (Schmit et al. 2016; Morel et al. 2017), the lack of profit motivation, combined with limited or no experience in agricultural or horticultural production, and limited investments in key labour-saving technologies and innovations, often leads to unprofitable business models or situations where producers subsidise their way of life by taking a minimal wage from the business, often less than the national legal minimum wage (NRC 2010; Blättel-Mink 2014), often being exposed to increasing levels of self-exploitation, namely, extremely long hours of hard physical labour for a limited financial reward (Galt 2013). For example, results from a study conducted by Morel et al. (2017) suggest that many microfarms are unable to earn the French minimum salary of €1000 per month, and many struggle to make €600 per month, often working many more hours per week than urban workers (Morel et al. 2017; Capt and Wavresky 2014).

In addition to the self-exploitation of farmers, an increased focus on small and ecologically diverse farms, run by amateur farmers, with less focus on the efficiency and cost of production, often results in increased unit costs, which are then, in turn, passed onto the consumer. This is not a problem when consumers have the financial resources and are willing to pay premiums. However, when consumers have limited financial resources, the price of foods supplied through SFSCs is often uncompetitive compared to large national food retailers who take advantage of cheap food products originating from LFSCs (Chiffoleau et al. 2019). For example, in the case of leafy greens, farmers involved in LFSCs receive around €0.20 per vegetable, whilst farmers in SFSCs receive around €0.60. Consumers in SFSCs were expected to pay a 50% premium compared to those buying from supermarkets supplied by LFSCs (i.e. €1.20 compared to €0.80) (Chiffoleau et al. 2019).

Issues such as price, limited food diversity and seasonality, and constraints in production capacity continue to constrain the growth of SFSCs. In Vermont, often seen as a vanguard city, SFSCs contribute less than 10% to food provision. In 2015, only 7.8% of farms in the USA sold food into SFSCs, with approximately 1.5% of the value of agricultural production in the USA (Low et al. 2015). This figure is similar to that estimated for the UK, which was around 3.5% of the value of agricultural production (DEFRA 2011).

7.7.3.3 Diversity and seasonality

The geographical and climatic boundedness of SFSCs also constrains the diversity of foods that can be produced in each urban food system. Whilst highly diverse urban food systems in tropical and sub-tropical latitudes can provide diverse local foods all year round, in temperate latitudes, such as northern Europe, only a limited number of commercial fruits, vegetables, etc., can be grown outdoors, and this range is significantly curtailed during the winter months (Vaarst et al. 2018). Whilst it might be possible to grow increasing amounts of vegetables, especially leafy vegetables, in controlled environments (glasshouses, polythene tunnels, vertical farms, etc.), it is unlikely that these will be able to meet the current demand for exotic fruits and vegetables (Vaarst et al. 2018). Whilst consumers are likely to be increasingly encouraged to adjust their consumption patterns to better match local supplies, it is likely that even the best-planned urban SFCSs will, to one degree or another, still rely on LFSCs both for difficult-to-give-up tropical and sub-tropical products (such as pineapples and coffee) and as a resilience mechanism to support urban SFSCs faced with climatic or other shocks (Vaarst et al. 2018).

7.7.3.4 Spatial limits

There is a growing consensus that, in isolation, many urban and peri-urban SFSCs do not have the physical capacity to meet urban food and nutrition security needs (Frayne et al. 2014; UN-Habitat 2018; Tacoli 2019). The estimated total area of urban land, including buildings and roads, is approximately 66 million ha, which equates to about 4% of cropland (Schneider et al. 2010). According to Ramankutty et al. (2019), many urbanisations across SSA, Central America, and South and Southeast Asia would not even be able to produce enough vegetables in urban areas to meet demand. Whilst vertical farms and other types of controlled environment agriculture (CEA), or alternative methods of protein production such as cultured meat) may be able to meet a significant proportion of vegetable and protein demand in some urban areas, the demand for cereals, pulses, roots, meat, etc., will need to be met in extensive hinterlands (Morel et al. 2017; Ramankutty et al. 2019; Edmondson et al. 2020). The extent of these hinterlands depends on several factors, including the absolute size of the urban population and population density, the quantity and quality of adjoining agricultural lands, and the proximity of neighbouring towns or cities (Zasada et al. 2019; Ramankutty et al. 2019). Ex ante food policy scenarios developed by Zasada et al. (2019) clearly depict the spatial range of urban foodsheds needed for the key European cities of London, Rotterdam, Milan, and Berlin.

Figure 103 illustrates similarities between the respective urbanisations with regard to the demand for food by individual urban inhabitants and the area of land required to supply this demand. As expected, meat production, especially

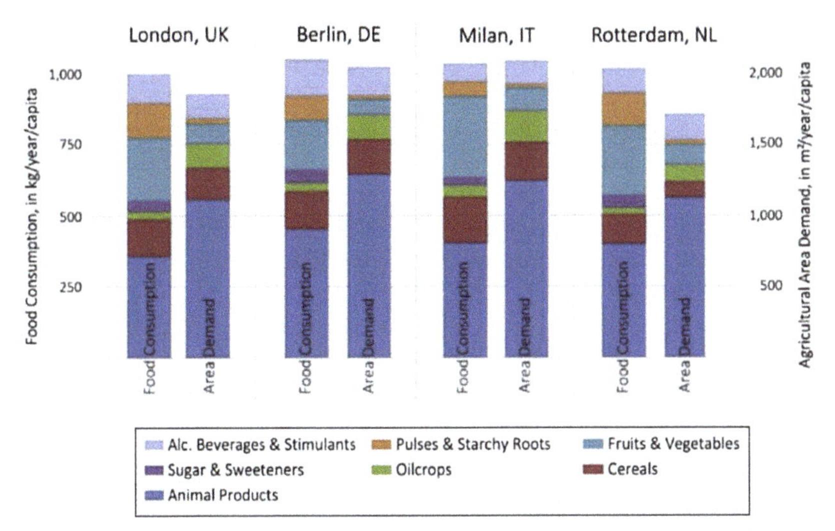

Figure 103 Annual food consumption and associated agricultural area demand. Source: Adapted from: Zasada et al. (2019).

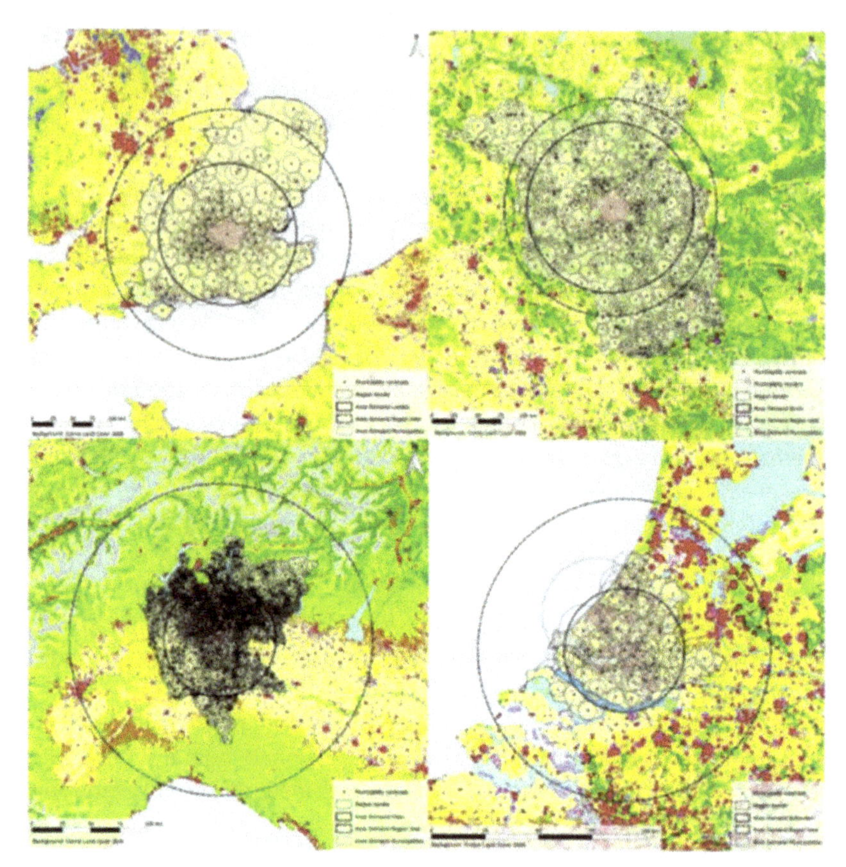

Figure 104 Metropolitan foodshed extent for London (upper left), berlin (upper right), Milan (lower left), and Rotterdam (lower right). Source: Adapted from: Zasada et al. (2019).

livestock and dairy, etc., requires the largest areas of land (between 59.6% and 65.5%). Food loss and food waste equate to 31% of the production area. When production is converted to organic, the area of land required increases considerably, in the case of London, by 41% (Zasada et al. 2019).

Figure 104 illustrates that whilst the foodsheds of Berlin, Milan, and Rotterdam are relatively small, the foodshed for London (top-right) is enormous (Zasada et al. 2019). Whilst London may be surrounded by fertile agricultural lands, its population size, and the size of its neighbours, spread the foodshed over an area of 42 176 km^2 (Zasada et al. 2019). Even then, the regional demand for food is double the regional supply. In the case of Rotterdam, regional demand outstrips regional supply by a factor of four (Zasada et al. 2019). Only Berlin, with its relatively low population density, can match regional demand with regional supply.

Part 3

The golden chalice of sustainability and the evolution of food systems

Chapter 8

The golden chalice of sustainability

8.1 Introduction: what do sustainable food systems look like?

In line with the 1987 Brundtland Report published by the World Commission on Environment and Development (WCED) (Brundtland 1987), the FAO defined sustainable development as:

'The management and conservation of the natural resource base, and the orientation of technological and institutional change in such a manner as to ensure the attainment and continued satisfaction of human needs for present and future generations'.

In the specific context of sustainable food systems, development must conserve natural resources, such as soil, water, nutrients, and genetic diversity, and be environmentally sensitive, profitable, and socially acceptable (Holden et al. 2018). This definition, and similar definitions, have generally been interpreted in two ways:

- As practices; and
- As outcomes.

Actors focussing on practices have generally taken the concept of sustainable food systems to mean opposite of the industrial agriculture model (Caira and Ferranti 2016; Martin-Guay et al. 2018). Here, alternative food systems, such as agroecological, organic, and permaculture agriculture, are built on traditional crop rotations, intercropping, agroforestry, pest and disease management, organic nutrient management, etc. (Migliorini et al. 2018).

Conversely, actors focussing on outcomes have generally taken the concept of sustainable food systems to mean the delivery of food and nutrition security, conservation of biodiversity and natural habitats and natural resources, the maintenance of economic viability, and enhanced social welfare (Lang and Barling 2012; Allen and Prosperi 2016; Gillespie and van der Bold 2017; GPAFS 2016; Haddad et al. 2016; Migliorini et al. 2018). This second approach generally internalises the concept of planetary boundaries. Both interpretations incorporate the concept of resilience (i.e. the system's ability to absorb shocks, maintain integrity, and regenerate post-shock) (Stefanovic et al. 2020).

In understanding debates about sustainability, the idea of a 'consensus frame' (Gamson 1995) is useful. A consensus frame relates to the distinct sets of meanings (sub-frames) that can be attributed to a concept that seems to attract a broad level of support. Sustainability is a concept that, at least superficially, attracts a broad consensus. After all, who would support 'unsustainable' food systems?

However, proponents of sustainable food systems often hold contradictory positions, arguing that their practices and approaches can alone deliver sustainable food systems (Candel et al. 2014). A common dichotomy, for example, is between 'land sparing' and 'land sharing' approaches. 'Land sparing' assumes that some land is set aside for conservation purposes whilst another land dedicated to agriculture is subject to 'sustainable intensification' to optimise food production (e.g. using 'reformist' practices such as precision agriculture for more efficient resource use). 'Land sharing' assumes that agricultural land is farmed using more nature-friendly, agroecological methods that balance food production with reducing environmental impacts and promoting conservation and biodiversity. Chapters 4–7 present alternative and competing visions of what sustainable food systems should or could look like. Because of the variety and often disparate nature of these vying approaches, an outcome interpretation of sustainable food systems has been adopted.

8.2 Comparing the performance of different food system paradigms against sustainability criteria

This chapter compares neo-productivist, reformist, progressive, and radical paradigms, using the available peer-reviewed literature. Comparisons are made through four different 'sustainability' lenses or criteria (Zurek et al. 2018):

- food and nutrition security;
- environmental sustainability;
- economic sustainability; and

- social sustainability.

This chapter will also review the current political sustainability of, and support for, these paradigms. Stability and resilience are internalised as cross-cutting themes (Béné et al. 2019).

The criterion of food and nutrition security also encapsulates the recent pivot towards improved health (Galli et al. 2018). Health is not treated as a separate lens because health outcomes are impacted by a wide range of factors, such as genetic disposition, wealth, water, and sanitation infrastructure (Béné et al. 2019). However, several studies do compare different approaches to sustainability based on their potential contribution to improved health.

8.3 Food and nutrition security

To date, there are no comparative studies that assess the contribution of neo-productivist, reformist, progressive, and radical food system approaches to food and nutrition security. However, there are numerous individual studies and some meta-analyses that compare neo-productivist/reformist (conventional agriculture) against progressive approaches, especially organic agriculture.

8.3.1 Conventional (neo-productivist and reformist) versus organic crop production

Crop productivity, expressed as crop yield per area, is the most used metric for measuring the food production capacity of comparative production systems. Comparisons are generally made between monocultures, namely fields producing only one crop species, e.g. wheat, maize, rice, or soybeans. Most short-term and long-term comparative studies have been undertaken in temperate regions, with less work being undertaken in tropical regions, especially in terms of long-term comparisons (Bationo et al. 2012; Seufert et al. 2012).

In these comparisons, yields of crops produced using conventional technologies and practices tend to consistently outperform those crops grown using organic technologies and approaches. The yield gap between conventional and organic crops varies from 15% (Meng et al. 2017) to 34% (Benton and Harwatt 2022), up to a significant 42% (Smith et al. 2018; OECD 2021; Seufert et al. 2012; De Ponti et al. 2012; Hansen 2019; Ponisio et al. 2015; Meng et al. 2017). On average, organic agriculture is generally perceived to yield approximately 20–25% less than conventional agriculture (Seufert et al. 2012; Reganold and Wachter 2016). However, studies have determined that yield gaps between conventionally and organically grown crops vary significantly by factors such as

- crop species;
- management practices;
- climate and soil conditions; and
- systems-level impacts.

(de Ponti et al. 2012; Seufert et al. 2012; Ponisio et al. 2015; Gabriel et al. 2013; Ramankutty et al. 2019; Joseph et al. 2019).

- **Crop species and management practices**

The performance of organic versus conventional varies significantly depending on crop species and management practices. For example, organically produced cereal crops (such as wheat, barley, and maize), generally produce around 75% of the yield of conventionally grown crops. However, studies have shown significant yield variation between organically and conventionally grown cereals, with organically grown crops in some studies producing only 42% of conventional yields, whilst in others, they have delivered yields of up to 85% of conventionally grown cereals (De Ponti et al. 2012; Smith et al. 2018; Joseph et al. 2019; Wachter et al. 2019). Spring-sown organically grown cereals have also been shown to yield 82% of conventionally grown spring cereals (Klima et al. 2020). Similar variance has been found in comparative studies involving fruit and vegetables, with organically farmed fruits and vegetables producing between 72% and 90% of conventionally produced fruits and vegetables (Meng et al. 2017; Joseph et al. 2019).

Other studies stress the enhanced performance of organic mixed cropping systems (often a cereal and legume crop grown in the same field at the same time) vis-à-vis the conventional production of monoculture crops (only one crop grown in the same field at the same time), especially under low fertility conditions (Adamtey et al. 2016). According to Klima et al. (2020), spring cereal mixtures (oats and triticale) produced 8.5% more than conventionally produced mono-culture spring cereals.

- **Climate and soil conditions**

High soil fertility (defined by high organic carbon and sufficient quantities of major nutrients), balanced pH, as well as ample amounts of well-distributed rainfall, have been demonstrated to dramatically reduce the yield gap between organically and conventionally produced crops (Tuomisto et al. 2012; Benton and Harwatt 2022). Indeed, in some cases, organically grown crops have produced similar, or even higher, yields than conventionally grown crops (Lotter et al. 2003; Mucheru-Muna et al. 2014; Seufert et al. 2012; Joseph et al. 2019). Organically produced crops grown on low-fertility soil have also been reported to generate comparable yields to conventionally grown crops (Gomiero et al. 2011; Forster et al. 2013).

Other studies have looked at food security more in terms of stability of supply, e.g. in adapting to more unpredictable and extreme weather events associated with climate change. Studies have documented higher performance of organically grown crops vis-à-vis conventionally grown crops when they have been exposed to water stress, both drought and excess rainfall. In part, this has been attributed to the higher water-holding and water infiltration rates of soils with high organic matter content (Reganold and Wachter 2016; Seufert et al. 2012). The high organic matter content of the soil, often associated with organic production systems, is also seen as providing enhanced soil stability and yield resilience (Pretty and Hine 2001; FAO 2002b, 2008; Halberg et al. 2006; Badgley and Perfecto 2007; Badgley et al. 2007; El-Hage Scialabba 2007; Niggli et al. 2007, 2008).

- **Systems-level impacts**

It is also essential to take account of system-level dimensions (i.e. field, farm, landscape/watershed, regional, and global) when making comparisons between organically and conventionally grown crops. Here, several studies have stressed the reliance of organic production on inputs such as farmyard manure, which, in turn, relies on additional land being dedicated to the production of livestock and livestock feed, or the need to include lower-yielding legumes in the farm's crop rotation to build soil fertility, suggesting more land is needed for organic crop production. When this additional land is factored into the comparison, the yield gap between organic and conventionally produced crops increases (De Ponti et al. 2012).

Lastly, in comparisons of organically vis-à-vis conventionally produced foods, it is important to consider quality alongside quantity. Here, organically grown crops are argued to have both a nutritional (secondary metabolites) and health advantage over conventionally produced crops that have been exposed to synthetic fertilisers and pesticides, and those that may have been genetically modified (Kesse-Guyot et al. 2013; Baudry et al. 2015; Ramankutty et al. 2019).

8.3.2 Conventional (neo-productivist and reformist) versus organic livestock production

In many respects, the yield gap between organically and conventionally farmed livestock, and livestock products, resembles the situation with crops. At just 33% of the conventional yield, poultry is the poorest performer, with pork production at 54%, beef production at 60%, and fish and fish products at 80% of conventional yields (Joseph et al. 2019). In a similar vein to organic crops, the quality of inputs, especially feed, is responsible for the reduced performance of organic livestock, especially housed livestock such as chickens and pigs (Smith et al. 2018). In outdoor grazing systems for beef and sheep, the reduced quantity of grazing/fodder limits yields (Smith et al. 2018). Again, the nutritional and health comparisons between organic and conventional livestock production systems produce similar results to crop comparisons. Here the highly restricted use of antibiotics, and the ban on growth hormones, in organic systems is seen by many consumers as significantly preferable (Muscanescu 2013).

8.3.3 Conventional (neo-productivist and reformist) versus agroecological agriculture

Whilst similar in many ways to organic agriculture, at the production level, agroecological agriculture tends to focus more explicitly on building soil biodiversity and fertility, as well as on often quite complex/intricate crop species mixes, crop rotations, green manuring, and composting regimes. However, unlike the numerous studies undertaken comparing organic versus conventional yields, few, if any rigorous comparisons exist for agroecological versus conventional agriculture (Bernstein 2014). Indeed, because of this, scientists wanting to compare the two approaches have often resorted to using surrogate comparisons of organic and conventional.

Such work that has been done often relies on comparisons between monocultures and polycultures (also known as mixed cropping), across different climatic zones. Consequently, this lack of like-for-like comparison has allowed proponents of conventional agriculture to declare an almost default position that agroecological systems significantly underperform when compared against conventionally produced crops (Holt Giménez and Altieri 2013; Bellamy and Ioris 2017). To this end, there does appear to be some agreement that conventionally grown crops, such as maize (grown in a monoculture), will generally outyield maize grown agroecologically (in a monoculture).

However, proponents of agroecological agriculture are quick to retort that crops are rarely ever grown in monoculture on agroecological farms. Where crops are produced in polycultures, several researchers have claimed that agroecological yields can match or even exceed conventional agriculture yields (Altieri and Toledo 2011; Rosset et al. 2011; Francis 1986; Hecht 1989; Pimentel and Pimentel 1979). For example, in Cuba, since the widespread adoption of agroecological agriculture, reported yields have tripled between 1998 and 2009 (Rosset et al. 2011). In Brazil, reports suggest that mixed cropping of maize and beans has resulted in yield increases of up to 28% when compared to monoculture yields of the same crops under conventional management (Francis 1986). Indeed, some comparisons in other parts of the Amazon have claimed a doubling of yields compared to conventional farming (Hecht 1989). There appears to be a growing acknowledgement that, under certain conditions (enabling soil, climate, local knowledge, etc.), polycultures outperform conventionally produced monocultures on a yield/area basis (Badgley et al. 2007). However, scaling these successful agroecological approaches has proved difficult and has constrained agroecological success to specific localities (Jansen 2015; Martínez-Torres and Rosset 2014). Concerns also abound regarding the significant requirement for farm labour needed to manage polycultures. High demand for relatively unskilled manual labourers may work in certain geographies, especially where indigenous peoples wrest a living from their environment in far-flung corners of the developing world, but it would struggle to work in developed food systems, where finding anyone to work the land is becoming increasingly challenging. Aside from the growing, yet proportionately small, number of lifestyle farmers, consciously adopting agroecological practices, the wholesale transition to agroecological food systems would most likely suffer from severe labour shortages (Jansen 1998; Bellamy and Ioris 2017; Pereira et al. 2018; Benton et al. 2021).

Whilst there have been limited comparisons between the nutritional outputs of agroecological and conventional approaches, at least in small and peasant-based food systems, the sheer diversity of agroecological polycultures, especially in tropical and sub-tropical geographies, should be able to deliver highly nutritious diets throughout most of the year (Altieri 2000). Lastly, agroecology's use of polycultures and focus on soil health claim to provide greater economic, and biophysical resilience, especially when exposed to the extreme weather conditions promulgated by climate change (Rosset et al. 2011; Pretty 1995; Holt-Giménez 2002; Badgley et al. 2009; Pretty and Hine 2000).

8.3.4 Conventional (neo-productivist and reformist) versus regenerative agriculture

To date, very few comparative analyses of regenerative agriculture versus conventional agriculture have been undertaken. Those that exist are often contested on the grounds of perceived robustness, bias, and incompatibility of comparisons. In a similar vein to both organic and agroecological approaches, under specific circumstances, such as enabling soils, climate, farmer knowledge, mixed cropping, etc., regenerative agriculture has been reported as delivering similar or even higher yields when compared to conventional farming approaches (EIT Food 2020). Other studies, however, suggest that, in line with both organic and agroecological approaches, regenerative agriculture generally produces around 71% of conventional yields (LaCanne and Lundgren 2018; Beyond Pesticides 2019; EIT Food 2020; Milinchuk 2020).

Recent research on the impact of increased soil organic matter (SOM) and associated microbial communities on crop nutrient density suggests that the conversion to regenerative agriculture could result in the production of food with higher micronutrients and phytochemical content (Montgomery and Biklé 2021). This compares to increased levels of Cd, Ni, and Na in conventionally farmed crops, which are harmful to human health. Regeneratively farmed livestock was also shown to have higher levels of conjugated linoleic acid (CLA), omega-3, and omega-6 fats (Montgomery et al. 2022).

8.3.5 *Conventional (neo-productivist and reformist) versus new technologies*

Lastly, it would be amiss not to explore the potential impact of rapidly evolving technologies within the neo-productivist and reformist food system paradigms. Whilst the potential of new neo-productivist technologies, such as cultured and plant-based meat analogues and vertical agriculture, have been acknowledged as increasing the speed and efficiency of producing food products, as well as the high level of productivity of yield/area, to date, no analyses have been undertaken to determine the performance of these new approaches vis-à-vis other production models or, indeed, the potential contributions of these new technologies to global food and nutrition security. Indeed, comparisons are made even more challenging in the case of vertical agriculture, as many operators are using organic or similar production methods. Ultimately, whilst these new approaches are energy-intensive, their productivity per unit of land holds promise, especially if production continues to be in areas of high population (Benton et al. 2021).

8.3.6 *What's needed to ensure food and nutrition security?*

Ultimately, whilst disagreements abound (Bagley et al. 2007; Halberg 2006; Seufert et al. 2012; Ponisio et al. 2015; Altieri 2002; Wezel et al. 2013; Jansen 2015), most analyses conclude that the replacement of conventional agriculture with alternative food system approaches (organic, agroecological, regenerative, etc.) would lead to reduced yields, and the consequent need to expand agricultural lands to balance global food production with food consumption (Candel et al. 2014; Feuerbacher et al. 2018; Desquilbet et al. 2018; Giller et al. 2021; Muller et al. 2017). Indeed, according to OECD (2021), a yield gap of 20% would necessitate an additional 25% of cropland being brought into production to produce the same yield as conventional approaches. An even larger yield gap of 30% would necessitate an additional 42% of cropland being brought into production, with an associated increase in GHGEs, biodiversity loss, and environmental pollution (Benton and Harwatt 2022).

To date, little analysis has been done to model the expected impacts of 100% conversion to alternative food systems (Feuerbacher et al. 2018). Where real-life examples exist of country-wide transitions to organic agriculture, there have been less than convincing results. On the one hand, the disastrous transition to organic agriculture in Sri Lanka sent alarm bells ringing across the globe. On the other hand, both Bhutan and Cuba have been hailed for their successful national transition to organic agriculture. However, in the case of Bhutan, whilst the government set a target of 100% transition between 2012 and 2020, recent estimates suggest that this target probably will not be achieved until at least 2035 (Böll 2022). Ultimately, whilst 80% of Bhutan's farmers are organic by default (i.e. no synthetic inputs, etc.), only 5.6% of the land area is certified as organic (Böll 2022). In the case of Cuba, whilst the government is promoting organic, only 20% of output is organic, with the remaining 80% of the country's output being produced on intensive/non-organic farms. Cuba imports between 60% and 80% of its food requirements (Guardian 2017). Also, population density is also a major consideration in the case of Bhutan and Cuba. The population density of Cuba (102 people per km^2), especially Bhutan (20 people per km^2), gives significant room for manoeuvre compared to other countries with much higher population densities, such as the UK (276 people per km^2) and the Netherlands (520 people per km^2). Indeed, the current trajectory of the European Union to produce at least 30% of food using organic farming methods has stimulated significant debate (Moudrý et al. 2018; Wezel et al. 2018). A lack of nitrogen is cited as probably the most limiting factor of the economy-wide transition to organic production. Pro-organic scientists have been accused of grossly overestimating the potential of organic systems to generate enough nitrogen to produce the high yields required at a global level (Muller et al. 2017). Indeed, several analysts suggest that the global conversion to organic agriculture could only support a global population of three to four billion (Buringh and van Heemst 1979; Smil 2000).

Depending on how much the global population increases before plateauing out, it is important to remember that dietary changes and a reduction in food losses and food waste could potentially reduce the pressure on the yield gaps that organic, agroecological, and regenerative agricultures need to fill (Benton and Harwatt 2022). Ultimately, the big question is, how much can the yield gap between alternative and conventional food systems be closed through ecological intensification, how much can diets be changed, and how much can food losses and waste be reduced? (Priefer et al. 2013; Doherty et al. 2017; von Koerber et al. 2017; Joseph et al. 2019; Hunter et al. 2017; Muller et al. 2017; Benton et al. 2021; Bellamy and Ioris 2017).

8.4 Environmental sustainability

To date, there are no comparative studies that assess the environmental sustainability of neo-productivist, reformist, progressive, and radical food system approaches. Where studies exist, they generally compare conventional versus organic approaches, or long food supply chains (LFSCs) against short food supply chains (SFSCs). Broad-based agroecological (including Organic and regenerative agriculture), or similar approaches, which are based on crop, livestock, and landscape diversity, reduced use of synthetic/industrial inputs, and active promotion of biodiversity (including subterranean diversity) are automatically attributed with possessing higher levels of biodiversity in farmed lands. However, as discussed in Chapter 6, due to generally lower yields, these approaches are open to criticism due to the associated need to expand farmed lands into areas of semi-natural and natural habitats.

This section summarises evidence on the ways in which neo-productivist, reformist, progressive, and radical food systems attempt to balance nature, environment, and food production, especially regarding greenhouse gas emissions (GHGEs), biodiversity, soil health, and resilience.

8.4.1 Greenhouse gases

Due to their reliance on cheap and plentiful fossil fuels, conventional food systems are traditionally seen as being highly polluting, especially regarding the production of GHGEs, and their subsequent exacerbation of climate change. However, there are several new neo-productivist technologies and innovations that claim to greatly reduce the amount of GHGEs produced per unit of food consumed. These include key advances, such as GMOs and new breeding techniques (NBTs), cultured and plant-based meat analogues, and controlled environment agriculture.

Studies suggest that compared to non-GMO conventional production, GMOs and NBTs reduce crop production-related GHGEs (Kovak et al. 2022). For example, it is suggested that the adoption of insect-resistant and herbicide-tolerant GMOs globally has reduced pesticide use by 775.4 million kg (8.3%), as well as reducing the amount land cultivation (Brookes and Barfoot 2018), with significant savings in fuel consumption, equivalent to the GHGEs of 15.27 million cars (Brookes and Barfoot 2020), and that this figure would be even higher if nano-additives were added to the fuel to further increase fuel economy. According to Kovak et al. (2022), if GMOs were adopted widely across Europe, total agriculture-related GHGEs could be reduced by an estimated 7.5%. Ultimately, GMOs' greatest contribution to the reduction of GHGEs is through higher yields and their land-sparing effects (Kovak et al. 2022). Other technological advances, such as the use of renewable/green energy to power farm machinery, food transportation, processing, and storage, as well as the production of agricultural inputs, especially fertilisers and pesticides, have the potential to dramatically reduce GHGEs, even to the point of the development of carbon-neutral, and potentially carbon-negative food systems (Brown 2018).

It is also claimed that, compared to the conventional approach to meat production, i.e. grazing sheep, beef yards, and poultry sheds, the production of both cultured and plant-based meat analogues reduces GHGEs between 30% and 95% (ETC Group 2019a; Burton 2019; Credit Suisse 2021; Future Meat Technologies 2022; Webb and Hessel 2022). See Chapter 4 for more details. Again, GHGEs would be even lower if the production of cultured and plant-based meat analogues can be undertaken using 100% renewable energy. Indeed, according to solar foods, which uses 100% solar energy to produce a plant-based meat substitute, their production of the natural protein Solein generates just 1% of the GHGEs produced by conventionally farmed animals (Credit Suisse 2021). Due to the rocketing cost of fossil fuels during 2021 and 2022, emergent vertical farms are being forced to explore cheaper sources of renewable energy (solar and wind, etc.) to remain profitable. Many now aspire to become carbon-negative (Agrizon 2021; Eastern Daily Press 2021). Additionally, when cultured and plant-based meat analogues, and vertically farmed produce, as well as organic and agroecological-based produce, are produced close to high consumption areas, SFSC can, if organised well, reduce GHGEs by extending food shelf life, reducing food waste, minimising transport costs, and reducing food miles (Born and Purcell 2006; van Delden et al. 2021).

Reformist technologies and innovations, such as precision agriculture and climate-smart agriculture, also lay claim to reducing GHGEs through the more precise management of agricultural inputs (fertiliser, pesticides, energy, etc.), which reduces the generation of GHGEs during production, transportation, and application of inputs, and reduces farm pollution and waste, whilst maintaining or even increasing yields. Reformist practices, such as minimal cultivation and Conservation Agriculture, can also contribute to CO_2 sequestration via increases in soil organic matter, at least in the top few centimetres of the soil profile (Nyasimi et al. 2014). Meta-analysis of almost 200 studies conducted during the past 15 years determined that minimal cultivation increased SOC by 0.06 g C 100 g^{-1} compared to conventional agricultural practices (Jordan et al. 2022).

Progressive technologies and innovations, such as regenerative agriculture, organic agriculture, and agroecological agriculture, tend not to focus on reducing GHGEs through increasing yields. Here, it is argued that GHGEs are lowered via increased carbon sequestration in the soil or in complex tree, livestock, and cropping systems, and reduced soil disturbance that inherently stores more carbon than conventional monoculture cropping or livestock systems. For example, progressive technologies, such as regenerative agriculture (no-till, cover crops, and diverse rotations), claim to sequester significant amounts of carbon and claim to be able to more than double SOM, especially on small vegetable farms (Montgomery et al. 2022). Whilst it is widely recognised that regenerative agriculture does increase soil organic carbon, the scale of the claims made is generally unsubstantiated and requires further research (Paustian et al. 2020). For example, whilst no-till practices increase SOC in the top few centimetres of the soil, it can lead to reduced SOC in lower layers of soil (Powlson et al. 2014).

In addition to the sequestration of CO_2, regenerative agriculture also reduces the generation of GHGEs by reducing input, machinery, and fuel use (e.g. by replacing ploughing with direct drilling), often without crop yield losses (Jordan et al. 2022). Organic agriculture also claims to reduce direct fossil fuel energy consumption (linked to the production of synthetic inputs), leading to a 30% reduction in GHGEs compared to conventional agriculture. However, it is important to note that, whilst organic farming generally generates lower levels of GHGEs, the fact that yields per hectare are so much lower, often results in increased GHGEs per tonne of yield produced (Clark and Tilman 2017). Indeed, on a yield basis, studies have demonstrated that organic generates an average of 23% more nitrous oxide emissions and 49% more methane emissions than conventional/industrial production systems (Ramankutty et al. 2019).

Studies have also demonstrated that radical food systems, such as SFSC, often do not lead to reduced GHGEs (Schmitt et al. 2017). This is based on the argument that large-scale bulk transportation of food along global supply chains (by cargo ship and rail) is often much more efficient per unit than SFSC transportation, which may be by small vans and cars (Coley et al. 2011, 2009; Wallgren 2006).

8.4.2 Biodiversity

Due to the reliance on intensive mono-culture production systems and their associated negative externalities, conventional industrial food systems are seen as highly destructive and destabilising for the conservation of biodiversity. For several decades, proponents of industrial food systems have claimed that their productivity has spared the conversion of both pristine virgin and highly biodiverse semi-natural habitats into farmland, either for arable cropping, livestock grazing, or plantation crops. There is broad agreement between ecologists that land spared from agricultural conversion or mono-cultured tree production, has the highest levels of biodiversity (Phalan et al. 2011).

Additionally, throughout the past 30 years or so, the most environmentally damaging and persistent pesticides have been significantly reduced or removed from production. Figure 105 demonstrates the notable reduction in application rate, toxicity, and persistence of pesticides in the USA between the late 1968 and 2008. More recently, a range of non-selective pesticides (especially insecticides) have been removed from production or restricted in their use (Watson 2018). In studies undertaken by Pretty and Bharucha (2013), pesticide use across target countries has been reduced by 44% in the UK, 38%

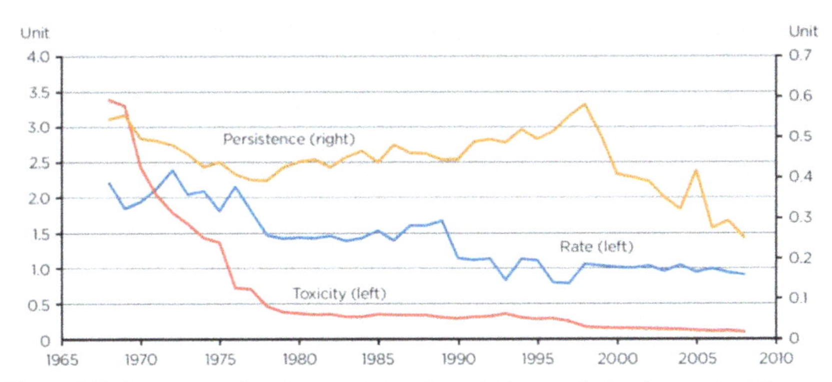

Figure 105 Average quality characteristics of pesticides applied to four major US crops, 1968–2008. Source: Adapted from: Phillips McDougall (2019).

in France, 32% in Japan, and 24% in Vietnam. According to Perry et al. (2016), whilst herbicide applications in soybean crops have increased (primarily due to the evolution of GM-resistant weeds), insecticide use has reduced by over 11.2% compared to non-GM production systems (Campos et al. 2019).

In addition to these traditional defensive positions, new neo-productivist approaches to food production, such as cultured meats and plant-based meat analogues, and controlled environment agriculture, especially vertical farming, are making even bolder claims regarding the conservation of natural habitats. For example, in Chapter 4, it is claimed that the production of plant-based meat analogues uses 47–99% less land compared to farmed livestock (Burton 2019; Credit Suisse 2021; Future Meat Technologies 2022). Furthermore, in Chapter 4, it is claimed that, due to highly conducive growing environments, crops produced in vertical farms grow 30–40% faster than those conventionally farmed (Little 2021) and could produce 350 times more output per unit of land than conventional farming (Credit Suisse 2021; Little 2021). If widely adopted, both the cultured and plant-based meat revolution and vertical farming have the potential to take the land-sparing debate to a completely new level.

Reformist approaches, especially the development and application of precision agriculture, have also reduced the impact on biodiversity, both within cropped fields as well as the surrounding environment. Precision agriculture, which increasingly involves micro-doses of pesticides (see Chapter 5), dramatically reduces the amount of active ingredient (pesticide agent or nutrient, etc.) applied to the soil, and emergent nanotechnologies significantly reduce and target applications of pesticides and fertilisers, etc. Precision agriculture also endeavours to ensure that the crop inputs are applied in the right amounts, in the right place, and at the right time to maximise crop growth whilst reducing waste and impact on the environment (Neo 2020; Pahpy 2020). Other reformist approaches have also demonstrated significant reductions in pesticide use. For example, in a study spanning a 20-year timeframe and 24 countries, integrated pest management (IPM) was demonstrated to have reduced pesticide use by 69%, whilst yields had risen by 41% (Pretty and Bharucha 2013).

Conversely, both progressive and radical approaches to sustainable food systems tend to promote a more land-sharing approach (see Chapter 6). This is achieved by increasing the diversity of crops, livestock, and trees integrated into the farming system, and reducing or eliminating the use of all synthetic inputs, such as pesticides and fertilisers, etc. See Chapter 6 for details. Whilst conventional/industrial food systems rely on a handful of crop species, agroecological food systems rely on a diverse range of crop species. In India, up to 1500 species are eaten. Globally, it is estimated that up to 300 000 plant species are potentially edible (Benton et al. 2021). In addition, agroecological farmers often maintain areas such as hedges, herbaceous areas, woodlands, and ponds and wet areas for beneficial organisms (predators of crop pests) and wildlife (Gomiero et al. 2011; Benton et al. 2021).

Most analyses that have been undertaken support the argument that food system approaches using a broad agroecological approach, such as organic, regenerative, permaculture, and agroecological agriculture, work with, rather than against, natural systems, have greater in-field and on-farm biodiversity, and reduce negative impacts on the environment (Gomiero et al. 2011; Reganold and Wachter 2016; Seufert and Ramankutty, 2017; Cisilino et al. 2019; Gaitán-Cremaschi et al. 2020; Seufert et al. 2019; Benton et al. 2021; Gabriel et al. 2013). For example, compared to conventional/industrial agriculture, Organic Agriculture has been demonstrated to increase organism abundance by between 40% and 153% and species diversity from between 1% and 105% (Fuller et al. 2005; Seufert and Ramankutty 2017; Ramankutty et al. 2019). Interestingly, it is suggested that the conventionalisation of organic agriculture towards large-scale mono-culture production could erode much of the floral and faunal diversity of this approach (Guthman 2000; Templer et al. 2018; Rosa-Schleich et al. 2019).

Plant diversity and pollinators benefit significantly under organic management practices (Seufert and Ramankutty 2017). Earlier studies, by Fuller et al. (2005), suggested that organic fields contained 5–48% more spiders, 16–62% more birds, and 6–75% more bats. However, biodiversity demonstrates significant spatial variability. In some situations, small-scale conventional/industrial agriculture in heterogeneous landscapes, where uncropped areas act as a refuge for weed populations, can deliver similar or higher levels of biodiversity compared to organic systems (Gomiero et al. 2011; Gabriel et al. 2010; Dainese et al. 2019). Much of the plant (weed) diversity is directly related to the reduced use of inputs and crop diversity (Gomiero et al. 2011). Indeed, plant diversity in organic systems is often like that found in low-input conventional/industrial production systems, which suggests that it is often the intensity of production (especially weed management) that determines both the floral and faunal diversity (Gabriel et al. 2013). The yield-biodiversity relationship has been well-studied (Seufert et al. 2012). Ultimately, in a similar vein to the generation of GHGs, when biodiversity is measured against crop yields, sensitively managed conventional/industrial farms, with patch-work areas and corridors set aside for biodiversity conservation, may indeed be more efficient than large areas of lower-yielding organic, regenerative, or agroecological cropping in conserving biodiversity (Green et al. 2005; Benton and Harwatt 2022).

Lastly, in a similar vein to the arguments above, SFSCs, especially based on local organic or explicit agroecological production methods, are generally argued to enhance local biodiversity (Lamine et al. 2012; Brunori et al. 2011). This is based on the premise that SFSCs allow closer monitoring of local food production, as well as easy and frequent communication between consumers and producers (Vittersø et al. 2019). Ultimately, whilst a closer and more collaborative relationship between producers and consumers could lead to increased local biodiversity, several studies have warned of what they term the 'local trap', whereas consumers automatically assume that local production is more biodiverse and environmentally sensitive than LCSCs. The case can often be the reverse (Purcell and Brown 2005; Born and Purcell 2006).

8.4.3 Soil health

The loss of SOM and soil organic carbon (SOC), along with increased soil compaction, erosion, and contamination, have been widely documented over the past decades. What is somewhat lacking in the literature is neo-productivist attempts to increase SOM and SOC. Some attempts, however, have been made under reformist approaches to address soil health, targeting soil structure, biological activity, and remediation. These approaches, which include the development of nanotechnologies to reduce soil erosion, increase soil fertility and water-holding capacity, and decrease SOM decomposition, as well as both minimal (Gomiero et al. 2011) and zero tillage (Conservation Agriculture), are outlined in Chapter 5. Soil health, however, has been a central focus of progressive food systems approaches, such as organic, regenerative (which are usually based on organic practices), and agroecological agricultures (Seufert and Ramankutty 2017; Reeve et al. 2016).

Whilst it often takes several years, the conversion to organic-based agricultures generally results in measurable increases in SOM (Clark et al. 1998; Reganold et al. 1987; Reganold 1995; Drinkwater et al. 1998; Siegrist et al. 1998; Fließbach et al. 2000, 2007; Glover et al. 2000; Stölze et al. 2000; Stockdale et al. 2001; Mäder et al. 2002a; Lotter et al. 2003; Delate and Cambardella 2004; Pimentel et al. 2005; Kasperczyk and Knickel 2006; Marriott and Wander 2006; Briar et al. 2007; Liu et al. 2007). Whilst results vary by soil type and climate, on average, soils under organic management have 11% more SOM compared to conventionally managed fields (Ramankutty et al. 2019). Much of the increased SOM is found in the uppermost parts of the soil profile (Gomiero et al. 2011). Trials conducted by the Rodale Institute, over a 22-year period, recorded a 27.9% and 15.1% increase in soil carbon from organic animal and organic legume production systems (Gomiero et al. 2011). In a 180-year-long trial (Broad balk) at Rothamsted in the UK, SOM has increased by over 120% under organic manured production (Gomiero et al. 2011). In turn, SOM has a direct effect on soil biodiversity, especially soil microbes, and soil micro and macro fauna (Paoletti et al. 1995, 1998; Gunapala and Scow 1998; Fließbach and Mäder 2000; Hansen et al. 2001; Mäder et al. 2002a; Marinari et al. 2006; Tu et al. 2006; Briar et al. 2007; Fließbach et al. 2007; Liu et al. 2007; Birkhofer et al. 2008; Phelan 2009). For example, the formation of mycorrhizae activity can be more than 40% higher under organic management compared to conventional management, and earthworm biomass and abundance can be between 30% and 320% higher under organic management compared to conventional management (Gomiero et al. 2011).

However, whilst organic management generally leads to higher levels of SOM compared to conventionally managed fields, some studies have found no significant differences between fields under organic and conventional management (Kirchmann et al. 2007). In some cases, organic fields were even more prone to the leaching of nitrates and phosphates when compared to conventionally managed fields (Migliorini et al. 2018; Seufert et al. 2017).

8.4.4 Resilience

The concept of resilience is widely recognised as the capacity of a given socio-ecological system to absorb an external shock/disturbance and to reorganize/adapt to retain, bounce back, or improve upon previous socio-ecological conditions (Walker et al. 2004; Barrett and Constas 2014). In recent years, climate change has been a major shock to global food systems. Neo-productivist food systems initially responded to climatic shocks through the development of drought, waterlogging, heat, pests, and diseases resilient crop varieties, and improved crop management (Watson 2019a,b). More recently, plant breeding has benefited from the advent of gene editing (see Chapter 4) and deep learning neural networks, which harness big data and machine learning to predict bespoke genotypes for specific environments affected by climate change. Vertical farming and both cultured and plant-based analogues rely much less on climatic variability and, whilst rarely championed in discussions on resilience, could also be seen as innovations that build food system resilience to climatic shocks. Reformist technologies such as climate-smart agriculture, Conservation Agriculture, and precision agriculture (using emergent nanotechnologies) also make a significant contribution to resilience (see Chapter 5).

All progressive innovations, such as regenerative, organic, and agroecological agricultures, claim to contribute to climate resilience. This claim is primarily linked to increased SOM associated with these approaches as well as the diverse crop, livestock, and tree crop rotations. For example, under severe drought conditions, organic fields containing higher SOM (and subsequent improved structure and higher water-holding capacity) have been shown to out-yield conventionally managed crops by up to 90% (Lockeretz et al. 1981; Stanhill 1990; Smolik et al. 1995; Teasdale et al. 2000; Lotter et al. 2003; Pimentel et al. 2005; EIT Food 2020; Mijatovi et al. 2013; Leippert et al. 2020). For every 1% increase in SOM, it has been estimated that a hectare of soil can hold an additional 10 000–11 000 litres of plant-available water (Sullivan 2002). Landscape complexity and high levels of agrobiodiversity, including the conservation of local crop varieties and crop wild relatives, increase resilience to pests and diseases, stabilize livestock production, and buffer against climatic shocks, such as drought, erratic or untimely rainfall, and floods (Altieri et al. 2015; Altieri and Manuel Toledo 2011; Tittonell 2015; Kremen and Merenlender 2018; Kerr et al. 2023; Cabell and Oelofse 2012; Rosa-Schleich et al. 2019; Benton et al. 2021) (see Chapter 6). Lastly, radical approaches, such as permaculture, incorporate multi-layered cropping and seasonal diversity by design to bolster overall system resilience (see Chapter 7).

The literature appears less aligned with regard to food system resilience to economic shocks. Neo-productivist and reformist food systems, which dominate much of the developed and developing world, rely almost exclusively on the capacity of the market to balance supply against demand and to provide an adequate return on investment for all food system actors. In cases where farmers' margins are squeezed too tight, existing or new, government subsidies, and legislation, are expected to buffer the economic impact until the market can re-balance. Economic shocks associated with recent food crises (2008/2009, 2011/2012, 2019 (COVID-19), 2022/2023) have generally quickly subsided, with normal conditions returning after just a couple of years. With government support, most food chain actors can weather economic shocks and continue to grow (Soubry and Sherren 2022). Where government support is not forthcoming, economic shocks, such as those associated with the COVID-19 pandemic, can force some food system actors to either temporarily or permanently close their businesses and focus on more rewarding opportunities in other sectors (Watson 2021). Progressive food system approaches, such as organic and regenerative, generally rely on price premiums to compensate for lower yields and high labour costs. Where these premiums and market share can be maintained, organic and regenerative farmers, who rely less on bought inputs and have a more diverse range of crops, are often able to survive protracted economic shocks (Reganold and Wachter 2016). Whilst approaches such as agroecological agriculture are generally seen as more resilient to economic shocks due to the limited reliance on input markets, recent research on the impact of COVID-19 on agroecological food systems in Guatemala has raised several concerns. According to Rice et al. (2023), whilst food self-sufficiency improved for many farmers, and local non-farming rural consumers, during the pandemic, with a more diverse range of crops planted, increased cropping intensity, and more cultivated land, nearly a third of farmers could not sell food to urban consumers. Even where sales to urban centres were possible, these tended to be restricted by distance, with only nearby centres being provided with food (Rice et al. 2023). Those farmers who depended on markets in distant urban centres for their livelihoods suffered significant economic hardship.

8.5 Economic sustainability

For decades, governments across the world have endeavoured to develop sustainable food systems that provide adequate amounts of low-cost staple foods. Large agri-food corporations have evolved that have invested heavily in the supply of low-cost foods alongside healthy profit margins. The provisioning of low-cost, healthy, and nutritious foods for predominantly highly urbanised populations remains a priority for all governments. This priority has been somewhat elevated after a run of food crises (2008, 2011/2012, 2022/2023), and the COVID-19 pandemic.

8.5.1 Neo-productivist systems

Whilst environmental externalities remain a significant challenge, industrial food systems dominate the provisioning of cheap food. Cheap food has and continues to be the trump card of industrial neo-productivist food systems. The adoption of neo-productivist technologies, such as GM and gene-edited crops and plant-based meat analogues, has further reduced the cost of production and increased food system profitability compared to conventional and other reformist, progressive, and radical food systems (Klümper and Qaim 2014; Méda et al. 2017; Gaitán-Cremaschi et al. 2020; Durham and Mizik 2021). Brookes (2022) estimates that between 1996 and 2020, the adoption of GMO crops has increased income by US$261.3 billion, or an average of US$112/ha/yr, compared to conventional farms, benefiting developed and developing countries almost equally. Figure 106 illustrates income gains associated with the use of herbicide-tolerant maize.

Increased incomes have been derived primarily from increased yields (330 million tonnes of soybeans and 595 million tonnes of maize), as well as reduced input costs. It is suggested that, without the use of GM crops, more than 23

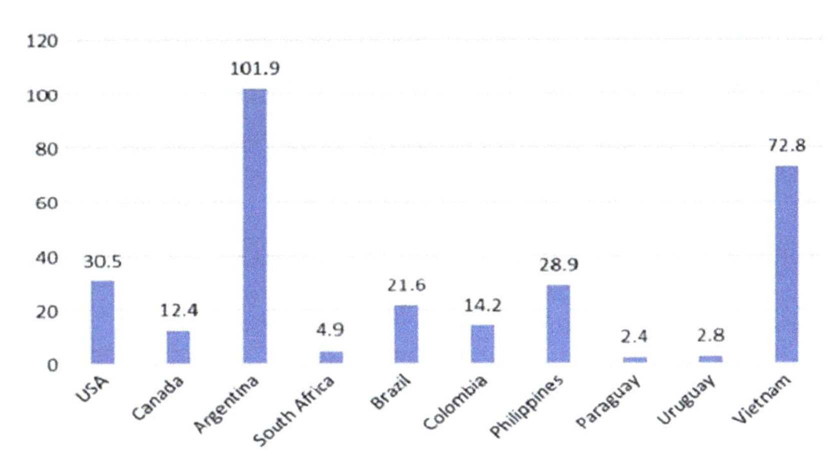

Figure 106 Average farm income gain (percentage) from using GM HT maize by country, 1997–2020. Source: Adapted from: Brookes (2022).

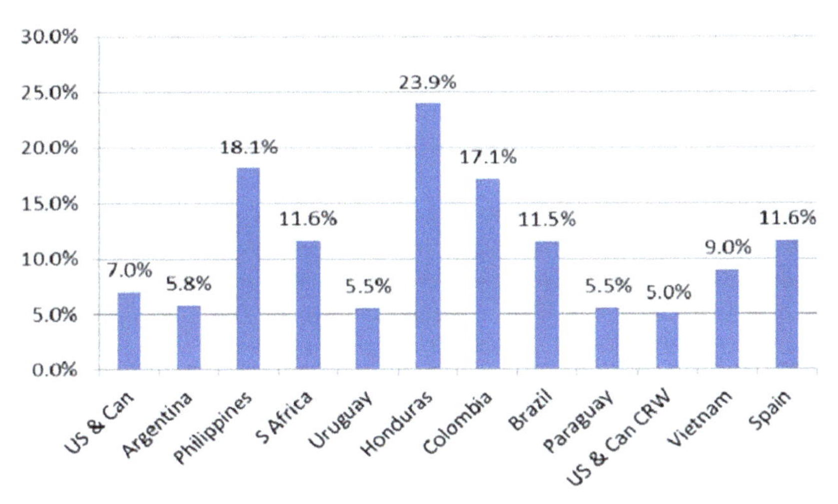

Figure 107 Average yield gains GM IR maize by country 1996–2020. Source: Adapted from: Brookes (2022).

million hectares of additional farmland would have been required to produce this additional yield (Brookes 2022). Figure 107 illustrates average yield gains associated with the adoption of insect-resistant maize.

Table 4 outlines the multiple benefits that can be achieved through the adoption of GM technologies compared to the use of conventional practices.

Other neo-productivist technologies, such as cultured meats and vertical farming, have yet to prove their competitiveness in the marketplace.

Table 4 The benefits of GM adoption versus conventional production (Durham and Mizik 2021)

Characteristics	Conventional	Conventional plus biotechnology
Yield	Normal	Up to 15% higher
Pesticide cost	Normal	Lower
Fertiliser cost	Normal	Normal
Labor cost	Normal	Lower
Product variety	Specialisation	Specialisation
Product price	Normal	Normal
Cross margin	Normal	Much higher
Pricing/ Business model	Volume	Volume
Environmental benefits	Normal	Higher

8.5.2 Reformist systems

Reformist approaches, such as precision agriculture, climate-smart agriculture, and Conservation Agriculture, face similar pressures to produce nutritious low-cost foods whilst at the same time increasing food system resilience, increasing resource use efficiency, and reducing negative environmental externalities.

In the case of precision agriculture (PA), there are several factors that have limited adoption (see Chapter 5 for details). Amongst these, the high cost of PA technologies is often the most important factor. Consequently, aside from some capital-intensive businesses, such as intensive livestock and viticulture, the adoption of PA has primarily been restricted to large and highly mechanised arable farms in countries such as the USA, Canada, and Brazil, where capital is available to fund the initial investment, significant intra-field variability exists, PA and GM technologies are combined, and costs can be spread over large areas of land (Yatribi 2020). Even under these conditions, whilst input costs can be cut (Figure 108), considering the cost and other challenges associated with PA, the commercial benefits of adopting PA are often extremely small (Schimmelpfennig 2016).

Indeed, in many situations, whilst PA technologies save inputs, these savings are often outweighed, especially by the costly investments in PA technologies, both hardware and software, that need to be made (Dhoubhadel 2020).

In many respects, the adoption of climate-smart agriculture (CSA) technologies/practices is similar to PA. Whilst CSA can lead to higher gross margins via higher yields (see Figure 109 below), CSA practices are generally associated with higher input costs (fertilisers, seeds, herbicides, and labour), and the need for a higher level of farmer skills (Branca et al. 2021). Figure 109 illustrates that CSA, in this example an approach based on the adoption of Conservation Agriculture, which incorporates minimum soil disturbance (MSD), in combination with cover crops, inclusion of legume, residue retention, and

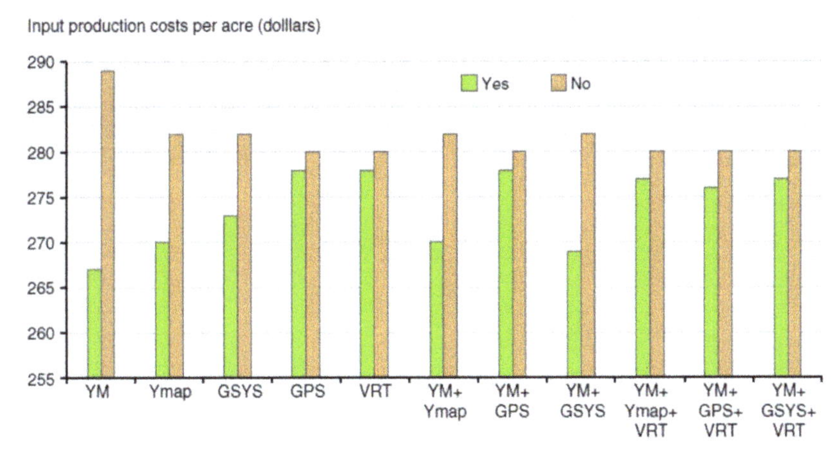

Figure 108 Input costs with and without PA technologies. Source: Adapted from: Schimmelpfennig (2016).

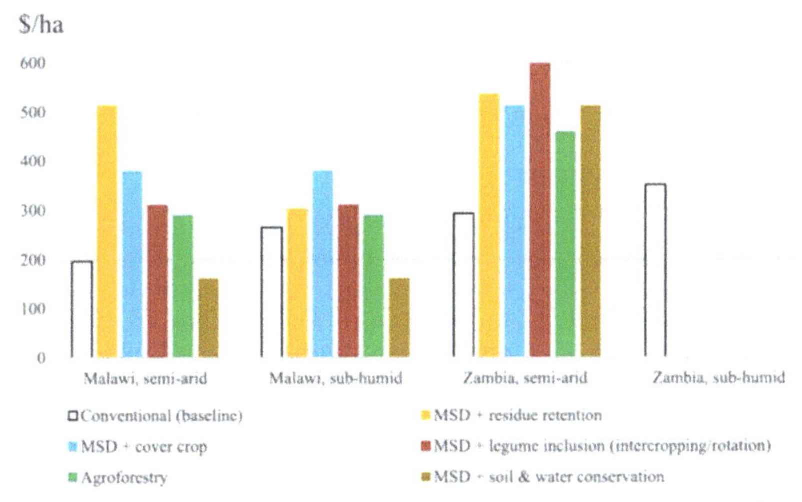

Figure 109 Gross margins of maize cropping in different agroecological zones of Malawi and Zambia. Source: Adapted from: Branca et al. (2021).

soil and water conservation measures, can significantly improve farmer gross margins (Branca et al. 2021).

Table 5 below provides a breakdown of the financial cost and returns of CSA in Malawi and Zambia in semi-arid test sites. Whilst input costs are higher in CSA compared to conventional agriculture, the returns to capital and labour in this example are higher under CSA.

However, it must be noted that there is mixed evidence regarding the performance of CA regarding both crop yield (Giller et al. 2015), and profitability (Jat et al. 2020).

Table 5 Capital and labour-related indicators of maize cropping by agroecological zones (Branca et al. 2021)

Indicator	Unit of measure	Semi-arid areas	
		Conventional	CSA
		A) Malawi	
Cash inputs	$/ha	310.30	347.16
Labour costs	$/ha	85.60	110.60
Returns to cash capital	$/$	2.13	2.78
Return to family labour	$/person day	3.29	5.95
Labour productivity	kg/person day	13.2	19.7
Capital intensity		3.6	3.1
		B) Zambia	
Cash inputs	$/ha	409.5	444.6
Labour costs	$/ha	78.8	108.7
Returns to cash capital	$/$	0.9	1.0
Return to family labour	$/person day	0.3	1.2
Labour productivity	Kg/person day	18.8	22.4
Capital intensity		5.2	4.1

8.5.3 Progressive systems

Whilst the economic sustainability of neo-productivist and reformist food systems is argued primarily on reduced unit costs of production (based primarily on higher yields and similar or reduced amounts of inputs), progressive food systems claim economic sustainability based on reduced expenditure on farm inputs, and often higher premiums, on what is claimed to be higher quality (safer, healthier, and more nutritious) outputs.

In the case of regenerative agriculture, evidence suggests that, due to increased soil organic matter, in some cases, yields can match or even beat yields from comparable conventional farms (LaCanne and Lundgren 2018). Under these conditions, reduced input costs can lead to increased profitability compared to conventional farms (Moyer et al. 2020; EIT Food 2020). However, in cases where crop yields are reduced under regenerative agriculture, the existence of premiums, especially where farmers are certified as organic, can lead to higher levels of profitability (Milinchuk 2020; LaCanne and Lundgren 2018; Impey 2021a). See Chapter 6 for more details. According to Khangura et al. (2023), more empirical studies are needed to determine to profitability of regenerative agriculture vis-à-vis conventional agriculture.

In many respects, the profitability of organic agriculture is similar to that of regenerative agriculture. The profitability of organic agriculture is achieved via lower overall input costs (despite higher labour costs Gomiero et al. 2011;

Wachter et al. 2019), receipt of subsidies for organic conversion, and premiums on the outputs produced under organic certification (Gomiero et al. 2011; Wachter et al. 2019; Joseph et al. 2019). In some contexts, the lower yields achieved under organic production are generally compensated for by lower production costs, leading to similar profits in both organic and conventional production systems (Gomiero et al. 2011). However, in many contexts, a premium for organic outputs is necessary to make up for lower yields in organic production systems. Without these premiums, organic agriculture would not be profitable (Crowder and Reganold 2015; Ramankutty et al. 2019). According to Crowder and Reganold (2015), organic premiums made organic agriculture 22–35% more profitable than conventional agriculture, and only a premium of 5–7% is required for organic to match the profitability of conventional agriculture. Conversely, in some instances, it is argued that organic agriculture can be more profitable vis-à-vis conventional agriculture, even without subsidies (Wachter et al. 2019; Klima et al. 2020). Table 6 below illustrates the profitability of organic versus conventional/integrated production in spring cereal production in mountainous Poland. Ultimately, when compared to conventional production, higher prices for organic products act as a barrier to achieving higher sales (Ramankutty et al. 2019).

Whilst there is considerable heterogeneity associated with the competitiveness of organic agriculture, it has also been demonstrated to be profitable in developing countries. For example, gross margins for cotton production in India were shown to be 30–40% higher under organic systems compared to conventional systems. This was due to lower production costs (10% to 20%) and the higher prices of organic products (20%) (Eyhorn et al. 2007). In Kenya, Adamtey et al. (2016) demonstrated the profitability of organic food systems compared to conventional food systems. Indeed, in some cases, organic food systems were even more profitable without organic premiums.

Compared to other food system approaches, the profitability of agroecological agriculture is extremely difficult to estimate. In part, this is because much agroecological production is aimed at self-sufficiency, with limited sales onto local and regional markets. In cases where agroecological farmers produce for the market, economic analyses are extremely mixed in their findings and often accused of lacking analytical rigour (D'Annolfo et al. 2017). Complexities especially arise when comparative analyses focus on the performance of single crops. Given that agroecological food systems are based on often complex mixes of arable, livestock, and tree production, it is often like comparing apples and oranges. In addition, unlike more industrial approaches, agroecological food systems explicitly focus on the generation of multiple objectives, such as the conservation of biodiversity and ecosystem services, timber for construction and fuel purposes, or to reduce soil erosion, and the generation of local employment and community sustainability, which

Table 6 Economic indicators of pure and mixed sowings of spring cereals in integrated or organic crop rotations (EUR ha^{-1}); prices for the year 2019 (Klima et al. 2020)

Source of cost	Pure sowing						Spring cereal mixture						Mean	
	Oats		Barley		Triticale		Oats +	Barley	Oats	Triticale	Triticale	+ Barley		
	I	O	I	O	I	O	I	O	I	O	I	O	I	O
Input costs	544.3	421.9	612.8	499.2	588.9	471.9	570.6	496.8	523.1	431.1	544.7	459.7	564.0	463.3
SGM without subsidies	57.7	123.7	138.6	214.4	103.7	175.1	89.9	204.7	37.2	133.6	64.9	168.3	82.1	169.8
SGM with subsidies	235.6	487.6	316.5	5783	281.6	539.0	267.8	568.6	215.1	4975	242.8	532.2	260.0	533.7
Share of subsidies in the SGM (%)	75	74	56	63	63	67	66	64	82	73	73	68	68	68
Direct profitability index														
Without subsidies	1.11	1.41	1.29	1.74	1.21	158	1.18	1.69	1.07	1.44	1.13	1.57	1.17	1.57
With subsidies	1.48	2.63	1.66	3.01	1.57	2.81	1.55	1.94	1.44	2.67	1.50	2.82	1.53	2.81

I, integrated; O, organic; SGM, standard gross margin. Conversion rate of €1 = 4.2585 PLN in accordance with the National Polish Bank exchange rate on 31 December, 2014.

often are not rewarded financially (Wezel and Silva 2017; Van der Ploeg et al. 2019). In this section, the science and practice of agroecological agriculture are being compared. The explicit social and political dimensions associated with agroecological agriculture are discussed in the next section.

The often-higher total production per area, and higher value addition, gained through the sales of specialised/niche products, either processed or unprocessed, generally in local markets, can often result in farmers receiving higher prices, and keeping a higher proportion of the total value-added, which, in turn, may result in agroecological farmers attaining a higher profitability per area of land cultivated (Ricciardi 2019). Whilst this potentially higher profitability per area farmed is encouraging, the generally small-scale nature of agroecological agriculture may ultimately result in insufficient farmer incomes (Dumont and Baret 2017; Ramankutty et al. 2019). The profitability of agroecological approaches may also be lower than conventional approaches when farmers invest disproportionate amounts of their labour into the integration of agroforestry into the farm and via the construction of highly labour-intensive soil and water conservation measures (Branca et al. 2021).

Conversely, other studies have demonstrated that agroecological food systems can deliver more profitable livelihoods compared to conventional food system approaches. Table 7 illustrates the profitability of agroecological grassland-based milk production versus conventional milk production in Bretagne, France.

Table 7 illustrates that whilst the gross value of production (GVP) is much higher on conventional farms €118 281 compared to €86 837 for grassland-based agroecological farms, the value addition (VA) per worker €44 179 and income per family worker €27 271 are much higher. Here, higher profitability is attained through reducing reliance on expensive external resources and replacing them with freely available or cheaper internal (farm) inputs. In the case above, whilst the grassland-based agroecological farms achieved a lower milk yield, the quality of the milk produced was higher quality, fetching a higher premium price at the market, and was cheaper per litre to produce (Van der Ploeg et al. 2019). Additionally, a higher proportion of the value-added captured at the farm level was spread across higher full-time labour,

Table 7 Gross value of production (in euros) for conventional and agroecological grassland farms (Van der Ploeg et al. 2019)

	Conventional farms	Grassland-based farms	Difference
G VP/worker	€118 281	€86 837	−27%
VA/GVP	33%	51%	+ 54%
VA/worker	€38 884	€44 179	+ 14%
Family income/family worker	€15 797	€27 271	+ 73%

thus contributing to local employment (Van der Ploeg et al. 2019). Reduced reliance on expensive external farm inputs, especially during times of high prices and market volatility, especially when output prices are low, has helped agroecological farmers to mitigate the growing cost/price squeeze (Wezel and Silva 2017; Van der Ploeg et al. 2019).

Similarly, the agroecological system of rice intensification (SRI) practiced in Asia has been shown to increase rice yields by 24% compared to conventional rice production systems (Zhao et al. 2010), and increased profitability by up to 206% (Cornell University 2023).

8.5.4 Radical systems

Given the dearth of information on the economic sustainability of both the slow food movement and permaculture, it is almost impossible to say much at this juncture. However, given that slow food and permaculture generally draw on agroecological-based approaches, it is probably fair to draw lessons and insights from the previous paragraphs. A body of literature does exist, however, on SFSCs. As with agroecological food systems, and the limited number of comparative analyses, the verdict is still out on the economic efficiency and sustainability of SFSCs (Kneafsey et al. 2014). Inherently, SFSCs are often seen as being more profitable than LFSCs, especially for producers. Indeed, there is significant evidence to suggest this. The shortness of the food chain (reduced logistical costs, higher capture of value-added, and often price premiums for local produce, whether or not it is certified as organic) is suggested to lead to higher incomes and job creation for local producers (Aubry and Kebir 2013; Mount 2012; Bereuter et al. 2016; Aubry and Kebir 2013; Duram and Oberholtzer 2010; Kneafsey et al. 2008, 2014; Mundler and Laughrea 2016; Gorton et al. 2014; Wittman et al. 2012; Malak-Rawlikowska et al. 2019). However, whilst SFSCs can be more profitable than LFSCs for farmers, the bigger picture is much more complicated. Whilst farmers can often receive higher prices for their produce through the direct provisioning of local markets, they often struggle to meet local demand, delivering only a fraction of what is required, and can often only supply a fraction of the range of produce demanded at the required time. Where local provisioning systems attempt to produce for the mainstream, as opposed to niche markets, the issue of price sensitivity becomes much more important (Vittersø et al. 2019). Here, only a sub-set of consumers can pay large premiums for local produce. Aggregation and distribution are also problematic, as most local aggregators, processors, and distributors have either grown large (serving LFSCs) or been forced out of business due to high-cost structures and lack of profitability. Indeed, where individual farmers and farmer cooperatives have sought to expand from supplying directly to the consumers, via box schemes or farmers' markets, etc., they have generally been obliged to work with the larger, more efficient, aggregators, processors, and distributors that service LFSCs (Izumi et al. 2010). In many cases, this has led to food producers provisioning both SFSCs and LFSCs. Conversely, many larger aggregators, processors, and distributors, have downscaled their operations to exploit new opportunities associated with the local provisioning of food.

8.6 Social sustainability

A socially sustainable food system can be understood as one that provides adequate amounts of affordable, healthy, nutritious, and culturally acceptable food in a just, equitable, and resilient manner, and promotes the well-being of both producers and consumers.

Since the 1960s, citizens across the developed and developing world have increasingly believed in the power of science and innovation to provide for their physical needs. Many accepted that new scientific discoveries and applied innovations would automatically enhance their individual and community well-being (Watson 2018). Since the 1980s, rapid globalisation and the evolving free market model promoted by the New Right have persuaded citizens to rely on the market to 'maximise human welfare': economically, markets efficiently distribute knowledge and resources; socially, liberal individualism would maximise moral worth; and politically, liberalism would maximise political freedoms, since it rests on the most efficient distribution of resources and wealth (Watson 2018).

8.6.1 Neo-productivist food systems

For many, neo-productivist innovations, such as plant-based meat analogues and cultured meats, genetically modified and gene-edited crops, and vertical agriculture, were just part of a continuous stream of modern agricultural innovations. Plant-based meat analogues are becoming more socially acceptable as consumers begin to turn away from the consumption of animal meat, on the grounds of animal welfare, resource, and environmental conservation, climate change mitigation, and other factors, which have been popularised by a variety of local, national, and international NGOs (van der Weele et al. 2019). However, it is suggested that the expansion of plant-based meat analogues has been limited due to issues related to the taste and smell of PBMA products (unpleasant flavour of soybean), concerns about potential allergenic effects of pulses, gluten intolerance of wheat-based products, and the perception of pulses being the food of the poor (van der Weele et al. 2019; Ismail et al. 2020). In this light, whilst cultured meat is seen by some consumers as unnatural and undesirable, to others, it exemplifies just what can be achieved by high science. Indeed, recent research has demonstrated that between 24%

and 32% of the public are open to consuming both PBMAs and cultured meats (Burton 2019; Credit Suisse 2021).

Cultured meats have especially caught the media's attention, in a similar way to GMOs during the 1990s and 2000s (van der Weele et al. 2019). It is yet to be seen whether this publicity has a positive or negative effect on the growth of the industry. Additionally, whilst currently only available in Singapore and the USA, cultured meats have certainly sparked interest from exclusive restaurants and their status-sensitive consumers, who are willing to pay high prices to eat cultured meat (Bryant and Barnett 2018; Stephens et al. 2018; Van der Weele and Driessen 2013).

GMOs were commercialised in the 1990s. To date, there have been few, if any, of the side effects once feared. Whilst there is still strong opposition to GMOs, and non-GM crops are generally preferred, consumer resistance to GMOs is slowly eroding, especially where adoption of GM crops leads to lower pesticide use and crops have higher nutraceutical properties (Edenbrandt et al. 2017).

Whilst technologically advanced, new developments in controlled agriculture, such as vertical farming, appear to be perceived as a natural evolution from the greenhouse and polythene tunnel production of fruits and vegetables and have not yet stimulated reaction from activist groups.

8.6.2 Reformist food systems

Reformist innovations, such as big data, computer-controlled machinery, nanotechnologies, climate-smart, and precision agriculture, also appear to be entering the marketplace without resistance. Indeed, where innovations increase input use efficiency and yield, at same time as reducing negative impacts on the environment, these technologies are generally quickly accepted by the media, consumers, and interest groups. Whilst perhaps only being incremental in nature, these innovations continue to nudge commercial agriculture in a positive direction.

8.6.3 Progressive food systems

Progressive food systems, such as regenerative, organic, and agroecological, on the other hand, proposed to approach social sustainability from a more socially inclusive, equitable, resilient, and culturally acceptable angle to that of neo-productivist and reformist food systems. Progressive food systems aim to rebuild the social fabric and trust relations of both rural communities that produce food and the predominantly urban communities that consume the food (Vittersø et al. 2019). Proponents of progressive food systems emphasise the duty of care to all food system actors. This duty of care extends to catering for producers' and consumers' physical and mental health and well-being through the provision of wholesome, healthy, and nutritious (often local) foods. At the producer level, this approach aims to ensure that farmers and their workers are appreciated by their communities for the fruits of their labour, that they enjoy their work, and earn a fair living wage (Doernberg et al. 2016; Dupré et al. 2017; Kim et al. 2018; Ramankutty et al. 2019; Rodale 1983; Beyond Pesticides 2019; Burgess et al. 2019; McCarthy et al. 2020; IDH 2020; Schreefel et al. 2020; Newton et al. 2020; Rhodes 2017; Sherwood and Uphoff 2000; Gaitán-Cremaschi et al. 2020). Additionally, from a progressive food systems perspective, it is important that the production, transportation, processing, and consumption of food meet evolving consumer demands, such as the local traceable quality foods, high animal welfare, safe, healthy, and nutritious foods, mitigation of and adaptation to climate change, environmental conservation (biodiversity, soil health, etc.), enhanced ecosystem services (clean water, etc.), and reduced pollution (Doernberg et al. 2016; Kim et al. 2018; Gaitán-Cremaschi et al. 2020).

However, whilst this idyllic image conjured by advocates of progressive food systems has significant appeal, others suggest that reality is often far different. At the producer level, life can often be far from idyllic. Working hours can be much longer than conventional farmers and often more manual and stressful, especially when managing a multitude of crops and livestock, and may also be less financially rewarding (Dupré et al. 2017). Additionally, rather than being accessible to all, the added expense of organic foods, etc., can reduce the affordability to poorer sections of the local community, and can indeed become almost status symbols. Here, consumers outwardly demonstrate their own social standing (wealth and social conscience) by consuming organically certified or similar products, which are often promoted by celebrities, making these goods elitist (Kim et al. 2018). Even the image of local and healthy food is questioned. This concern is especially linked to the conventionalisation and globalisation of organic and regenerative approaches (Doernberg et al. 2016).

8.6.4 Radical food systems

Given that agroecology is dealt with above, and that little has been written about the social sustainability of permaculture, this section focusses exclusively on the social sustainability of SFSCs. In many respects, much of what needs to be said about the social sustainability of radical food systems has already been discussed above. After all, SFSCs predominately focus on the provision of organic, regenerative, or agroecological foods. However, this is not always the case, as SFSCs focus on the provisioning of local foods, which may be produced on conventional farms or in vertical farms, etc. For example, in the case of SFSCs, especially those supplying regional specialty food products (SFPs), whilst often organic, many may also have been produced using synthetic fertilisers and pesticides and may have been subject to significant levels of traditional processing. Their claim to social sustainability is primarily linked to value-laden attributes such as tradition, authenticity, nostalgia, spiritual, emotional, healthy, fresh, and sustainability (especially linked to reduced food miles) (Andress et al. 2020; Johnson et al. 2016; Sacchi et al. 2018). In many respects, it is more about the place, history, and tradition than it is about the production process (Ilbery and Kneafsey 2000). Consuming local or regional specialty foods is perceived as wholesome. Something that both supports and binds communities through a collective act (Andress et al. 2020). In part, the values that are endowed in SFSCs also help to create shared social obligations, both on the part of consumers and producers. For example, Vittersø et al. (2019) write about the case of a Norwegian Fish shop, which was originally established in 1948. Over the years, the shop had become much-loved. It was relied upon as a community resource, a beloved institution, and a place where long-term friendships had been established between generations of consumers and the family members who owned and ran the fish shop. Here, the owner's son felt a strong obligation to maintain the shop, both as a livelihood for his family and as a resource for the community.

Conversely, others warn about what has come to be known as the local trap. The term local trap refers to the widely held assumption that buying locally is inherently better than buying food supplied via LFSCs (Born and Purcell 2006). Caution should be exercised in this quarter, as well-organised and socially conscious LFSCs may be more just and sustainable than SFSCs. Hinrichs (2000) suggests that, depending on the locality's comparative or absolute advantage, buying locally may lead to economic losses compared to buying nationally or internationally, where the same food products can be produced significantly cheaper. Likewise, even when local produce can be produced more competitively than in other parts of the world, this does not always lead to economic benefits being shared fairly. Such was the case of the elitist capture of expensive local organic produce, which prices out consumption by the poorest in the local economy (Murdoch et al. 2000; Sacchi et al. 2018).

8.7 Political sustainability

The political sustainability of food systems can be envisaged as the stable collective support for a course of actions aimed at maintaining or promoting food systems that provide adequate amounts of nutritious and culturally appropriate food, operate within planetary boundaries, efficiently manage biophysical resources, whilst underpinning both rural and urban livelihoods.

8.7.1 Neo-productivist and reformist food systems

In many respects, political support for both neo-productivist and reformist food systems builds on decades of political, economic, social, organisational, and technical support for the Third Food Regime (also known as the Neo-Productivist Regime, Corporate Food Regime, or 'Food From Nowhere' Regime), which was part of the Neo-Fordist/Neo-Liberal Regime (McMichael 2016). What this means in practice is that, almost by default, government and public sector institutions, including government-funded research institutions, land grant universities, global institutions (World Bank, IMF, WTO, UN agencies), as well as agri-food corporations and key philanthropic actors, such as BMGF, are guided by both the principles and rules of the Regime (Holt-Giménez and Altieri 2013).

Investment in agricultural research and development is a prime example of the institutionalised support for industrial agriculture. Key agricultural policies, such as the EU 2007–2013 R&D Framework Programme and the US National Innovation Act, remained focussed on the development of scientific innovations for industrial agriculture. Public sector resources were invested in molecular biology research (Vanloqueren and Baret 2009). For example, between 1970 and 1997, France's Institut National de la Recherche Agronomique (INRA) molecular biologists increased from less than 10% of staff to 20% of the INRA's total staff (Mignot and Poncet 2001). Furthermore, due to the Neo-Fordist Regime, molecular technologies had become patentable and promised huge profits for the private corporations that owned these patents (Miles et al. 2017). Indeed, the Neo-Fordist Regime was so enshrined that public goods research increasingly churned out outputs that would quickly be commercialised as private goods (Busch and Bain 2004). Through constant lobbying via industry groups such as Bio in the USA and Europabio in the EU, and a plethora of public–private partnerships, global corporations have continually influenced the governments of developed and developing nations to invest in blue-sky and proof-of-concept research, eagerly swooping in to take ownership of commercially promising emergent technologies, via IPR arrangements brokered with the WTO (ETC and HBF 2018). Examples of these partnerships include Novartis and the University of California and the French Government's investment in Genoplante (Vanloqueren and Baret 2009). Recent global food crises (2008/2009, 2011/2012, 2022/2023) have further rallied support for increased agricultural production driven by the adoption of 'magic-bullet' GM and gene-edited crops (ETC and HBF 2018; Jacobs 2020). According to ETC and HBF (2018), Emerging Ag Inc. was paid US$1.6 million by BMGF to promote gene editing as a solution to increased food production in UN meetings. Philanthropic institutions, such as the BMGF, have also invested heavily in developing GM technologies for developing countries. This has primarily materialised through support for the CGIAR-System, which has developed several biotechnologies for key staple crops such as maize, rice, and wheat. Biotechnologies have slowly become integral parts of government agricultural strategies in both developed and developing countries (IPES-Food

2020). Ultimately, research into agroecological-based solutions has been grossly underfunded (Vanloqueren and Baret 2009).

In many respects, the political economy of PBMAs and cultured meat is like that of GM and gene-edited crops. Namely, the private sector, protected by strong IPR, expects significant profits to ensue from the development and sale of these high-tech foods (van der Weele et al. 2019). Whilst, as novel products, cultured meats, algae, and insects face stringent regulatory health and safety tests, PBMAs have quickly entered to market. Opposition to cultured and PBMAs is primarily coming from livestock producer groups, who see their livelihoods threatened by these developments. Other parts of the meat industry, such as processors, distributors, and wholesalers/retailers, are already expanding their product portfolios, especially in the PBMA sector (van der Weele et al. 2019).

To date, part from the broader question of how R&D funding is allocated, vertical farming in Europe and the USA has surprisingly avoided the attention of activists (Orsini et al. 2020; Armanda et al. 2019; Benke and Tomkins 2017). This has allowed public institutions to actively support vertical farming. For example, the USA provides R&D support to vertical farming, and the USDA has established a National Committee on Urban Agriculture, which provides 'guidance on policy formulation and outreach for urban, indoor and other emerging agricultural production practices and addresses obstacles to urban agriculture' (Hu et al. 2011). Asian support for both vertical farming and cultured meat is more advanced. Countries such as Singapore, with limited arable land, built on initial R&D support for vertical farms with a suite of proposed tax breaks (Kozai et al. 2020). In Singapore, vertical farms have already been awarded organic certification. After several years of discussion and debate, some vertical farms are even being awarded organic certification in the USA. However, in the EU, vertically farmed crops using aquaponics and other soil-free approaches cannot be certified as organic because 'only crops grown in soil can qualify for such certification' (Vertical Farming Planet 2023).

CSA receives a significant amount of political support. Many public and private sector institutions have rallied around CSA. Intergovernmental support for CSA arises from UN institutions such as the FAO, IFAD, WFP, United Nations Development Programme (UNDP), UNFCCC, and the World Bank (Newell and Taylor 2018). In 2014, the FAO established the Global Alliance for Climate Smart Agriculture platform (Newell and Taylor 2018). Other key supporters of CSA include the OECD, CGIAR, and DFID (now FCDO) (Pimbert and Moeller 2018). Given that CSA furthers capitalist and free market solutions, support comes from a range of private sector organisations, from crop breeding (including GM), fertiliser, pesticide, and machinery manufacturers, providers of climate insurance, and crop storage solutions, etc. (Fairhead et al. 2012). CSA has also been centre-stage in recent COP meetings, especially COP22 in Marrakech (Newell and Taylor 2018), and COP 27, in Sharm El-Sheikh, Egypt (Reuters

2022b). African national governments, such as Senegal, regional economic commissions, such as ECOWAS, and continental bodies such as NEPAD, have also officially embraced CSA (IPES-Food 2020).

However, given its focus on more corporate and high-tech solutions to climate adaptation and mitigation, CSA remains contested and faces organised political opposition. Indeed, CSA is regularly accused of corporate greenwashing by NGO groups such as La Vía Campesina, Greenpeace, ActionAid International, and ETC Group (Newell and Taylor 2018).

8.7.2 Progressive food systems

Regenerative agriculture has recently experienced significant political and financial support, especially from downstream corporate food system actors such as food processors, wholesalers, and retailers. In a similar vein to CSA, regenerative agriculture has been able to create greater legitimacy through the production of food products that address at least some of the broader food system concerns expressed by NGOs and the general public, such as the quality of the end product (much of regenerative agriculture is organic certified), an increased focus on enhancing the incomes of smallholder farming families, a central focus on improving soil health, and claims to address climate change.

Organic agriculture also benefits from support from NGOs, corporate food system actors, and the public. Additionally, organic agriculture has significant support from several national governments (e.g. Bhutan and Sweden) (Feuerbacher et al. 2018; IPES-Food 2019), as well as supranational institutions such as the EU (Cisilino et al. 2019; IPES-Food 2019) and key organs of the United Nations, such as the FAO (Aschemann-Witzel and Zielke 2017). Unlike regenerative agriculture, political support for organic agriculture has grown slowly since the 1950s and 1960s but has shown regional and sub-regional disparities. For example, biodynamic farming (similar to organic) received strong support in the northern and central zones of Italy during the 1950s. Conversely, organic agriculture received political support in southern Italy during the 1990s (Darnhofer et al. 2019). Political support for organic agriculture grew significantly during the 1980s (e.g. Austria) and 1990s (e.g. EU) (Darnhofer et al. 2019).

Political support for agroecological agriculture, at least the science and practice of agroecology, has grown steadily since the 1970s. Initially, political support for agroecological agriculture was much less pronounced than support for organic agriculture. However, whilst agroecological agriculture has not really been able to galvanise support from the corporate sector in the way that organic agriculture has, due to limited opportunities for the development of proprietary technologies (Vanloqueren and Baret 2009), the advent of regenerative agriculture, which incorporates principles of agroecology, has helped to secure broader political support for agroecological agriculture. Given the similarity of approaches, private and public sector investments in organic and agroecological agriculture during the past 40 years have also managed to identify opportunities for ecological intensification in agroecology-based food systems (Vanloqueren and Baret 2009; Miles et al. 2017). However, both public and private sector R&D investments in agroecological agriculture remain woefully inadequate (Miles et al. 2017). For example, in the USA, agroecological agriculture R&D received less than 4% of USDA research funds (DeLonge et al. 2016).

Political support for Agroecological Agriculture (science and practice) has grown significantly during the past 10–15 years (IPES-Food 2018, 2019, 2020). This is primarily due to the accelerated search for sustainable food systems. Agroecological agriculture is increasingly being seen as a way to produce food, whilst improving food quality, conserving biodiversity, adapting to mitigating climate change, and promoting social justice (Wezel et al. 2018). The International Assessment of Agricultural Science and Technology for Development (IAASTD 2009a) called for greater support for agroecological agriculture. As of 2018, only France and the Czech Republic in the EU had officially embraced explicit agroecological policy interventions (Wezel et al. 2018).

Both organic and agroecological agriculture have become embroiled in the conventionalisation debate (see Chapter 6). Whilst this continues to raise concerns from traditionalists, some governments, universities, and philanthropic institutions welcome the appropriation of elements from organic and agroecological agriculture into mainstream neo-productivist and reformist food system approaches, suggesting that this can create highly productive but ecologically sensitive hybrid food system solutions (Holt-Giménez and Altieri 2013; Giraldo and Rosset 2018). Examples of hybridisation of agroecological elements include CSA, sustainable intensification, drought and heat-resilient GMOs, and precision agriculture (Holt-Giménez and Altieri 2013; Loos et al. 2014; Pimbert 2015). For example, the FAO is convinced that 'biotechnologies and their products can be used in production systems, based on agroecological principles, to enhance productivity while ensuring sustainability, conservation of genetic resources and use of indigenous knowledge' (FAO 2016c). Ultimately, this appropriation is likely to assist in mainstreaming these elements into neo-productivist and reformist food systems, where political support and corporate food system infrastructure are entrenched.

8.7.3 Radical food systems

Interest in and support for the degrowth agenda, especially from academic and social activists, has grown significantly during the past 20 years or so. However, that said, the degrowth debate remains peripheral in both political circles and media coverage (Buch-Hansen 2018; Koch 2018). Several academics attribute this marginalisation to the resilience of the dominant growth paradigm attendant in capitalism, which has provided prosperity and stability for hundreds of millions of working-class families, especially in the developed world (Göpel 2016). Indeed, many academics see growth as locked into developed and transitional economies. To date, no political party has been able to win elections based on promoting a degrowth agenda. Whilst politicians generally are too politically savvy to address degrowth head-on in the contemporary period, things may begin to change if public attitudes begin prioritising well-being and environmental conservation over economic growth based on the unsustainable exploitation of natural resources and environmental destruction.

Unlike the degrowth agenda, several countries, including Bolivia, Brazil, and Colombia (in selected provinces), and Cuba, Ecuador, and Venezuela have endorsed food systems based on agroecological principles and deployed a mix of food sovereignty elements, including protectionist and redistributive policies (see Chapter 7 for details) (Acevedo-Osorio and Chohan 2020). Political support/buy-in has taken several decades to galvanise, and much of this can be attributed to the establishment of La Via Campesina in 1993 and the respective national food sovereignty movements that had been compounding political capital since the 1960s and 1970s (Holt-Giménez and Altieri 2013; Wezel et al. 2018).

In the case of Cuba, the collapse of the Soviet Union and the ensuing isolation from international markets, coupled with an evolving state of emergency, led to the adoption of agroecological agriculture more as a *fait accompli* than a conscious policy decision based on weighing alternatives (Holt Giménez and Altieri 2013; Bellamy and Ioris 2017).

The case of Brazil presents a very different picture. One side of this bifurcated country owes its political ownership of food sovereignty to the persistent efforts of civil society (Latin American Consortium on Agroecology (CLADES), the National Agroecology Alliance (ANA), and the Brazilian Agroecological Association (ABA-Agroecology), research and extension (Brazilian Agricultural Research Organisation, and National Policy for Technical Assistance and Rural Extension (PNATER)), public procurement schemes (National School Feeding Program (PNAE)) (Nicholls and Altieri 2018), religious organisations, and farmers' movements over past decades (Bellamy and Ioris 2017). Brazilian academic institutions offer more than 100 degree-level and post-graduate courses in agroecology (Bellamy and Ioris 2017).

Over the years, support for agroecology, and to some extent food sovereignty, has grown within the FAO. However, given that the FAO's vision for agroecology includes its integration with sustainable intensification (climate-smart agriculture and biotechnologies), rather than ecological intensification, its vision for agroecology often better fits the appropriationist 'reformist' model and not the 'radical' or even 'progressive' food system approaches (Giraldo and Rosset 2018).

Whilst still somewhat politically marginalised, Slow Food's focus on traditional food production (local production, using local varieties and livestock breeds, organic fertilisers, and pest and disease control), and traditional food consumption (local recipes, and the cultural and social importance of eating) has garnered significant support from increasing numbers of those weary of successive food safety scares, bland foods, and driven by the rhetoric of healthy eating (Born and Purcell 2006; Wezel et al. 2018; Darnhofer et al. 2019).

SFSCs have garnered substantial support from local and regional governments during the past couple of decades. More recently, however, especially after a succession of global food crises and the recent COVID-19 pandemic, food sovereignty-styled approaches have increasingly been adopted by a growing number of local and regional authorities in both developed and developing countries (see Chapter 7).

Chapter 9

Food system evolution

9.1 Introduction

The initial chapters of this book highlighted how the global food system that we see today evolved. It also endeavoured to highlight its notable strengths and its underlying weaknesses. Subsequent chapters outlined a range of approaches that have been proposed to either build on the existing successes of the so-called Third Food Regime, whilst addressing inherent weaknesses, or consider alternative or disruptive approaches to building food systems that more equitably address humanity's need for food, fibre, and fuel, whilst rebuilding, refocussing, or resetting our relationship with the natural world. This chapter explores how the global food system is likely to evolve. In keeping with the conceptual framework of regimes of accumulation, this chapter takes a Multi-Level Perspective (MLP), to consider how change occurs in food systems and how likely it is that the pressures for change will result in the transformation of the global food system, or at least parts of it.

9.1.1 MLP concept

The MLP framework is a widely used approach for understanding the change process in complex socio-technical systems, especially with sustainability transitions (Vandeventer et al. 2019). One of the central tenets of the MLP is that actors are embedded in political, economic, social, and technical structures that both shape their worldview and enable or limit their agency and disposition to act (Geels 2010; Dumont et al. 2020). Although not without criticism (Hermans 2011; Hargreaves et al. 2013; Fuenfschilling and Truffer 2014; Ingram et al. 2015), MLP, in combination with Actor-Network Theory, can provide a more nuanced understanding of both macro-structural changes as well as the underpinning micro-social network processes that bring about such change. According to the MLP, change can be conceptualised as the result of interactions between three levels landscape, regime, and niche (Geels 2002, 2004; Geels and Schot 2007).

Figure 110 depicts the relationship between landscape, regime, and niche. These levels are delineated on the Y-axis by their level of structuration (stability of societal structures), with the landscape depicted as the most fixed/stable configuration of biophysical, economic, political, and socio-cultural dimensions. Landscapes are defined as the relatively stable context in which regimes, and in turn, niches are framed. Niches are depicted as the least fixed/stable configuration (Vandeventer et al. 2019).

Landscape features include macro-economic patterns, deep-seated cultural traditions and behaviours, demography, climate, etc. (Geels 2002; Hermans 2011; Smith et al. 2010; Van Driel and Schot 2005; Dumont et al. 2020). Whilst landscape-level patterns, such as the capitalist mode of production and

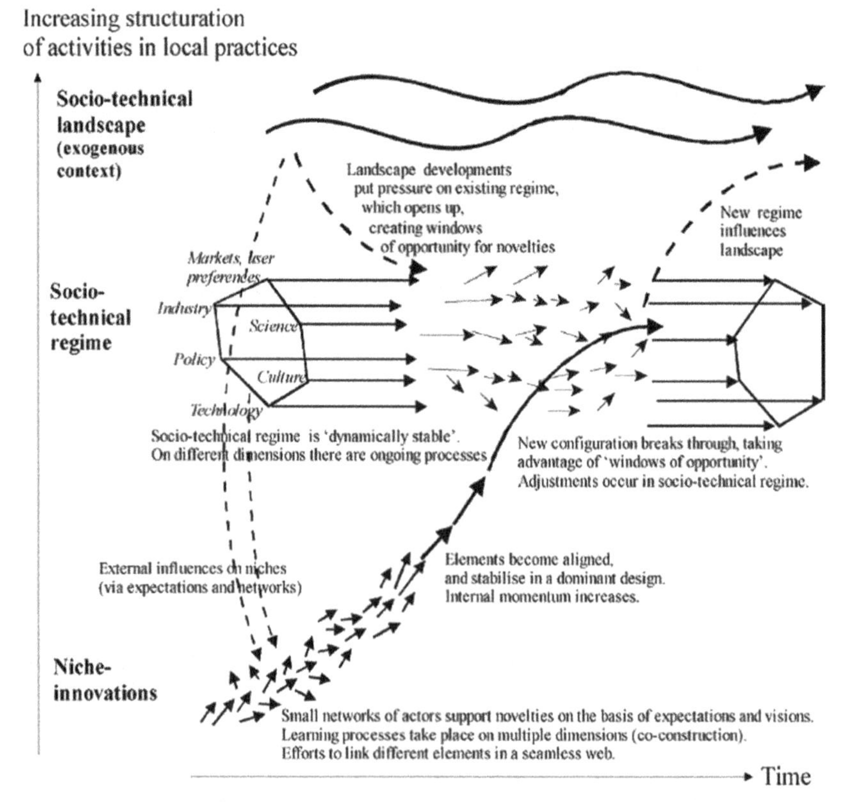

Figure 110 Multi-Level Perspective on system change. Source: Adapted from: Geels and Schot (2007).

the free market economy, are stable and deeply entrenched, gradual changes at the landscape level, such as the increasing human population, dietary shifts, climate change, environmental pollution, and the loss of biodiversity, as well as rapid changes associated with wars, pandemics (such as Covid-19), and the transition away from fossil fuels, can create pressures for change at the regime and/or niche levels (Hargreaves et al. 2013; Marsden 2013; Maye 2018; Kuokkanen et al. 2018; Dumont et al. 2020).

The concept of the socio-technical regime, in this case, a food regime, refers to the deep-structured complex of institutions, rules (both formal and informal), norms, and practices that both stabilise and legitimise the way food is produced and consumed (Hermans 2011; Hargreaves et al. 2013; Kuokkanen et al. 2018; EEA 2018; Geels 2004). On the one hand, whilst regimes are resilient and entrenched, and relatively resistant to wholesale change, they are also durable and adaptive, whereby incremental change generally occurs along accepted pathways (Hargreaves et al. 2013; Ingram 2015). For instance, in the case of crop breeding, new higher-yielding or nutritionally enhanced varieties emerge and replace existing varieties (Geels 2004). Regimes can also exert significant influence on each other (Hargreaves et al. 2013). For example, changes in the dominant energy regime (i.e. the transition from fossil fuels to renewables) are having a profound impact on the food regime.

Lastly, using the example of food systems, the niche concept can be defined as a space for disruptive experimentation, innovation, and configuration of alternative food system approaches. This space is protected from the constraints of the dominant regime by either political and social interventions or by geography (Geels and Raven 2006). For example, the EU's Farm to Fork Strategy provides regulatory protection and financial incentives for the expansion of organic farming in the EU. In the same light, local and regional authorities that are committed to localising, often organic-based forms of production, create both institutions, regulations, and economic incentives that provide a safe space for experimentation with SFSCs. For example, this includes community-based action, where land is acquired for local (often organic) food production, which is destined to be consumed (usually in a relatively unprocessed form) by local consumers. Geographical spaces that exist outside the dominant industrial food regime also have the freedom to experiment, not only with the promotion of food sovereignty based on alternative agroecological production systems but also with the re-configuration of political and economic governance of national food systems. This protective space is essential if disruptive regime-breaking innovation is to be developed (Smith and Raven 2011). Niche actors are often outside or marginalised within the socio-technical regime, often living by different rules, and demonstrating different practices and behaviours (Van de Poel 2000).

According to the MLP, change in the dominant regime can occur when tensions arise between landscape rules, norms, practices, and those of the dominant regime. Tension destabilises the regime and creates an opportunity for change (Vandeventer et al. 2019). Likewise, alternative food systems developed in niches can threaten regimes as they gain higher levels of support from the public and politically influential leaders, and values and opinions begin to shift at the landscape level (Marsden 2013; Maye 2018; Dumont et al. 2020).

There are basically two different types of niche innovations, those that fit and conform (also known as transitional innovations) and those that stretch and transform (also known as transformational innovations (Smith and Raven 2011; Weber et al. 2020). Here are some examples of the first type of fit and conform innovations. The development of cultured meat and plant-based meat analogues was, in part, stimulated by the need to produce animal protein or plant-based analogue protein that meets consumer needs without the production of high levels of GHGEs, destruction of forests and wetlands for grazing animals, and which eliminates the unnecessary suffering of domesticated animals. Likewise, vertical farming evolved to increase yields per area and provide highly nutritious foods in, or close to, areas of high population density, reducing food miles and increasing freshness and naturalness. GMOs and gene-edited crop varieties are examples of innovations that improve yields, reduce costs, and/or include additional nutritional benefits. Precision agriculture innovations, such as nano-fertilisers, nano-pesticides, nano-sensors, and machinery fitted with GIS and GPS capacities, developed adjacent to the dominant socio-technical regime, and are generally quickly and easily incorporated into the dominant food regime (Kuokkanen et al. 2018). Examples of stretch and transform innovations include regenerative, organic, and agroecological agriculture, which are responding to landscape-level cultural changes that demand more natural, healthy, and wholesome foods, produced in traditional ways that reduce the types of negative environmental externalities associated with the dominant industrial regime and improve farmer livelihoods. To go to scale, these niche innovations require transformational change in beliefs, personalities, attitudes, preferences, values, and motives, as well as behavioural change, adoption of radically new technologies, and institutional and political reforms (O'Brien and Sygna 2013; Dessart et al. 2019).

More extreme innovations such as degrowth and food sovereignty not only look to replace the dominant food regime but seek to further dismantle, replace, or restructure broader landscape-level structures, such as free market capitalism (Geels and Schot 2007; Vandeventer et al. 2019).

9.2 Path-dependency and socio-technological regime lock-ins

Unless significant changes occur at the landscape level, fit and conform niche innovations are much more likely to go to scale and be institutionalised compared to stretch and transform niche innovations. This is due to two interrelated phenomena known as technological and institutional path dependency and socio-technological regime lock-ins. Technological and institutional path dependency relates to the inherent rigidity, embeddedness, and self-reinforcing resistance of the dominant regime to radical and transformational change (Unruh 2000; Foray 1997; Cowan 1990; Smith and Stirling 2010). The rigidity and resistance have been adequately demonstrated in the First, Second, and Third Food Regimes discussed in Chapter 1. In addition to agricultural innovation (Vanloqueren and Baret 2009) and food systems (Kuokkanen et al. 2017), path dependencies have also been well-documented in the fossil fuel energy regime (Unruh 2000), nuclear power (Cowan 1990), and transportation sector (Klitkou et al. 2015). In practice, this means that regimes tend to change quite slowly, incrementally, and along almost predetermined trajectories (Feyereisen et al. 2017). In the case of the Second and Third Food Regimes, the drive for even cheaper food, facilitated by the global free market and liberal financial regulation, has provided signals that have guided farms to grow, consolidate, specialise, and industrialise to remain profitable (IPES-Food 2020). Farmers' identities have increasingly been linked to the production of ever higher yields, whilst, at the same time, driving down the unit cost of production (IPES-Food 2016). To attain higher yields and increased competitiveness, farmers have needed to industrialise, consolidate, and specialise even further to service the high levels of investments that increasingly need to be made in yield-enhancing and cost-reducing technologies, especially labour-saving technologies such as high-powered and high-tech machinery. Similar investments have been made by food logistic companies that ship food (processed or raw, refrigerated, or unrefrigerated) around the globe, as well as processors, wholesalers, and retailers, investments in processing, storage, and sales centres (Reganold et al. 2011; IPES-Food 2020). Consumers not only rely on cheap food but on a wide range of blemish-free, out-of-season, and exotic fruits and vegetables that have often been transported halfway around the world (Vanloqueren and Baret 2009; IPES-Food 2016; Béné et al. 2019b).

De facto, path dependency results in, and is the result of, the 'lock-in' of certain beliefs, values, institutions, behaviours and practices, technologies, policies, laws, research and education, and financial incentives that perpetuate the existing dominant regime (Vanloqueren and Baret 2009; Ingram 2015; IPES-Food 2016; Bellamy and Ioris 2017; Frison and Clément 2020). According to IPES (2016), there are eight key lock-ins of Industrial Agriculture, namely, the expectation of cheap food, feed-the-world narratives, compartmentalised thinking, short-term thinking, measures of success, path dependency, and export orientation (see Fig. 111). The concentration of corporate power is seen to choreograph path dependency and lock-ins.

Ultimately, what is not locked into the dominant regime is, by default, 'locked out' (Seyfang and Smith 2007; Hermans 2011). Agroecology is one of the most often cited examples of locked-out innovations.

Power, exercised by global input, aggregation, logistics, processing, and retail corporations, favours a narrow range of crop species, transformed, and traded along international supply chains. Value is added at every juncture. For example, corporate seed, pesticide, fertiliser, and machinery manufacturers search for even greater profits based on sales of proprietary technologies/innovations, such as biotech seeds, or intelligent machinery. In agroecological food systems, traditional local crop varieties are planted by farmers. This means that farmers can use farm-saved seeds and do not have to purchase seeds from corporate seed companies. Simple small-scale traditional machinery is often used by agroecological farmers, which negates the need to invest in large/hi-tech machinery manufactured by large corporations (BFED and IPES-Food 2020). This reduced opportunity for profit-making by global corporations not only makes agroecological agriculture uninteresting to them, but they often see agroecology as a competitive model that could put them out of business. To this end, corporations, and often the governments that they influence, overtly target agroecological agriculture. They insist that, due to low levels of productivity and production, agroecological agriculture cannot feed the world (IPES-Food 2016, 2017a; HLPE 2017; Bellamy and Ioris 2017). Because of this bias, alternative agriculture receives limited public sector support for research and 'extension, storage, distribution and processing facilities, affordable credit and insurance

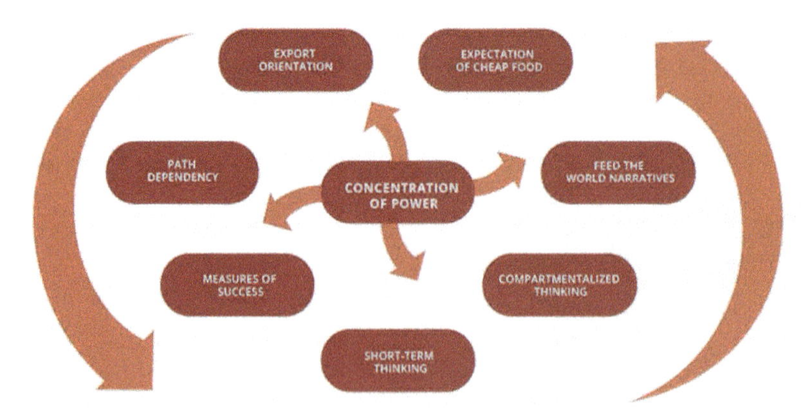

Figure 111 The eight key lock-ins of Industrial Agriculture . Source: Adapted from: IPES-Food (2016).

policies' (Carolan 2005; Gaitán-Cremaschi et al. 2020). In the case of fertiliser corporations, investments have been made in the production of synthetic nitrogenous fertilisers, and the opening of new phosphate and potash mines. This gives global fertiliser corporations an opportunity to expand their linear markets for proprietary fertilisers, which are shipped and sold across the world. This model is pursued by the corporate sector to the detriment of promoting more circular approaches to nutrient management, including nitrogen fixation by leguminous crops and plants, and circulating phosphates, potash, and micronutrients through the recycling of nutrients in animal manure, plant wastes, and human sewage. Corporations have also exercised their power in the sphere of agricultural research for development. Here, through organised lobbying of governments and the provision of public and private research grants, the development of new proprietary technologies has dominated agricultural research, at the expense of research into agroecological agriculture, which remains woefully under-funded (Crowder and Reganold 2015; Willer and Lernoud 2016; Watson 2018). According to Miles et al. (2017), agroecological research received less than 15% of USDA's competitive research funding. In turn, this has influenced investments in research infrastructure and biased the number of scientists and their disciplines in favour of Industrial Agriculture.

9.3 Incremental or radical change?

Given what we know from history about the First, Second, and Third Food Regimes, and the phenomena of lock-ins, lockouts, and path-dependency, what will future food systems look like? Can alternative food systems take root and replace, or provide real competition for, the global Neo-Productivist Food Regime? Indeed, does it make sense for them to do so? Well, let's first be clear about one thing. Whilst both the First, Second, and Third Food Regimes have dominated the way we grow, transport, process, trade, wholesale, retail, and consume food, alternative modalities, such as indigenous/traditional food systems, have continued to operate in parallel. These include food systems based on traditional practices and locally important crops, livestock, and fish species, which often operate in tandem with hunting and gathering. These include communities living in remote areas, which include many parts of Asia, Africa, Central America, South America, etc. Even where industrial agriculture has dominated, most of these geographies witnessed continued pockets of traditional food systems, parallel systems of food sharing and bartering, as well as the emergence of alternative food systems, such as organic, biodynamic, and agroecological agriculture, which arose in the early twentieth century when industrial agriculture was beginning to dominate much of the developed world. This pluriverse of food systems across the globe also includes those arguing for the necessity of degrowth (Kaul et al. 2022).

Ultimately, due to the structural tendencies of lock-in and path dependency, most food system analysts expect incremental, as opposed to radical, change in the global, regional, national, and local food systems (Loos et al. 2014; Marsden 2016; Gaitán-Cremaschi et al. 2020). For example, in response to landscape-level concerns over animal welfare, healthy diets, natural resource depletion, climate change, energy consumption, food miles, etc., the production and consumption of plant-based meat alternatives (PBMAs) is expected to grow incrementally by approximately 10% CAGR over the next decade or so (Wild et al. 2014; Ismail et al. 2020). The market for PBMAs in developing countries is also expected to grow by 73% by 2050 (FAO 2011d). But these are not the only drivers of change. High-tech innovations such as PBMA have only materialised due to significant R&D investments by both food and non-food corporations. These corporations have made substantial investments in proprietary infrastructure due to projected lucrative income streams that can be protected by robust intellectual property regimes (IPRs). These are the same types of IPRs that evolved alongside the development of improved crop varieties, pesticides, biotechnologies, and food processing technologies (van der Weele et al. 2019).

Within this pluriverse, neo-productivist, reformist, progressive, and radical approaches will continue to compete. However, to complicate the situation further, it is increasingly likely that the world will witness greater levels of appropriation, substitution, and the creation of hybrid food systems. As discussed in Chapter 2, appropriation refers to the phenomena of industrial inputs and processes replacing organic inputs and traditional farm practices. Examples of appropriation and substitutionism include the production of cultured meat, PBMAs, and vertical farming. Here, in the case of cultured meat, aside from the initial use of animal stem cells, the whole production process has been shifted from the farm to the laboratory/factory. Furthermore, the evolution of food systems is likely to be complicated by the advent of increased conventionalisation and hybridisation of food systems. Chapter 2 discussed the conventionalisation of organic food systems (Renting et al. 2003; Doernberg et al. 2016). Due to the lure of higher profits and the downward pressure on output prices imposed by large organic food processors, wholesalers, and retailers, many traditional small-scale, artisanal, organic producers have been replaced by large-scale organic farming. Here, aside from adhering to the bare minimum organic standards, farms more closely resemble industrial rather than organic farms in many parts of the developed world. Indeed, corporations responsible for driving the conventionalisation of organic agriculture have been accused of 'greenwashing' industrial agriculture to make it more socially and ecologically acceptable (Lamine and Dawson 2018; Lamine 2011; Giménez Cacho et al. 2018; Therond et al. 2019). In other cases, small-scale organic producers have been obliged to utilise existing industrial food system infrastructure to scale their operations and access more distant, or more diverse, markets. Examples include the increased utilisation of industrial plants (e.g. abattoirs), storage, and logistical networks for the competitive post-farm gate processing, storage, and distribution of organic food. Indeed, many organic farmers supply both local and regional SFSCs, as well as LFSC (Ilbery and Maye 2005b; Doernberg et al. 2016).

The final section of this book examines the likely evolution of competing food system paradigms, set against some of the key landscape-level drivers. Table 8 provides a high-level and relatively subjective overview of both the strengths and weaknesses of competing food system paradigms when attempting to address landscape-level drivers of change. It can clearly be seen that neo-productivist, reformist, progressive, and radical food systems paradigms address most landscape-level drivers to a greater or lesser extent.

9.3.1 Increasing food supply

Regarding the pressure to increase food supplies and feed the world, new approaches that fall under the neo-productivist paradigm score highly compared to all other paradigms. Whilst GM and gene-edited crops have already demonstrated their contribution to increasing yields, neo-productivist technologies such as cultured meats, PBMAs, and vertical farming score highly due to their unfolding potential. Indeed, unless these technologies face significant resistance during scaling, they are likely to play an increasingly important role in the future provisioning of food. Conversely, it is argued that with increased R&D investments into agroecological intensification, yields from both the progressive and radical paradigms could dramatically increase (Titonell 2014; Gonzalez de Molina and Guzman Casado 2017). Indeed, there remains significant scope to narrow the yield gap between progressive food systems and neo-productivist and reformist food systems. For example, computational agroecology, and a fresh more mainstream look at Permaculture, could produce key opportunities to narrow the yield gap. It is also likely that greater hybridisation will occur between the four food system paradigms. For example, Reformist approaches such as CSA could increasingly appropriate elements from agroecological agriculture, such as crop diversification, the inclusion of livestock, fish, and agroforestry, and the focus on building soil organic matter. Likewise, progressive food system approaches may integrate GM and gene-edited crops and livestock, especially those that embody pest and disease resistance, and climate resilience.

9.3.2 Increasing resource use efficiency

Neo-productivist approaches also perform well with respect to resource use efficiency, as do Reformist approaches, especially precision agriculture. However, whilst the improved efficiency of resources is important, in the future,

Table 8 Food system paradigms' responses to landscape level food system drivers

	Food System Paradigm												
	Land Sparing							Land Sharing					
	Neo-Productivist					Reformist		Progressive			Radical		
	Cultured meat	PBMAs	GMOs and gene editing	Vertical farming	Nano-technology	Precision agriculture	Climate-smart agriculture	Regenerative agriculture	Organic agriculture	Agroecological agriculture	Degrowth	Food Sovereignty	Regional food systems
Increasing food supply	✓✓✓	✓✓✓	✓✓✓	✓✓✓	✓	✓	✓✓	✓	✓	✓	✓	✓	✓
Increasing resource use efficiency	✓✓✓	✓✓✓	✓✓	✓✓✓	✓✓	✓✓✓	✓✓	✓✓✓	✓✓	✓✓✓	✓✓	✓✓	✓
Reducing food loss and waste	✓✓	✓✓	✓✓	✓✓✓	✓✓	✓✓	✓✓	✓✓	✓✓	✓✓✓	✓✓✓	✓✓	✓✓
Improving diets	✓✓✓	✓✓✓	✓✓	✓✓✓	✓		✓	✓	✓✓	✓✓✓	✓	✓	✓✓
Mitigating and adapting to climate change	✓✓	✓✓	✓✓✓	✓✓	✓	✓✓	✓✓✓	✓✓✓	✓	✓✓✓	✓✓✓	✓✓	✓✓
Reducing habitat and species loss	✓✓✓	✓✓✓	✓✓	✓✓✓					✓	✓✓	✓✓	✓	✓✓
Reducing environmental pollution	✓✓✓	✓✓	✓	✓✓✓	✓✓	✓✓	✓✓	✓✓	✓✓	✓✓✓	✓✓	✓	✓✓
Increasing food system equity								✓✓	✓	✓✓	✓✓	✓✓	✓
Increasing access to food for the poor							✓	✓	✓	✓✓	✓✓✓	✓✓✓	✓✓

there will be increasing pressure to recycle resources, especially key crop nutrients such as potassium and phosphate, fresh water, and organic matter. The linear approach to resource use is unsustainable in the long term. Progressive approaches also score well against this driver by explicitly taking a more circular approach to nutrient recycling and reducing the need for synthetic inputs. In the future, Progressive approaches may increasingly borrow emerging Reformist technologies such as biostimulants, and nanotechnologies.

9.3.3 Reducing food loss and waste

There is mixed performance with regard to reducing food loss and waste. Ultimately, both food loss and waste are imperatives that all food system approaches need to tackle. In the case of neo-productivist and reformist approaches, technologies such as cultured meat, PMBAs, GM/gene-edited crops, nanotechnologies, and precision agriculture aim to reduce food losses. Progressive and radical approaches also aim to reduce losses via crop diversification, natural control of pests and diseases, etc. To date, however, aside from limited inroads made by radical approaches such as food banks and food sharing, and food gleaning, the competing paradigms have yet to make headway in this important area. Whilst organic waste (farmyard manure, spoilt crops, and other biomass) at the farm level in both progressive and radical approaches tend to be recycled, very little is currently being done to recycle food waste post-farm gate.

9.3.4 Improving diets

There is a mixed performance with regard to improving diets. Currently, both neo-productivist and reformist do little to improve diets. Whilst they may contribute to improving access to large quantities of cheap calories, much of this tends to be ultra-processed, often unhealthy, and of limited nutritional value. However, with the right incentives, these food systems could easily pivot to deliver healthy and nutritious diets. Progressive and radical approaches are generally seen as more diverse, nutritious, and healthy. However, at scale, the inherent capacities to produce and supply adequate quantities of nutritious foods, especially exotic or out-of-season fruits and vegetables, have yet to be tested. It is likely that a hybrid of production paradigms, using both SFSCs and LFSCs, will continue to meet this demand. More research is needed to better understand the most effective and cost/resource-efficient approaches to changing diets at scale and ensuring both physical and economic access to healthy and nutritious diets for the poor.

9.3.5 Mitigating and adapting to climate change

Most paradigms actively contribute to mitigating and adapting to climate change. Neo-productivist and reformist approaches do this by generating high yields and using fossil fuels more efficiently. Innovations such as cultured meats, PBMAs, vertical farms, electric farm vehicles, and food chain vehicles show great promise. In the future, if these approaches can harness green energy, they may well become carbon-neutral, or even carbon-negative. Progressive and Reformist approaches score well because they rely less on fossil-fuel-derived energy sources and farm inputs. Ultimately, however, if lower yielding approaches are widely adopted, this would generate demand for new agricultural land to be brought into production. This would potentially release significant amounts of GHGs through deforestation, conversion of grasslands, and drainage of wetlands.

9.3.6 Reducing habitat and species loss

Neo-productivist approaches also score well with regard to reducing species and habitat loss. This is primarily achieved through higher yields per unit area (i.e. land sparing). However, the application of machine learning and computational ecology could lead to the design of heterogeneous landscapes that produce high agricultural yields as well as deliver species richness. Progressive and radical approaches take a more land-sharing approach, which is inherently more biodiverse than monoculture crop production and intensively managed grasslands, etc. However, if the yield gap cannot be closed, the lower yields associated with these approaches would potentially lead to the need for more agricultural land. This land would be carved out of biodiverse natural and semi-natural habitats, impacting heavily on biodiversity.

9.3.7 Reducing environmental pollution

All paradigms have mixed scores with regard to environmental pollution.

9.3.8 Increasing food system equity

In response to food system equity and access to food, both neo-productivist and reformist paradigms perform poorly, whilst progressive and radical paradigms perform well.

9.3.9 Increasing access to food for the poor

Due to an explicit focus on social justice, radical approaches, such as degrowth and food sovereignty, perform well with regard to increasing access to food for the poor. However, this does rely on the capacity of radical food systems to produce enough food for urban and rural populations alike. It is important that some of these more radical food system models, and approaches to degrowth, are rigorously tested. This could be at either national or sub-national scales.

9.4 Summary

For some, the Neo-Productivist Regime is viewed as simply hanging in and limping forwards, providing nothing more than superficial band-aids to mask the cuts and bruises it is being dealt. Others, however, imply that, whilst battered and bruised, the current regime is at worst still somewhat functional. Namely, it can feed unprecedented numbers of both human and animal mouths, whilst at the same time preserving remnants of wilderness, reducing food system pollution, and recycling limited resources. Whilst numerous alternative food system models are being forwarded, many still see the current model as the only tried and tested approach at scale. Concerns are especially raised regarding the capacity of alternative food systems to produce adequate amounts of nutritious food, at affordable prices, for 10 or 11 billion people. Lastly, for the ardent believers - pursuing 'food security' and 'free trade' as agribusiness projects, new technologies and innovations, many of which are highlighted in this book, are seen as having the inherent capacity to provide long-term and sustainable solutions to the problems facing the current food system.

Chapter 10

References

Abad, E., Palacio, F., Nuin, M., Zárate, A. Gd., Juarros, A., Gómez, J. M. and Marco, S. (2009). RFID smart tag for traceability and cold chain monitoring of foods: demonstration in an intercontinental fresh fish logistic chain. *Journal of Food Engineering* 93(4), 394–399.

ABC News (2019). CRISPR editing of plants and animals gets green light in Australia. Now what? *ABC News-Australia*. Available at: https://www.abc.net.au/news/science/2019-04-30/crispr-gene-editing-in-the-food-chain/11053622.

Abdollahdokh, D., Gao, Y., Faramarz, S., Poustforoosh, A., Abbasi, M., Asadikaram, G. and Nematollahi, M. H. (2022). Conventional agrochemicals towards nano-biopesticides: an overview on recent advances. *Chemical and Biological Technologies in Agriculture* 9(1), 13. DOI: 10.1186/s40538-021-00281-0.

Abdullahi, H. S., Mahieddine, F. and Sheri, R. E. (2015). Technology impact on agricultural productivity: a review of precision agriculture using unmanned aerial vehicles. In: Proceedings of the International Conference on Wireless and Satellite Systems. Springer, Cham, pp. 388–400.

Abegunde, V. O., Sibanda, M. and Obi, A. (2019). The dynamics of climate change adaptation in Sub-Saharan Africa: a review of climate-smart agriculture among small-scale farmers. *Climate* 7(11), 132. DOI: 10.3390/cli7110132. Available at: www.mdpi.com/journal/climate.

Abegunde, V. O., Sibanda, M. and Obi, A. (2020a). Determinants of the adoption of climate-smart agricultural practices by small-scale farming households in King Cetshwayo district municipality, South Africa. *Sustainability* 12(1), 195.

Abegunde, V. O., Sibanda, M. and Obi, A. (2020b). Mainstreaming climate-smart agriculture in small-scale farming systems: a holistic nonparametric applicability assessment in South Africa. *Agriculture* 10, 52.

Acevedo-Osorio, Á. and Chohan, J. K. (2020). Agroecology as social movement and practice in Cabrera's peasant reserve zone, Colombia. *Agroecology and Sustainable Food Systems* 44(3), 331–351. DOI: 10.1080/21683565.2019.1623359.

ACF (2017). Fiche étude: gestion de l'eau en agroécologie. Action contre la Faim [ACF]. Available at: https://www.coalition-eau.org/wp-content/uploads/fiche-etude-gestion-de-leau-en-agroecologieacf-vf.pdf.

Action Aid (2012). *Fed Up: Now's the Time to Invest in Agroecology*. ActionAid, Johannesburg.

Adamides, G. (2020). A review of climate-smart agriculture applications in Cyprus. *Atmosphere* 11(9), 898. DOI: 10.3390/atmos11090898.

Adams, K. M., Benzie, M., Croft, S. and Sadowski, S. (2021). Climate change, trade, and global food security: a global assessment of transboundary climate risks in agricultural commodity flows. SEI Report. Stockholm Environment Institute, Stockholm. DOI: 10.51414/sei2021.009.

Adamtey, N., Musyoka, M. W., Zundel, C., Cobo, J. G., Karanja, E., Fiaboe, K. K. M., Muriuki, A., Mucheru-Muna, M., Vanlauwe, B., Berset, E., Messmer, M. M., Gattinger, A., Bhullar, G. S., Cadisch, G., Fliessbach, A., Mäder, P., Niggli, U. and Foster, D. (2016). Productivity, profitability, and partial nutrient balance in maize-based conventional and organic farming systems in Kenya. *Agriculture, Ecosystems and Environment* 235, 61–79.

Adebayo, M. A., Kolawole, A. O. and Adebayo, T. A. (2015). Assessment of new generation of drought-tolerant maize (*Zea mays* L.) hybrids for agronomic potential and adaptation in the dried savanna agro-ecologies of Nigeria. *International Journal of Agronomy and Agricultural Research* 7, 45–54.

Adebayo, M. A. and Menkir, A. (2014). Assessment of hybrids of drought tolerant maize (*Zea mays* L.) inbred lines for grain yield and other traits under stress managed conditions. *Nigerian Journal of Genetics* 28(2), 19–23.

AFSA (2019). A common food policy for Africa. Available at: https://afsafrica.org/towards-acommon-food-policy-for-africa/.

Afshin, A. A., Sur, P. J., Fay, K. A., Cornaby, L. and Ferrara, G. (2019). Health effects of dietary risks in 195 countries, 1990–2017: a systematic analysis for the Global Burden of Disease Study 2017. *The Lancet* 393(10184), 1958–1972. DOI: 10.1016/S0140-6736(19)30041-8.

AgbioInvestor (2023). Global GM crop area increased by 3.3% in 2022. Available at: https://gm.agbioinvestor.com/news/global-gm-crop-area-increased-by-33-in-2022# Accessed 30 November 2023.

Aggarwal, P. K., Jarvis, A., Campbell, B. M., Zougmoré, R. B., Khatri-Chhetri, A., Vermeulen, S. J., Loboguerrero, A. M., Sebastian, L. S., Kinyangi, J., Bonilla-Findji, O., Radeny, M., Recha, J., Martinez-Baron, D., Ramirez-Villegas, J., Huyer, S., Thornton, P., Wollenberg, E., Hansen, J., Alvarez-Toro, P., Aguilar-Ariza, A., Arango-Londoño, D., Patiño-Bravo, V., Rivera, O., Ouedraogo, M. and Tan Yen, B. T. (2018). The climate-smart village approach: framework of an integrative strategy for scaling up adaptation options in agriculture. *Ecology and Society* 23(1), 14.

Agriculture Outlook (2000). U.S. farm policy: the first 200 years. *Agriculture Outlook*. USDA Economic Research Service, Washington, DC.

Agrimonti, C., Lauro, M. and Visioli, G. (2021). Smart agriculture for food quality: facing climate change in the 21st century. *Critical Reviews in Food Science and Nutrition* 61(6), 971–981. DOI: 10.1080/10408398.2020.1749555.

Agrizon (2021). Vertical farming is headed for the 'trough of disillusionment.' Here's why that's a good thing. Publicado el 31 Diciembre 2021.

Ahmad, M., Masih, I. and Giordano, M. (2014). Constraints and opportunities for water savings and increasing productivity through Resource Conservation Technologies in Pakistan. *Agriculture, Ecosystems and Environment* 187, 106–115.

Ahmar, S., Mahmood, T., Fiaz, S., Mora-Poblete, F., Shafique, M. S., Chattha, M. S. and Jung, K. H. (2021). Advantage of nanotechnology-based genome editing system and its application in crop improvement. *Frontiers in Plant Science* 12, 663849. DOI: 10.3389/fpls.2021.663849.

Aktar, M. W., Sengupta, D. and Chowdhury, A. (2009). Impact of pesticides use in agriculture: their benefits and hazards. *Interdisciplinary Toxicology* 2(1), 1–12. DOI: 10.2478/v10102-009-0001-7.

Al Jazeera (2023). 'A new era': lab-grown meat gets approval for sale in US. Available at: https://www.aljazeera.com/news/2023/6/21/a-new-era-lab-grown-meat-gets-approval-for-sale-in-us. Accessed 21 June 2023.

Alameen, A. A., Al-Gaadi, K. A. and Tola, E. (2019). Development and performance evaluation of a control system for variable rate granular fertilizer application. *Computers and Electronics in Agriculture* 160, 31–39.

Albright, R. K., Kram, B. W. and White, R. P. (1983). Malathion exposure associated with acute renal failure. *Journal of the American Medical Association* 250(18), 2469. DOI: 10.1001/jama.1983.03340180031010.

Alexander, P., Brown, C., Arneth, A., Finnigan, J., Moran, D. and Rounsevell, M. D. A. (2017). Losses, inefficiencies, and waste in the global food system. *Agricultural Systems* 153, 190–200. DOI: 10.1016/j.agsy.2017.01.014.

Alexandratos, N. and Bruinsma, J. (2012). *World Agriculture: Towards 2015/2030. The 2012 Revision*. Food and Agriculture Organization of the United Nations, Rome.

Alkon, A. H. and McCullen, C. G. (2011). Whiteness and farmers markets: performances, perpetuations... contestations? *Antipode* 43(4), 937–959.

Alkon, A. H. and Norgaard, K. M. (2009). Breaking the food chains: an investigation of food justice activism. *Sociological Inquiry* 79(3), 289–305.

Allen, P. and Kovach, M. (2000). The capitalist composition of organic: the potential of markets in fulfilling the promise of organic agriculture. *Agriculture and Human Values* 17(3), 221–232.

Allen, R. (1980). *How to Save the World: Strategy for World Conservation*. Kogan Page, London.

Allen, T. and Prosperi, P. (2016). Modelling sustainable food systems. *Environmental Management* 57(5), 956–975.

Allied Analytics LLP (2022). Biopesticides market value to cross $33,638.90 million by 2031. Available at: https://www.einnews.com/pr_news/566066297/biopesticides-market-value-to-cross-33-638-90-million-by-2031-top-companies-and-industry-growth-insights. Accessed 21 March 2022.

Allison, K. (2005). Investors dabble with living off the land. *Financial Times*, 5 April 2005.

Alston, J. M., MacEwan, J. P. and Okrent, A. M. (2016). The economics of obesity and related policy. *Annual Review of Resource Economics* 8(1), 443–465. DOI: 10.1146/annurev-resource-100815-095213.

Altieri, M. A. (1989). Agroecology: a new research and development paradigm for world agriculture. *Agriculture, Ecosystems and Environment* 27(1–4), 37–46.

Altieri, M. A. (1993). Ethnoscience and biodiversity: key elements in the design and of sustainable pest management systems for small farmers in developing countries. *Agriculture, Ecosystems and Environment* 46(1–4), 257–272.

Altieri, M. A. (1995). *Agroecology: The Science of Sustainable Agriculture*. Westview Press, Boulder, CO, 433 p.

Altieri, M. A. (2000). Multifunctional dimensions of ecologically based agriculture in Latin America. *International Journal of Sustainable Development and World Ecology* 7(1), 62–75.

Altieri, M. A. (2002). Agroecology: the science of natural resource management for poor farmers in marginal environments. *Agriculture, Ecosystems and Environment* 93(1–3), 1–24.

Altieri, M. A. (2009). Agroecology, small farms, and food sovereignty. *Monthly Review* 61(3), 102–113.

Altieri, M. A. (2010). Scaling up agroecological approaches to food sovereignty in Latin America. In: Wittman, Desmarais and Wiebe (Eds), pp. 120–133.

Altieri, M. A. (2012). *Agroecologia: Bases Científicas para Uma Agricultura Sustentável* (3rd edn). Expressão Popular, São Paulo; AS-PTA, Rio de Janeiro. ISBN 978-85-7743-191-5.

Altieri, M. A. and Nicholls, C. I. (2004). *Biodiversity and Pest Management in Agroecosystems* (2nd edn). CRC Press, Boca Raton, FL.

Altieri, M. A. and Nicholls, C. I. (2005). *Agroecology and the Search for a Truly Sustainable Agriculture*. Primera, México; University of California, Berkeley, CA.

Altieri, M. A. and Nicholls, C. I. (2020). Agroecology and the reconstruction of a post-COVID-19 agriculture. *The Journal of Peasant Studies* 47(5), 881–898. DOI: 10.1080/03066150.2020.1782891.

Altieri, M. A., Nicholls, C. I., Henao, A. and Lana, M. A. (2015). Agroecology and the design of climate change-resilient farming systems. *Agronomy for Sustainable Development* 35(3), 869–890. DOI: 10.1007/s13593-015-0285-2.

Altieri, M. A. and Rosset, P. (1996). Agroecology and the conversion of large-scale conventional systems to sustainable management. *International Journal of Environmental Studies* 50(3–4), 165–185.

Altieri, M. A., Rosset, P. and Thrupp, L. R. (1998). *The Potential of Agroecology to Combat Hunger in the Developing World. 2020 Brief.* International Food Policy Research Institute, Washington, DC.

Altieri, M. A. and Toledo, V. M. (2011). The agroecological revolution in Latin America: rescuing nature, ensuring food sovereignty, and empowering peasants. *Journal of Peasant Studies* 38(3), 587–612. DOI: 10.1080/03066150.2011.582947.

Amate, J. I. and González de Molina, M. (2013). Sustainable de-growth in agriculture and food: an agro-ecological perspective on Spain's agri-food system (year 2000). *Journal of Cleaner Production* 38, 27–35.

Amede, T. (2003). Opportunities and challenges in reversing land degradation: the regional experience. In: Amede, T. (Ed.) *Natural Resource Degradation and Environmental Concerns in the Amhara National Regional State; Impact on Food Security*. Ethiopian Soil Science Society, Addis Ababa, pp. 173–183.

Amekawa, Y., Sseguya, H., Onzere, S. and Carranza, I. (2010). Delineating the multifunctional role of agroecological practices: toward sustainable livelihoods for smallholder farmers in developing countries. *Journal of Sustainable Agriculture* 34(2), 202–228. DOI: 10.1080/10440040903433079.

Anderson, C. (2004). The long tail. *Wired Magazine*. Available at: https://www.wired.com/2004/10/tail/. Accessed 10 January 2004.

Anderson, C., Pimbert, M. and Kiss, C. (2015). *Building, Defending and Strengthening Agroecology: A Global Struggle for Food Sovereignty*. ILEIA, Centre for Learning on Sustainable Agriculture, Wageningen.

Anderson, F. R., Glicksman, R. L., Mandelker, D. R. and Tarlock, A. D. (1999). *Environmental Protection: Law and Policy* (3rd edn). Aspen Law and Business, New York.

Anderson, M. E. and Sobsey, M. D. (2006). Detection and occurrence of antimicrobially resistant *E. coli* in groundwater on or near swine farms in eastern North Carolina. *Water Science and Technology* 54(3), 211–218. DOI: 10.2166/wst.2006.471.

Andrée, P., Clark, J. K., Levkoe, C. Z. and Lowitt, K. (Eds) (2019). *Civil Society and Social Movements in Food System Governance*. Routledge, Oxon. ISBN 9780429503597.

Andress, L., Shanks, C. B., Hardison-Moody, A., Prewitt, T. E., Kinder, P. and Haynes-Maslow, L. (2020). The curated food system: a limiting aspirational vision of what constitutes "good" food. *International Journal of Environmental Research and Public Health 2020* 17, 6157. DOI: 10.3390/ijerph17176157.

Alonso-Fradejas, A., Borras Jr, S. M., Holmes, T., Holt-Giménez, E. and Robbins, M. J. (2015). Food sovereignty: convergence and contradictions, conditions, and challenges. *Third World Quarterly* 36(3), 431–448. DOI: 10.1080/01436597.2015.1023567.

Antonelli, A. (2023). Indigenous knowledge is key to sustainable food systems. *Nature* 613, , 239.

AQUASTAT (2016). *Water Uses*. Food and Agriculture Association of the United Nations. FAO, Rome. Available at: http://www.fao.org/nr/water/aquastat/water_use/index.stm.

Arabian Business (2022). *Group Amana Constructs World's Largest Vertical Farm in Dubai South*. 26 August. Available at: https://www.arabianbusiness.com/industries/construction/group-amana-constructs-worlds-largest-vertical-farm-in-dubai-south.

Arad, B., Balendonck, J., Barth, R., Ben-Shahar, O., Edan, Y., Hellström, T., Hemming, J., Kurtser, P., Ringdahl, O., Tielen, T. and van Tuijl, B. (2020). Development of a sweet pepper harvesting robot. *Journal of Field Robotics* 37(6), 1027–1039.

Araújo, R., Castro, A. C. M. and Fiúza, A. (2015). The use of nanoparticles in soil and water remediation processes. *Materials Today: Proceedings* 2(1), 315–320. DOI: 10.1016/j.matpr.2015.04.055.

ARC2020 (2015). *Transitioning Towards Agroecology: Using the CAP to Build New Food Systems*. ARC2020, Friends of the Earth Europe, and IFOAM, Berlin.

Arfini, F., Mancini, M. C. and Donati, M. (Eds) (2012). *Local Agri-Food Systems in a Global World: Market, Social and Environmental Challenges*. Cambridge Scholars Publishing, Newcastle Upon Tyne.

Arguelles Gonzalez, V. (2022). Can we raise livestock sustainably? A win-win solution for climate change, deforestation, and biodiversity loss. *The Conversation*. 14 March 2022.

Armanda, D. T., Guinée, J. B. and Tukker, A. (2019). The second green revolution: innovative urban agriculture's contribution to food security and sustainability–a review. *Global Food Security* 22, 13–24.

Armstrong, D. (2000). A survey of community gardens in upstate New York: implications for health promotion and community development. *Health and Place* 6(4), 319-327.

Arnall, A., Oswald, K., Davies, M., Mitchell, T. and Coirolo, C. (2010). *Adaptive Social Protection: Mapping the Evidence and Policy Context in the Agriculture Sector in South Asia; IDS Working Papers*. IDS, Wivenhoe Park, pp. 1–92.

Arrebola, J. P., Fernández, M. F., Olea, N., Ramos, R. and Martin-Olmedo, P. (2013). Human exposure to p,p'-dichlorodiphenyldichloroethylene (p,p'-DDE) in urban and semi-rural areas in southeast Spain: a gender perspective. *Science of the Total Environment* 458–460, 209–216.

Arrighi, G. (2009). The winding paths of capital: interview by David Harvey. *New Left Review* 56, 61-96.

Aschemann-Witzel, J. and Zieke, S. (2017). Can't buy me green? A review of consumer perceptions of and behaviour toward the price of organic food. *The Journal of Consumer Affairs* Spring 2017, 211–251. DOI: 10.1111/joca.1209.

Asfaw, A. (2011). Does consumption of processed foods explain disparities in the body weight of individuals? The case of Guatemala. *Health Economics* 20(2), 184–195.

Aslani, F., Bagheri, S., Muhd Julkapli, N., Juraimi, A. S., Hashemi, F. S. G. and Baghdadi, A. (2014). Effects of engineered nanomaterials on plants growth: an overview. *The Scientific World Journal* 2014, 641759.

Asseng, S. and Asche, F. (2019). Future farms without farmers. *Science Robotics* 4(27), eaaw1785.

Asseng, S., Guarin, J. R., Raman, M., Monje, O., Kiss, G., Despommiere, D. D., Meggers, F. M. and Gauthier, P. P. G. (2020). Wheat yield potential in controlled-environment vertical farms. *PNAS* 117(32), 19131–19135.

Astier, M., Argueta, J. Q., Orozco-Ramírez, Q., González, M. V., Morales, J., Gerritsen, P. R. W., Escalona, M. A., Rosado-May, F. J., Sánchez-Escudero, J., Martínez Saldaña, T., Sánchez-Sánchez, C., Arzuffi Barrera, R., Castrejón, F., Morales, H., Soto, L., Mariaca, R., Ferguson, B., Rosset, P., Ramírez, H., Jarquin, R., García-Moya, F., Ambrosion Montoya, M. and González-Esquivel, C. (2017). Back to the roots: understanding current agroecological movement, science, and practice in Mexico. *Agroecology and Sustainable Food Systems* 41(3–4), 329–348.

Atkins, P. and Bowler, I. (2001). *Food in Society. Economy, Culture, Geography*. Arnold, London.

Atkins, P. J. (2007). A tale of two cities: a comparison of food supply in London and Paris in the 1850s. In: Atkins, P. J., Lummel, P. and Oddy, D. J. (Eds) *Food and the city in Europe Since 1800*. Ashgate, Aldershot, pp. 25–38.

Aubry, C. and Kebir, L. (2013). Shortening food supply chains: a means for maintaining agriculture close to urban areas? The case of the French metropolitan area of Paris. *Food Policy* 41, 85–93. DOI: 10.1016/j.foodpol.2013.04.006.

Augére-Granier, L. M. (2016). *Short Food Supply Chains and Local Food Systems in the EU*. European Parliamentary Research Service (EPRS), Brussels.

Austin, J. E. and Esteva, G. (1987). Final reflections. In: Austin, J. E. and Esteva, G. (Eds) *Food Policy in Mexico: the Search for Self-Sufficiency*. Cornell University Press, Ithaca, NY, pp. 353–373.

AZoNano.com (2013). *Nanofibers to Be Used in Drug Delivery, Gene Therapy, Crop Engineering and Environmental Monitoring*. AzoM.com Pty. Ltd, Manchester. Available at: http://www.azonano.com/article.aspx?ArticleID=114. Accessed 19 April 2014.

Azupogo, F., Aurino, E., Gelli, A., Bosompem, K. M., Ayi, I., Osendarp, S. J. M., Brouwer, I. D. and Folson, G. (2019). Agro-ecological zone and farm diversity are factors associated with haemoglobin and anaemia among rural school-aged children and adolescents in Ghana. *Maternal and Child Nutrition* 15(1). DOI: 10.1111/mcn.12643.

Babel, M. S. and Wahid, S. W. (2008). Freshwater under threat in South Asia. UNEP Report. United Nations Environment Programme (UNEP). ISBN 978-92-807-2949-8, 29 pp.

Bac, C. W., van Henten, E. J., Hemming, J. and Edan, Y. (2014). Harvesting robots for high-value crops: state-of-the-art review and challenges ahead. *Journal of Field Robotics* 31(6), 888–911.

Bacco, M., Barsocchi, P., Ferro, E., Gotta, A. and Ruggeri, M. (2019). The digitisation of agriculture: a survey of research activities on smart farming. *Array* 2019, 3–4, 100009.

Bacco, M., Berton, A., Ferro, E., Gennaro, C., Gotta, A., Matteoli, S., Paonessa, F., Ruggeri, M., Virone, G. and Zanella, A. (2018). Smart farming: opportunities, challenges, and technology enablers. In: Proceedings of the 2018 IoT Vertical and Topical Summit on Agriculture–Tuscany (IOT Tuscany), Tuscany, Italy, 8–9 May 2018, pp. 1–6.

Badgley, C., Moghtader, J., Quintero, E., Zakem, E., Chappell, M. J., Avilés-Vázquez, K., Samulon, A. and Perfecto, I. (2007). Organic agriculture and the global food supply. *Agriculture and Food Systems* 22(2), 86–108. DOI: 10.1017/S1742170507001640.

Badgley, C. and Perfecto, I. (2007). Can organic agriculture feed the world? *Renewable Agriculture and Food Systems* 22(2), 80–86.

Bahadur, K. C. K., Dias, G. M., Veeramani, A., Swanton, C. J., Fraser, D., Steinke, D., Lee, E., Wittmen, H., Farber, J. M., Dunfield, K., McCann, K., Anand, M., Campbell, M., Rooney, N., Raine, N. E., Van Acker, R., Hanner, R., Pascoel, S., Sherif, S., Benton, T. G. and Fraser, E. D. G. (2018). When too much isn't enough: does current food production meet global nutritional needs? *PLoS ONE* 13(10), 16. DOI: 10.1371/journal.pone.0205683.

Bai, Y. and Gao, J. (2021). Optimization of the nitrogen fertilizer schedule of maize under drip irrigation in Jilin, China, based on DSSAT and GA. *Agricultural Water Management* 244, 106555.

Bailey, L. H. (1909). *Cyclopedia of American Agriculture: A Popular Survey of Agricultural Conditions, Practices and Ideals in the United States and Canada* (Vol. 2). Macmillan Company, New York.

Bailey, R., Benton, T. G., Challinor, A., Elliott, J., Gustafson, D., Hiller, B., Jones, A., Jahn, M., Kent, C., Lewis, K., Meacham, T., Rivington, M., Robson, D., Tiffin, R. and Wuebbles, D. J.(2015). Extreme weather and resilience of the global food system. Final Project Report from the UKUS Taskforce on Extreme Weather and Global Food System Resilience. The Global Food Security Programme, UK.

Baines, J. (2015). Fuel, feed, and the corporate restructuring of the food regime. *Journal of Peasant Studies* 42(2), 295–321.

Bajželj, B., Allwood, J. M. and Cullen, J. M. (2013). Designing climate change mitigation plans that add up. *Environmental Science and Technology* 47(14), 8062–8069. DOI: 10.1021/es400399h.

Baker, L. E. (2004). Tending cultural landscapes and food citizenship in Toronto's community gardens. *Geographical Review* 94(3), 305–325.

Baker, P. and Friel, S. (2014). Processed foods and the nutrition transition: evidence from Asia. *Obesity Reviews* 15(7), 564–577.

Baker, P. and Friel, S. (2016). Food systems transformations, ultra-processed food markets and the nutrition transition in Asia. *Globalization and Health* 12(1), 80. DOI: 10.1186/s12992-016-0223-3.

Balafoutis, A. T., Van Evert, F. K. V. and Fountas, S. (2020). Smart farming technology trends: economic and environmental effects, labor impact, and adoption readiness. *Agronomy* 10(5), 743. DOI: 10.3390/agronomy10050743.

Balmford, A., Amano, T., Bartlett, H., Chadwick, D., Collins, A., Edwards, D., Field, R., Garnsworthy, P., Green, R., Smith, P., Waters, H., Whitmore, A., Broom, D. M., Chara, J., Finch, T., Garnett, E., Gathorne-Hardy, A., Hernandez-Medrano, J., Herrero, M., Hua, F., Latawiec, A., Misselbrook, T., Phalan, B., Simmons, B. I., Takahashi, T., Vause, J., zu Ermgassen, E. and Eisner, R. (2018). The environmental costs and benefits of high yield farming. *Nature Sustainability* 1(9), 477–485. Available at: www.nature.com/natsustain.

Balmford, A., Green, R. E. and Scharlemann, J. P. W. (2005). Sparing land for nature: exploring the potential impact of changes in agricultural yield on the area needed for crop production. *Global Change Biology* 11(10), 1594–1605. DOI: 10.1111/j.1365-2486.2005.001035.x.

Balwinder-Singh, , Humphreys, E., Gaydonca, D. S. and Yadavb, S. (2015). Options for increasing the productivity of the rice-wheat system of northwest India while reducing groundwater depletion. Part 2. Is conservation agriculture the answer? *Field Crops Research* 173, 81–94.

Bane, P. (2012). *The Permaculture Handbook: Garden Farming for Town and Country*. New Society, New York.

Bannari, A., Mohamed, A. M. A. and El-Battay, A. (2017). Water stress detection as an indicator of red palm weevil attack using worldview-3 data. In: Proceedings of the 2017 IEEE International Geoscience and Remote Sensing Symposium (IGARSS), Fort Worth, TX, USA, 23 July 2017, pp. 4000–4003.

Baqué, P. (2012). *La Bio Entre Business et Projet de Société*. Agone, Marseille.

Baragwanath, T. (2021). Digital opportunities for demand-side policies to improve consumer health and the sustainability of food systems. OECD Food, Agriculture and Fisheries Papers, No. 148. OECD Publishing, Paris. DOI: 10.1787/bec87135-en.

Barański, M., Srednicka-Tober, D., Volakakis, N., Seal, C., Sanderson, R., Stewart, G. B., Benbrook, C., Biavati, B., Markellou, E., Giotis, C., Gromadzka-Ostrowska, J., Rembiałkowska, E., Skwarło-Sońta, K., Tahvonen, R., Janovská, D., Niggli, U., Nicot, P. and Leifert, C. (2014). Higher antioxidant and lower cadmium concentrations and lower incidence of pesticide residues in organically grown crops: a systematic literature re- view and meta-analyses. *British Journal of Nutrition* 112(5), 794–811. DOI: 10.1017/S0007114514001366.

Bàrberi, P. (2019). Ecological weed management in sub-Saharan Africa: prospects and implications on other agroecosystem services. *Advances in Agronomy* 156(4), 219–264.

Barchowsky, A. and O'Hara, K. A. (2003). Metal-induced cell signalling and gene activation in lung diseases. *Free Radical Biology and Medicine* 34(9), 1130–1135. DOI: 10.1016/S0891-5849(03)00059-5.

Barham, E. (2003). Translating terroir: the global challenge of French AOC labelling. *Journal of Rural Studies* 19(1), 127–138.

Barles, S. (2007). Feeding the city: food consumption and flow of nitrogen, Paris, 1801–1914. *Science of the Total Environment* 375(1–3), 48–58. DOI: 10.1016/j.scitotenv.2006.12.003.

Barnes, A. P., Soto, I., Eory, V., Beck, B., Balafoutis, A., Sánchez, B., Vangeyte, J., Fountas, S., van der Wal, T. and Gómez-Barbero, M. (2019). Exploring the adoption of precision agricultural technologies: a cross regional study of EU farmers. *Land Use Policy* 80, 163–174. DOI: 10.1016/j.landusepol.2018.10.004.

Barnett, B. J., Barrett, C. B. and Skees, J. R. (2008). Poverty traps and index-based risk transfer products. *World Development* 36(10), 1766–1785.

Barrett, C. B. and Constas, M. A. (2014). Toward a theory of resilience for international development applications. *Proceedings of the National Academy of Sciences of the United States of America* 111(40), 14625–14630.

Barrett, H. and Rose, D. C. (2020). Perceptions of the fourth agricultural revolution: what's in, what's out, and what consequences are anticipated? *Sociologia Ruralis* 62(2). DOI: 10.1111/soru.12324.

Basso, B. and Antle, J. (2020). Digital agriculture to design sustainable agricultural systems. *Nature Sustainability* 3(4), 254–256.

Bationo, A., Pinto-Toyi, K. A., Ayuk, E. and Mokwunye, A. U. (2012). Overview of long-term experiments in Africa. In: Bationo, A. (Ed.) *Lessons Learned from Long-Term Soil Fertility Management Experiment in Africa*. Springer, pp. 1–26.

Baud, C. and Durand, C. (2012). Financialization, globalization, and the making of profits by leading retailers. *Socio-Economic Review* 10(2), 241–266.

Baudry, J., Méjean, C., Allès, B., Péneau, S., Touvier, M., Hercberg, S., Lairon, D., Galan, P. and Kesse-Guyot, E. (2015). Contribution of organic food to the diet in a large sample of French adults (the NutriNet-Santé cohort study). *Nutrients* 7(10), 8615–8632.

Baulcombe, D., Crute, I., Davies, B., Dunwell, J., Gale, M., Jones, J., Pretty, J., Sutherland, W., Toulmin, C. and Green, N. (2009). *Reaping the Benefits: Science and the Sustainable Intensification of Global Agriculture*. The Royal Society, London.

Bauwens, T., Hekkert, M. and Kirchherr, J. (2020). Circular futures: what will they look like? *Ecological Economics* 175, 106703.

Bayford, K. (2023). Half UK consumers who cut back on animal products find plant-based diet more expensive. Available at: https://www.grocerygazette.co.uk/2023/02/22/uk-adults-plant-based-expense/. Accessed 22 February 2023.

BBC (2021). Anti-obesity drive: junk food TV adverts to be banned before 9 pm. Available at: https://www.bbc.com/news/uk-politics-57593599. Accessed 10 July 2021.

BBC (2023). Lean times hit the vertical farming business. Available at: https://www.bbc.com/news/business-66173872. Accessed 18 July 2023.

BCERF (1998). Pesticides and breast cancer risk, an evaluation of heptachlor. Cornell University Program on Breast Cancer and Environmental Risk Factors in New York State. Fact Sheet 12, March 1998.

BCFN (2012). Food waste: causes, impacts and proposals. Barilla Center for Food and Nutrition. Codice Edizioni. Available at: https://www.barillacfn.com/en/publications/foodwaste-causes-impacts-and-proposals/.

Beach, R. H., Creason, J., Ohrel, S. B., Ragnauth, S., Ogle, S., Li, C., Ingraham, P. and Salas, W. (2015). Global mitigation potential and costs of reducing agricultural non-CO_2 greenhouse gas emissions through 2030. *Journal of Integrative Environmental Sciences* 12, 87–105.

Beauregard, S. (2009). *Food Policy for People: Incorporating Food Sovereignty Principles into State Governance. Case Studies of Venezuela, Mali, Ecuador, and Bolivia*. Urban and Environmental Policy Department, Occidental College, Los Angeles, CA. Available at: http://www.oxy.edu/sites/default/files/assets/UEP/Comps/2009/Beauregard%20Food%20Policy%20for%20People.pdf.

Bell, G. (2005). *The Permaculture Garden*. Chelsea Green, White River Junction.

Bellamy, A. S. and Ioris, A. A. R. (2017). Addressing the knowledge gaps in agroecology and identifying guiding principles for transforming conventional agri-food systems. *Sustainability* 9(3), 330. DOI: 10.3390/su9030330.

Bellon, M. R., Ntandou-Bouzitou, G. D. and Caracciolo, F. (2016). On-farm diversity and market participation are positively associated with dietary diversity of rural mothers in southern Benin, West Africa. *PLoS ONE* 11(9). DOI: 10.1371/journal.pone.0162535.

Bellon, S. and de Abreu, L. (2006). Rural social development: small-scale horticulture in Sao Paulo, Brazil. In: Holt, G. C. and Reed, M. (Eds) *Sociological Perspectives of Organic Agriculture: From Pioneer to Policy*. CABI Publishing, pp. 243–259. DOI: 10.1079/9781845930387.0000.

Bellon-Maurel, V. and Huyghe, C. (2017). Putting agricultural equipment and digital technologies at the cutting edge of agroecology. *Oilseeds Fats Crops Lipids* 24(3), D307.

Belmonte-Ureña, L. J., Plaza-Úbeda, J. A., Vazquez-Brust, D. and Yakovleva, N. (2021). Circular economy, degrowth and green growth as pathways for research on sustainable development goals: a global analysis and future agenda. *Ecological Economics* 185, 107050.

Béné, C., Oosterveer, P., Lamotte, L., Brouwer, I. D., de Haan, S., Prager, S. D., Talsma, E. F. and Khoury, C. K. (2019a). When food systems meet sustainability - current narratives and implications for actions. *World Development* 113, 116-130.

Béné, C., Prager, S. D., Achicanoy, H. A. E., Toro, P. A., Lamotte, L., Cedrez, C. B. and Mapes, B. R. (2019b). Understanding food systems drivers: a critical review of the literature. *Global Food Security* 23, 149-159.

Bengtsson, J., Ahnström, J. and Weibull, A. C. (2005). The effects of organic agriculture on biodiversity and abundance: a meta-analysis. *Journal of Applied Ecology* 42(2), 261-269.

Benis, K. and Ferrão, P. (2018). Commercial farming within the urban built environment–taking stock of an evolving field in northern countries. *Global Food Security* 17, 30-37.

Benke, K. and Tomkins, B. (2017). Future food-production systems: vertical farming and controlled-environment agriculture. *Sustainability: Science, Practice and Policy* 13(1), 13-26.

Bennetzen, E. H., Smith, P. and Porter, J. R. (2016). Decoupling of greenhouse gas emissions from global agricultural production: 1970-2050. *Global Change Biology* 22(2), 763-781.

Benoit, M., Rizzo, D., Marraccini, E., Moonen, A. C., Galli, M., Lardon, S., Rapey, H., Thenail, C. and Bonari, E. (2012). Landscape agronomy: a new field for addressing agricultural landscape dynamics. *Landscape Ecology* 27(10), 1385-1394. DOI: 10.1007/s10980-012-9802-8.

Bentley, J. W., Rodríguez, G. and González, A. (1994). Science and people: Honduran campesinos and natural pest control inventions. *Agriculture and Human Values* 11(2-3), 178-182.

Benton, T. G. (2012). Managing agricultural landscapes for multiple services: the policy challenge. *International Agricultural Policy* 1, 7-18.

Benton, T. G. (2015). Chapter 6: Sustainable intensification. In: Pritchard, B., Ortiz, R. and Shekar, M. (Eds) *Routledge Handbook of Food and Nutrition Security*. Routledge. ISBN 9781138343498.

Benton, T. G. and Bailey, R. (2019). The paradox of productivity: agricultural productivity promotes food system inefficiency. *Global Sustainability* 2, 1-8. DOI: 10.1017/sus.2019.3.

Benton, T. G., Bieg, C., Harwatt, H., Pudasaini, R. and Wellesley, L. (2021). Food system impacts on biodiversity loss: Three levers for food system transformation in support of nature. Research Paper. Energy, Environment and Resources Programme. Chatham House - The Royal Institute of International Affairs.

Benton, T. G. and Harwatt, H. (2022). Sustainable agriculture and food systems: comparing contrasting and contested versions. Research Paper. Environment and Society Programme. Chatham House, London.

Benyezza, H., Bouhedda, M. and Rebouh, S. (2021). Zoning irrigation smart system based on fuzzy control technology and IoT for water and energy saving. *Journal of Cleaner Production* 302, 127001.

Berckmans, D. (Ed) (2022). *Advances in Precision Livestock Farming*. Burleigh Dodds Science Publishing, Cambridge.

Berdegué, J. A., Proctor, F. J. and Cazzuffi, C. (2014). Inclusive rural-urban linkages. Working Paper Series No. 123. Rimisp, Santiago, Chile.

Berenbaum, M. R. (1995). *Bugs in the System: Insects and Their Impact on Human Affairs*. Addison-Wesley, Reading, MA.

Bereuter, D., Glickman, D. and Reardon, T. A. (2016). Transforming food systems in an urbanizing world. Chicago Council on Global Affairs or the Global Food and Agriculture Program. Chicago, IL.

Bernardes, M. F. F., Pazin, M., Pereira, L. C. and Dorta, D. J. (2015). Impact of pesticides on environmental and human health. In: Andreazza, A. C. and Scola, G. (Eds) *Toxicology Studies - Cells, Drugs and Environment*. InTech. DOI: 10.5772/58714. DOI: 10.5772/59710.

Berners-Lee, M., Hoolohan, C., Cammack, H. and Hewitt, C. N. (2012). The relative greenhouse gas impacts of realistic dietary choices. *Energy Policy* 43, 184–190.

Bernstein, H. (2014). Food sovereignty via the 'peasant way': a sceptical view. *The Journal of Peasant Studies* 41(6), 1031–1063. DOI: 10.1080/03066150.2013.852082.

Beste, A. (2021). Comparing organic, agroecological and regenerative farming part 3 - regenerative. Available at: https://www.resilience.org/stories/2021-02-12/comparing-organic-agroecological-and-regenerative-farming-part-3-regenerative/.

Beuchelt, T. D. and Virchow, D. (2012). Food sovereignty or the human right to adequate food: which concept serves better as international development policy for global hunger and poverty reduction? *Agriculture and Human Values* 29(2), 259–273.

Beudou, J., Martin, G. and Ryschawy, J. (2017). Cultural and territorial vitality services play a key role in livestock agroecological transition in France. *Agronomy for Sustainable Development* 37(4), 1–11. DOI: 10.1007/s13593-017-0436-8.

Beukes, P. C., McCarthy, S., Wims, C. M., Gregorini, P. and Romera, A. J. (2019). Regular estimates of herbage mass can improve profitability of pasture-based dairy systems. *Animal Production Science* 59(2), 359.

Beyond Meat (2022). Home Page. Available at: https://www.beyondmeat.com/en-US/products/. Accessed 29 January 2022.

Beyond Pesticides (2019). Organic systems: the path forward. *Pesticides and You* Summer 2019. Available at: www.BeyondPesticides.org.

BFED and IPES-Food (2020). *Money Flows: What Is Holding Back Investment in Agroecological Research for Africa?* Biovision Foundation for Ecological Development (BFED) and International Panel of Experts on Sustainable Food Systems (IPES). Available at: www.biovision.ch and www.ipes-food.org.

Bhat, B. A., Ahmad Bhat, I., Vishwakarma, S., Verma, A. and Saxena, G. (2012). A comparative study on the toxicity of a synthetic pesticide, dichlorvos and a neem-based pesticide, neem on to *Labeo rohita* (Hamilton). *Current World Environment* 7(1), 157–161.

Bhatt, V. and Farah, L. M. (2016). Cultivating montreal: a brief history of citizens and institutions integrating urban agriculture in the city. *Urban Agriculture and Regional Food Systems* 1(1). DOI: 10.2134/urbanag2015.01.1511.

Biel, R. (2016). *Sustainable Food Systems*. UCL Press, London.

Bijman, J. (2001). Advanta: worldwide challenges. *AgBioForum* 4, 34–39.

Billen, G. (2011). Grain, meat, and vegetables to feed Paris: where did and do they come from? Localising Paris food supply areas from the eighteenth to the twenty-first century. *Regional Environmental Change* 12(2), 325–335. DOI: 10.1007/s10113-011-0244-7.

Billen, G., Barles, S., Garnier, J., Rouillard, J. and Benoit, P. (2009). The food-print of Paris: long-term reconstruction of the nitrogen flows imported into the city from its rural hinterland. *Regional Environmental Change* 9(1), 13–24. DOI: 10.1007/s10113-008-0051-y.

Bindraban, P. S., van der Velde, M., Ye, L. M., van den Berg, M., Materechera, S., Kiba, D. I., Tamene, L., Ragnarsdottir, K. V., Jongschaap, R., Hoogmoed, M., Hoogmoed, W., van Beek, C. and van Lynden, G. (2012). Assessing the impact of soil degradation on food production. *Current Opinion in Environmental Sustainability* 4(5), 478–488.

Birch, N. and Glare, T. (Eds) (2020). *Biopesticides for Sustainable Agriculture*. Burleigh Dodds Science Publishing, Cambridge.

Birkhofer, K., Bezemer, T. M., Bloem, J., Bonkowski, M., Christensen, S., Dubois, D., Ekelund, F., Fließbach, A., Gunst, L., Hedlund, K., Mäder, P., Mikola, J., Robin, C., Setälä, H., Tatin-Froux, F., Van der Putten, W. H. and Scheu, S. (2008). Long-term organic farming fosters below and aboveground biota: implications for soil quality, biological control, and productivity. *Soil Biology and Biochemistry* 40(9), 2297–2308.

BIS (2009). *Statistics on Amounts Outstanding of OTC Equity-Linked and Commodity Derivatives, by Instrument and Counterparty*. Bank for International Settlements, Basel. Available at: http://www.bis.org/statistics/otcder/dt21c22a.pdf.

Bisht, I. S., Mehta, P. S., Negi, K. S., Verma, S. K., Tyagi, R. K. and Garkoti, S. C. (2018). Farmers' rights, local food systems, and sustainable household dietary diversification: a case of Uttarakhand Himalaya in north-western India. *Agroecology and Sustainable Food Systems* 42(1), 77–113. DOI: 10.1080/21683565.2017.1363118.

BlackRock (2013). Black Rock world agricultural fund fact sheet. Available at: http://www.blackrock.com.hk/content/groups/hongkongsite/documents/literature/1111121111.pdf.

Blandford, D. and Hassapoyannes, K. (2018). The role of agriculture in global GHG mitigation. OECD Food, Agriculture and Fisheries Papers, No. 112. OECD Publishing, Paris. DOI: 10.1787/da017ae2-en.

Blasch, G., Spengler, D., Hohmann, C., Neumann, C., Itzerott, S. and Kaufmann, H. (2015). Multitemporal soil pattern analysis with multispectral remote sensing data at the field scale. *Computers and Electronics in Agriculture* 113, 1–13.

Blättel-Mink, B. (2014). Active consumership as a driver towards sustainability? *GAIA - Ecological Perspectives for Science and Society* 23(3), 158–165. DOI: 10.14512/gaia.23.S1.3.

Blay-Palmer, A. (2009). The Canadian pioneer: the genesis of urban food policy in Toronto. *International Planning Studies* 14(4), 401–416. DOI: 10.1080/13563471003642837..

Blay-Palmer, A., Santini, G., Dubbeling, M., Renting, H., Taguchi, M. and Giordano, T. (2018). Validating the city region food system approach: enacting inclusive, transformational city region food systems. *Sustainability* 10(5), 1680. DOI: 10.3390/su10051680.

Bleich, S., Cutler, D., Murray, C. and Adams, A. (2008). Why is the developed world obese? *Annual Review of Public Health* 29, 273–295. DOI: 10.1146/annurev.publhealth.29.020907.090954.

Blesh, J. and Wolf, S. A. (2014). Transitions to agroecological farming systems in the Mississippi River basin: toward an integrated socioecological analysis. *Agriculture and Human Values* 31(4), 621–635.

Blok, P. M., van Evert, F. K., Tielen, A. P. M., van Henten, E. J. and Kootstra, G. (2021). The effect of data augmentation and network simplification on the image-based detection of broccoli heads with Mask R-CNN. *Journal of Field Robotics* 38(1), 85–104.

Bloom, J. (2015). Something to shoot for: US sets food waste reduction goal. *Wasted Food*. Available at: http://www.wastedfood.com/2015/09/16/something-toshoot-for-us-sets-food-waste-reduction-goal/. Accessed 18 December 2015.

Bloomberg (2022). Impossible foods plans to lay off about 20% of workers. Available at: https://www.bloomberg.com/news/articles/2023-01-30/impossible-foods-plans-to-lay-off-about-20-of-employees. Accessed 30 January 2023.

Bloomberg (2023a). Insect farming startup raises $175million for food expansion. Available at: https://www.bloomberg.com/news/articles/2023-04-16/insect-farming-startup-ynsect-raises-175-million-for-expansion. Accessed 16 April 2023.

Bloomberg (2023b). Lab-grown meat has a bigger problem than the lab. Available at: https://www.bloomberg.com/news/features/2023-02-07/lab-grown-meat-has-bigger-challenges-than-the-fda. Accessed 7 February 2023.

Bloomberg (2023c). Funding is drying up for AI-run vertical farms. Available at: https://www.bloomberg.com/news/newsletters/2023-06-16/from-appharvest-to-aerofarms-funding-is-drying-up-for-ai-run-vertical-farms. Accessed 16 June 2023.

Blythman, J. (2020). If gene editing to 'improve' food sounds too good to be true, that's because it is. *The Herald Newspaper* 22 August 2020.

Boardman, R. (1986). *Pesticides in World Agriculture: The Politics of International Regulation*. St Martin's Press, New York.

Böll, H. K. (2022). Bhutan's challenges and prospects in becoming a 100% organic country. Available at: https://hk.boell.org/en/2022/09/08/bhutans-challenges-and-prospects-becoming-100-organic-country. Accessed 12 March 2023.

Bommarco, R., Kleijn, D. and Potts, S. G. (2013). Ecological intensification: harnessing ecosystem services for food security. *Trends in Ecology and Evolution* 28(4), 230–238.

Bonaiuti, M. (Eds) (2011). *From Bioeconomics to Degrowth: Georgescu-Roegen's 'New Economics' in Eight Essays*. Routledge, New York.

Bootsman, R. and Rogers, R. (2010). *What's Mine Is Yours: The Rise of Collaborative Consumption*. HarperCollins Publishers, HarperCollins. ISBN 0062014056, 9780062014054.

Boraeve, F., Dendoncker, N., Cornélis, J. T., Degrune, F. and Dufrêne, M. (2020). Contribution of agroecological farming systems to the delivery of ecosystems services. *Journal of Environmental Management* 260, 109576.

Born, B. and Purcell, M. (2006). Avoiding the local trap scale and food systems in planning research. *Journal of Planning Education and Research* 26(2), 195–207. DOI: 10.1177/0739456X06291389.

Borras, S. M. Jr., Edelman, M. and Kay, C. (2008). Transnational agrarian movements: origins and politics, campaigns, and impact. *Journal of Agrarian Change* 8(2–3), 169–204. DOI: 10.1111/j.1471-0366.2008.00167.x.

Borras, S. M. Jr. and Franco, J. C. (2012). Global land grabbing and trajectories of agrarian change: a preliminary analysis. *Journal of Agrarian Change* 12(1), 34–59.

Borras, S. M. Jr, Franco, J. C., Gómez, S., Kay, C. and Spoor, M. (2012). Land grabbing in Latin America and the Caribbean. *The Journal of Peasant Studies* 39(3–4), 845–872.

Boursianis, A. D., Papadopoulou, M. S., Diamantoulakis, P., Liopa-Tsakalidi, A., Barouchas, P., Salahas, G., Karagiannidis, G., Wan, S. and Goudos, S. K. (2020). Internet of Things (IoT) and Agricultural Unmanned Aerial Vehicles (UAVs) in smart farming: a comprehensive review. *IEEE Internet Things*. DOI: 10.1016/j.iot.2020.100187.

Bouwman, A. F., Bierkens, M. F. P., Griffioen, J., Hefting, M. M., Middelburg, J. J., Middelkoop, H. and Slomp, C. P. (2013). Nutrient dynamics, transfer, and retention along the aquatic continuum from land to ocean: towards integration of ecological and biogeochemical models. *Biogeosciences* 10(1), 1–22. DOI: 10.5194/bg-10-1-2013.

Bowler, I. R. (1990). Agricultural geography. *Progress in Human Geography* 14(4), 569–578.

Bowler, I. R. (1992). *The Geography of Agriculture in Developed Market Economies*. Longman Scientific and Technical, New York.

Boyer, J. (2010). Food security, food sovereignty, and local challenges for transnational agrarian movements: the Honduras case. *Journal of Peasant Studies* 37(2), 319–351.

Brady, M., Kellermann, K., Sahrbacher, C. and Jelinek, L. (2009). Impacts of decoupled agricultural support on farm structure, biodiversity, and landscape mosaic: some EU results. *Journal of Agricultural Economics* 60(3), 563–585. DOI: 10.1111/j.1477-9552.2009.00216.x.

Branca, G., Arslan, A., Paolantonio, A., Grewer, U., Cattaneo, A., Cavatassi, R., Lipper, L., Hillier, J. and Vetter, S. (2021). Assessing the economic and mitigation benefits of climate-smart agriculture and its implications for political economy: a case study in Southern Africa. *Journal of Cleaner Production* 285, 125161.

Brandmeier, J., Reininghaus, H. and Scherber, C. (2023). Multispecies crop mixtures increase insect biodiversity in an intercropping experiment. *Ecological Solutions and Evidence*. DOI: 10.1002/2688-8319.12267.

Brandt, L. and Erixon, F. (2013). The prevalence and growth of obesity and obesity-related illnesses in Europe. Available at: http://ecipe.org/app/uploads/2014/12/Think_piece_obesity_final.pdfn.

Brass, T. (2015). Peasants, academics, populists: forward to the past? *Critique of Anthropology* 35(2), 187–204.

Breger Bush, S. (2012). *Derivatives and Development: A Political Economy of Global Finance, Farming, and Poverty*. Palgrave Macmillan, New York.

Brewster, C., Roussaki, I., Kalatzis, N., Doolin, K. and Ellis, K. (2017). IoT in agriculture: designing a Europe-wide large-scale pilot. *IEEE Communications Magazine* 55(9), 26–33.

Briar, S. S., Grewal, P. S., Somasekhar, N., Stinner, D. and Miller, S. A. (2007). Soil nematode community, organic matter, microbial biomass, and nitrogen dynamics in field plots transitioning from conventional to organic management. *Applied Soil Ecology* 37(3), 256–266.

Briske, D. D., Ash, A. J., Derner, J. D. and Huntsinger, L. (2014). Commentary: a critical assessment of the policy endorsement for holistic management. *Agricultural Systems* 125, 50–53.

Britannica (2015). Britannica, The Editors of Encyclopaedia. Agricultural Revolution. *Encyclopaedia Britannica*. Available at: https://www.britannica.com/topic/agricultural-revolution. Accessed 4 December 2015.

Brittain, C. A., Vighi, M., Bommarco, R., Settele, J. and Potts, S. G. (2010). Impacts of a pesticide on pollinator species richness at different spatial scales. *Basic and Applied Ecology* 11(2), 106–115. DOI: 10.1016/J.BAAE.2009.11.007.

Britton, D. (1990). *Agriculture in Britain: Changing Pressures and Policies*. CAB International, Wallingford.

Britwum, K., Yiannaka, A. and Kastanek, K. (2018). Public perceptions of genetically engineered nutraceuticals. *AgBioForum* 21(1), 13–24.

Broad, G. M. (2019). Plant-based and cell-based animal product alternatives: an assessment and agenda for food tech justice. *Geoforum* 107, 223–226.

Broad, G. M. (2020a). Know Your indoor farmer: square roots, techno-local food, and transparency as publicity. *American Behavioral Scientist* 64(11). DOI: 10.1177/0002764220945349.

Broad, G. M. (2020b). Making meat, better: the metaphors of plant-based and cell-based meat innovation. *Environmental Communication* 14(7), 919–932. DOI: 10.1080/17524032.2020.1725085.

Brock, D. A., Douglas, T. E., Queller, D. C. and Strassmann, J. E. (2011). Primitive agriculture in a social amoeba. *Nature* 469(7330), 393–396.

Bródy, L. S. and de Wilde, M. (2020). Cultivating food or cultivating citizens? On the governance and potential of community gardens in Amsterdam. *Local Environment* 25(3), 243–257. DOI: 10.1080/13549839.2020.1730776.

Brondizio, E. S., Settele, J., Díaz, S. and Ngo, H. T. (Eds) (2019). Global assessment report on biodiversity and ecosystem services. Intergovernmental Science-Policy Platform on Biodiversity and Ecosystem Services (IPBES Secretariat, 2019).

Bronson, K. (2018). Smart farming: including rights holders for responsible agricultural innovation. *Technology Innovation Management Review* 8(2), 7–14. DOI: 10.22215/timreview/1135.

Bronson, K. (2019). The digital divide and how it matters for Canadian food system equity. *Canadian Journal of Communication* 44(2), 63–68.

Bronson, K. and Knezevic, I. (2016). Big Data in food and agriculture. *Big Data and Society* 2016, 3. DOI: 10.1177/2053951716648174.

Brookes, G. (2022). Farm income and production impacts from the use of genetically modified (GM) crop technology 1996-2020. *GM Crops and Food* 13(1), 171-195. DOI: 10.1080/21645698.2022.2105626.

Brookes, G. and Barfoot, P. (2018). Environmental impacts of genetically modified (GM) crop use 1996–2016: impacts on pesticide use and carbon emissions. *GM Crops and Food* 9(3), 109–139. DOI: 10.1080/21645698.2018.1476792. PMID: 29883251; PMCID: PMC6277064.

Brookes, G. and Barfoot, P. (2020). Environmental impacts of genetically modified (GM) crop use 1996–2018: impacts on pesticide use and carbon emissions. *GM Crops and Food* 11(4), 215–241. DOI: 10.1080/21645698.2020.1773198. PMID: 32706316; PMCID: PMC7518756.

Brouder, S. M. and Gomez-Macpherson, H. (2014). The impact of conservation agriculture on smallholder agricultural yields: a scoping review of the evidence. *Agriculture, Ecosystems and Environment* 187, 11–32.

Brown (2018). *Innovations in Ammonia*. US Department of Energy, H2@Scale R&D Consortium Kick-Off Meeting Chicago, 1 August 2018.

Brown, O. (2008). Migration and climate change. IOM Migration Research Series, No. 31. International Organization for Migration, Geneva. Available at: https://www.iom.cz/files/Migration_and_Climate_Change_-_IOM_Migration_Research_Series_No_31.pdf.

Brown, P., Daigneault, A. and Dawson, J. (2019). Age, values, farming objectives, past management decisions, and future intentions in New Zealand agriculture. *Journal of Environmental Management* 231, 110–120. DOI: 10.1016/j.jenvman.2018.10.018.

Brown, P. and Roper, S. (2017). Innovation and networks in New Zealand farming. *Australian Journal of Agricultural and Resource Economics* 61(3), 422–442. DOI: 10.1111/1467-8489.12211.

Brundtland, G. (1987). Report of the World Commission on Environment and Development: Our Common Future. United Nations General Assembly document, A/42/427.

Brunori, G., Rossi, A. and Malandrin, V. (2011). Co-producing transition: innovation processes in farms adhering to solidarity-based purchase groups (GAS) in Tuscany, Italy. *International Journal of Sociology of Agriculture and Food* 18, 28–53.

Bryan, D. and Rafferty, M. (2006). *Capitalism with Derivatives*. Palgrave Macmillan, Basingstoke.

Bryant, C. and Barnett, J. (2018). Consumer acceptance of cultured meat: a systematic review. *Meat Science* 143, 8–17.

Bryant, C. R., Peña Díaz, J., Keraita, B., Lohrberg, F. and Yokohari, M. (2016). Urban agriculture from a global perspective. In: Lohrberg, F. and Timpe, A. (Eds) *Urban Agriculture Europe*. Jovis, Berlin, pp. 30–37.

Bryden, J. and Hart, K. (2001). *Dynamics of Rural Areas (DORA): The International Comparison. The Arkleton Centre for Rural Development Research*. University of Aberdeen, St. Mary's, Aberdeen.

Bryngelsson, D., Wirsenius, S., Hedenus, F. and Sonesson, U. (2016). How can the EU climate targets be met? A combined analysis of technological and demand-side changes in food and agriculture. *Food Policy* 59, 152–164.

Brynjolfsson, E., Hu, Y. J. and Smith, M. D. (2010). The longer tail: the changing shape of Amazon's sales distribution curve. *SSRN Electronic Journal*. DOI: 10.2139/ssrn.1679991.

Buch-Hansen, H. (2018). The prerequisites for a degrowth paradigm shift: insights from critical political economy. *Ecological Economics* 146, 157–163.

Buck, D., Getz, C. and Guthman, J. (1997). From farm to table: the organic vegetable commodity chain of Northern California. *Sociologia Ruralis* 37(1), 3–20.

Bugge, M., Hansen, T. and Klitkou, A. (2016). What is the bioeconomy? A review of the literature. *Sustainability* 8(7), 691.

Bull, D. (1982). *A Growing Problem: Pesticides and the Third World Poor*. Oxfam, Oxford.

Bunge (2023). Bunge launches program to drive regenerative agriculture in Brazil. Available at: https://www.bunge.com/Press-Releases/Bunge-Launches-Program-to-Drive-Regenerative-Agriculture-in-Brazil. Accessed 21 April 2023.

Burch, D. and Lawrence, G. (2009). Towards a third food regime: behind the transformation. *Agriculture and Human Values* 26(4), 267–279.

Burch, D. and Lawrence, G. (2013). Financialization in agri-food supply chains: private equity and the transformation of the retail sector. *Agriculture and Human Values* 30(2), 247–258.

Burgess, P. J., Harris, J., Graves, A. R. and Deeks, L. K. (2019). Regenerative agriculture: identifying the impact; enabling the potential. Report for SYSTEMIQ. Cranfield University, Bedfordshire.

Buringh, P. and van Heemst, H. D. (1979). Potential world food production. In: Linnemann, H., de Hoogh, J., Keyser, M. A. and van Heemst, H. D. (Eds) *MOIRA: A Model of International Relations in Agriculture*. Elsevier, Amsterdam, pp. 19-72.

Burke, K. (2021). Soil carbon sequestration on farms alone won't absolve our daily emission sins. *The Guardian*. Available at: https://www.theguardian.com/australia-news/2021/dec/19/soil-carbon-sequestration-on-farms-alone-wont-absolve-our-daily-emission-sins. Accessed 18 December 2021.

Burnett, K. and Murphy, S. (2014). What place for international trade in food sovereignty? *The Journal of Peasant Studies* 41(6), 1065–1084.

Burton, R. J. F. (2019). The potential impact of synthetic animal protein on livestock production: the new "war against agriculture"? *Journal of Rural Studies* 68, 33–45.

Busch, L. and Bain, C. (2004). New! improved? the transformation of the global agrifood system. *Rural Sociology* 69(3), 321–346.

Bush, R. and Martiniello, G. (2017). Food riots and protest: agrarian modernizations and structural crises. *World Development* 91, 193–207.

Bush, S. B. (2012). *Derivatives and Development*. Palgrave Macmillan, New York.

Business Insider (2021). See how these rare, $50 'luxury' strawberries are grown in vertical farms designed to replicate the Japanese Alps with bees and AI robots. Available at: https://www.businessinsider.com/rare-omakase-strawberries-luxury-berries-price-online-vertical-farms-oishii-2021-11?amp.

Butland, B., Jebb, S., Kopelman, P., McPherson, K., Thomas, S., Mardell, J. and Parry, V. (2007). *Tackling Obesities: Future Choices - Project Report*. Government Office for Science of the United Kingdom, London.

Butler, R. A. (2021). Amazon destruction. *Mongabay*, November 2021. Available at: https://rainforests.mongabay.com/amazon/amazon_destruction.html. Accessed 10 November 2023.

Buxton, A., Campanale, M. and Cotula, L. (2012). Farms and funds: investment funds in the global land rush. *IIED Briefing*, January. IIED, London. Available at: http://pubs.iied.org/pdfs/17121IIED.pdf.

Byé, P. and Font, M. (1991). *Technical Change in Agriculture and New Functions for Rural Spaces in Europe*. Paper presented at the American Sociological Association. Cincinnati, OH.

Byé, P., Schmidt, V. D. B. and Schmidt, W. (2002). Transferência de dispositivos de reconhecimento da agricultura orgânica e apropriação local: uma análise sobre a Rede Ecovida. Desenvolvimiento e Meio Ambiente. *Curitiba* 6, 81–93.

Cabell, J. F. and Oelofse, M. (2012). An indicator framework for assessing agroecosystem resilience. *Ecology and Society* 17(1), 18-13. DOI: 10.5751/ES-04666-170118.

Caira, S. and Ferranti, P. (2016). Innovation for sustainable agriculture and food production. In: *Reference Module in Food Science*. Elsevier. DOI: 10.1016/B978-0-08-100596-5.21018-4.

Calegari, F., Tassi, D. and Vincini, M. (2013). Economic and environmental benefits of using a spray control system for the distribution of pesticides. *Journal of Agricultural Engineering* 44(2s), 163–165. DOI: 10.4081/jae.2013.274.

Callaway, R. M. (1995). Positive interactions among plants. *Botanical Review* 61(4), 306-349. DOI: 10.1007/BF02912621.

Calo, J. R., Crandall, P. G., O'Bryan, C. A. and Ricke, S. C. (2015). Essential oils as antimicrobials in food systems: a review. *Food Control* 54, 111–119.

Camara, O. (2013). Seasonal price variability and the effective demand for nutrients: evidence from cereals markets in Mali. *African Journal of Food, Agriculture, Nutrition and Development* 13(58), 7640–7661.

Camargo, C. and Beduschi, L. C. (2013). Agroecologia e seus sistemas de garantia: construindo confiança e participação entre os agricultores. *Cadernos de Agroecologia*. Available at: http://revistas.abaagroecologia.org.br/index.php/cad/article/view/14996. Accessed 28 December 2013.

Camargo, J. A. and Alonso, A. (2006). Ecological and toxicological effects of inorganic nitrogen pollution in aquatic ecosystems: a global assessment. *Environment International* 32(6), 831–849.

Campbell, B. M., Hansen, J., Rioux, J., Stirling, C. M., Twomlow, S. and Wollenberg, E. (2018). Urgent action to combat climate change and its impacts (SDG 13): transforming agriculture and food systems. *Current Opinion in Environmental Sustainability* 34, 13–20. DOI: 10.1016/j.cosust.2018.06.005.

Campos, E. V. R., Proença, P. L. F., Oliveira, J. L., Bakshic, M., Abhilash, P. C. and Fraceto, L. F. (2019). Use of botanical insecticides for sustainable agriculture: future perspectives. *Ecological Indicators* 105, 483–495.

Candel, J. J. L., Breeman, G. E., Stiller, S. J., Termeer, C. J. A. M. and Termeer, J. A. M. (2014). Disentangling the consensus frame of food security: the case of the EU Common Agricultural Policy reform debate. *Food Policy* 44, 47–58.

Cao, L., Zhou, Z., Niu, S., Cao, C., Li, X., Shan, Y. and Huang, Q. (2018). Positive-charge functionalized mesoporous silica nanoparticles as nanocarriers for controlled 2,4-dichlorophenoxy acetic acid sodium salt release. *Journal of Agricultural and Food Chemistry* 66(26), 6594–6603. DOI: 10.1021/acs.jafc.7b01957.

Capt, D. and Wavresky, P. (2014). Determinants of direct-to-consumer sales on French farms. *Revue d'Études en Agriculture et Environnement* 95(3), 351–377.

Carey, J. (2011). *Who Feeds Bristol? Towards a Resilient Food Plan; A Baseline Study of the Food System That Serves Bristol and the Bristol City Region*. Bristol City Council, Bristol.

Carlisle, L. and Miles, A. (2013). Closing the knowledge gap: how the USDA could tap the potential of biologically diversified farming systems. *Journal of Agriculture, Food Systems, and Community Development* 3, 219–225.

Carlson, R. (2016). Estimating the biotech sector's contribution to the US economy. *Nature Biotechnology* 34(3), 247–255.

Carolan, M. (2005). Barriers to the adoption of sustainable agriculture on rented land: an examination of contesting social fields. *Rural Sociology* 70(3), 387–413.

Carolan, M. (2017). Publicising food: big data, precision agriculture, and co-experimental techniques of addition. *Sociologia Ruralis* 57(2), 135–154. DOI: 10.1111/soru.12120.

Carolan, M. (2018). 'Smart' farming techniques as political ontology: access, sovereignty and the performance of neoliberal and not-so-neoliberal worlds sociologia ruralis VC 2018 European Society for Rural Sociology. *Sociologia Ruralis* 58(4). DOI: 10.1111/soru.12202.

Caron, P., de Loma-Osorio, G. F., Nabarro, D., Hainzelin, E., Guillou, M., Andersen, I., Arnold, T., Astralaga, M., Beukeboom, M., Bickersteth, S., Bwalya, M., Caballero, P., Campbell, B. M., Divine, N., Fan, S., Frick, M., Friis, A., Gallagher, M., Halkin, J. P., Hanson, C.,

Lasbennes, F., Ribera, T., Rockstrom, J., Schuepbach, M., Steer, A., Tutwiler, A. and Verburg, G. (2018). Food systems for sustainable development: proposals for a profound four-part transformation. *Agronomy for Sustainable Development* 38(4), 41. DOI: 10.1007/s13593-018-0519-1.

Carpenter, S. R. (2005). Eutrophication of aquatic ecosystems: bistability and soil phosphorus. *Proceedings of the National Academy of Sciences of the United States of America* 102(29), 10002–10005. DOI: 10.1073/pnas.0503959102.

Carrington, D. (2020). No-kill, lab-grown meat to go on sale for first time. *The Guardian Newspaper*, 02 December.

Carroll, M. (2017). The sticky materiality of neo-liberal neonatures: GMOs and the agrarian question. *New Political Economy* 22(2), 203–218. DOI: 10.1080/13563467.2016.1214696.

Carson, R. (1964). *The Silent Spring*. Fawcett, New York.

Carter, M., de Janvry, A., Sadoulet, E. and Sarris, A. (2014). Index-based weather insurance for developing countries: a review of evidence and a set of propositions for up-scaling; Background document for the workshop: "Microfinance products for weather risk management in developing countries: State of the arts and perspectives". FERDI, Paris, France. Available at: https://econpapers.repec.org/paper/fdiwpaper/1800.htm.

Castaldi, F., Hueni, A., Chabrillat, S., Ward, K., Buttafuoco, G., Bomans, B., Vreys, K., Brell, M. and Van Wesemael, B. (2019). Evaluating the capability of the Sentinel 2 data for soil organic carbon prediction in croplands. *ISPRS Journal of Photogrammetry and Remote Sensing* 147, 267–282.

Castle, M. H., Lubben, B. D. and Luck, J. D. (2016). Factors influencing the adoption of PA technologies by Nebraska Producers. Presentations, Working Papers, and Gray Literature: Agricultural Economics. Available at: http://digitalcommons.unl.edu/ageconworkpap/49.

Castro, J., Krajter Ostoi'c, S., Cariñanos, P., Fini, A. and Sitzia, A. (2018). "Edible" urban forests as part of inclusive, sustainable cities. *Unasylva* 69, 59–65.

Cavazza, L. (1996). Agronomia aziendale e agronomia del territorio.*Rivista di Agronomia* 30, 310–319.

CBAN (2021). Product profile – GM waxy corn. Available at: https://cban.ca/wp-content/uploads/GM-Waxy-Corn-Corteva-product-profile-CBAN.pdf. Canadian Biotechnology Action Network Updated 12 April 2021.

Chagnon, M., Kreutzweiser, D., Mitchell, E. A., Morrissey, C. A., Noome, D. A. and Van der Sluijs, J. P. (2015). Risks of large-scale use of systemic insecticides to ecosystem functioning and services. *Environmental Science and Pollution Research* 22(1), 119-134. DOI: 10.1007/s11356-014-3277-x.

Chakraborty, I. and Deshmukh, R. (2022). Biopesticides market by product type (microbial, predators, others), by formulation (dry form, liquid form), by crop type (orchards, grazing and dry land, field crops): global opportunity analysis and industry forecast, 2020-2031. Available at: https://www.alliedmarketresearch.com/biopesticides-market.

Chalker-Scott, L. (2010). Permaculture–my final thoughts. *Gard. Profr.–WSU Ext*. Available at: https://sharepoint.cahnrs.wsu.edu/blogs/urbanhort/archive/2010/05/26/permaculture-my-final-thoughts.aspx. Accessed 29 May 2013.

Challinor, A. J., Müller, C., Asseng, S., Deva, C., Nicklin, K. J., Wallach, D., Vanuytrecht, E., Whitfield, S., Ramirez-Villegas, J. and Koehler, A. K. (2018). Improving the use of crop models for risk assessment and climate change adaptation. *Agricultural Systems* 159, 296–306.

Chang, H. S. and Zepeda, L. (2005). Consumer perceptions and demand for organic food in Australia: focus group discussions. *Renewable Agriculture and Food Systems* 20(3), 155–167.

Chaplain, V., Mamy, L., Vieublé-Gonod, L., Mougin, C., Benoit, P., Barriuso, E. and Nélieu, S. (2011). Fate of pesticides in soils: toward an integrated approach of influential factors. In: Stoytcheva, M. (Ed.) *Pesticides in the Modern World – Risks and Benefits*. InTech. DOI: 10.5772/949.

Chaplin-Kramer, R., O'Rourke, M. E., Blitzer, E. J. and Kremen, C. (2011). A meta-analysis of crop pest and natural enemy response to landscape complexity. *Ecology Letters* 14(9), 922–932. DOI: 10.1111/j.1461-0248.2011.01642.x. Accessed 6 November 2020.

Chappell, M. J. and LaValle, L. A. (2011). Food security and biodiversity: can we have both? An agroecological analysis. *Agriculture and Human Values* 28(1), 3–26. DOI: 10.1007/s10460-009-9251-4.

Chaudhry, Q. (2020). *Nanomaterials in Food Products: New Prospects, New Challenges and Regulation of Risk*. Food Standards Agency Food for Thought Seminar Series 11 November 2020.

Chaudhry, Q., Scotter, M., Blackburn, J., Ross, B., Boxall, A., Castle, L., Aitken, R. and Watkins, R. (2008). Applications and implications of nanotechnologies for the food sector. *Food Additives and Contaminants. Part A, Chemistry, Analysis, Control, Exposure and Risk Assessment* 25(3), 241–258. DOI: 10.1080/02652030701744538.

Chemistry World (2022). Cultured meat flexes its muscles. Available at: https://www.chemistryworld.com/features/cultured-meat-flexes-its-muscles/4016057.article.

Chen, D., Rocha-Mendoza, D., Shan, S., Smith, Z., García-Cano, I., Prost, J., Jimenez-Flores, R. and Campanella, O. (2022). Characterization and cellular uptake of peptides derived from in vitro digestion of meat analogues produced by a sustainable extrusion process. *Journal of Agricultural and Food Chemistry* 70(26), 8124–8133. DOI: 10.1021/acs.jafc.2c01711. Epub 2022 June 22.

Chen, H. and Yada, R. (2011). Nanotechnologies in agriculture: new tools for sustainable development. *Trends in Food Science and Technology* 22(11), 585-594. DOI: 10.1016/j.tifs.2011.09.004.

Cherlet, M., Hutchinson, C., Reynolds, J., Hill, J., Sommer, S. and von Maltitz, G. (Eds) (2018). *World Atlas of Desertification*. Publication Office of the European Union, Luxembourg.

Chhipa, H. (2017). Nanofertilizers and nanopesticides for agriculture. *Environmental Chemistry Letters* 15(1), 15–22.

Chi, P. Y., Chen, J. H., Chu, H. H. and Lo, J. L. (2008). Enabling calorie-aware cooking in a smart kitchen. In: Oinas-Kukkonen, H., Hasle, P., Harjumaa, M., Segerståhl, K. and Øhrstrøm, P. (Eds) *Persuasive Technology*. Springer, Berlin, Hiedelberg, pp. 116–127.

Chiffoleau, Y., Millet-Amrani, S. and Canard, A. (2017). From short food supply chains to sustainable agriculture in urban food systems: food democracy as a vector of transition. *Agriculture* 6(4), 57. DOI: 10.3390/agriculture6040057.

Chiffoleau, Y., Millet-Amrani, S., Rossi, A., Rivera-Ferre, M. G. and Merino, P. L. (2019). The participatory construction of new economic models in short food supply chains. *Journal of Rural Studies* 68, 182–190.

Chiles, R. M. (2013). If they come, we will build it: in vitro meat and the discursive struggle over future agrofood expectations. *Agriculture and Human Values* 30(4), 511-523.

Chin, J. H., Gamuyao, R., Dalid, C., Bustamam, M., Prasetiyono, J., Moeljopawiro, S., Wissuwa, M. and Heuer, S. (2011). Developing rice with high yield under phosphorus deficiency: Pup1 sequence to application. *Plant Physiology* 156(3), 1202–1216.

Chivenge, P. P., Murwira, H. K., Giller, K. E., Mapfumo, P. and Six, J. (2007). Long-term impact of reduced tillage and residue management on soil carbon stabilization: implications for conservation agriculture on contrasting soils. *Soil and Tillage Research* 94(2), 328–337.

Chopra, M. and Darnton-Hill, I. (2004). Tobacco and obesity epidemics: not so different after all? *BMJ* 328(7455), 1558–1560.

Chowdhury,. A., Pradhan, S., Saha, M. and Sanyal, N. (2008). Impact of pesticides on soil microbiological parameters and possible bioremediation strategies. *Indian Journal of Microbiology* 48(1), 114–127. DOI: 10.1007/s12088-008-0011-8.

Chriki, S. and Hocquette, J. F. (2020). The myth of cultured meat: a review. *Frontiers in Nutrition* 7, 7. DOI: 10.3389/fnut.2020.00007.

Chung, E. (2014). Lake Erie's algae explosion blamed on farmers. *CBC News*. Available at: https://www.cbc.ca/news/science/lake-erie-s-algae-explosion-blamed-on-farmers-1.2729327. Accessed 7 August 2014.

Cioffi, N., Torsi, L., Ditaranto, N., Sabbatini, L., Zambonin, P. G., Tantillo, G., Ghibelli, L., D'Alessio, M., Bleve-Zacheo, T. and Traversa, E. (2004). Antifungal activity of polymer-based copper nanocomposite coatings. *Applied Physics Letters* 85(12), 2417-2419.

Cisilino, F., Bodini, A. and Zanoli, A. (2019). Rural development programs' impact on environment: an ex-post evaluation of organic farming. *Land Use Policy* 85, 454–462.

Cisternas, I., Velásquez, I., Caro, A. and Rodríguez, A. (2020). Systematic literature review of implementations of precision agriculture. *Computers and Electronics in Agriculture* 176, 105626.

Civil Eats (2020). Does overselling regenerative Ag's climate benefits undercut its potential? *Civil Eats*, 1 October 2020.

Clapp, J. (2002). *Transnational Corporate Interests and Global Environmental Governance: Negotiating Rules for Agricultural Biotechnology and Chemicals*. Trent University, Ontario.

Clapp, J. (2012). *Food*. Polity Press, Malden, MA.

Clapp, J. (2014). Financialization, distance and global food politics. *Journal of Peasant Studies* 41(5), 797–814. DOI: 10.1080/03066150.2013.875536.

Clapp, J. and Fuchs, D. (Eds) (2009). *Corporate Power in Global Agrifood Governance*. MIT Press, Cambridge, MA.

Clapp, J. and Helleiner, E. (2012). International political economy and the environment: back to the basics? *International Affairs* 88(3), 485–501.

Clapp, J., Newell, P. and Brent, Z. W. (2018). The global political economy of climate change, agriculture and food systems. *The Journal of Peasant Studies* 45(1), 80–88. DOI: 10.1080/03066150.2017.1381602.

Clapp, J. and Ruder, S. L. (2020). Precision technologies for agriculture: digital farming, gene edited crops, and the politics of sustainability. *Global Environmental Politics* 20(3), 49–69.

Clark, M. and Tilman, D. (2017). Comparative analysis of environmental impacts of agricultural production systems, agricultural input efficiency, and food choice. *Environmental Research Letters* 12(6), 64016. DOI: 10.1088/1748-9326/aa6cd5.

Clark, M. S., Horwath, W. R., Shennan, C. and Scow, K. M. (1998). Changes in soil chemical properties resulting from organic and low-input farming practices. *Agronomy Journal* 90(5), 662–671.

Clark, P. (2017). Neo-developmentalism and a 'Vía campesina' for rural development: unreconciled projects in Ecuador's citizen's revolution. *Journal of Agrarian Change* 17(2), 348–364.

Cloke, P. and Goodwin, M. (1992). Conceptualizing countryside change: from post-Fordism to rural structural coherence. *Transactions of the Institute of British Geographers* 17(3), 321–336.

Clunies-Ross, T. (1990). Agricultural change and politics of organic farming. Unpublished PhD thesis. University of Bath.

Cocozza, P. (2013). 'If I shop here I've got money for gas': inside the UK's first social supermarket. *The Guardian*. Available at: www.theguardian.com/society/2013/dec/09/inside-britains-first-social-supermarket-goldthorpe-yorkshire.

Cohen, J. (2019). *To Feed Its 1.4 Billion, China Bets Big on Genome Editing of cropsScienceMag.org*. American Association for the Advancement of Science, Washington, DC.

Colchero, M. A., Rivera-Dommarco, J., Popkin, B. M. and Ng, S. W. (2017). In Mexico, evidence of sustained consumer response two years after implementing a sugar-sweetened beverage tax. *Health Affairs* 36(3), 564–571.

Coley, D., Howard, M. and Winter, M. (2009). Local food, food miles and carbon emissions: a comparison of farm shop and mass distribution approaches. *Food Policy* 34(2), 150–155. DOI: 10.1016/j.foodpol.2008.11.001.

Coley, D., Howard, M. and Winter, M. (2011). Food miles: time for a re-think? *British Food Journal* 113(7), 919–934. DOI: 10.1108/00070701111148432.

Collective Green (2022). Foodsharing and Foodsaving worldwide – a global and distributed grassroots movement against food waste. Available at: https://www.collectivegreen.de/foodsharing-and-foodsaving-worldwide/. Accessed 23 October 2022.

Colonna, P., Fournier, S., Touzard, J.-M., Abécassis, J., Broutin, C., Chabrol, D., Champenois, A., Deverre, C., François, M., Stimolo, D., Méry, V., Moustier, P. and Trystram, G. (2013). Food systems. In: Esnouf, C., Russel, M. and Bricas, N. (Eds) *Food System Sustainability: Insights from duALIne*. Cambridge University Press, Cambridge, pp. 69–100.

Commoner, B. (1971). *The Closing Circle*. Random House, New York.

Commoner, B. (1975). *Making Peace with the Planet*. Pantheon, New York.

Conijn, J. G., Bindraban, P. S., Schröder, J. J. and Jongschaap, R. E. E. (2018). Can our global food system meet food demand within planetary boundaries? *Agriculture, Ecosystems and Environment* 251, 244–256.

Constance, D. H., Choi, J. Y. and Lyke-Ho-Gland, H. (2008). Conventionalization, bifurcation, and quality of life: certified and non-certified organic farmers in Texas. South. *Rural Sociology* 23(1), 208–234.

Constance, D. H., Friedland, W. H., Renard, M. C. and Rivera-Ferre, M. G. (2014). The discourse on alternative agrifood movements. In: Constance, D. H., Renard, M.-C., Rivera-Ferre, M. G. (Eds) *Alternative Agrifood Movements: Patterns of Convergence and Divergence*. Binkley, Emerald.

Conti, B., Flamini, G., Cioni, P. L., Ceccarini, L., Macchia, M. and Benelli, G. (2014). Mosquitocidal essential oils: are they safe against non-target aquatic organisms? *Parasitology Research* 113(1), 251-259. DOI: 10.1007/s00436-013-3651-5.

Conway, G. (2012). *One Billion Hungry: Can We Feed the World?* Cornell University Press, Cornell.

Conway, G. R. (1987). The properties of agroecosystems. *Agricultural Systems* 24(2), 95–117.

Cooper, P. J. M., Dimes, J., Rao, K. P. C., Shapiro, B., Shiferaw, B. and Twomlow, S. (2008). Coping better with current climatic variability in the rain-fed farming systems of sub-Saharan Africa: an essential first step in adapting to future climate change? *Agriculture, Ecosystems and Environment* 126(1–2), 24–35.

Corbeels, M., Graaff, J., Ndah, T. H., Penot, E., Baudron, F., Naudin, K., Andrieu, N., Chirat, G., Schuler, J., Nyagumbo, I., Rusinamhodzi, L., Traore, K., Mzoba, H. D. and Adolwa, I. S. (2014a). Understanding the impact and adoption of conservation agriculture in Africa: a multi-scale analysis. *Agriculture, Ecosystems and Environment* 187, 155–170.

Cordell, D., Drangert, J. O. and White, S. (2009). The story of phosphorus: global food security and food for thought. *Global Environmental Change* 19(2), 292–305.

Cordero, P., Li, J., Temple, J. L., Nguyen, V. and Oben, J. A. (2016). Epigenetic mechanisms of maternal obesity effects on the descendants. In: Green, L. R. and Hester, R. L. (Eds) *Parental Obesity: Intergenerational Programming and Consequences*. Springer, New York, pp. 355–368.

Cornell University (2023). *System of Rice Intensification (SRI)*. Sri International Network and Resources Center, Ithaca, NY. Available at: http://sri.ciifad.cornell.edu/index.html. Accessed 24 June 2023.

Cosby, A. M., Falzon, G. A., Trotter, M. G., Stanley, J. N., Powell, K. S. and Lamb, D. W. (2016). Risk mapping of redheaded cockchafer (*Adoryphorus couloni*) (Burmeister) infestations using a combination of novel k-means clustering and on-the-go plant and soil sensing technologies. *Precision Agriculture* 17(1), 1–17.

Cosgrove, E. (2017). Cultured meat will cost start-ups $150m–$370m (and take at least 4 more years) to bring to market. *Agfunder News*, 10 October. Available at: https://agfundernews.com/cultured-meat-will-cost-startups-150m-370m-and-take-at-least-4-more-years-to-bring-to-market.html. Accessed 12 August 2018.

Cosme, I., Santos, R. and O'Neill, D. W. (2017). Assessing the degrowth discourse: a review and analysis of academic degrowth policy proposals. *Journal of Cleaner Production* 149, 321-334.

Costanza, R., d'Arge, R., de Groot, R., Farberk, S., Grasso, M., Hannon, B., Limburg, K., Naeem, S., O'Neill, R. V., Paruelo, J., Raskin, R. G., Sutton, P. and van den Belt, M. (1997). The value of the world's ecosystem services and natural capital. *Nature* 387 | 15 May 1997.

Cotula, L. (2012). The international political economy of the global land rush: a critical appraisal of trends, scale, geography, and drivers. *The Journal of Peasant Studies* 39(3–4), 649–680.

Cowan, R. (1990). Nuclear power reactors: a study in technological lock-in. *Journal of Economic History* 50(3), 541–567.

Credit Suisse (2021). *The Global Food System: Identifying Sustainable Solutions*. Credit Suisse Group AG, London. research.institute@credit-suisse.com.

CropLife (2019). Lettuce and gene editing: increasing heat resistance to combat changing climate. Available at: https://croplife.org/wp-content/uploads/2019/12/CRP17043_CaseStudy_Lettuce_FINAL.pdf.

Crotty, J. (2009). Structural causes of the global financial crisis: a critical assessment of the 'new financial architecture'. *Cambridge Journal of Economics* 33(4), 563–580.

Crowder, D. W. and Reganold, J. P. (2015). Financial competitiveness of organic agriculture on a global scale. *PNAS* 112(24), 7611–7616.

CTA (2016). COP 22 action for agriculture press communique. Available at: http://www.cta.int/en/article/2016-11-14/cop22-n-action-for-agriculture.html.

Cui, Z. L., Zhang, H., Chen, X., Zhang, C., Ma, W., Huang, C., Zhang, W., Mi, G., Miao, Y., Li, X., Gao, Q., Yang, J., Wang, Z., Ye, Y., Guo, S., Lu, J., Huang, J., Lv, S., Sun, Y., Liu, Y., Peng, X., Ren, J., Li, S., Deng, X., Shi, X., Zhang, Q., Yang, Z., Tang, L., Wei, C., Jia, L., Zhang, J., He, M., Tong, Y., Tang, Q., Zhong, X., Liu, Z., Cao, N., Kou, C., Ying, H., Yin, Y., Jiao, X., Zhang, Q., Fan, M., Jiang, R., Zhang, F. and Dou, Z. (2018). Pursuing sustainable productivity with millions of smallholder farmers. *Nature* 555(7696), 363–366.

Cultivated Food (2023). Cutting out the cow: new project ams to make vegan meat directly from grass proteins. Available at: https://cultivatedmeats.org/2023/02/11/making-vegan-meat-directly-from-grass-proteins-new-project-cuts-out-the-cow/.

Cunha, D. B., da Costa, T. H. M., da Veiga, G. V., Pereira, R. A. and Sichieri, R. (2018). Ultra-processed food consumption and adiposity trajectories in a Brazilian cohort of adolescents: ELANA study. *Nutrition and Diabetes* 8(1), 28.

Curtis, P. G., Slay, C. M., Harris, N. L., Tyukavina, A. and Hansen, M. C. (2018). Classifying drivers of global forest loss. *Science* 361(6407), 1108–1111. DOI: 10.1126/science.aau3445.

D'Alisa, G., Demari, F. and Kallis, G. (Eds) (2015). *Degrowth – A Vocabulary for a New Era*. Routledge, New York/London.

D'Annolfo, R., Gemmill-Herren, B., Graeub, B. and Garibaldi, L. A. (2017). A review of social and economic performance of agroecology. *International Journal of Agricultural Sustainability* 15(6), 632–644. DOI: 10.1080/14735903.2017.1398123.

D'Antoni, J. M., Mishra, A. K. and Joo, H. (2012). Farmers' perception of precision technology: the case of autosteer adoption by cotton farmers. *Computers and Electronics in Agriculture* 87, 121–128. DOI: 10.1016/j.compag.2012.05.017.

da Silveira, S. M., Luciano, F. B., Fronza, N., Cunha, A., Scheuermann, G. N. and Vieira, C. R. W. (2014). Chemical composition and antibacterial activity of *Laurus nobilis* essential oil towards foodborne pathogens and its application in fresh *Tuscan sausage* stored at 7 degrees C. *LWT – Food Science and Technology* 59(1), 86–93.

DagangHalal (2022). P3 nano sugar. Available at: https://www.daganghalal.com/Product/p3_nano_sugar_37227. Accessed 13 February 2022.

Dahm, M. J., Samonte, A. V. and Shows, A. R. (2009). Organic foods: do eco-friendly attitudes predict eco-friendly behaviours? *Journal of American College Health* 58(3), 195–202.

Dainese, M., Isaac, N. J. B., Powney, G. D., Bommarco, R., Öckinger, E., Kuussaari, M., Pöyry, J., Benton, T. G., Gabriel, D., Hodgson, J. A., Kunin, W. E., Lindborg, R., Sait, S. M. and Marini, L. (2016). Landscape simplification weakens the association between terrestrial producer and consumer diversity in Europe. *Global Change Biology* 23(8), 3040–3051. DOI: 10.1111/gcb.13601.

Dainese, M., Martin, E. A., Aizen, M. A., Albrecht, M., Bartomeus, I., Bommarco, R., Carvalheiro, L. G., Chaplin-Kramer, R., Gagic, V., Garibaldi, L. A., Ghazoul, J., Grab, H., Jonsson, M., Karp, D. S., Kennedy, C. M., Kleijn, D., Kremen, C., Landis, D. A., Letourneau, D. K., Marini, L., Poveda, K., Rader, R., Smith, H. G., Tscharntke,

T., Andersson, G. K. S., Badenhausser, I., Baensch, S., Bezerra, A. D. M., Bianchi, F. J. J. A., Boreux, V., Bretagnolle, V., Caballero-Lopez, B., Cavigliasso, P., Ćetković, A., Chacoff, N. P., Classen, A., Cusser, S., da Silva E Silva, F. D., de Groot, G. A., Dudenhöffer, J. H., Ekroos, J., Fijen, T., Franck, P., Freitas, B. M., Garratt, M. P. D., Gratton, C., Hipólito, J., Holzschuh, A., Hunt, L., Iverson, A. L., Jha, S., Keasar, T., Kim, T. N., Kishinevsky, M., Klatt, B. K., Klein, A. M., Krewenka, K. M., Krishnan, S., Larsen, A. E., Lavigne, C., Liere, H., Maas, B., Mallinger, R. E., Martinez Pachon, E., Martínez-Salinas, A., Meehan, T. D., Mitchell, M. G. E., Molina, G. A. R., Nesper, M., Nilsson, L., O'Rourke, M. E., Peters, M. K., Plećaš, M., Potts, S. G., Ramos, D. L., Rosenheim, J. A., Rundlöf, M., Rusch, A., Sáez, A., Scheper, J., Schleuning, M., Schmack, J. M., Sciligo, A. R., Seymour, C., Stanley, D. A., Stewart, R., Stout, J. C., Sutter, L., Takada, M. B., Taki, H., Tamburini, G., Tschumi, M., Viana, B. F., Westphal, C., Willcox, B. K., Wratten, S. D., Yoshioka, A., Zaragoza-Trello, C., Zhang, W., Zou, Y. and Steffan-Dewenter, I. (2019). A global synthesis reveals biodiversity-mediated benefits for crop production. *Science Advances* 5(10), eaax0121. DOI: 10.1126/sciadv.aax0121.

Dalgaard, T., Hutchings, N. J. and Porter, J. R. (2003). Agroecology, scaling and interdisciplinarity. *Agriculture, Ecosystems and Environment* 100(1), 39–51.

Dalin, C., Wada, Y., Kastner, T. and Puma., M. J. (2017). Groundwater depletion embedded in international food trade. *Nature* 543(7647), 700–704.

Daniel, A. C. (2017). Fonctionnement et durabilité des microfermes urbaines, une observation participative sur le cas des fermes franciliennes. Chaire Eco-Conception, AgroParisTech, INRA, UMR SADAPT, France. Available at: http://www.cityfarmer.org/2017DanielACD.pdf.

Daniel, S. (2012). Situating private equity capital in the land grab debate. *Journal of Peasant Studies* 39(3–4), 703–729.

Dara, S. K. (2017). Insect resistance to biopesticides. *UCANR E-journal - Strawberries and Vegetables*. Accessed 6 December 2017. Available at: http://ucanr.edu/blogs/blogcore/postdetail.cfm?postnum=25819.

Dardonville, M., Urruty, N., Bockstaller, C. and Therond, O. (2020). Influence of diversity and intensification level on vulnerability, resilience, and robustness of agricultural systems. *Agricultural Systems* 184, 102913.

Darnhofer, I., D'Amico, S. and Fouilleux, E. (2019). A relational perspective on the dynamics of the organic sector in Austria, Italy, and France. *Journal of Rural Studies* 68, 200–212.

Dasgupta (2020). The Dasgupta review - independent review on the economics of biodiversity interim report. Available at: https://www.gov.uk/government/publications/interim-report-the-dasgupta-review-independent-review-on-the-economics-of-biodiversity. Accessed April 2020.

Dasgupta, N., Ranjan, S., Chakraborty, A. R., Ramalingam, C., Shanker, R. and Kumar, A. (2016). Nano agriculture and water quality management. In: Ranjan, S., Nandita, D. and Lichtfouse, E. (Eds) *Sustainable Agriculture Reviews: Nanoscience in Food and Agriculture*. Springer, Berlin, Heidelberg, pp. 1–42.

Dauvergne, P. (2008). *The Shadows of Consumption*. MIT Press, Cambridge, MA.

Dauvergne, P. and Neville, K. J. (2010). Forests, food, and fuel in the tropics: the uneven social and ecological consequences of the emerging political economy of biofuels. *The Journal of Peasant Studies* 37(4), 631–660.

Davies, A., Titterington, A. J. and Cochrane, C. (1995). Who buys organic food? A profile of the purchasers of organic food in Northern Ireland. *British Food Journal* 97(10), 17-23.

Davies, A. R. and Legg, R. (2018). Fare sharing: interrogating the nexus of ICT, urban food sharing, and sustainability. *Food, Culture and Society* 21(2), 233–254. DOI: 10.1080/15528014.2018.1427924.

Davies, M., Béné, C., Arnall, A., Tanner, T., Newsham, A. and Coirolo, C. (2013). Promoting resilient livelihoods through adaptive social protection: lessons from 124 programmes in South Asia. *Development Policy Review* 31(1), 27–58.

Davies, M., Guenther, B. and Leavy, J. (2009). Climate change adaptation, disaster risk reduction and social protection: complementary roles in agriculture and rural growth? *IDS Working Papers* 320, 39. IDS, Wivenhoe Park .

Davies, M., Guenther, B., Leavy, J., Mitchell, T. and Tanner, T. (2008). 'Adaptive social protection': synergies for poverty reduction. *IDS Bulletin* 39(4), 105–112.

Davies, S. (2019). Time to solve childhood obesity. An Independent Report by the Chief Medical Officer.

Davignon, L. F., St-Pierre, J., Charest, G. and Tourangeau, F. J. (1965). A study of the chronic effects of insecticides in man. *Canadian Medical Association Journal* 92(12), 597–602.

Daviron, B., Dembele, N. N., Murphy, S. and Rashid, S. (2011). Price volatility and food security. A Report by the High-Level Panel of Experts on Food Security and Nutrition of the Committee on World Food Security. Available at: http://www.fao.org/fileadmin/user_upload/hlpe/hlpe_documents/HLPE-price-volatility-and-food-security-report-July-2011.pdf.

Davis, A. (2011). Primavera takes stake in Chinese fertilizer maker. *Asian Venture Capital Journal*. Available at: http://www.avcj.com/avcj/news/2127979/primaveratakes-stake-chinese-fertilizer-maker. Accessed 28 November 2011.

Davis, J., Sonesson, U., Baumgartner, D. U. and Nemecek, T. (2010). Environmental impact of four meals with different protein sources: case studies in Spain and Sweden. *Food Research International* 43(7), 1874–1884.

de Boer, I. J. M. and van Ittersum, M. K. (2018). Circularity in agricultural production. Available at: https://www.wur.nl/upload_mm/7/5/5/14119893-7258-45e6-b4d0-e514a8b6316a_Circularity-in-agriculturalproduction-20122018.pdf.

De Castro, A. I., Peña, J. M., Torres-Sánchez, J., Jiménez-Brenes, F. and López-Granados, F. (2017). Mapping *Cynodon dactylon* in vineyards using UAV images for site-specific weed control. *Advances in Animal Biosciences* 8(2), 267–271.

De Molina, M. G. (2012). Agroecology and politics. How to get sustainability? About the necessity for a political agroecology. Agroecology and Sustainable Food Systems 37, 45–59. DOI: 10.1080/10440046.2012.705810.

de Oliveira, J. L., Campos, E. V. R., Bakshi, M., Abhilash, P. C. and Fraceto, L. F. (2014). Application of nanotechnology for the encapsulation of botanical insecticides for sustainable agriculture: prospects and promises. *Biotechnology Advances* 32(8), 1550–1561.

de Ponti, T., Rijk, B. and van Ittersum, M. K. (2012). The crop yield gap between organic and conventional agriculture. *Agricultural Systems* 108, 1-9.

de Sartre, A., Charbonneau, X.M. and Charrier, O. (2019). How ecosystem services and agroecology are greening French agriculture through its reterritorialization. *Ecology and Society* 24(2), 2. DOI: 10.5751/ES-10711-240202.

De Schutter, O. (2010). Food commodities speculation and food price crises. UN Special Rapporteur on the Right to Food. Briefing Note 02 September. Available at: http://www.srfood.org/images/stories/pdf/otherdocuments/20102309_briefing_note_02_ee_ok.pdf.

De Schutter, O. (2014). Final report: the transformative potential of the right to food. United Nations Human Rights Council, New York, p. 27.

De Schutter, O., Jacobs, N. and Clément, C. (2020). A 'Common Food Policy' for Europe: how governance reforms can spark a shift to healthy diets and sustainable food systems. *Food Policy* 96, 101849.

De Vogli, R., Kouvonen, A., Elovainio, M. and Marmot, M. (2014). Economic globalization, inequality, and body mass index: a cross-national analysis of 127 countries. *Critical Public Health* 24(1), 7–21. DOI: 10.1080/09581596.2013.768331.

De Vogli, R., Kouvonen, A. and Gimeno, D. (2011). 'Globesization': ecological evidence on the relationship between fast food outlets and obesity among 26 advanced economies. *Critical Public Health* 21(4), 395–402.

de Vries, F. T., Thebault, E., Liiri, M., Birkhofer, K., Tsiafouli, M. A., Bjørnlund, L., Bracht Jørgensen, H., Brady, M. V., Christensen, S., de Ruiter, P. C., d'Hertefeldt, T., Frouz, J., Hedlund, K., Hemerik, L., Hol, W. H., Hotes, S., Mortimer, S. R., Setälä, H., Sgardelis, S. P., Uteseny, K., van der Putten, W. H., Wolters, V. and Bardgett, R. D. (2013). Soil food web properties explain ecosystem services across European land use systems. *Proceedings of the National Academy of Sciences of the United States of America* 110(35), 14296–14301.

Debatewise (2021). Junk food should be banned. Available at: https://debatewise.org/31023-junk-food-should-be-banned/. Accessed 10 July 2021.

Deep Green (2022). Permaculture design principle 4 – zones and sectors, efficient energy planning. Available at: https://deepgreenpermaculture.com/permaculture/permaculture-design-principles/4-zones-and-sectors-efficient-energy-planning/. Accessed 25 September 2022.

DEFRA (2000). *Design of a Tax or Charge Scheme for Pesticides*. Department for Environment, Food and Rural Affairs, London.

DEFRA (2011). *Food Statistics Pocketbook 2010*. Department for Environment, Food and Rural Affairs, London.

DEFRA (2014). *Food Statistics Pocketbook*. Department for Environment, Food and Rural Affairs, London.

Degré, A., Debouche, C. and Verhève, D. (2007). Conventional versus alternative pig production assessed by multicriteria decision analysis. *Agronomy for Sustainable Development* 27(3), 185–195.

Degrowth (2022). Degrowth conferences. Available at: https://degrowth.info/en/conferences. Accessed 4 November 2022.

Dela Rue, B. T. and Eastwood, C. R. (2017). Individualised feeding of concentrate supplement in pasture-based dairy systems: practices and perceptions of New Zealand dairy farmers and their advisors. *Animal Production Science* 57(7), 1543-1549. DOI: 10.1071/AN16471.

Delate, K. and Cambardella, C. A. (2004). Agroecosystem performance during transition to certified organic grain production. *Agronomy Journal* 96(5), 1288–1298.

Deléage, E. (2011). Les mouvements agricoles alternatifs. *Informacao e Sociedade* 164, 44–50.

DeLonge, M. S., Miles, A. and Carlisle, L. (2016). Investing in the transition to sustainable agriculture. *Environmental Science and Policy* 55, 266–273. DOI: 10.1016/j.envsci.2015.09.013.

Demaria, F. (2019). Degrowth: A call for radical socio-ecological transformation. March 14, 2019. Available at: https://globaldialogue.isa-sociology.org/articles/degrowth-a-call-for-radical-socio-ecological-transformation. Accessed 14 November 2022.

Demaria, F., Schneider, F., Sekulova, F. and Martinez-Alier, J. (2013). What is degrowth? From an activist slogan to a social movement. *Environmental Values* 22(2), 191–215. DOI: 10.3197/096327113X13581561725194.

Demaria, F., Kallis, G. and Bakker, K. (2019). Geographies of degrowth: nowtopias, resurgences and the decolonization of imaginaries and places. *Environment and Planning E: Nature and Space* 2(3), 431–450. DOI: 10.1177/2514848619869689.

Demmler, K. M., Ecker, O. and Qaim, M. (2018). Supermarket shopping and nutritional outcomes: a panel data analysis for urban Kenya. *World Development* 102, 292–303. DOI: 10.1016/j.worlddev.2017.07.018.

Dent, D. R. (2000). *Insect Pest Management* (2nd edn). CABI Publishing, Wallingford.

Despommier, D. (2010). *The Vertical Farm: Feeding the World in the 21st Century*. Macmillan, London.

Desquilbet, M., Maigné, E. and Monier-Dilhan, S. (2018). Organic food retailing and the conventionalisation debate. *Ecological Economics* 150, 194–203.

Dessart, F. J., Barreiro-Hurle, J. and van Bavel, R. (2019). Behavioural factors affecting the adoption of sustainable farming practices: a policy-oriented review. *European Review of Agricultural Economics* 46(3), 417–471.

Deutsche Bank (1999). GMOs are dead, 12 July. Available at: http://www.biotech-info.net/Deutsche.html. Accessed on 20 September 2003.

Dey, M. M., Paraguas, F. J., Kambewa, P. and Pemsl, D. E. (2010). The impact of integrated aquaculture-agriculture on small-scale farms in Southern Malawi. *Agricultural Economics* 41(1), 67–79. DOI: 10.1111/j.1574-0862.2009.00426.x.

Dharmasena, S. and Capps Jr., O. (2012). Intended and unintended consequences of a proposed national tax on sugar-sweetened beverages to combat the US obesity problem. *Health Economics* 21(6), 669–694.

Dhoubhadel, S. P. (2020). Precision agriculture technologies and farm profitability. *Journal of Agricultural and Resource Economics* 46(2), 256–268 ISSN 1068-5502. DOI: 10.22004/ag.econ.303598.

Diabetes UK (2022). New low-cost sweetener developed to help diabetics. Available at: https://www.diabetes.co.uk/news/2010/sep/new-low-cost-sweetener-developed-to-help-diabetics-98802891.html. Accessed 13 February 2022.

Díaz, S., Settele, J., Brondízio, E., Ngo, H. T., Guèze, M., Agard, J. and Zayas, C. (2019). *Summary for Policymakers of the Global Assessment Report on Biodiversity and Ecosystem Services of the Intergovernmental Science-Policy Platform on Biodiversity and Ecosystem Services*. University of the West of England, Bristol.

Dimitri, C., Oberholtzer, L. and Pressman, A. (2016). Urban agriculture: connecting producers with consumers. *British Food Journal* 118(3), 603–617.

Dimkpa, C. O. and Bindraban, P. S. (2018). Nanofertilizers: new products for the industry? *Journal of Agricultural and Food Chemistry* 66(26), 6462–6473. DOI: 10.1021/acs.jafc.7b02150.

Dion, C. (2013). Le bio dans les Supermarchés, c'est Pire ou c'est Mieux? *Kaizen Magazine*, 2 Juin. Available at: http://www.kaizen-magazine.com/le-bio-dans-les-supermarches-cest-pireou-cest-mieux/.

Dixson-Declève, S., Gaffney, O., Ghosh, J., Randers, G., Rockström, J. and Espen Stokne, P. (2022). Earth for all: a survival guide for humanity. A Report to the Club of Rome (2022) Fifty Years after the Limits to Growth (1972). New Society Publishers.

DJI Enterprise (2022). Precision agriculture with drone technology. Available at: https://enterprise-insights.dji.com/blog/precision-agriculture-drones. Accessed 14 March 2022.

Doernberg, A., Zasada, I., Bruszewska, K., Skoczowski, B. and Piorr, A. (2016). Potentials and limitations of regional organic food supply: a qualitative analysis of two food chain types in the Berlin metropolitan region. *Sustainability* 8(11), 1125. DOI: 10.3390/su8111125.

Doherty, R., Benton, T. G., Fastoso, F. J. and Gonzalez Jimenez, H. (2017). *British Food-What Role Should UK Food Producers Have in Feeding the UK?* Morrisons Supermarket, Bradford.

Doll, P. and Siebert, S. (2002). Global modeling of irrigation water requirements. *Water Resources Research* 38(4), 1-10. DOI: 10.1029/2001WR000355.

Donnelly, L. (2018). Cancer warning over processed foods that make up half of UK diet. *Telegraph*, 14 February. Available at: https://www.telegraph.co.uk/news/2018/02/14/cancer-warning-processed-foods-make-half-uk-diet/.

Donovan, M. (2020). *What Is Cconservation Agriculture?* CIMMYT, El Batan. Available at: https://www.cimmyt.org/news/what-is-conservation-agriculture/#.

Doré, T., Makowski, D., Malézieux, E., Munier-Jolain, N., Tchamitchian, M. and Tittonell, P. (2011). Facing up to the paradigm of ecological intensification in agronomy: revisiting methods, concepts, and knowledge. *European Journal of Agronomy* 34(4), 197-210.

Dorward, A. and Chirwa, E. (2011). The Malawi agricultural input subsidy programme: 2005/06 to 2008/09. *International Journal of Agricultural Sustainability* 9, 232-247.

Dos Santos, N. and Gottschalk Nolasco, L. (2017). A ênfase sobre conduta ética e os fatores incerteza e a condição humana inerentes às nanotecnologias.. *Revista da Faculdade de Direito da UFMG*. DOI: 10.12818/P.0304-2340.2016v69p441.

Doshi, R., Braida, W., Christodoulatos, C., Wazne, M. and O'Connor, G. (2008). Nano-aluminium: transport through sand columns and environmental effects on plants and soil communities. *Environmental Research* 106(3), 296-303.

Dou, Z. and Toth, J. D. (2020). Global primary data on consumer food waste: rate and characteristics - a review. *Resources, Conservation and Recycling*, 105332. DOI: 10.1016/j.resconrec.2020.105332.

Douglass, G. (Ed.) (1984). *Agricultural Sustainability in a Changing World Order*. Westview Press, Boulder, CO, 282 p.

Dowd, A. M., Marshall, N., Fleming, A., Jakku, E., Gaillard, E. and Howden, M. (2014). The role of networks in transforming Australian agriculture. *Nature Climate Change* 4(7), 558-563.

Doyle, M. P., Erickson, M. C., Alali, W., Cannon, J., Deng, X., Ortega, Y., Smith, M. A. and Zhao, T. (2015). The food industry's current and future role in preventing microbial foodborne illness within the United States. *Clinical Infectious Diseases*, 252-259. DOI: 10.1093/cid/civ253.

Drinkwater, L. E., Wagoner, P. and Sarrantonio, M. (1998). Legume-based cropping systems have reduced carbon and nitrogen losses. *Nature* 396(6708), 262-265.

Druilhe, Z. and Barreiro-Hurlé, J. (2012). Fertilizer subsidies in sub-Saharan Africa. ESA Working Paper No. 12-04. FAO.

Duah-Yentumi, S. and Johnson, D. B. (1986). Changes in soil microflora in response to repeated applications of some pesticides. *Soil Biology and Biochemistry* 18(6), 629-635. DOI: 10.1016/0038-0717(86)90086-6.

Dubbeling, M., Campbell, M. C., Hoekstra, F. and Veenhuize, R. V. (2009). Building resilient cities. *Urban Agriculture Magazine* 22, 3-11.

Dubbeling, M. and Zeeuw, H. (2011). Urban agriculture and climate change adaptation: ensuring food security through adaptation. In: Otto-Zimmermann, K. (Ed.) *Resilient Cities* (Vol. 1). Springer, Dordrecht, The Netherlands, pp. 441-449.

Dubock, A. (2017). Biofortified golden rice: an additional intervention for vitamin A deficiency. In: Sasaki, T. (Ed.). *Achieving Sustainable Cultivation of Rice Volume 1: Breeding for Higher Yield and Quality*. Burleigh Dodds Science Publishing, Cambridge.

Dubrovsky, N. M., Burow, K. R., Clark, G. M., Gronberg, J. M., Hamilton, P. A., Hitt, K. J., Mueller, D. K., Munn, M. D., Nolan, B. T. and Puckett, L. J. (2010). *The Quality of Our Nation's Waters–Nutrients in the Nation's Streams and Groundwater, 1992-2004*. U.S. Geological Survey, Reston, VA.

Duhan, J. S., Kumar, R., Kumar, N., Kaur, P., Nehra, K. and Duhan, S. (2017). Nanotechnology: the new perspective in precision agriculture. *Biotechnology Reports* 15, 11-23. DOI: 10.1016/j.btre.2017.03.002.

Dumont, A. M. and Baret, P. V. (2017). Why working conditions are a key issue of sustainability in agriculture? A comparison between agroecological, organic and conventional vegetable systems. *Journal of Rural Studies* 56, 53-64.

Dumont, A. M., Gasselin, P. and Baret, P. V. (2020). Transitions in agriculture: three frameworks highlighting coexistence between a new agroecological configuration and an old, organic, and conventional configuration of vegetable production in Wallonia (Belgium). *Geoforum* 108, 98-109.

Duncan, J. and Pascucci, S. (2017). Mapping the organisational forms of networks of alternative food networks: implications for transition. Sociologia ruralis VC 2017 European society for rural sociology. *Sociologia Ruralis* 57(3, July), 316-339.

Dunlap, A., Hammelbo Søyland, L. and Shokrgozar, S. (Eds) (2021). *Debates in Post-Development and Degrowth: Volume 1*. TvergasteinBlindern. ISSN 1893-5834.

Dunlap, T. R. (1981). *DDT: Scientists, Citizens, and Public Policy*. Princeton University Press, Princeton, NJ.

Dupré, L., Lamine, C. and Navarrete, M. (2017). Short food supply chains, long working days: active work and the construction of professional satisfaction in French diversified organic market gardening. *Sociologia Ruralis* 57(3, July). DOI: 10.1111/soru.12178.

Duram, L. and Oberholtzer, L. (2010). A geographic approach to place and natural resource use in local food systems. *Renewable Agriculture and Food Systems* 25(2), 99-108. DOI: 10.1017/S1742170510000104.

Durham, T. C. and Mizik, T. (2021). Comparative economics of conventional, organic, and alternative agricultural production systems. *Economies* 9(2): 64. DOI: 10.3390/economies9020064.

Dwivedi, S. L., van Bueren, T. L., Ceccarelli, S., Grando, S., Upadhyaya, H. D. and Ortiz, R. (2017). Diversifying food systems in the pursuit of sustainable food production and healthy diets. *Trends in Plant Science* 22(10). DOI: 10.1016/j.tplants.2017.06.011.

Eastern Daily Press (2021). World's largest vertical farm is being built in Norfolk. Available at: https://www.edp24.co.uk/news/business/fischer-farms-builds-vertical-farm-in-norfolk-8465252. Accessed 5 November 2021.

Eastwood, C., Klerkx, L., Ayre, M. and Dela Rue, B. (2019). Managing socio-ethical challenges in the development of smart farming: from a fragmented to a comprehensive approach for responsible research and innovation. *Journal of Agricultural and Environmental Ethics* 32(5–6), 741–768.

Eberle, U. and Fels, J. (2014). Environmental impacts of German food consumption and food losses. In: Proceedings of the 9th International Conference Life Cycle Assessment Agri-Food Sector. DOI: 10.1007/s11367-015-0983-7.

EC (2002). *Implementation of Council Directive 91/676/EEC Concerning the Protection of Waters against Pollution Caused by Nitrates from Agricultural Sources; Synthesis from Year 2000 Member States Reports*. European Commission Office for Official Publications of the European Communities, Luxembourg, p. 44.

EC (2004). European action plan for organic food and farming. Commission Staff Working Document SEC739. Annex to the Communication from the Commission COM(2004)415final. Available at: http://ec.europa.eu/agriculture/organic, 33 pp.

EC (2022). Bioeconomy. Available at: https://ec.europa.eu/info/research-and-innovation/research-area/environment/bioeconomy_en#:~:text=The%20bioeconomy%20means%20using%20renewable,produce%20food%2C%20materials%20and%20energy.&text=As%20action%20under%20the%202018,been%20launched%20in%20June%202021. Accessed 21 January 2022.

Echegoyen, Y. and Nerin, C. (2015). Performance of an active paper based on cinnamon essential oil in mushrooms quality. *Food Chemistry* 170, 30–36.

Economist (2010). Economic growth: the solution to all problems. Available at: http://www.economist.com/blogs/freeexchange/2010/06/economic_growth.

Ecumenical Advocacy Alliance (2012). *Nourishing the World Sustainably: Scaling up Agroecology*. Ecumenical Advocacy Alliance, Geneva.

Edan, Y. (1999). Food and agriculture robotics. In: Nof, S. Y. (Ed.) *Handbook of Industrial Robotics* (2nd edn). John Wiley & Sons, Inc., Hoboken, NJ.

Eddleston, M., Karalliedde, L., Buckley, N., Fernando, R., Hutchinson, G., Isbister, G., Konradsen, F., Murray, D., Piola, J. C., Senanayake, N., Sheri, R., Singh, S., Siwach, S. B. and Smit, L. (2002). Pesticide poisoning in the developing world: a minimum pesticides list. *The Lancet* 360(9340), 1163–1167.

Edelman, M. (2005). Bringing the moral economy back in…to the study of 21st-century transnational peasant movements. *American Anthropologist* 107(3), 331–345.

Edelman, M. (2014). Food sovereignty: forgotten genealogies and future regulatory challenges. *Journal of Peasant Studies* 41(6), 959–978. DOI: 10.1080/03066150.2013.876998.

Edenbrandt, E. K., Gamborg, C. and Thorsen, B. J. (2017). Consumers' preferences for bread: transgenic, cisgenic, organic or pesticide-free? *Journal of Agricultural Economics* 69(1), 121–141. DOI: 10.1111/1477-9552.12225.

Edmondson, J. L., Cunningham, H., Densley Tingley, D. O., Dobson, M. C., Grafius, D. R., Leake, J. R., McHugh, N., Nickles, J., Phoenix, G. K., Ryan, A. J., Stovin, V., Buck, N. T.,

Warren, P. H. and Cameron, D. D. (2020). The hidden potential of urban horticulture. *Nature Food* 1(3), 155–159. Available at: www.nature.com/natfood.

EEA (2003). *Europe's Water: An Indicator-Based Assessment*. European Environment Agency, Copenhagen, p. 97. Available at: http://www.eea.europa.eu/publications/topic_report_2003_1.

EEA (2015). *The European Environment–State and Outlook 2015*. European Environment Agency, Copenhagen.

EEA (2017). Environmental indicator report 2017. Support to the Monitoring of the Seventh Environment Action Programme. European Environment Agency, Copenhagen.

EEA (2018). Perspectives on transitions to sustainability. European Environment Agency Report No 25/2017. European Environment Agency, Copenhagen. ISSN 1725-9177.

EEA (2019). Sustainability transitions: policy and practice. European Environment Agency. LU, Publications Office. DOI: 10.2800/641030. ISBN 9789294800862.

Efron, E. (1985). *The Apocalyptics*. Touchstone/Simon & Schuster, New York, p. 268.

Ehrlich, P. R. (1968). *The Population Bomb*. Buccaneer Books, Cutchogue, NY.

Ehrlich, P. R. and Holdren, J. P. (1971). Impact of population growth. *Science* 171, 212–213.

Eigenbrod, C. and Gruda, N. (2015). Urban vegetable for food security in cities: a review. *Agronomy for Sustainable Development* 35(2), 483–498.

EIT (2023). Can regenerative agriculture replace conventional farming? Available at: https://www.eitfood.eu/blog/can-regenerative-agriculture-replace-conventional-farming. Accessed 27 October 2023.

EIT Food (2020). Can regenerative agriculture replace conventional farming? 25 August. EIT Food iVZW, Belgium. info@eitfood.eu.

El Diario (2008). Presidente Correa fija precio de quintal de arroz. 25 August. Available at: https://www.eldiario.ec/noticias-manabi-ecuador/89347-presidente-correa-fija-precio-de-quintal-de-arroz/.

El Universo (2008). Correa fija precio de la leche y controla los de la harina y el arroz. Available at: https://www.eluniverso.com/2008/01/03/0001/9/8AE02227F7364215A3E65A64A62EF4E4.html.

ELD Initiative (2015). *Report for Policy and Decision Makers: Reaping Economic and Environmental Benefits from Sustainable Land Management*. Economics of Land Degradation Initiative, Bonn.

Eldridge, B. M., Manzoni, L. R., Graham, C. A., Rodgers, B., Farmer, J. R. and Dodd, A. N. (2020). Getting to the roots of aeroponic indoor farming. *New Phytologist* 228(4), 1183–1192.

Electris, C., Humphreys, J., Lang, K., LeZaks, D. and Silverstein, J. (2019). Soil wealth: investing in regenerative agriculture across asset classes. *Croatan Institute*. Available at: https://croataninstitute.org/wp-content/uploads/2021/03/soil-wealth-2019.pdf.

El-Hage Scialabba, N. (2007). Organic agriculture and food security. In: International Conference on Agriculture and Food Security. FAO, Rome. Available at: ftp://ftp.fao.org/paia/organicag/ofs/OFS-2007-5.pdf.

El-Hage Scialabba, N. (2013). Organic agriculture's contribution to sustainability. *Crop Management*. DOI: 10.1094/CM-2013-0429-09-PS.

Elijah, O., Rahman, T. A., Orikumhi, I., Leow, C. Y. and Hindia, M. N. (2018). An overview of Internet of Things (IoT) and data analytics in agriculture: benefits and challenges. *IEEE Internet Things* 5(5), 3758–3773.

Ellen MacArthur Foundation (2015). Growth within: a circular economy vision for a competitive Europe. Available at: https://www.ellenmacarthurfoundation.org/publications/growth-within-a-circulareconomy-vision-for-a-competitive-Europe.

Elmqvist, T., Folke, C., Nyström, M., Peterson, G., Bengtsson, J., Walker, B. and Norberg, J. (2003). Response diversity, ecosystem change, and resilience. *Frontiers in Ecology and the Environment* 1(9), 488–494. DOI: 10.1890/1540-9295(2003)001[0488:RDECAR]2.0.CO;2.

Elver, H. (2017). Report of the Special *rapporteur* on the right to food. UN doc. A/HRC/34/48. United Nations Human Rights Council, Geneva.

Enclona, E. A., Thenkabail, P. S., Celis, D. and Diekmann, J. (2004). Within-field wheat yield prediction from IKONOS data: a new matrix approach. *International Journal of Remote Sensing* 25(2), 377–388.

Engelen, E. (2008). The case for financialization. *Competition and Change* 12(2), 111–119.

Engels, J. M. M. (1995). In-situ conservation and use of plant genetic resources for food and agriculture in developing countries. Report of a DSE/ATSAF/IPGRI Workshop, 2–4 May 1995, Bonn-Röttgen. Germany.

EPA (1975). Train stops manufacture of heptachlor/chlordane. Cites imminent cancer risk. *EPA*, press release, July 1975. Available at: epa.gov/history/topics/legal/01.htm. Accessed 12 January 2004.

EPA (2003). DDT regulatory history: a brief survey (to 1975). *EPA*. Available at: epa.gov/history/topics/ddt/02.htm. Accessed 12 January 2004.

EPA (2022). What are biopesticides? The United States Environmental Protection Agency. Available at: https://www.epa.gov/ingredients-used-pesticide-products/what-are-biopesticides#. Accessed 11 April 2022.

EPHA (2016). *Agriculture and Public Health: Impacts and Pathways for Better Coherence*. Brussels, Belgium.

Epstein, G. (2001). Financialization, rentier interests, and central bank policy (version 1.2, June 2002). Paper prepared for PERI Conference on the 'Financialization of the World Economy', 7–8 December 2001,. University of Massachusetts Amherst.

Epstein, G. (2005). Introduction: financialization and the world economy. In: Epstein, G. (Ed.) *Financialization and the World Economy*. Edwar Elgar, Cheltenham, pp. 3–16.

Epstein, G. and Jayadev, A. (2005). The rise of rentier incomes in OECD countries: financialization, central bank policy and labour solidarity. In: Epstein, G. (Ed.) *Financialization and the World Economy*. Edward Elgar Publishing, Cheltenham, pp. 46–58.

Erb, K. H., Lauk, C., Kastner, T., Mayer, A., Theurl, M. C. and Haberl, H. (2016). Exploring the biophysical option space for feeding the world without deforestation. *Nature Communications* 7, 11382.

Ercsey-Ravasz, M., Toroczkai, Z., Lakner, Z. and Baranyi, J. (2012). Complexity of the international agro-food trade network and its impact on food safety. *PLoS ONE* 7(5), e37810.

Esteban-Tejeda, L., Malpartida, F., Esteban-Cubillo, A., Pecharromán, C. and Moya, J. S. (2009). Antibacterial and antifungal activity of a soda-lime glass containing copper nanoparticles. *Nanotechnology* 20(50), 505701.

Esteva, G. (1984). *Por una nueva política alimentaria*.Sociedad Mexicana de Planificación, Mexico.

Eswaran, H., Lal, R. and Reich, P. F. (2001). Land degradation: an overview. Responses to Land degradation. In: Proceedings of the 2nd International Conference on Land Degradation and Desertification, Khon Kaen, Thailand. Oxford Press, pp. 20–35.

ETC and HBF (2018). *Forcing the Farm: How Gene Drive Organisms Could Entrench Industrial Agriculture and Threaten Food Sovereignty*. ETC Group and Heinrich Boll Foundation (HBF). Available at: www.etcgroup.org and www.boell.de/en. Accessed October 2018.

ETC Group (2017). *Who Will Feed Us? The Industrial Food Chain vs. The Peasant Food Web* (3rdedn). Available at: https://www.etcgroup.org/content/who-will-feed-us-industrial-food-chain-vs-peasant-food-web.

ETC Group (2019a). *Lab-Grown Meat and Other Petri-Protein Industries*. ETC Group, May 2019. Available at: https://www.etcgroup.org/content/lab-grown-meat-and-other-petri-protein-industries.

ETC Group (2019b). *Plate Tech-Tonics: Mapping Corporate Power in Big Food*. ETC Group, November 2019. Available at: https://www.etcgroup.org/content/plate-tech-tonics.

EU (2013). *The Impact of EU Consumption on Deforestation: Comprehensive Analysis of the Impact of EU Consumption on Deforestation*. European Union. Available at: http://ec.europa.eu/environment/forests/pdf/1.%20Report%20analysis%20of%20impact.pdf.

EU (2017). Precision farming: sowing the seeds of a new agricultural revolution. European Commission. The Community Research and Development Information Service (CORDIS), Luxembourg. ISBN 978-92-78-41485-6.

EU (2020). EU biodiversity Strategy for 2030: bringing nature back into our lives. Communication From the Commission to the European Parliament, the Council, the European. Economic and Social Committee and the Committee of the Regions, Brussels, 20 May 2020. COM (2020) 380 final.

EU (2021). *Organic Farming Statistics in the EU*. European Commission, EUROSTAT. Available at: https://ec.europa.eu/eurostat/statistics-explained/index.php/Organic_farming_statistics. Accessed 5 April 2021.

Euronews (2022). Future of food: the tech tool that precisely measures vitamins in fruit and veg. Available at: https://www.euronews.com/my-europe/2022/08/01/food-for-thought-the-tech-tool-that-can-precisely-measure-vitamins-in-fruit-and-veg. Accessed 1 August 2022.

European Commission (2020). Farm to fork strategy. Available at: https://food.ec.europa.eu/horizontal-topics/farm-fork-strategy_en.

European CSA Research Group (2016). *Overview of Community Supported Agriculture in Europe*. Urgenci, Aubagne.

Ewert, F., Baatz, R. and Finger, R. (2023). Agroecology for a sustainable agriculture and food system: from local solutions to large-scale adoption. *Annual Review of Resource Economics* 15(1), 351–381. DOI: 10.1146/annurev-resource-102422-090105.

Eyhorn, F., Ramakrishnan, M. and Mäder, P. (2007). The viability of cotton-based organic agriculture systems in India. *International Journal of Agricultural Sustainability* 5(1), 25–38.

Fairbairn, M. (2014). 'Like gold with yield': evolving intersections between farmland and finance. *The Journal of Peasant Studies* 41(5), 777–795. DOI: 10.1080/03066150.2013.873977.

Fairhead, J., Leach, M. and Scoones, I. (2012). Green grabbing: a new appropriation of nature? *Journal of Peasant Studies* 39(2), 237–261.

Falco, G., Nicola, M., Pini, M., Marucco, G., Wilde, W. D., Popugaev, A., Gay, P. and Aimonino, D. R. (2019). Investigation of performance of GNSS-based devices for precise positioning in harsh agriculture environments. In: Proceedings of the 2019 IEEE International Workshop on Metrology for Agriculture and Forestry (MetroAgriFor), Portici, Italy, 24–26 October 2019, pp. 1–6.

FAO (1995). *Dimensions of Need: an Atlas of Food and Agriculture* (1st edn). Food and Agriculture Organization of the United Nations, Santa Barbara, CA.

FAO (1997). *State of the World's Forests*. United Nations-Food and Agriculture Organization, Rome, Italy.

FAO (2002a). *The State of Food Insecurity in the World 2001*. FAO, Rome.

FAO (2002b). Organic agriculture, environment and food security. Environment and Natural Resources Service Sustainable Development Department, FAO, Rome. Available at: http://www.fao.org/DOCREP/005/Y4137E/y4137e00.htm#TopOfPage.

FAO (2008). The state of food insecurity in the developing world. FAO, Rome. Available at: http://www.fao.org/docrep/012/i0876e/i0876e00.HTM.

FAO (2010). Food, agriculture and cities-challenges of food and nutrition security, agriculture and ecosystem management in an urbanizing world. Food and Agriculture Organization of the United Nations, Rome. Available at: http://www.fao.org/fileadmin/templates/FCIT/PDF/FoodAgriCities_Oct2011.

FAO (2011a). *The State of Food Insecurity in the World 2011*. Food and Agriculture Organization, Rome.

FAO (2011b). *Energy Smart Food for People and Climate*. - Issue Paper. Food and Agriculture Organization of the United Nations, Rome.

FAO (2011c). World livestock 2011: livestock in food security. Food and Agriculture Organization of the United Nations, Rome. Available at: http://www.fao.org/docrep/014/i2373e/i2373e.pdf.

FAO (2012). In: Alexandratos, N. and Bruinsma, J. (Eds) *World Agriculture Towards 2030/2050* (2012 revision). Global Perspective Studies Team, Food and Agriculture Organization of the United Nations, Rome.

FAO (2013a). *Statistical Yearbook of World Food and Agriculture for the United Nations*. Food and Agriculture Organization, Rome.

FAO (2013b). *Resilient Livelihoods: Disaster Risk Reduction for Food and Nutrition Security, Updated New Edition* Emergency and Rehabilitation Division, Food and Agriculture Organization of the United Nations, Rome.

FAO (2013c). FAOSTAT. Available at: http://faostat3.fao.org/home/index.html.

FAO (2013d). *Climate-Smart Agriculture Sourcebook*. Food and Agriculture Organisation, Rome. Available at: http://www.fao.org/docrep/018/i3325e/i3325e.pdf.

FAO (2014). *Agriculture, Forestry and Other Land Use Emissions by Sources and Removals by Sinks*. Food and Agriculture Organization, Rome.

FAO (2015a). *World Agriculture: Towards 2015/2030: An FAO Perspective: Livestock Commodities*. Food and Agriculture Organization of the United Nations, Rome. Available at. http://www.fao.org/docrep/005/y4252e/y4252e05b.htm.

FAO (2015b) *GLADIS*. Food and Agriculture Organization of the United Nations, Rome. Available at: www.fao.org/nr/lada/gladis/gladis/.

FAO (2015c). Report on the regional meeting on agroecology in Sub-Saharan Africa. Dakar, 5–6 November 2015. Available at: http://www.fao.org/3/a-i6364e.pdf.

FAO (2015d). Addressing social and economic burden of malnutrition through nutrition sensitive agricultural and food policies in the region of Europe and Central Asia.

Agenda item #6. 39th Session, European Commission on Agriculture. Budapest, Hungary.

FAO (2016a). Agriculture and food security at heart of climate change action. Available at: http://www.fao.org/news/story/en/item/453416/icode/.

FAO (2016b). *Influencing Food Environment for Healthy Diet*. Food and Agriculture Organization of the United Nations, Rome.

FAO (2016c). Summary Report of the FAO International Symposium "The Role of Agricultural Biotechnologies in Sustainable Food Systems and Nutrition". Food and Agriculture Organization, Rome.

FAO (2016d). *Energy, Agriculture and Climate Change: Towards Energy-Smart Agriculture*. Food and Agriculture Association of the United Nations, Rome. Available at: http://www.fao.org/3/a-i6382e.pdf.

FAO (2017a). *Water for Sustainable Food and Agriculture*. Food and Agriculture Organization of the United Nations, Rome. Available at. http://www.fao.org/3/a-i7959e.pdf.

FAO (2017b). *Energy for Agriculture*. Food and Agriculture Organization of the United Nations, Rome. Available at: http://www.fao.org/docrep/003/x8054e/x8054e05.htm.

FAO (2018a). *Sustainable Food Systems: Concept and Framework*. Food and Agriculture Organization of the United Nations, Rome. Available at: http://www.fao.org/sustainable-food-value-chain.

FAO (2018b). 2nd International Symposium on Agroecology: Scaling up Agroecology to Achieve the Sustainable Development Goals (SDGs), 3-5 April 2018, Rome. Chair's Summary. http://www.fao.org/3/CA0346EN/ca0346en.pdf.

FAO (2018c). *Global Livestock Environmental Assessment Model*, Version 2.0 (5th revision), July 2018. Food and Agriculture Organization of the United Nations, Rome.

FAO (2018d). 1st International Symposium on Agroecology for Food Security and Nutrition 2014. Available at: http://www.fao.org/about/meetings/afns/en/?__scoop_post=2964fe60-3d9a-11e4-daf6-90b11c3d2b20&_scoop_topic=2055030.

FAO (2018e). *FAO's Work on Agricultural Innovation: Sowing the Seeds of Transformation to Achieve the SDGs*. Food and Agriculture Organization, Rome.

FAO (2018f). *Scaling up Agroecology Initiative: Transforming Food and Agricultural Systems in Support of the SDGs*. Food and Agriculture Organization of the United Nations, Rome.

FAO (2018g). *Sustainable Agriculture and Food Systems in Europe and Central Asia in a Changing Climate* (31st Session). FAO Regional Conference for Europe, Voronezh, Russian Federation, 16–18 May 2018. ERC/18/2.

FAO (2019). The state of food security in the world. Available at: http://www.fao.org/3/ca5162en/ca5162en.pdf.

FAO (2020). The state of food security and nutrition in the world 2020. Available at: http://www.fao.org/publications/sofi/2020/en/.

FAO (2021). Unlocking the potential of protected agriculture in the countries of the Gulf Cooperation Council – saving water and improving nutrition. Cairo. DOI: 10.4060/cb4070en.

FAO (2022). *The State of Agricultural Commodity Markets 2022. Part One: Global and Regional Trade Networks*. Available at: https://www.fao.org/3/cc0471en/online/state-of-agricultural-commodity-markets/2022/food-agricultural-trade-globalization.html. Accessed 25 November 2023.

FAO, IFAD, UNICEF, WFP and WHO (2017). *The State of Food Security and Nutrition in the World: Building Resilience for Peace and Food Security*. Food and Agriculture Organization, Rome.

FAO, IFAD, UNICEF, WFP and WHO (2020). *The State of Food Security and Nutrition in the World 2020. Transforming Food Systems for Affordable Healthy Diets*. FAO, Rome. DOI: 10.4060/ca9692en.

FAO, IFAD, UNICEF, WFP and WHO (2023). *The State of Food Security and Nutrition in the World 2023. Urbanization, Agrifood Systems Transformation and Healthy Diets across the Rural-Urban Continuum*. Food and Agriculture Organization, Rome. DOI: 10.4060/cc3017en.

FAO-AU (2020). *Measures for Supporting Domestic Markets during the Covid-19 Outbreak in Africa*. Food and Agriculture Organisation and African Union, Rome.

Farmindustrynews.com (2017). Precision planting sold to AGCO. Available at: https://www.farmprogress.com/business/precision-planting-sold-agco. Accessed 26 July 2017.

Farooq, M. and Siddique, K. H. M. (Eds) (2015). *Conservation Agriculture: Concepts, Brief History, and Impacts on Agricultural Systems*. Springer, New York.

Farrell, D., Hoover, M., Chen, H. and Friedersdorf, L. (2013). *Overview of Resources and Support for Nanotechnology for Sensors and Sensors for Nanotechnology: Improving and Protecting Health, Safety, and the Environment*. US National Nanotechnology Initiative, Arlington, VA. Available at: http://nano.gov/sites/default/files/pub_resource/nsi_nanosensors_resources_for_web.pdf.

Fast Company (2022). *How Nanotechnology Is Changing How You Eat and Taste*. Available at: https://www.fastcompany.com/1680380/how-nanotechnology-is-changing-how-you-eat-and-taste. Accessed 13 February 2022.

Fastler, J. (2021). Regenerative agriculture needs a reckoning. *The Counter*. Available at: https://thecounter.org/regenerative-agriculture-racial-equity-climate-change-carbon-farming-environmental-issues/. Accessed 5 March 2021.

Faulkner, E. H. (1943). *Plowman's Folly*. Micheal Joseph, London.

Feagan, R. (2007). The place of food: mapping out the local in local food systems. *Progress in Human Geography* 31(1), 23–42.

Feenstra, G. W. (1997). Local food systems and sustainable communities. *American Journal of Alternative Agriculture* 12(1), 28–36.

Feldmann, C. and Hamm, U. (2015). Consumers' perceptions and preferences for local food: a review. *Food Quality and Preference* 40, 152–164.

Feliciano, D., Ledo, A., Hillier, J. and Nayak, D. R. (2018). Which agroforestry options give the greatest soil and above ground carbon benefits in different world regions? *Agriculture, Ecosystems and Environment* 254, 117–129.

Feng, Y., Zeng, Z. Searchinger, T. D., Ziegler, A. D., Wu, J., Wang, D., He, X., Elsen, P. R., Ciais, P., Xu, R., and Guo, Z. (2022). Doubling of annual forest carbon loss over the tropics during the early twenty-first century. *Nature Sustainability*. Available at: www.nature.com/natsustain. DOI: 10.1038/s41893-022-00854-3.

Ferguson, R. S. and Taylor Lovell, S. T. (2014). Permaculture for agroecology: design, movement, practice, and worldview: a review. *Agronomy for Sustainable Development* 34(2), 251–274. DOI: 10.1007/s13593-013-0181-6.

Fernandez, M., Goodall, K., Olson, M. and Méndez, E. (2013). Agroecology and alternative agri-food movements in the United States: toward a sustainable agri-food system. *Agroecology and Sustainable Food Systems* 37(1), 115–126. DOI: 10.1080/10440046.2012.735633.

Fernandez, M. and Méndez, V. E. (2019). Subsistence under the canopy: agrobiodiversity's contributions to food and nutrition security amongst coffee communities in Chiapas, Mexico. *Agroecology and Sustainable Food Systems* 43(5), 579–601. DOI: 10.1080/21683565.2018.1530326.

Ferris, J., Norman, C. and Sempik, J. (2001). People, land and sustainability: community gardens and the social dimension of sustainable development. *Social Policy and Administration* 35(5), 559–568.

Feskens, E. J. M., Sluik, D. and van Woudenbergh, G. J. (2013). Meat consumption, diabetes, and its complications. *Current Diabetes Reports* 13(2), 298–306. DOI: 10.1007/s11892-013-0365-0.

Feuerbacher, A., Luckmann, J., Boysen, O., Zikeli, S. and Grethe, H. (2018). Is Bhutan destined for 100% organic? Assessing the economy-wide effects of a large-scale conversion policy. *PLoS ONE* 13(6), e0199025. DOI: 10.1371/journal.pone.0199025.

Feyereisen, M., Stassart, P. M. and Mèlard, F. (2017). Fair trade milk initiative in Belgium: bricolage as an empowering strategy for change. Sociologia ruralis VC 2017 European society for rural sociology. *Sociologia Ruralis* 57(3, July), 297–315.

Fi Global Insights (2021). Vertical farms are part of 'the new agricultural revolution'. But how sustainable are they? Available at: https://insights.figlobal.com/sustainability/vertical-farms-are-part-new-agricultural-revolution-how-sustainable-are-they.

Field, C. B. (2014). *Climate Change: Impacts, Adaptation and Vulnerability: Regional Aspects*. Cambridge University Press, Cambridge; New York, pp. 1–32.

Field, T. and Bell, B. (2013). *Harvesting Justice: Transforming Food, Land, and Agricultural Systems in the Americas*. Other Worlds & U.S. Food Sovereignty Alliance, New York.

Filippini, R., Marraccini, E., Houdart, M., Bonari, E. and Lardon, S. (2016). Food production for the city: hybridization of farmers' strategies between alternative and conventional food chains. *Agroecology and Sustainable Food Systems* 40(10), 1058–1084. DOI: 10.1080/21683565.2016.1223258.

Financial Times (2022a). Cultivated meat start-ups race to add products to the menu. Available at: https://www.ft.com/content/5cba69a1-860e-4f56-b5f8-26f50fb35abe. Accessed 17 December 2022.

Financial Times (2022b). Gene-edited crops: 'frankenfoods' or a fix for climate change? Available at: https://www-ft-com.ezproxy.depaul.edu/content/337ba132-cc33-42a9-9df3-cd53114200bc.

Financial Times (2023a). Agritech and foodtech sectors suffer funding slump. 28 December 2022.

Financial Times (2023b). EU plans to relax GMO restrictions to help farmers adapt to climate change. Available at: https://www.ft.com/content/5c799bc0-8196-466e-b969-4082e917dbe6. Accessed 23 June 2023.

Financial Times (2023c). Losses on meat alternatives leave investors with sour aftertaste. Available at: https://www.ft.com/content/f92a8bfe-9ec1-4423-a05a-e2f0d9047f03. Accessed 8 June 2023.

Finger, R., Swinton, S. M., El Benni, N. and Walter, A. (2019). Precision farming at the nexus of agricultural production and the environment. *Annual Review of Resource Economics* 11(1), 313–335.

Finless Foods (2022). Home Page. Available at: https://finlessfoods.com/about/#products. Accessed 27 January 2022.

Firth, C., Maye, D. and Pearson, D. (2011). Developing "community" in community gardens. *Local Environment* 16(6), 555–568.

Fischer, J., Brosi, B., Daily, G. C., Ehrlich, P. R., Goldman, R., Goldstein, J., Lindenmayer, D. B., Manning, A. D., Mooney, H. A., Pejchar, L., Ranganathan, J. and Tallis, H. (2008). Should agricultural policies encourage land sparing or wildlife-friendly farming? *Frontiers in Ecology and the Environment* 6(7), 380–385.

Fisher, L. (2022). Regenerative agriculture has a big race and equity issue, and it's not going away anytime soon. Available at: https://www.wellandgood.com/equity-issues-regenerative-agriculture/. Accessed 11 April 2022.

Fisher, M., Abate, T., Lunduka, R. W., Asnake, W., Alemayehu, Y. and Madulu, R. B. (2015). Drought tolerant maize for farmer adaptation to drought in sub-Saharan Africa: determinants of adoption in eastern and southern Africa. *Climatic Change* 133(2), 283–299.

Fitzpatrick, I. (2015). *From the Roots Up: How Agroecology Can Feed Africa*. Global Justice Now, London.

Fletcher, B. (2020). Blue fletcher investigates projects that are using vertical farming as a developing resource for our future. Available at: https://greenkode.net/vertical-farming/.

Fließbach, A. and Mäder, P. (2000). Microbial biomass and size-density fractions differ between soils of organic and conventional agricultural systems. *Soil Biology and Biochemistry* 32(6), 757–768.

Fließbach, A., Mäder, P. and Niggli, U. (2000). Mineralization and microbial assimilation of 14C-labeled straw in soils of organic and conventional agricultural systems. *Soil Biology and Biochemistry* 32(8–9), 1131–1139.

Fließbach, A., Oberholzer, H. R., Gunst, L. and Mäder, P. (2007). Soil organic matter and biological soil quality indicators after 21 years of organic and conventional farming. *Agriculture, Ecosystems and Environment* 118(1–4), 273–284. DOI: 10.1016/j.agee.2006.05.022.

Folberth, C., Yang, H., Gaiser, T., Liu, J., Wang, X., Williams, J. and Schulin, R. (2014). Effects of ecological and conventional agricultural intensification practices on maize yields in sub-Saharan Africa under potential climate change. *Environmental Research Letters* 9(4), 044004. DOI: 10.1088/1748-9326/9/4/044004.

Foley, J. A., Ramankutty, N., Brauman, K. A., Cassidy, E. S., Gerber, J. S., Johnston, M., Mueller, N. D., O'Connell, C., Ray, D. K., West, P. C., Balzer, C., Bennett, E. M., Carpenter, S. R., Hill, J., Monfreda, C., Polasky, S., Rockstrom, J., Sheehan, J., Siebert, S., Tilman, D. and Zaks, D. P. M. (2011). Solutions for a cultivated planet. *Nature* 478(7369), 337–342.

FoodCloud (2021). 2021 annual report: food waste hurts our planet. Help us fight it. Available at: https://food.cloud/about.

Food Navigator (2022a). Are consumers ready for plant-based and cultivated meat hybrids? FoodNavigator.com. 9 December 2022.

Food Navigator (2022b). Big meat retrenches as meat alternatives lose their lustre. FoodNavigator-usa.com. 13 October 2022.

Food Navigator (2023a). 'The race is on': Is Japan's commitment to cultivated meat another sign Europe is lagging? FoodNavigator.com. 14 March 2023.

Food Navigator (2023b). As vertical farms topple, can tech sow sustainability into controlled environment agriculture? FoodNavigator.com. 27 July 2023.

Food Processing-Technology (2022). *Future Meat Technologies' Cultured Meat Production Facility*. Rehovot, Israel. Available at: https://www.foodprocessing-technology.com/projects/future-meat-technologies-cultured-meat-production-facility-israel/. Accessed 29 January 2022.

Foray, D. (1997). The dynamic implications of increasing returns: technological change and path dependent inefficiency. *International Journal of Industrial Organization* 15(6), 733–752.

Forney, J. and Häberli, I. (2015). Introducing "seeds of change" into the food system? Localisation strategies in the Swiss Dairy industry: introducing seeds of change into the food system? *Sociologia Ruralis*. DOI: 10.1111/soru.12072.

Forsberg, E. M. and de Lauwere, C. (2013). Integration needs in assessments of nanotechnology in food and agriculture. *Etikk i Praksis – Nordic Journal of Applied Ethics* 1(1), 38–54.

Forster, A. C. and Church, G. M. (2007). Synthetic biology projects in vitro. *Genome Research* 17(1), 1–6.

Forster, D., Andres, C., Verma, R., Zundel, C., Messmer, M. M. and Mäder, P. (2013). Yield and economic performance of organic and conventional cotton-based farming systems – results from a field trial in India. *PLoS ONE* 8(12), e81039. DOI: 10.1371/journal.pone.0081039.

Forster, T., Egal, F., Getz Escudero, A., Dubbeling, M. and Renting, H. (2015). Milan urban food policy pact. Selected good practices from cities. Utopie 29. *Globalizzazione* (Ebook). Available at: http://www.fondazionefeltrinelli.it/article/ebook-utopie-milan-urban-food-policy-pact/.

Fortune Business Insights (2020). Biopesticides market size. Available at: https://www.fortunebusinessinsights.com/industry-reports/biopesticides-market-100073.

Fortune Business Insights (2023a). Vertical farming market. Available at: https://www.fortunebusinessinsights.com/industry-reports/vertical-farming-market-101958. Vertical, Farming Market.

Fortune Business Insights (2023b). Agricultural drone market size. Available at: https://www.fortunebusinessinsights.com/agriculture-drones-market-102589.

Fountain, E. D. and Wratten, S. D. (2013). Conservation biological control and biopesticides in agricultural. In: Elias, S. A. (Ed.) *Reference Module in Earth Systems and Environmental Sciences*. Elsevier, New York. DOI: 10.1016/B978-0-12-409548-9.00539-X.

Fournier, V. (2008). Escaping from the economy: the politics of degrowth. *International Journal of Sociology and Social Policy* 28(11/12), 528–545.

Fraceto, L. F., Grillo, R., de Medeiros, G. A., Scognamiglio, V., Rea, G. and Bartolucci, C. (2016). Nanotechnology in agriculture: which innovation potential does it have? *Frontiers in Environmental Science* 4, 20. DOI: 10.3389/fenvs.2016.00020.

Francis, C., Lieblein, G., Gliessman, S., Breland, T. A., Creamer, N., Harwood, R., Salomonsson, L., Helenius, J., Rickerl, D., Salvador, R., Wiedenhoeft, M., Simmons, S., Allen, P., Altieri, M., Flora, C. and Poincelot, R. (2003). Agroecology: the ecology of food systems. *Journal of Sustainable Agriculture* 22(3), 99–118.

Francis, C. A. (1986). *Multiple Cropping Systems*. MacMillan, New York.

Francis, C. A., Harwood, R. R. and Parr, J. F. (1986). The potential for regenerative agriculture in the developing world. *American Journal of Alternative Agriculture* 1(2), 65–74.

Frayne, B., McCordic, C. and Shilomboleni, H. (2014). Growing out of poverty: does urban agriculture contribute to household food security in southern African cities? *Urban Forum* 25(2), 177–189. DOI: 10.1007/s12132-014-9219-3.

Freijer, K., Tan, S. S., Koopmanschap, M. A., Meijers, J. M., Halfens, R. J. and Nuijten, M. J. (2013). The economic costs of disease related malnutrition. *Clinical Nutrition* 32(1), 136–141. DOI: 10.1016/j.clnu.2012.06.009.

French, S. and Morris, P. (2006). Assessing the evidence for sugar-sweetened beverages in the aetiology of obesity: a question of control. *International Journal of Obesity* 30(S3), S37–S39. DOI: 10.1038/sj.ijo.0803490.

Fresh Plaza (2023). Future crops declared bankrupt: all equipment in online auction at Troostwijk Auctions. Accessed 6 April 2023. Available at: https://www.freshplaza.com/europe/article/9518097/future-crops-declared-bankrupt-all-equipment-in-online-auction-at-troostwijk-auctions/.

Fresh Plaza (2022). Vertical future raises £21 million to stabilize production in the UK. Available at: https://www.freshplaza.com/article/9391522/vertical-future-raises-ps21-million-to-stabilize-production-in-the-uk/.

Friedland, J. (2011). Morgan Stanley closes $50 million investment in Chinese fertilizer producer Yongye International. *Big Emerging Economies (BEEs)*, 16 June. Available at: http://bigemergingeconomies.wordpress.com/2011/06/16/morgan-stanley-closes-50-millioninvestment-in-chinese-fertilizer-producer-yongye-international/.

Friedmann, H. (1994). Distance and durability: shaky foundations of the world food economy. In: McMichael, P. (Ed.) *The Global Restructuring of Agro-Food Systems*. Cornell University Press, Ithaca, NY, pp. 258–276.

Friedmann, H. (2005). From colonialism to green capitalism: social movements and the emergence of food regimes. In: Buttel, F. H. and McMichael, P. D. (Eds) *New Directions in the Sociology of International Development: Research in Rural Sociology and Developments* (Vol. 11). Elsevier, Amsterdam, pp. 227–264.

Friedmann, H. and McMichael, P. (1989). Agriculture and the state system: the rise and decline of national agricultures, 1870 to the present. *Sociologia Ruralis* 29(2), 93-117. DOI: 10.1111/j.1467-9523.1989.tb00360.x.

Frison, E. and Clément, C. (2020). The potential of diversified agroecological systems to deliver healthy outcomes: making the link between agriculture, food systems & health. *Food Policy* 96, 101851.

Fromartz, S. (2007). *Organic Inc: Natural Foods and How They Grew*. Houghton Mifflin Harcourt, Orlando, FL.

Fuchs, D. and Glaab, K. (2011). Material power and normative conflict in global and local agrifood governance: the lessons of 'Golden Rice' in India. *Food Policy* 36(6), 729–735.

Fuenfschilling, L. and Truffer, B. (2014). The structuration of socio-technical regimes–conceptual foundations from institutional theory. *Research Policy* 43(4), 772–791.

Fuller, R. J., Norton, L. R., Feber, R. E., Johnson, P. J., Chamberlain, D. E., Joys, A. C., Mathews, F., Stuart, R. C., Townsend, M. C., Manley, W. J., Wolfe, M. S., Macdonald, D. W. and Firbank, L. G. (2005). Benefits of organic farming to biodiversity vary among taxa. *Biology Letters* 1(4), 431–434.

Future Meat (2022). Home Page. Available at: https://future-meat.com/. Accessed 29 January 2022.

Gabel, M. (1979). *Ho-Ping: A World Scenario for Food Production*. World Game Institute, Philadelphia, PA.

Gabriel, D., Sait, S. M., Hodgson, J. A., Schmutz, U., Kunin, W. E. and Benton, T. G. (2010). Scale matters: the impact of organic farming on biodiversity at different spatial scales. *Ecology Letters* 13(7), 858–869. DOI: 10.1111/j.1461-0248.2010.01481.x.

Gabriel, D., Sait, S. M., Kunin, W. E. and Benton, T. G. (2013). Food production vs. biodiversity: comparing organic and conventional agriculture. *Journal of Applied Ecology* 50(2), 355–364. DOI: 10.1111/1365-2664.12035.

Gaitán-Cremaschi, D., Klerkx, L., Duncan, J., Trienekens, J. H., Huenchuleo, C., Dogliotti, S., Contesse, M. E., Benitez, F. J. and Rossing, W. A. H. (2020). Sustainability transition pathways through ecological intensification: an assessment of vegetable food systems in Chile. *International Journal of Agricultural Sustainability* 18(2), 131-150. DOI: 10.1080/14735903.2020.1722561.

Galli, F. and Brunori, G. (2013). Short food supply chains as drivers of sustainable development. Evidence document. FP7 Project Food Links–European Commission, Brussels, Belgium.

Galli, F., Favilli, E., D'Amico, S. and Brunori, G. (2018). *A Transition Towards Sustainable Food Systems in Europe. Food Policy Blueprint Scoping Study*. Laboratorio di Studi Rurali Sismondi, Pisa. ISBN 9788890896040.

Galt, R. E. (2013). The moral economy is a double-edged sword: explaining farmers' earnings and self-exploitation in community-supported agriculture. *Economic Geography* 89(4), 341–365.

Gamson, W. A. (1995). Constructing social protest. In: Johnston, H. and Klandermands, B. (Eds) *Social Movements and Culture: Social Movements, Protest, and Contention*. University of Minnesota Press, Minneapolis, MN, pp. 85–106.

Gao, B., Huang, T., Ju, X., Gu, B., Huang, W., Xu, L., Rees, R. M., Powlson, D. S., Smith, P. and Cui, S. (2018). Chinese cropping systems are a net source of greenhouse gases despite soil carbon sequestration. *Global Change Biology* 24(12), 5590–5606.

GAP (2020). Artificial Intelligence is in a path to reduce herbicide use by 90%. Available at: https://globalagriculturalproductivity.org/artificial-intelligence-is-in-a-path-to-reduce-herbicide-use-by-90/. Accessed 14 April 2022.

Garibaldi, L. A., Carvalheiro, L. G., Vaissière, B. E., Gemmill-Herren, B., Hipólito, J., Freitas, B. M., Ngo, H. T., Azzu, N., Sáez, A., Åström, J., An, J., Blochtein, B., Buchori, D., Chamorro García, F. J., Oliveira da Silva, F., Devkota, K., Ribeiro, Mde F., Freitas, L., Gaglianone, M. C., Goss, M., Irshad, M., Kasina, M., Pacheco Filho, A. J., Kiill, L. H., Kwapong, P., Parra, G. N., Pires, C., Pires, V., Rawal, R. S., Rizali, A., Saraiva, A. M., Veldtman, R., Viana, B. F., Witter, S. and Zhang, H. (2016). Mutually beneficial pollinator diversity and crop yield outcomes in small and large farms. *Science* 351(6271), 388–391.

Garnett, T. (2010). The food miles debate: is shorter better? In: McKinnon, A., Cullinane, S., Browne, M. and Whiteing, A. (Eds) *Green Logistics: Improving the Environmental Sustainability of Logistics*. Kogan Page, London, pp. 265–281.

Garnett, T., Appleby, M. C., Balmford, A., Bateman, I. J., Benton, T. G., Bloomer, P., Burlingame, B., Dawkins, M., Dolan, L., Fraser, D., Herrero, M., Hoffmann, I., Smith, P., Thornton, P. K., Toulmin, C., Vermeulen, S. J. and Godfray, H. C. (2013). Sustainable intensification in agriculture: premises and policies. *Science* 341(6141), 33–34.

Garnett, T. and Dodfray, C. H. (2012). Sustainable intensification in agriculture Navigating a course through competing food system priorities. A report on a workshop, funded by the UK Government's Foresight Programme as part of its follow up activities to the Future of Food and Farming Project, 51.

Garnett, T., Godde, C. and Muller, A. (2017). *Grazed and Confused? Ruminating on Cattle, Grazing Systems, Methane, Nitrous Oxide, the Soil Carbon Sequestration Question - and What It All Means for Greenhouse Gas Emissions*. Food Climate Research Network, University of Oxford, Oxford, p. 127.

Garnett, T. and Wilkes, A. (2014). *Appetite for Change: Social, Economic, and Environmental Transformations in China's Food System*. Food Climate Research Network, University of Oxford, Oxford.

Garrison, G. L., Biermacher, J. T. and Brorsen, B. W. (2022). How much will large-scale production of cell-cultured meat cost? *Journal of Agriculture and Food Research* 10, 100358.

Gascón, J. (2010). ¿Del paradigma de la industrialización al de la soberanía alimentaria? Una comparación entre los gobiernos nacionalistas latinoamericanos del siglo XX y los pos-neoliberales a partir de sus políticas agrarias. In: Gascón, J. and Montagut, X. (Eds) *Cambio de rumbo en las políticas agrarias latinoamericanas? Estado, movimientos sociales campesinos y soberanía alimentaria*. Icaria Editorial, Barcelona, pp. 215–259.

Gascuel-Odoux, C., Lescourret, F., Dedieu, B., Detang-Dessendre, C., Faverdin, P., Hazard, L., Litrico-Chiarelli, I., Petit, S., Roques, L., Reboud, X., Tixier-Boichard, M., de Vries, H. and Caquet, T. (2022). A research agenda for scaling up agroecology in European countries. *Agronomy for Sustainable Development* 42(3), 53.

Gathala, M. K., Kumar, V., Sharma, P. C., Saharawat, Y. S., Jat, H. S., Singh, M., Kumar, A., Jat, M. L., Humphreys, E., Sharma, D. K., Sharma, S. and Ladha, J. K. (2014). Reprint of "Optimizing intensive cereal-based cropping systems addressing current and future drivers of agricultural change in the North-western Indo-Gangetic Plains of India". *Agriculture, Ecosystems and Environment* 187, 33–46.

Gattinger, A., Muller, A., Haeni, M., Skinner, C., Fliessbach, A., Buchmann, N., Mäder, P., Stolze, M., Smith, P., Scialabba, N. E.-H. and Niggli, U. (2012). Enhanced topsoil carbon stocks under organic farming. *Proceedings of the National Academy of Sciences of the United States of America* 109(44), 18226–18231. DOI: 10.1073/pnas.1209429109.

GBNews (2023). Plant-based meat substitute manufacturer suffers 30.5 per cent net revenue nosedive in just three months. *GBNews*, 08 August 2023.

Gebbers, R. and Adamchuk, V. I. (2010). Precision agriculture and food security. *Science* 327(5967), 828–831.

Geels, F. W. (2002). Technological transitions as evolutionary reconfiguration processes: a multi-level perspective and a case-study. *Research Policy* 31(8–9), 1257–1274.

Geels, F. W. (2004). From sectoral systems of innovation to socio-technical systems: insights about dynamics and change from sociology and institutional theory. *Research Policy* 33(6–7), 897–920.

Geels, F. W. (2005). *Technological Transitions and System Innovations: A Co-evolutionary and Socio-technical Analysis*. Edward Elgar Publishing, Cheltenham.

Geels, F. W. (2010). Ontologies, socio-technical transitions (to sustainability), and the multi-level perspective. *Research Policy* 39(4), 495–510.

Geels, F. W. (2011). The multi-level perspective on sustainability transitions: responses to seven criticisms. *Environmental Innovation and Societal Transitions* 1(1), 24–40.

Geels, F. W. (2014). Regime resistance against low-carbon transitions: introducing politics and power into the multi-level perspective. *Theory, Culture and Society* 31(5), 21–40.

Geels, F. W. and Raven, R. (2006). Non-linearity and expectations in niche-development trajectories: ups and downs in Dutch biogas development (1973–2003). *Technology Analysis and Strategic Management* 18(3–4), 375–392.

Geels, F. W. and Schot, J. (2007). Typology of sociotechnical transition pathways. *Research Policy* 36(3), 399–417.

GEF (2021). Land degradation. Available at: https://www.thegef.org/topics/land-degradation. Accessed 3 June 2021.

Geiger, F., Bengtsson, J., Berendse, F., Weisser, W. W., Emmerson, M., Morales, M. B., Ceryngier, P., Liira, J., Tscharntke, T., Winqvist, C., Eggers, S., Bommarco, R., Pärt, T., Bretagnolle, V., Plantegenest, M., Clement, L. W., Dennis, C., Palmer, C., Oñate, J. J., Guerrero, I., Hawro, V., Aavik, T., Thies, C., Flohre, A., Hänke, S., Fischer, C., Goedhart, P. W. and Inchausti, P. (2010). Persistent negative effects of pesticides on biodiversity and biological control potential on European farmland. *Basic and Applied Ecology* 11(2), 97–105. DOI: 10.1016/J.BAAE.2009.12.001.

Genentech (2003). Genetech's Corporate History. Available at: http://www.gene.com/gene/about/corporate/history/.

Gentilini, U. (2013). Banking on food: the state of food banks in high-income countries. *IDS Working Papers* 415, 1–18. DOI: 10.1111/j.2040-0209.2013.00415.x.

George, D. R., Sparagano, O. A. E., Port, G., Okello, E., Shiel, R. S. and Guy, J. H. (2010). Toxicity of plant essential oils to different life stages of the poultry red mite, Dermanyssus gallinae, and non-target invertebrates. *Medical and Veterinary Entomology* 24(1), 9–15. DOI: 10.1111/j.1365-2915.2009.00856.x.

Georgescu-Roegen, N. (1975). Energy and economic myths. *Southern Economic Journal* 41(3), 347–381.

Georghiou, G. P. (1986). The magnitude of the problem. In: Glass, E. H. (Ed.) *Pesticide Resistance: Strategies and Tactics for Management*. National Academy Press, Washington, DC.

Georghiou, G. P. and Taylor, C. (1976). Pesticide resistance as an evolutionary phenomenon. In: Proceedings of the XVth International Congress of Entomology, 19–27 August 1976, Washington, DC, pp. 759–785.

Gerber, P. J., Steinfeld, H., Henderson, B., Mottet, A., Opio, C., Dijkman, J., Falcucci, A. and Tempio, G. (2013). *Tackling Climate Change through Livestock - A Global Assessment of Emissions and Mitigation Opportunities*. Food and Agriculture Organization of the United Nations, Rome.

Germain, M. (2017). Optimal versus sustainable degrowth policies. *Ecological Economics* 136, 266–281. DOI: 10.1016/j.ecolecon.2017.02.001.

Ghafar, A. S. A., Hajjaj, S. S. H., Gsangaya, K. R., Sultan, M. T. H., Mail, M. F. and Hua, L. S. (2023). Design and development of a robot for spraying fertilizers and pesticides for agriculture. *Materials Today: Proceedings* 81(2), 242–248.

Ghisellini, P., Cialani, C. and Ulgiati, S. (2016). A review on circular economy: the expected transition to a balanced interplay of environmental and economic systems. *Journal of Cleaner Production* 114, 11–32.

Ghosh, J. (2010). The unnatural coupling: food and global finance. *Journal of Agrarian Change* 10(1), 72–86.

Ghosh, J., Heintz, J. and Pollin, R. (2012). Speculation on commodities futures markets and destabilization of global food prices: exploring the connections. *International Journal of Health Services: Planning, Administration, Evaluation* 42(3), 465–483.

Gibbs, H. K., Ruesch, A. S., Achard, F., Clayton, M. K., Holmgren, P., Ramankutty, N. and Foley, J. A. (2010). Tropical forests were the primary sources of new agricultural land in the 1980s and 1990s. *Proceedings of the National Academy of Sciences of the United States of America* 107(38), 16732–16737. DOI: 10.1073/pnas.0910275107.

Gibson-Graham, J. K. (1996). *The End of Capitalism (As We Knew It)*. University of Minnesota Press, Minneapolis, MN.

Gibson-Graham, J. K. (2006). *A Postcapitalist Politics*. University of Minnesota Press, Minneapolis, MN.

Gill, M., Feliciano, D., Macdiarmid, J. and Smith, P. (2015). The environmental impact of nutrition transition in three case study countries. *Food Security* 7(3), 493–504.
Giller, K. E., Andersson, J. A., Corbeels, M., Kirkegaard, J., Mortensen, D., Erenstein, O. and Vanlauwe, B. (2015). Beyond conservation agriculture. *Frontiers in Plant Science* 6, 870. DOI: 10.3389/fpls.2015.00870.
Giller, K. E., Hijbeek, R., Andersson, J. A. and Sumberg, J. (2021). Regenerative agriculture: an agronomic perspective. *Outlook on Agriculture* 50(1), 13–25.
Gillespie, S. and van der Bold, M. (2017). Agriculture, food systems, and nutrition: meeting the challenge. *Global Challenges* 1(3). DOI: 10.1002/gch2.201600002.
Gilligan, D. O., Hoddinott, J. and Seyoum Taffesse, A. S. (2009). The impact of Ethiopia's productive safety net programme and its linkages. *Journal of Development Studies* 45(10), 1684–1706.
Giménez Cacho, M., Giraldo, O. F., Aldasoro, M., Morales, H., Ferguson, B. G., Rosset, P., Khadse, A. and Campos, C. (2018). Bringing agroecology to scale: key drivers and emblematic cases. *Agroecology and Sustainable Food Systems* 42(6), 637–665. DOI: 10.1080/21683565.2018.1443313.
Giraldo, O. F. and Rosset, P. M. (2018). Agroecology as a territory in dispute: between institutionality and social movements. *The Journal of Peasant Studies* 45(3), 545–564. DOI: 10.1080/03066150.2017.1353496.
Giroto, A. S., Guimarães, G. G. F., Foschini, M. and Ribeiro, C. (2017). Role of slow-release nanocomposite fertilizers on nitrogen and phosphate availability in soil. *Scientific Reports* 7, 46032.
Gleeson, T., Wada, Y., Bierkens, M. F. and van Beek, L. P. (2012). Water balance of global aquifers revealed by groundwater footprint. *Nature* 488(7410), 197–200. DOI: 10.1038/nature11295.
Gleick, P. H., Burns, W., Chalecki, E., Cohen, M., Cushing, K., Mann, A., Reyer, R., Wolff, G. and Wong, A. (2002). *The World's Water 2002-2003. The Biennial Report on Freshwater Resources*. Island Press, Washington, DC.
Gliessman, S. (2016). Transforming food systems with agroecology. *Agroecology and Sustainable Food Systems* 40(3), 187–189. DOI: 10.1080/21683565.2015.1130765.
Gliessman, S. (2018a). Transforming our food systems. *Agroecology and Sustainable Food Systems* 42(5), 475–476. DOI: 10.1080/21683565.2018.1412568.
Gliessman, S. (2018b). Defining agroecology. *Agroecology and Sustainable Food Systems* 42(6), 599–600.
Gliessman, S., Friedmann, H. and Howard, P. H. (2019). Agroecology and food sovereignty. The IDS Bulletin is published by Institute of Development Studies, Library Road, Brighton BN1 9RE, UK. *IDS Bulletin* 50(2), 91–110.
Gliessman, S. R. (1989). *Agroecology: Researching the Ecological Basis for Sustainable Agriculture*. Springer-Verlag, Berlin.
Gliessman, S. R. (1990). *Agroecology: Researching the Ecological Basis for Sustainable Agriculture*. *Ecological Studies Series* No. 78. Springer, New York.
Gliessman, S. R. (1997). *Agroecology: Ecological Processes in Sustainable Agriculture*. CRC Press, Boca Raton, FL, 384 p.
Gliessman, S. R. (1998). *Agroecology: Ecological Processes in Sustainable Agriculture*. Ann Arbor Press, Ann Arbor, MI.
Gliessman, S. R. (2007). *Agroecology: The Ecology of Sustainable Food Systems*. CRC Press, Taylor & Francis, New York, 384 p.

Gliessman, S. R. (2016). Agroecology: roots of resistance to industrialized food systems. In: Méndez, V. E., Bacon, C. M., Cohen, R. and Gliessman, S. R. (Eds) *Agroecology as a Transdisciplinary, Participatory, and Action-Oriented Approach, Advances in Agroeocology*. CRC Press/Taylor and Francis Group, Boca Raton, FL.

Gliessman, S. R., Garcia-Espinosa, R. E. and Amador, M. A. (1981). The ecological basis for the application of traditional agricultural technology in the management of tropical agroecosystems. *Agro-Ecosystems* 7(3), 173–185.

Global News Wire (2021). Big da analytics market | 2021 size, growth insights, share, COVID-19 impact, emerging technologies, key players, competitive landscape, regional and global forecast to 2028. Available at: https://www.globenewswire.com/news-release/2021/12/16/2353210/0/en/Big-Data-Analytics-Market-2021-Size-Growth-Insights-Share-COVID-19-Impact-Emerging-Technologies-Key-Players-Competitive-Landscape-Regional-and-Global-Forecast-to-2028.html.

GLOPAN (2016). *Food Systems and Diets: Facing the Challenges of the 21st Century*. Global Panel on Agriculture and Food Systems for Nutrition, London.

Glover, J. D., Reganold, J. P. and Andrews, P. K. (2000). Systematic method for rating soil quality of conventional, organic, and integrated apple orchards in Washington State. *Agriculture, Ecosystems and Environment* 80(1–2), 29–45.

GLP (2022). *10 Years into the Lab Grown Meat Revolution Yet There Are Few Approved Foods. What Are the Barriers to Scaling Cell Meat into Blockbusters?* Genetic Literacy Project. 22 September 2022.

Godfray, H. C. J., Beddington, J. R., Crute, I. R., Haddad, L., Lawrence, D., Muir, J. F., Pretty, J., Robinson, S., Thomas, S. M. and Toulmin, C. (2010). Food security: the challenge of feeding 9 billion people. *Science* 327(5967), 812–818.

Gomiero, T., Pimentel, D. and Paoletti, M. G. (2011a). Is there a need for a more sustainable agriculture? *Critical Reviews in Plant Sciences* 30(1–2), 6–23.

Gomiero, T., Pimentel, D. and Paoletti, M. G. (2011b). Environmental impact of different agricultural management practices: conventional vs. organic agriculture. *Critical Reviews in Plant Sciences* 30(1–2), 95–124. DOI: 10.1080/07352689.2011.554355.

Gomiero, T. (2016). Soil degradation, land scarcity and food security: reviewing a complex challenge. *Sustainability* 8(3), 281. DOI: 10.3390/su8030281.

Gomiero, T. (2017). Agriculture and degrowth: state of the art and assessment of organic and biotech-based agriculture from a degrowth perspective. *Journal of Cleaner Production*. DOI: 10.1016/j.jclepro.2017.03.237.

Gomiero, T. (2018). Food quality assessment in organic vs. conventional agricultural produce: findings and issues. *Applied Soil Ecology* 123, 714–728. DOI: 10.1016/j.apsoil.2017.10.014.

Gonçalves, D., Coelho, P., Martinez, L. F. and Monteiro, P. (2021). Nudging consumers toward healthier food choices: a field study on the effect of social norms. *Sustainability* 13(4), 1660. DOI: 10.3390/su13041660.

Gonzales de Molina, M. (2012). Agro-ecology and politics. How to get sustainability? About the necessity for a political agroecology. *Agroecology and Sustainable Food Systems* 37, 45–59.

Gonzalez de Molina, M. and Guzman Casado, G. (2017). Agroecology and ecological intensification: a discussion from a metabolic point of view. *Sustainability* 9(1), 86. DOI: 10.3390/su9010086.

González Esteban, A. L. (2017). Patterns of world wheat trade 1945–2010: the long hangover from the second food regime. *Journal of Agrarian Change* 18(1), 87–111. DOI: 10.1111/joac.12219.

Goodland, R. (1991). The case that the world has reached its limits: more precisely that the current throughput growth in the global economy cannot be sustained. In: Goodland, R., Daly, H., El Serafy, S. and von Droste, B. (Eds) *Environmentally Sustainable Economic Development: Building on Brundtland*. UNESCO, Paris.

Goodman, D. (2000). Organic and conventional agriculture: materializing discourse and agro-ecological managerialism. *Agriculture and Human Values* 17(3), 215–219.

Goodman, D. (2003). The quality turn and alternative food practices: reflections and agenda. *Journal of Rural Studies* 19(1), 1–7.

Goodman, D. (2004). Rural Europe redux? Reflections on alternative agro-food networks and paradigm change. *Sociologia Ruralis* 44(1), 3–16.

Goodman, D., DuPuis, E. M. and Goodman, M. K. (2012). *Alternative Food Networks: Knowledge, Practice, and Politics*. Taylor and Francis, Hoboken, NJ.

Goodman, D. and Goodman, M. (2009). Alternative food networks. In: Kitchen, R. and Thrift, N. (Eds) *International Encyclopaedia of Human Geography* (vol. 8). Elsevier, Oxford. ISBN 978-0-08-044919-7.

Goodman, D. and Redclift, M. (1991). *Refashioning Nature: Food, Ecology and Culture*. Routledge, New York.

Goodman, D., Sorj, B. and Wilkinson, J. (1987). *From Farming to Biotechnology*. Blackwell Publishing, Oxford.

Goodman, D. and Watts, M. J. (Eds) (1997). *Globalising Food: Agrarian Questions and Global Restructuring*. Routledge, London and New York.

Göpel, M. (2016). *The Great Mind Shift: How a New Economic Paradigm and Sustainability Transformations Go Hand In Hand*. Springer, New York.

Gorton, M., Salvioni, C. and Hubbard, C. J. E. (2014). Semi-subsistence farms and alternative food supply chains. *EuroChoices* 13(1), 15–19.

Gorz, A. (1972). *Proceedings of the from a Public Debate Organized in Paris by the Club du Nouvel Observateur*. Nouvel Observateur, Paris, p. IV, 397.

Gottlieb, R. (2009). Where we live, work, play... and eat: expanding the environmental justice agenda. *Environmental Justice* 2(1), 7–8. DOI: 10.1089/env.2009.0001.

GovGrant (2023). How well is Europe playing the cultured meat game? What the numbers tell us. Available at: https://www.govgrant.co.uk/sector-research/how-well-is-europe-playing-the-cultured-meat-game/. Accessed 28 February 2023.

GPAFS (2016). *Food Systems and Diets: Facing the Challenges of the 21st Century*. Global Panel on Agriculture and Food Systems for Nutrition, London.

GPAFS (2020). Rethinking trade policies to support healthier diets. Global Panel on Agriculture and Food Systems for Nutrition. Policy Brief No. 13. February 2020.

Graamans, L., Baeza, E., van den Dobbelsteen, A., Tsafaras, I. and Stanghellini, C. (2018). Plant factories versus greenhouses: comparison of resource use efficiency. *Agricultural Systems* 160, 31–43.

Graamans, L., Tenpierik, M., van den Dobbelsteen, A. and Stanghellini, C. (2020). Plant factories: reducing energy demand at high internal heat loads through façade design. *Applied Energy* 262, 114544.

Graf, S. and Cecchini, M. (2017). Diet, physical activity, and sedentary behaviours: analysis of trends, inequalities and clustering in selected OECD countries. OECD Health Working Papers No. 100. OECD Publishing, Paris. DOI: 10.1787/54464f80-en.

Grain (2011). *Food and Climate Change: The Forgotten Link*. Grain. Available at: https://www.grain.org/e/4357. Accessed 28 September 2011.

Grain and IATP (2018). *Emissions Impossible: How Big Meat and Dairy Are Heating up the Planet*. Grain and the Institute for Agriculture and Trade Policy. Available at: https://www.iatp.org/sites/default/files/2018-07/Emissions%20impossible%20EN%2012-page.pdf.

Grantham, J. (2011). GMO quarterly letter, April 2011. Available at: http://www.scribd.com/doc/54681895/Jeremy-Grantham-Investor-Letter-1Q-2011.

Gray, H. and Nuri, K. R. (2020). Differing visions of agriculture: industrial-chemical vs. small farm and urban organic production. *American Journal of Economics and Sociology*, 79 (3). DOI: 10.1111/ajes.12344. © 2020 American Journal of Economics and Sociology, Inc.

Grayson, R. (2010). *Permaculture Papers 7: Into the New Century*. Pac. Edge. Available at: http://pacific-edge.info/2010/09/permaculture-papers6/. Accessed 30 May 2013.

Green Queen (2023a). Cultivated meat regulation: the 10 most supportive countries, from funding to policy. Available at: https://www.greenqueen.com.hk/cultivated-meat-government-support/. Accessed 23 January 2023.

Green Queen (2023b). Beef lite: impossible foods launches new meat alternative with 45% less fat than animal version. Available at: https://vegconomist.com/products-launches/impossible-foods-beef-lite/. Accessed 8 March 2023.

Green, M. (1976). *Pesticides, Boon, or Bane?* Electra Books, London.

Green, R. E., Cornell, S. J., Scharlemann, J. P. and Balmford, A. (2005). Farming and the fate of wild nature. *Science* 307(5709), 550–555. DOI: 10.1126/science.1106049.

Greenland, D. J. and Szabolcs, I. (Eds) (1994). *Soil Resilience and Sustainable Land Use*. CAB International, Wallingford.

Greenpeace (2020a). *Farming for failure: How European Animal Farming Fuels the Climate Emergency*. Greenpeace EU.

Greenpeace (2020b). False sense of security: why European food systems lack resilience. *Greenpeace European Unit Rue Belliard* 199, 1040. Brussels, Belgium. Available at: www.greenpeace.eu.

Grewal, S. S. and Grewal, P. S. (2012). Can cities become self-reliant in food? *Cities* 29(1), 1–11.

Griffin, M., Sobal, J. and Lyson, T. A. (2009). An analysis of a community food waste stream. *Agriculture and Human Values* 26(1–2), 67–81.

Griffin, T. W., Miller, N. J., Bergtold, J., Shanoyan, A., Sharda, A. and Ciampitti, I. A. (2017). Farm's sequence of adoption of information-intensive precision agricultural technology. *Applied Engineering in Agriculture* 33(4), 521–527. DOI: 10.13031/aea.12228.

Grisa, C., Schmitt, C. J. and Mattei, L. (2011). Brazil's PAA: policy-driven food systems. *Farming Matters* 27(3), 34–36.

Griscom, B. W., Adams, J., Ellis, P. W., Houghton, R. A., Lomax, G., Miteva, D. A., Schlesinger, W. H., Shoch, D., Siikamäki, J. V., Smith, P., Woodbury, P., Zganjar, C., Blackman, A., Campari, J., Conant, R. T., Delgado, C., Elias, P., Gopalakrishna, T., Hamsik, M. R., Herrero, M., Kiesecker, J., Landis, E., Laestadius, L., Leavitt, S. M., Minnemeyer, S., Polasky, S., Potapov, P., Putz, F. E., Sanderman, J., Silvius, M., Wollenberg, E. and Fargione, J. (2017). Natural climate solutions. *Proceedings of the National Academy of Sciences of the United States of America* 114(44), 11645–11650.

Guan, H., Chi, D., Yu, J. and Li, X. (2008). A novel photodegradable insecticide: preparation, characterization, and properties evaluation of nano-imidacloprid. *Pesticide Biochemistry and Physiology* 92(2), 83-91.

Guardian (2017). Organic or starve: can Cuba's new farming model provide food security? *The Guardian*, 28 October.

Guardian (2020a). Lords seek to allow gene-editing in UK 'to produce healthy, hardier crops'. Available at: https://www.theguardian.com/science/2020/jun/14/lords-seek-to-allow-gene-editing-in-uk-to-produce-healthy-hardier-crops.

Guardian (2020b). UN under fire over choice of 'corporate puppet' as envoy at key food summit. Available at: https://www.theguardian.com/global-development/2020/mar/12/un-under-fire-over-choice-of-corporate-puppet-as-envoy-at-key-food-summit.

Guardian (2020c). Unhealthy snacks to be banned from checkouts at supermarkets in England. Available at: https://www.theguardian.com/business/2020/dec/28/unhealthy-snacks-to-be-banned-from-checkouts-supermarkets-in-england. Accessed 10 July 2021.

Guardian (2021). 'It's not as carbon-hungry': UK's largest sunlit vertical farm begins harvest. *Guardian*. Available at: https://www.theguardian.com/environment/2021/oct/18/its-not-as-carbon-hungry-uks-largest-sunlit-vertical-farm-begins-harvest.

Guardian (2022a). EU wastes 153m tonnes of food a year – much more than it imports, says report. *Guardian*, 20 September 2022.

Guardian (2022b). 'Fishless fish': the next big trend in the seafood industry, 22 October.

Guardian (2022c). Why the ancient art of gleaning is making a comeback across England. . Available at: https://www.theguardian.com/environment/2022/feb/19/harvest-for-all-why-ancient-art-of-gleaning-is-making-a-comeback-food-banks-food-waste. Accessed 19 February 2022.

Guardian (2022d). The rise of vertical farms: could indoor plant factories be the norm in 10 years? Available at: https://www.theguardian.com/environment/2022/aug/21/the-rise-of-vertical-farms-could-indoor-plant-factories-be-the-norm-in-10-years. Accessed 24 August 2022.

Guardian (2023). Global freshwater demand will outstrip supply by 40% by 2030, say experts. *The Guardian*. Available at: https://www.theguardian.com/environment/2023/mar/17/global-fresh-water-demand-outstrip-supply-by-2030. Accessed 17 March 2023.

Guerra, F. D., Attia, M. F., Whitehead, D. C. and Alexis, F. (2018). Nanotechnology for environmental remediation: materials and applications. *Molecules* 23(7), 1760.

Gunapala, N. and Scow, K. M. (1998). Dynamics of soil microbial biomass and activity in conventional and organic farming systems. *Soil Biology and Biochemistry* 30(6), 805–816.

Gupta, V., Roper, M. and Thompson, J. (2019). Harnessing the benefits of soil biology in conservation agriculture. In: Pratley, J. and Kirkegaard, J. (Eds) *Australian Agriculture in 2020: From Conservation to Automation*. Agronomy Australia and Charles Sturt University, Wagga Wagga, pp. 237–253.

Gustavsson, J., Cederberg, C., Sonesson, U., Van Otterdijk, R. and Meybeck, A. (2011). *Global Food Losses and Food Waste. Extent, Causes and Prevention*. Food and Agriculture Organization of the United Nations, Rome and Swedish Institute for Food and Biotechnology (SIK).

Guthman, J. (2000). Raising organic: an agro-ecological assessment of grower practices in California. *Agriculture and Human Values* 17(3), 257-266. DOI: 10.1023/A:1007688216321.

Guthman, J. (2004). *Agrarian Dreams: The Paradox of Organic Farming in California*. University of California Press, Berkeley, CA.

Haddad, L., Hawkes, C., Webb, P., Thomas, S., Beddington, J., Waage, J. and Flynn, D. (2016). A new global research agenda for food. *Nature* 540(7631), 30-32.

Hadi Ishak, A., Hajjaj, S. S. H., Rao Gsangaya, K., Thariq Hameed Sultan, M., Fazly Mail, M. and Seng Hua, L. (2023). Autonomous fertilizer mixer through the Internet of Things (IoT). *Materials Today: Proceedings*. 81(2), 295-301

Haggblade, S. and Tembo, G. (2003). Conservation farming in Zambia. EPTD Discussion Paper Number 108. International Food Policy Research Institute (IFPRI), Washington, DC.

Hajjaj, S. S. H. and Sahari, K. S. M. (2016). Review of agriculture robotics: practicality and feasibility. In: Proceedings of the 2016 IEEE International Symposium on Robotics and Intelligent Sensors (IRIS), Tokyo, Japan, 17-20 December 2016, pp. 194-198.

Halberg, N., Alroe, H. F., Knudsen, M. T. and Kristensen, E. S. (Eds) (2006). *Global Development of Organic Agriculture: Challenges and Prospects*. CABI Publishing, Wallingford.

Hall, K. D., Ayuketah, A., Brychta, R., Cai, H., Cassimatis, T., Chen, K. Y., Chung, S. T. (2019). Ultra-processed diets cause excess calorie intake and weight gain: an inpatient randomized controlled trial of ad libitum food intake. *Cell Metabolism* 30(1), 67-77. DOI: 10.1016/j.cmet.2019.05.008e63.

Hallmann, C. A., Foppen, R. P., van Turnhout, C. A., de Kroon, H. and Jongejans, E. (2014). Declines in insectivorous birds are associated with high neonicotinoid concentrations. *Nature* 511(7509), 341-343. DOI: 10.1038/nature13531.

Hallmann, C. A., Sorg, M. and Jongejans, E. (2017). More than 75 percent decline over 27 years in total flying insect biomass in protected areas. *PLoS ONE* 12, 0185809-0185821.

Halweil, B. (2004). *Eat Here: Reclaiming Homegrown Pleasures in a Global Supermarket*. Norton, New York.

Hansen, B., Alrøe, H. F. and Kristensen, E. S. (2001). Approaches to assess the environmental impact of organic farming with particular regard to Denmark. *Agriculture, Ecosystems and Environment* 83(1-2), 11-26.

Hansen, H. O. (2019). The agricultural treadmill - a way out through differentiation? An empirical analysis of organic farming and the agricultural treadmill. *Journal of Tourism, Heritage and Services Marketing* 5(2), 20-26. ISSN 2529-1947. DOI: 10.5281/zenodo.3601667.

Hansen, J., Mason, S. J., Sun, L. and Tall, A. (2011). Review of seasonal climate forecasting for agriculture in sub-Saharan Africa. *Experimental Agriculture* 47(2), 205-240.

Hansen, J., Hellin, J., Rosenstock, T., Fisher, E., Cairns, J., Stirling, C., Lamanna, C., van Etten, J., Rose, A. and Campbell, B. (2019). Climate risk management and rural poverty reduction. *Agricultural Systems* 172, 28-46.

Hansen-Kuhn, K. and Suppan, S. (2013). *Promises and Perils of the TTIP. Negotiating a Transatlantic Agricultural Market*. Institute for Agriculture and Trade Policy, Minneapolis, MN and Heinrich Böll Foundation, Berlin.

Hardin, G. (1968). The tragedy of the commons. *Science* 162(3859), 1243-1248.

Hardin, G. (1993). *Living within Limits: Ecology, Economics, and Population Taboos*. Oxford University Press, New York.

Hargreaves, T., Longhurst, N. and Seyfang, G. (2013). Up, down, round and round: connecting regimes and practices in innovation for sustainability. *Environment and Planning A* 45(2), 402–420.

Häring, A. M., Dabbert, S., Aurbacher, J., Bichler, B., Eichert, C., Gambelli, D., Lampkin, N., Offermann, F., Olmos, S., Tuson, J. and Zanoli, R. (2004). Organic farming and measures of European agricultural policy. *Organic Farming in Europe.: Economics & Policy* 11. University of Hohenheim, Department of Farm Economics, Hohenheim.

Hart, A. K., McMichael, P., Milder, J. C. and Scherr, S. J. (2015). Multi-functional landscapes from the grassroots? The role of rural producer movements. *Agriculture and Human Values* 33(2), 305–322.

Hart, R. (1996). *Forest Gardening*. Green Books, Cambridge, p. 41. ISBN 978-1-60358050-2.

Hartmann, M., Frey, B., Mayer, J., Mäder, P. and Widmer, F. (2015). Distinct soil microbial diversity under long-term organic and conventional farming. *ISME Journal* 9(5), 1177–1194. DOI: 10.1038/ismej.2014.210.

Harwatt, H., Sabaté, J., Eshel, G., Soret, S. and Ripple, W. (2017). Substituting beans for beef as a contribution toward US climate change targets. *Climatic Change* 143(1–2), 261–270. DOI: 10.1007/s10584-017-1969-1.

Hassanein, N. (2003). Practicing food democracy: a pragmatic politics of transformation. *Journal of Rural Studies* 19(1), 77–86.

Hatfield, J. (2007). Beyond the edge of the field. *Presidents Message-Arch. Soil Science Soceity of America*. Available at: https://www.soils.org/about-society/presidentsmessage/archive/13. Accessed 29 May 2013.

Hattab, S., Bougattass, I., Hassine, R. and Dridi-Al-Mohandes, B. (2019). Metals and micronutrients in some edible crops and their cultivation soils in eastern-central region of Tunisia: a comparison between organic and conventional farming. *Food Chemistry* 270, 293–298.

Hatton, T. J. and Nulsen, R. A. (1999). Towards achieving functional ecosystem mimicry with respect to water cycling in southern Australian agriculture. *Agroforestry Systems* 45(1/3), 203–214. DOI: 10.1023/A:1006215620243.

Hauser, M. and Lindtner, M. (2017). Organic agriculture in postwar Uganda: emergence of pioneer-led niches between 1986 and 1993. *Renewable Agriculture and Food Systems* 32(2), 169–178.

Hawkes, C. (2005). The role of foreign direct investment in the nutrition transition. *Public Health Nutrition* 8(4), 357–365.

Hawkes, C. (2006). Uneven dietary development: linking the policies and processes of globalization with the nutrition transition, obesity, and diet-related chronic diseases. *Globalization and Health*, 2(1), 4.

Hawkes, C. (2007). *Globalization, Food, and Nutrition Transitions*. Globalization and Health Knowledge Network, Commission on the Social Determinants of Health, World Health Organization, Geneva.

Hawkes, C. (2008). Dietary implications of supermarket development: a global perspective. *Development Policy Review* 26(6), 657–692.

Hawkes, C. (2010). The influence of trade liberalization and global dietary change: the case of vegetable oils, meat, and highly processed foods. In: Hawkes, C. (Ed.) *Trade, Food, Diet and Health: Perspectives and Policy Options*. John Wiley & Sons Ltd, Chichester, pp. 35–59.

Hawkes, C. (2015). *Nutrition in the Trade and Food Security Nexus*. Food and Agriculture Organization, Rome.

Hawkes, C., Chopra, M. and Friel, S. (2009). Globalization, trade, and the nutrition transition. In: Labonte, R., Schrecker, T., Packer, C. and Runnels, V. (Eds) *Globalization and Health: Pathways, Evidence and Policy*. Routledge, New York.

Hawkes, C., Eckhardt, C., Ruel, M. and Minot, N. (2005). Diet quality poverty, and food policy: a new research agenda for obesity prevention in developing countries. *SCN News* 29, 20–22.

Hayes (1991). Pesticide regulation. Available at: http://feql.wsu.edu/entom558/110303ppt.pdf.

Haysom, G., Olsson, E. G. A., Dymitrow, M., Opiyo, P., Buck, N. T., Oloko, M., Spring, C., Fermskog, K., Ingelhag, K., Kotze, S. and Agong, S. G. (2019). Food systems sustainability: an examination of different viewpoints on food system change. *Sustainability* 11(12), 3337. DOI: 10.3390/su11123337.

Hazra, D. K., Karmakar, R., Poi Rajlakshmi, B. S. and Mondal, S. (2017). Recent advances in pesticide formulations for eco-friendly and sustainable vegetable pest management: a review. *Archives of Agriculture and Environmental Science* 2(3), 232-237.

He, X., Deng, H. and Hwang, H. M. (2019). The current application of nanotechnology in food and agriculture. *Journal of Food and Drug Analysis* 27(1), 1-21.

Headey, D. (2013). Developmental drivers of nutritional change: a cross-country analysis. *World Development* 42, 76–88. DOI: 10.1016/j.worlddev.2012.07.002.

Health Canada (2018). Canadian ban on trans fats comes into force today - Canada.ca . Available at: https://www.canada.ca/en/health-canada/news/2018/09/canadian-ban-on-trans-fats-comes-into-force-today.html. Accessed 31 January 2020.

Heap, I. (2014). Herbicide resistant weeds. In: Pimentel, D. and Peshin, R. (Eds) *Integrated Pest Management*. Springer Netherlands, Dordrecht, pp. 281-301.

Heath, J. R. (1985). El Programa Nacional de Alimentación y la crisis de alimentos. *Revista Mexicana de Sociología* 47(3), 115–135.

Hecht, S. B. (1989). The sacred cow in the green hell: livestock and forest conversion in the Brazilian Amazon. *Ecologist* 19, 229–234.

Hecht, S. B. (1995). The evolution of agroecological thought. In: Altieri, M. A. (Ed.) *Agroecology: The Science of Sustainable Agriculture*. Westview Press, Boulder, CO, pp. 1–19.

Heckenmüller, M., Narita, D. and Klepper, G. (2014). Global availability of phosphorus and its implications for global food supply: an economic overview. Kiel Working Paper, No. 1897. Kiel Institute for the World Economy (IfW). Kiel.

Heckmann, L. H., Hovgaard, M. B., Sutherland, D. S., Autrup, H., Besenbacher, F. and Scott-Fordsmand, J. J. (2011). Limit-test toxicity screening of selected inorganic nanoparticles to the earthworm *Eisenia fetida*. *Ecotoxicology* 20(1), 226-233.

Hedberg, R. C. and Zimmerer, K. S. (2020). What's the market got to do with it? Social-ecological embeddedness and environmental practices in a local food system initiative. *Geoforum* 110, 35–45.

Hedenus, F., Wirsenius, S. and Johansson, D. J. A. (2014). The importance of reduced meat and dairy consumption for meeting stringent climate change targets. *Climatic Change* 124(1-2), 79–91. DOI: 10.1007/s10584-014-1104-5.

Heisler, J., Glibert, P., Burkholder, J., Anderson, D., Cochlan, W., Dennison, W., Gobler, C., Dortch, Q., Heil, C., Humphries, E., Lewitus, A., Magnien, R., Marshall, H., Sellner, K., Stockwell, D., Stoecker, D. and Suddleson, M. (2008). Eutrophication and harmful algal blooms: a scientific consensus. *Harmful Algae* 8(1), 3–13.

Heltberg, R., Hossain, N., Reva, A. and Turk, C. (2013). Coping and resilience during the food, fuel, and financial crisis. *The Journal of Development Studies* 49(5), 705–718.
Hemenway, T. (2009). *Gaia's Garden: A Guide to Home-Scale Permaculture* (2nd edn). Chelsea Green, White River Junction.
Henao, J. and Baanante, C. (2006). Agricultural production and soil nutrient mining in Africa: implications for resource conservation and policy development. IFDC Tech. Bull. International Fertiliser Development Centre, Muscle Shoals, AL.
Hénin, S. (1967). Les acquisitions techniques en production végétale et leurs applications. *Économie Rurale*, 74, 37–44. SFER, Paris.
Herforth, A., Bai, Y., Venkat, A., Mahrt, K., Ebel, A. and Masters, W. A. (2020). Cost and affordability of healthy diets across and within countries. Background paper for The State of Food Security and Nutrition in the World 2020. FAO Agricultural Development Economics Technical Study No. 9. Food and Agriculture Organization, Rome. DOI: 10.4060/cb2431en.
Hermans, F. (2011). Social learning in innovation networks: how multisectoral collaborations shape discourses of sustainable agriculture. PhD thesis. Wageningen University, The Netherlands.
Hernandez, M. A. and Torero, M. (2010). Examining the dynamic relationship between spot and future prices of agricultural commodities. IFPRI Discussion Paper 00988. International Food Policy Research Institute, Washington, DC.
Hernandez, M. A. and Torero, M. (2013). Market concentration and pricing behaviour in the fertilizer industry: a global approach. *Agricultural Economics* 44(6), 723–734.
Hernández, M. Y., Macario, P. A. and L´opez-Martínez, J. O. (2017). Traditional agroforestry systems and food supply under the food sovereignty approach. *Ethnobiology Letters* 8(1), 125–141. DOI: 10.14237/ebl.8.1.2017.941.
Herren, H. R., Haerlin, B. and IAASTD+10 Advisory Group (Eds) (2020). Transformation of our food systems. Zukunftsstiftung Landwirtschaft/Biovision, Berlin/Zurich. Available at: https://www.globalagriculture.org/fileadmin/files/weltagrarbericht/IAASTD-Buch/PDFBuch/BuchWebTransformationFoodSystems.pdf.
Herrero, M., Havlík, P., Valin, H., Notenbaert, A., Rufino, M. C., Thornton, P. K., Blümmel, M., Weiss, F., Grace, D. and Obersteiner, M. (2014). Biomass use, production, feed efficiencies, and greenhouse gas emissions from global livestock systems. *Proceedings of the National Academy of Sciences of the United States of America* 110(52), 20888–20893.
Herrero, M., Hugas, M., Lele, U., Wira, A. and Torero, M. (2020a). Shift to healthy and sustainable consumption patterns – a paper on action track 2 – October 26th, 2020. United Nations Food Systems Summit 2021 Scientific Group. Available at: https://www.un.org/en/food-systems-summit/leadership#scientific-group.
Herrero, M., Thornton, P. K., Mason-D'Croz, D., Palmer, J., Benton, T. G., Bodirsky, B. L., Bogard, J. R., Hall, A., Lee, B., Nyborg, K., Pradhan, P., Bonnett, G. D., Bryan, B. A., Campbell, B. M., Christensen, S., Clark, M., Cook, M. T., de Boer, I. J. M., Downs, C., Dizyee, K., Folberth, C., Godde, C. M., Gerber, J. S., Grundy, M., Havlik, P., Jarvis, A., King, R., Loboguerrero, A. M., Lopes, M. A., McIntyre, C. L., Naylor, R., Navarro, J., Obersteiner, M., Parodi, A., Peoples, M. B., Pikaar, I., Popp, A., Rockström, J., Robertson, M. J., Smith, P., Stehfest, E., Swain, S. M., Valin, H., van Wijk, M., van Zanten, H. H. E., Vermeulen, S., Vervoort, J. and West, P. C. (2020b). Innovation can accelerate the transition towards a sustainable food system. *Nature Food* 1(5), 266-272. ISSN 2662-1355. DOI: 10.1038/s43016-020-0074-1.

Hertel, T. W., Ramankutty, N. and Baldos, U. L. (2014). Global market integration increases likelihood that a future African Green Revolution could increase crop land use and CO_2 emissions. *Proceedings of the National Academy of Sciences of the United States of America* 111(38), 13799-13804.

Hicks, J. (1992). DDT - friend or foe? *Pesticide News, Journal of the Pesticide Trust*, 17 September.

Hielscher, H. (2017). German groceries: the organic retail revolution. *The European Business Daily Handelsblatt Global*, 22 January.

Higgins, C. (2022). You know what a greenhouse and vertical farm are, but what is controlled environment agriculture (CEA)? *Urban AG News*, 20 February.

Higgins, V., Bryant, M., Howell, A. and Battersby, J. (2017). Ordering adoption: materiality, knowledge, and farmer engagement with precision agriculture technologies. *Journal of Rural Studies* 55, 193-202.

HighQuest Partners (2010). Private financial sector investment in farmland and agricultural infrastructure. Organisation for Economic Co-operation and Development (OECD), Paris. Available at: http://www.oecd.org/officialdocuments/publicdisplaydocumentpdf/?cote=TAD/CA/APM/WP(2010)11/FINAL&docLanguage=En.

Hinrichs, C. C. (2000). Embeddedness and local food systems: notes on two types of direct agricultural market. *Journal of Rural Studies* 16(3), 295-303.

Hinrichs, C. C. (2003). The practice and politics of food system localization. *Journal of Rural Studies* 19(1), 33-45.

Hirschfeld, S. and Van Acker, R. (2020). Permaculture farmers consistently cultivate perennials, crop diversity, landscape heterogeneity and nature conservation. *Renewable Agriculture and Food Systems* 35(3), 342-351.

Hirvonen, K., Bai, Y., Headey, D. and Masters, W. A.(2019). Affordability of the EAT - lancet reference diet: a global analysis. *The Lancet Global Health* 19, 1-8. DOI: 10.1016/S2214-109X(19)30447-4.

HLPE (2013). Investing in smallholder agriculture for food security. A Report by the High-Level Panel of Experts on Food Security and Nutrition of the Committee on World Food Security. Food and Agriculture Organization of the United Nations, Rome.

HLPE (2014). Food losses and waste in the context of sustainable food systems. A Report by the High-Level Panel of Experts on Food Security and Nutrition of the Committee on World Food Security. HLPE, Rome.

HLPE (2016). Sustainable agricultural development for food security and nutrition: what roles for livestock? A Report by the High Level Panel of Experts on Food Security and Nutrition.

HLPE (2017). High level panel of experts on food security and nutrition. Nutrition and food systems. A Report by the High-Level Panel of Experts on Food Security and Nutrition. Committee on World Food Security, Rome, Italy.

HLPE (2019). Agroecological and other innovative approaches for sustainable agriculture and food systems that enhance food security and nutrition. A Report by the High-Level Panel of Experts (HLPE) on Food Security and Nutrition of the Committee on World Food Security, Rome.

Hoddinott, J., Berhane, G., Gilligan, D. O., Kumar, N. and Taffesse, A. S. (2012). The impact of Ethiopia's productive safety net programme and related transfers on agricultural productivity. *Journal of African Economies* 21(5), 761-786.

Hodgson, J. A., Kunin, W. E., Thomas, C. D., Benton, T. G. and Gabriel, D. (2010). Comparing organic farming and land sparing: optimizing yield and butterfly populations at a landscape scale. *Ecology Letters* 13(11), 1358–1367.

Hoechst (2003). Archive. Available at: hoechst.com/English/index.html. Accessed 21 November 2003.

Hoekstra, A. J. (2012). The hidden water resource use behind meat and dairy. Twente Water Centre, University of Twente, The Netherlands, p. 7.

Hoekstra, A. Y. and Mekonnen, M. M. (2012). The water footprint of humanity. *Proceedings of the National Academy of Sciences of the United States of America* 109(9), 3232–3237. DOI: 10.1073/pnas.1109936109.

Hoey, L. and Sponseller, A. (2018). It's hard to be strategic when your hair is on fire: alternative food movement leaders' motivation and capacity to act. *Agriculture and Human Values* 35(3), 595–609.

Hofman, K. (2021). GrowUp Farms aims to change food systems. Available at: https://www.agri-tech-e.co.uk/2021/01/29/. Accessed 29 January 2021.

Holden, N. M., White, E. P., Lange, M. C. and Oldfield, T. L. (2018). Review of the sustainability of food systems and transition using the internet of food. *Science of Food* 2(18). DOI: 10.1038/s41538-018-0027-3.

Holland, L. (2004). Diversity and connections in community gardens: a contribution to local sustainability. *Local Environment* 9(3), 285–305.

Holleran, C. (2012). *Shopping in Ancient Rome: The Retail Trade in the Late Republic and the Principate*. Oxford University Press, Oxford.

Holling, C. S., Berkes, F. and Folke, C. (1998). Science, sustainability, and resource management. In: Berkes, F. (Ed.) *Linking Social and Ecological Systems*. Cambridge University Press, Cambridge, pp. 346–362.

Holloway, L., Kneafsey, M., Venn, L., Cox, R., Dowler, E. and Tuomainen, H. (2007). Possible food economies: a methodological framework for exploring food production–consumption relationships. *Sociologia Ruralis* 47(1), 1–19. DOI: 10.1111/j.1467-9523.2007.00427.x.

Holmes, H. (2021). Regenerative agriculture: why PepsiCo, McDonald's and Waitrose are jumping on regen ag. *The Crocer*. Available at: https://www.thegrocer.co.uk/sustainability-and-environment/regenerative-agriculture-why-pepsico-mcdonalds-and-waitrose-are-jumping-on-regen-ag/659119.article. Accessed 23 August 2021.

Holmes, M. (2001). *Melancholy Consequences: Britain's Long Relationship with Agricultural Chemicals Since the Mid-Eighteenth Century*. White Horse Press, Cambridge.

Holmgren, D. (1992). Uncommon sense: development of the permaculture concept. *Permaculture International Journal* 44, 26–30.

Holmgren, D. (2002). *Permaculture: Principles and Pathways Beyond Sustainability* (1st edn). Holmgren Design Services, Hepburn. ISBN-13 978-0646418445.

Holmgren, D. (2004). *Permaculture: Principles and Pathways Beyond Sustainability*. Holmgren Design, Hepburn.

Holt-Giménez, E. (2001). Measuring farms' agroecological resistance to Hurricane Mitch. *Leisa* 17, 7–10.

Holt-Giménez, E. (2002). Measuring farmers' agroecological resistance after Hurricane Mitch in Nicaragua: a case study in participatory, sustainable land management impact monitoring. *Agriculture, Ecosystems and Environment* 93(1–3), 87–105.

Holt-Giménez, E. (2006). *Campesino a Campesino: Voices from Latin America's Farmer to Farmer Movement for Sustainable Agriculture*. Food First Books, Oakland.

Holt-Giménez, E. (2009). Linking farmers movements for advocacy and practice. *Journal of Peasant Studies* 37(1), 203-236.

Holt-Giménez, E. (2019). Capitalism, food, and social movements: the political economy of food system transformation. *Journal of Agriculture, Food Systems, and Community Development* 9 (Suppl. 1), 23–35. DOI: 10.5304/jafscd.2019.091.043.

Holt-Giménez, E. and Altieri, M. A. (2013). Agroecology, food sovereignty, and the new green revolution. *Agroecology and Sustainable Food Systems* 37(1), 90–102. DOI: 10.1080/10440046.2012.716388.

Holt-Giménez, E. and Shattuck, A. (2011). Food crises, food regimes and food movements: rumblings of reform or tides of transformation? *Journal of Peasant Studies* 38(1), 109–144. DOI: 10.1080/03066150.2010.538578.

Holweg, C., Lienbacher, E. and Schnedlitz, P. (2010). Social supermarkets: typology within the spectrum of social enterprises. In: Miller, K. W. and Mills, M. K. (Eds) *Doing More with Less*. ANZMAC, Christchurch.

Holweg, C. and Lienbacher, E. (2011). Social marketing innovation: new thinking in retailing. *Journal of Nonprofit and Public Sector Marketing* 23(4), 307-326.

Honrado, J. L. E., Solpico, D. B., Favila, C. M., Tongson, E., Tangonan, G. L. and Libatique, N. J. C. (2017). UAV Imaging with low-cost multispectral imaging system for precision agriculture applications. In: Proceedings of the 2017 IEEE Global Humanitarian Technology Conference (GHTC), San Jose, CA, USA, 19 October 2017.

Horlings, L. G. and Marsden, T. K. (2011). Towards the real green revolution? Exploring the conceptual dimensions of a new ecological modernisation of agriculture that could 'feed the world'. *Global Environmental Change* 21(2), 441–452.

Hosonuma, N., Herold, M., De Sy, V., De Fries, R. S., Brockhaus, M., Verchot, L., Angelsen, A. and Romijn, E. (2012). An assessment of deforestation and forest degradation drivers in developing countries. *Environmental Research Letters* 7(4), 044009, DOI: 10.1088/1748-9326/7/4/044009.

Hough, P. (1998). *The Global Politics of Pesticides. Forging Consensus from Conflicting Interests*. Earthscan Publications Ltd, London.

Howard, A. (1940). *An Agricultural Testament*. Oxford University Press, London.

Howard, A. (1943). *An Agricultural Testament*. Oxford University Press, London. Available at: http://journeytoforever.org/farm library/howardAT/ATtoc.html.

HSE (2013). *Health Survey for England – 2012*. Web Master, London. Available at: http://www.hscic.gov.uk/catalogue/PUB13219.

Hu, R., Liang, Q., Pray, C., Huang, J. and Jin, Y. H.(2011). Privatization, public R & D policy, and private R & D investment in China's agriculture. *Journal of Agricultural and Resource Economics* 36, 416–432.

Hu, W., Wan, L., Jian, Y., Ren, C., Jin, K., Su, X., Bai, X., Haick, H., Yao, M. and Wu, W. (2019). Electronic noses: from advanced materials to sensors aided with data processing. *Advanced Materials Technologies* 4(2), 1800488.

Huambachano, M. (2018). Enacting food sovereignty in Aotearoa New Zealand and Peru: revitalizing Indigenous knowledge, food practices and ecological philosophies. *Agroecology and Sustainable Food Systems* 42(9), 1003–1028. DOI: 10.1080/21683565.2018.1468380.

Huang, H., Lan, Y., Yang, A., Zhang, Y., Wen, S. and Deng, J. (2020). Deep learning versus Object-based Image Analysis (OBIA) in weed mapping of UAV imagery. *International Journal of Remote Sensing* 41(9), 3446–3479.

Huang, Q., Yu, H. and Ru, Q. (2010). Bioavailability and delivery of nutraceuticals using nanotechnology. *Journal of Food Science* 75(1), R50–R57. DOI: 10.1111/j.1750-3841.2009.01457.x.

Hugenschmidt, C. E. (2016). Type 2 Diabetes, obesity, and risk for dementia: recent insights into brain insulin resistance and hypometabolism. *Current Behavioral Neuroscience Reports* 3(4), 293–300. DOI: 10.1007/s40473-016-0093-2.

Hughner, R. S., McDonagh, P., Prothero, A., Shultz, C. J. and Stanton, J. (2007). Who are organic food consumers? A compilation and review of why people purchase organic food. *Journal of Consumer Behaviour* 6(2–3), 94–110.

Humberto, M. B., de Souza, E. G., Wendel, K. O., Moreira, R. S., Bazzi, C. L. and Rodrigues, M. (2022). Fertilizer recommendation methods for precision agriculture – a systematic literature study. *Engenharia Agrícola, Jaboticabal* 42, e20210185.

Hunt, J. R., Celestina, C. and Kirkegaard, J. A. (2020). The realities of climate change, conservation agriculture and soil carbon sequestration. *Global Change Biology* 26(6), 3188–3189. DOI: 10.1111/gcb.15082.

Hunter, D., Foster, M., McArthur, J. O., Ojha, R., Petocz, P. and Samman, S. (2011). Evaluation of the micronutrient composition of plant foods produced by organic and conventional agricultural methods. *Critical Reviews in Food Science and Nutrition* 51(6), 571–582.

Hunter, M. C., Smith, R. G., Schipanski, M. E., Atwood, L. W. and Mortensen, D. A. (2017). Agriculture in 2050: recalibrating targets for sustainable intensification. *BioScience* 67(4), 386–391.

Hutchison, W. D., Burkness, E. C., Mitchell, P. D., Moon, R. D., Leslie, T. W., Fleischer, S. J., Abrahamson, M., Hamilton, K. L., Steffey, K. L., Gray, M. E., Hellmich, R. L., Kaster, L. V., Hunt, T. E., Wright, R. J., Pecinovsky, K., Rabaey, T. L., Flood, B. R. and Raun, E. S. (2010). Areawide suppression of European corn borer with Bt maize reaps savings to non-Bt maize grower. *Science* 330(6001), 222–225. DOI: 10.1126/science.1190242.

Hvitsand, C. (2016). Community supported agriculture (CSA) as a transformational act: distinct values and multiple motivations among farmers and consumers. *Agroecology and Sustainable Food Systems* 40(4), 333–351. DOI: 10.1080/21683565.2015.1136720.

IAASTD (2009a). *International Assessment of Agricultural Knowledge, Science and Technology for Development*. Global Report. Island Press, Washington, DC.

IAASTD (2009b). *International Assessment of Agricultural Knowledge, Science and Technology for Development*. Executive Summary of the Synthesis Report. Island Press, Washington, DC.

IATP (2008). *Commodities Market Speculation: The Risk to Food Security and Agriculture*. Institute for Agriculture and Trade Policy, Minneapolis, MN.

IATP (2009). *Betting against Food Security: Futures Market Speculation*. Trade and Global Governance Programme Paper (January). Institute for Agriculture and Trade Policy, Minneapolis, MN.

IATP (2013). *Scaling up Agroecology. Toward the Realization of the Right to Food*. Institute for Agriculture and Trade Policy, Minneapolis, MN.

ICIS (2007). Abraaj Capital buys Egyptian fertilizer firm EFC. *ICIS*. Available at: http://www.icis.com/Articles/2007/06/08/4503061/abraaj+capital+buys+egyptian+fertilizer+firm+efc.html. Accessed 11 June 2007.

IDDRI (2018). *Une Europe agroécologique en 2050: une agriculture multifonctionnelle pour une alimentation saine*. Institute de Développement Durable et des Relations

Internationales, Paris. Available at: https://www.iddri.org/sites/default/files/PDF/Publications/Catalogue%20Iddri/Etude/201809-ST0918-tyfa_1.pdf.

IDH (2020). Transitioning to regenerative agriculture with smallholders. *Blog27*, November.

IFA (2015). *Declaration of the International Forum for Agroecology*. Nyéléni, Mali. International Forum for Agroecology. Available at: https://www.fao.org/family-farming/detail/en/c/341387/. Accessed 9 April 2019.

IFAD (2011). Climate-smart smallholder agriculture: what's different? Available at: https://www.ifad.org/Documents/10180/65e06cd3-5b59-4192-8416-a7089d91630c.

IFC (2011). *African Entrepreneur, Backed by IFC Private Equity, Wins Award from Ernst and Young*. International Finance Corporation (IFC), New York. Available at: http://www.ifc.org/wps/wcm/connect/region__ext_content/regions/sub-saharan+africa/news/african_entrepreneur_backed_by_ifc_private_equity_wins_award_from_ernst_young.

IFPRI (2011). Global hunger index: the challenge of hunger: taming price spikes and excessive food price volatility. International Food Policy Research Institute, Washington, DC. Available at: http://www.ifpri.org/sites/default/files/publications/ghi11.pdf.

IFPRI (2017). *Global Food Policy Report*. International Food Policy Research Institute, Washington, DC.

IGS (2019). *Global Sustainable Development Report 2019: The Future Is Now – Science for Achieving Sustainable Development*. Independent Group of Scientists appointed by the Secretary-General. United Nations, New York.

Ilbery, B. W. and Bowler, I. R. (1998). From agricultural productivism to post-productivism. In: Ilbery, B. (Ed.) *The Geography of Rural Change*. Longman, Harlow, pp. 57–84.

Ilbery, B. W. and Kneafsey, M. (1998). Product and place: promoting quality products and services in the lagging regions of the EU. *European Urban and Regional Studies* 5(4), 329–341.

Ilbery, B. W. and Kneafsey, M. (2000). Producer constructions of quality in regional speciality foods production: a case study from southwest England. *Journal of Rural Studies* 16(2), 217–230.

Ilbery, B. W. and Maye, D. (2005a). Alternative (shorter) food supply chains and specialist livestock products in the Scottish–English borders. *Environment and Planning A* 37(5), 823–844.

Ilbery, B. W. and Maye, D. (2005b). Food supply chains and sustainability: evidence from specialist food producers in the Scottish/English borders. *Land Use Policy* 22(4), 331–344.

Ilbery, B. W. and Maye, D. (2006). Retailing local food in the Scottish–English borders: a supply chain perspective. *Geoforum* 37(3), 352–367. DOI: 10.1016/j.geoforum.2005.09.003.

Illich, I. (1975). *Tools for Conviviality*. Fontana/Collins, Glasgow.

Impey, L. (2021a). Regenerative farming data show variable costs cut by 18%. *Farmers Weekly*. Available at: https://www.fwi.co.uk/arable/regenerative-farming-data-show-variable-costs-cut-by-nearly-20. Accessed 3 July 2021.

Impey, L. (2021b). Regenerative agriculture: tips on building carbon-rich soils. *Farmers Weekly*. Available at: https://www.fwi.co.uk/arable/land-preparation/soils/regenerative-agriculture-tips-on-building-carbon-rich-soils. Accessed 24 November 2021.

Impossible Foods (2022). Home Page. Available at: https://impossiblefoods.com/products. Accessed 27 January 2022.

Independent (2017). Vertical farming: is this the answer to the world's food shortage? *Independent*, 3 March 2017, 19:01.

Independent (2018). Why vertical farming isn't a miracle solution to food security. *Independent*, 25 September 2018. Available at: https://www.independent.co.uk/climate-change/news/vertical-farming-solution-food-security-hydroponics-aquaponics-urban-agriculture-a8547561.html.

Independent (2022). Restrictions on shop junk food displays come into effect. *Independent*, 30 September 2022, 12:17 BST. Available at: https://www.independent.co.uk/business/restrictions-on-shop-junk-food-displays-come-into-effect-b2183284.html.

Ingram, J. (2011). A food systems approach to researching food security and its interactions with global environmental change. *Food Security* 3(4), 417–431.

Ingram, J. (2015). Framing niche-regime linkage as adaptation: an analysis of learning and innovation networks for sustainable agriculture across Europe. *Journal of Rural Studies* 40, 59–75. ISSN 07430167.

Ingram, J., Maye, D., Kirwan, J., Curry, N. and Kubinakova, K. (2015). Interactions between niche and regime: an analysis of learning and innovation networks for sustainable agriculture across Europe. *The Journal of Agricultural Education and Extension* 21(1), 55–71. DOI: 10.1080/1389224X.2014.991114.

INIC (2014). *First Nano-Organic Iron Chelated Fertilizer Invented in Iran*. Iran Nanotechnology Initiative Council, Tehran. Available at: http://www.iranreview.org/content/Documents/Iranians_Researchers_Produce_Nano_Organic_Fertilizer.htm. . Accessed 11 April 2014.

Internet Society (2020). Internet of food. Available at: https://www.internetsociety.org/blog/2020/10/the-internet-of-food/.

IPBES (2019). *Summary for Policymakers of the Global Assessment Report on Biodiversity and Ecosystem Services of the Intergovernmental Science-Policy Platform on Biodiversity and Ecosystem Services*. IPBES Secretariat, Bonn.

IPCC (2014). Climate change 2014: synthesis report. Contribution of Working Groups I, II and III to the Fifth Assessment Report of the Intergovernmental Panel on Climate Change. Intergovernmental Panel on Climate Change, Geneva.

IPCC (2019). *Climate Change and Land – An IPCC Special Report on Climate Change, Desertification, Land Degradation, Sustainable Land Management, Food Security, and Greenhouse Gas Fluxes in Terrestrial Ecosystems – Summary for Policymakers*. Intergovernmental Panel on Climate Change. Available at: https://www.ipcc.ch/site/assets/uploads/2019/08/4.-SPM_Approved_Microsite_FINAL.pdf. Intergovernmental Panel on Climate Change.

IPCC (2022). Climate change 2022: mitigation of climate change. In: Shukla, P. R., Skea, J., Slade, R., Al Khourdajie, A., van Diemen, R., McCollum, D., Pathak, M., Some, S., Vyas, P., Fradera, R., Belkacemi, M., Hasija, A., Lisboa, G., Luz, S. and Malley, J. (Eds) *Contribution of Working Group III to the Sixth Assessment Report of the Intergovernmental Panel on Climate Change*. Cambridge University Press, Cambridge and New York. DOI: 10.1017/9781009157926.025.

IPES-Food (2016). *From Uniformity to Diversity: A Paradigm Shift from Industrial Agriculture to Diversified Agroecological Systems*. International Panel of Experts on Sustainable Food Systems, Brussels. Available at: www.ipes-food.org.

IPES-Food (2017a). *Unravelling the Food-Health Nexus: Addressing Practices, Political Economy, and Power Relations to Build Healthier Food Systems*. The Global Alliance for the Future of Food and IPES-Food, Brussels.

IPES-Food (2017b). Too big to feed: exploring the impacts of mega-mergers, concentration, concentration of power in the agri-food sector. Available at: www.ipes-food.org.

IPES-Food (2018). Breaking away from industrial food and farming systems: Seven case studies of agroecological transition. Available at: www.ipes-food.org.

IPES-Food (2019). *Towards a Common Food Policy for the European Union: The Policy Reform and Realignment That Is Required to Build Sustainable Food Systems in Europe*. International Panel of Experts on Sustainable Food Systems, Brussels. Available at: www.ipes-food.org.

IPES-Food (2020). *The Added Value(S) of Agroecology: Unlocking the Potential for Transition in West Africa*. International Panel of Experts on Sustainable Food Systems, Brussels. Available at: www.ipes-food.org.

IPES-Food and Frison, E. (2016). *From Uniformity to Diversity: A Paradigm Shift from Industrial Agriculture to Diversified Agroecological Systems*. International Panel of Experts on Sustainable Food Systems, Brussels.

IPNI (2008). Global potassium reserves and potassium fertiliser use. Symposium on Global Nutrient Cycling, 6 October 2008. 2008 Joint Annual Meeting. International Plant Nutrition Institute (IPNI).

Iqbal, M., Umar, S. and Na, M. (2020). Nano-fertilization to enhance nutrient use efficiency and productivity of crop plants. In: Husen, A. and Iqbal, M. (Eds) *Nanomaterials and Plant Potential*. Springer, New York. DOI: 10.1007/978-3-030-05569-1_19.

Irfanoglu, Z., Baldos, B., Lantz, U., Hertel, T. and van der Mensbrugghe, D. (2014). Impacts of reducing global food loss and waste on food security, trade, GHG emissions and land use. Selected Paper prepared for presentation at the 17th Annual Conference on Global Economic Analysis, Dakar, Senegal, 18–20 June 2014.

Irving, J. and Ceriani, S. (2013). *Slow Food Companion*. Slow Food, Bra. Available at: http://slowfood.com/filemanager/AboutUs/Companion13ENG.pdf.

ISAAA (2016). Global status of commercialized biotech/GM crops: 2016. ISAAA Brief No. 52. ISAAA, Ithaca, NY.

Isakson, S. R. (2014). Food and finance: the financial transformation of agro-food supply chains. *Journal of Peasant Studies* 41(5), 749–775. DOI: 10.1080/03066150.2013.874340.

Isgren, E. and Ness, B. (2017). Agroecology to promote just sustainability transitions: analysis of a civil society network in the Rwenzori Region, Western Uganda. *Sustainability* 9(8), 1357. DOI: 10.3390/su9081357.

Isgren, E. (2018). *Between Nature and Modernity: Agroecology as an Alternative Development Pathway: The Case of Uganda*. Lund University, Lund.

Ismail, I., Hwang, Y. H. and Joo, S. T. (2020). Meat analog as future food: a review. *Journal of Animal Science and Technology* 62(2), 111–120. DOI: 10.5187/jast.2020.62.2.111.

Isman, M. (2013). Botanical Insecticides in Modern Agriculture and an Increasingly Regulated World Conference at National Center for Animal and Plant Health (CENSA), Mayabeque, Cuba, 51.

ISPA (2019). *Precision ag Definition*. International Society of Precision Agriculture, Monticello, IL. Available at: https://www.ispag.org/about/definition. Accessed 13 July 2021.

IUCN (1991). *Caring for the Earth: A Strategy for Sustainable Living*. The World Conservation Union, UNEP (United Nations Environment Program) and WWF (World Wildlife Fund), Gland. Available at: http://www.iucn.org/about/index.htm.

IUCN (2016). The IUCN red list of threatened species. Version 2016-2. Available at: http://www.iucnredlist.org.

IUCN (2018). *The IUCN Red List of Threatened Species, Version 2018-1*. IUCN, Gland.

Izumi, B. T., Wright, D. W. and Hamm, M. W. (2010). Farm to school programs: exploring the role of regionally based food distributors in alternative agrifood networks. *Agriculture and Human Values* 27(3), 335–350.

Jacke, D. and Toensmeier, E. (2005). *Edible Forest Gardens: Ecological Design and Practice for Temperate-Climate Permaculture*. Chelsea Green, White River Junction.

Jacks, D. S., O'Rourke, K. H. and Williamson, J. G. (2011). Commodity price volatility and world market integration since 1700. *Review of Economics and Statistics* 93(3), 800–813.

Jackson, G., McNamara, K. E. and Witt, B. (2020). "System of hunger": understanding causal disaster vulnerability of indigenous food systems. *Journal of Rural Studies* 73, 163–175.

Jackson, T. (2009). *Prosperity Without Growth. Adapting to Our New Economic Reality*. New Society Publishers, Gabriola Island.

Jacobs (2020). *Outlook 2020: The Future of the Global Crop Protection Industry*. Agri-Business Global. 28 January 2020. Available at: https://edisciplinas.usp.br/pluginfile.php/5167743/mod_resource/content/0/Outlook%202020_%20The%20Future%20of%20the%20Global%20Crop%20Protection%20Industry%20-%20AgriBusiness%20Global.pdf. Accessed 15 November 2023.

Jägermeyr, J., Gerten, D., Heinke, J., Schaphoff, S., Kummu, M. and Lucht, W. Water savings potentials of irrigation systems: global simulation of processes and linkages. *Hydrology and Earth System Sciences* 19(7), 3073–3091.

Jahnke, S., Roussel, J., Hombach, T., Kochs, J., Fischbach, A., Huber, G. and Scharr, H. (2016). *pheno*Seeder–a robot system for automated handling and phenotyping of individual seeds. *Plant Physiology* 172(3), 1358–1370.

Jain, A., Ranjan, S., Dasgupta, N. and Ramalingam, C. (2018). Nanomaterials in food and agriculture: an overview on their safety concerns and regulatory issues. *Critical Reviews in Food Science and Nutrition* 58(2), 297–317. DOI: 10.1080/10408398.2016.1160363.

Jakku, E., Taylor, B., Fleming, A., Mason, C., Fielke, S., Sounness, C. and Thorburn, P. (2019). If they don't tell us what they do with it, why would we trust them? Trust, transparency, and benefit-sharing in smart farming. *NJAS-Wagening Journal of Life Sciences* 90, 100285.

Jalava, M., Kummu, M., Porkka, M., Siebert, S. and Varis, O. (2014). Diet change–a solution to reduce water use? *Environmental Research Letters* 9(7), 074016.

James, C. (2015). Special issue on agri-biotech studies from policy and regulatory perspectives: preface. *AgBioForum* 18(1), 1-2.

Jamil, A., Riaz, S., Ashraf, M. and Foolad, M. R. (2011). Gene expression profiling of plants under salt stress. *Critical Reviews in Plant Sciences* 30(5), 435–458. DOI: 10.1080/07352689.2011.605739.

Jansen, K. (1998). *Political Ecology, Mountain Agriculture, and Knowledge in Honduras*. Thela Publishers, Amsterdam.

Jansen, K. (2015). The debate on food sovereignty theory: agrarian capitalism, dispossession and agroecology. *Journal of Peasant Studies* 42(1), 213–232. DOI: 10.1080/03066150.2014.945166.

Janzen, D. H. (1973). Tropical agroecosystems. *Science* 182(4118), 1212–1219.

Jarosz, L. (2008). The city in the country: growing alternative food networks in Metropolitan areas. *Journal of Rural Studies* 24(3), 231–244.

Jasinski, S. M., Kramer, D. A., Ober, J. A. and Searls, J. P. (1999). Fertilizers–sustaining global food supplies. United States Geological Survey Fact Sheet FS-155-99.

Jat, M. L., Chakraborty, D., Ladha, J. K., Rana, D. S., Gathala, M. K., McDonald, A. and Gerard, B. (2020). Conservation agriculture for sustainable intensification in South Asia. *Nature Sustainability* 3(4), 336–343. Available at: www.nature.com/natsustain.

Jayakumar, M., Janapriya, S. and Surendran, U. (2017). Effect of drip fertigation and polythene mulching on growth and productivity of coconut (*Cocos nucifera* L.), water, nutrient use efficiency and economic benefits. *Agricultural Water Management* 182, 87–93. DOI: 10.1016/j.agwat.2016.12.012.

Jehlička, P. and Daněk, P. (2017). Rendering the actually existing sharing economy visible: home-grown food and the pleasure of sharing. Sociologia ruralis VC 2017 European Society for Rural Sociology. *Sociologia Ruralis* 57(3), 274–296.

Jellason, N. P., Conway, J. S. and Baines, R. N. (2021). Understanding impacts and barriers to adoption of climate-smart agriculture (CSA) practices in North-Western Nigerian drylands. *Journal of Agricultural Education and Extension* 27(1), 55–72.

Jensen, N. D. and Christopher, B. (2017). Agricultural index insurance for development. *Applied Economic Perspectives and Policy* 39(2), 199–219.

Jeunes Agriculteurs (2013). Enquête nationale sur les hors cadres familiaux en agriculture, qui sont-ils et quels sont leurs besoins? Available at: http://www.jeunes-agriculteurs.fr/deveniragriculteur/item/677-demain-je-serai-paysan-?-etat-des-lieux-des-installations-deshors-cadres-familiaux. Accessed 10 October 2015.

Jha, K., Doshi, A., Patel, P. and Shah, M. (2019). A comprehensive review on automation in agriculture using artificial intelligence. *Artificial Intelligence in Agriculture* 2, 1–12.

Ji, Y., Nuñez Ocaña, D., Choe, D., Larsen, D. H., Marcelis, L. F. M. and Heuvelink, E. (2020). Far-red radiation stimulates dry mass partitioning to fruits by increasing fruit sink strength in tomato. *New Phytologist* 228(6), 1914–1925.

Jo, Y. K., Kim, B. H. and Jung, G. (2009). Antifungal activity of silver ions and nano-particles on phytopathogenic fungi. *Plant Disease* 93(10), 1037–1043.

Jobbágy, E. G. and Jackson, R. B. (2004). The uplift of soil nutrients by plants: biogeochemical consequences across scales. *Ecology* 85(9), 2380–2389. 10.1890/03-0245.

John Deere (2020). Sustainability report. Available at: https://www.deere.com/assets/pdfs/common/our-company/sustainability/sustainability-report-2020.pdf.

John Deere (2021). Sustainability report. Available at: https://www.deere.com/assets/pdfs/common/our-company/sustainability/sustainability-report-2021.pdf.

Johnson, C., Bansha Dulal, H., Prowse, M., Krishnamurthy, K. and Mitchell, T. (2013). Social protection, and climate change: emerging issues for research, policy, and practice. *Development Policy Review* 31(s2), 2–18.

Johnson, R., Fraser, E. D. G. and Hawkins, R. (2016). Overcoming barriers to scaling up sustainable alternative food systems: a comparative case study of two Ontario-based wholesale produce auctions. *Sustainability* 8(4), 328. DOI: 10.3390/su8040328.

Johnston, A. E., Poulton, P. R. and Coleman, K. (2009). Chapter 1: Soil organic matter: its importance in sustainable agriculture and carbon dioxide fluxes. In: Sparks, D. L. (Ed.) *Advances in Agronomy*. Academic Press, Cambridge, MA, pp. 1-57.

Johnston, G. W., Vaupel, S., Kegel, F. R. and Cadet, M. (1995). Crop and farm diversification provide social benefits. *California Agriculture* 49(1), 10–16.

Jonas, T. (2021). *Regenerative Agriculture and Agroecology - What's in a Name?* Australian Food Sovereignty Movement, Sydney. Available at: https://afsa.org.au/blog/2021/06/28/13699/. Accessed 28 June 2021.

Jones, A. D., Creed-Kanashiro, H., Zimmerer, K. S., De Haan, S., Carrasco, M., Meza, K., Cruz-Garcia, G. S., Tello, M., Plasencia Amaya, F., Marin, R. M. and Ganoza, L. (2018). Farm-level agricultural biodiversity in the Peruvian Andes is associated with greater odds of women achieving a minimally diverse and micronutrient adequate diet. *Journal of Nutrition* 148(10), 1625–1637. DOI: 10.1093/jn/nxy166.

Jones, B. A., Grace, D., Kock, R., Jonas, S., Rushton, J. and Said, M. Y. (2013). Zoonosis emergence linked to agricultural intensification and environmental change. *PNAS* 110, 8399–8404. 10.1073/pnas.1208059110.

Jones, P. B. C. (2006). *A Nanotech Revolution in Agriculture and the Food Industry*. Information Systems for Biotechnology, Blacksburg, VA. Available at: http://www.isb.vt.edu/articles/jun0605.htm.

Jordon, M. W., Willis, K. J., Bürkner, P. C., Haddaway, N. R., Smith, P. and Petrokofsky, G. (2022). Temperate regenerative agriculture practices increase soil carbon but not crop yield—a meta-analysis. *Environmental Research Letters* 17(9), 093001.

Joseph, S., Peters, I. and Friedrich, H. (2019). Can regional organic agriculture feed the regional community? A case study for Hamburg and North Germany. *Ecological Economics* 164, 106342.

Joseph, T. and Morrison, M. (2006). Nanoforum report: nanotechnology in agriculture and food. *European Nanotechnology Gateway*. Available at: ftp://ftp.cordis.europa.eu/pub/nanotechnology/docs/nanotechnology_in_agriculture_and_food.pdf.

Juhn, T. (2007). Wal-Mart: Mexico's new bank. *Latin Business Chronicle*, 23 July.

Jurgilevich, A., Birge, T., Kentala-Lehtonen, J., Korhonen-Kurki, K., Pietikäinen, J., Saikku, L. and Schösler, H. (2016). Transition towards circular economy in the food system. *Sustainability* 8(1), 69.

Kah, M. (2015). Nanopesticides and nanofertilizers: emerging contaminants or opportunities for risk mitigation? *Frontiers in Chemistry* 3, 64.

Kah, M. and Hofmann, T. (2014). Nanopesticide research: current trends and future priorities. *Environment International* 63, 224–235.

Kah, M., Kookana, R. S., Gogos, A. and Bucheli, T. D. (2018). A critical evaluation of nanopesticides and nanofertilizers against their conventional analogues. *Nature Nanotechnology* 13(8), 677–684.

Kalambukattu, J. G., Kumar, S. and Raj, R. A. (2018). Digital soil mapping in a Himalayan watershed using remote sensing and terrain parameters employing artificial neural network model. *Environmental Earth Sciences* 77(5), 203.

Kaliba, A. R., Mushi, R. J., Gongwe, A. G. and Mazvimavi, K. (2020). A typology of adopters and nonadopters of improved sorghum seeds in Tanzania: a deep learning neural network approach. *World Development* 127. DOI: 10.1016/j.worlddev.2019.104839.

Kallis, G., Kerschner, C. and Martinez-Alier, J. (2012). The economics of degrowth. *Ecological Economics* 84, 172–180. DOI: 10.1016/j.ecolecon.2012.08.017.

Kallis, G., Demaria, F. and D'Alisia, G. (2015). Introduction: degrowth. In: D'Alisia, G., Demaria, F. and Kallis, G. (Eds) *Degrowth: A Vocabulary for a New Era*. Routledge, New York, p. 1e17.

Kallis, G., Kostakis, V., Lange, S., Muraca, B., Paulson, S. and Schmelzer, M. (2018). Research on degrowth. *Annual Review of Environment and Resources* 43(1), 291–316. DOI: 10.1146/annurev-environ-102017-025941.

Kamilaris, A., Kartakoullis, A. and Prenafeta-Boldú, F. X. (2017). A review on the practice of big data analysis in agriculture. *Computers and Electronics in Agriculture* 143, 23–37.

Kasperczyk, N. and Knickel, K. (2006). Environmental impact of organic agriculture. In: Kristiansen, P., Taji, A. and Reganold, J. (Eds) *Organic Agriculture: A Global Perspective*. CSIRO Publishing, Collingwood, pp. 259–294.

Kaur, H. and Garg, H. (2014). Pesticides: environmental impacts and management strategies. In: Soloneski, S. (Ed.) *Pesticides - Toxic Aspects*. InTech, London. DOI: 10.5772/57399.

Kaul, S., Akbulut, B., Demaria, F. and Gerber, J. F. (2022). Alternatives to sustainable development: what can we learn from the pluriverse in practice? *Sustainability Science* 17(4), 1149–1158. 10.1007/s11625-022-01210-2.

Kaut, A. H. E. E., Mason, H. E., Navabi, A., O'Donovan, J. T. and Spaner, D. (2008). Organic and conventional management of mixtures of wheat and spring cereals. *Agronomy for Sustainable Development* 28(3), 363–371.

Kearney, J. (2010). Food consumption trends and drivers. *Philosophical Transactions of the Royal Society of London. Series B, Biological Sciences* 365(1554), 2793–2807. DOI: 10.1098/rstb.2010.0149.

Kehoe, L., Romero-Muñoz, A., Polaina, E., Estes, L., Kreft, H. and Kuemmerle, T. (2017.). Biodiversity at risk under future cropland expansion and intensification. *Nature Ecology and Evolution* 1(8), 1129–1135. 10.1038/s41559-017-0234-3.

Kernecker, M., Knierim, A., Wurbs, A., Kraus, T. and Borges, F. (2020). Experience versus expectation: farmers' perceptions of smart farming technologies for cropping systems across Europe. *Precision Agriculture* 21(1), 34–50.

Kerr, R. B., Nyantakyi-Frimpong, H., Lupafya, E., Dakishoni, L., Shumba, L. and Luginaah, I. (2016). Building resilience in African smallholder farming communities through farmer-led agroecological methods. In: Nagothu, U. S. (Ed.) *Climate Change and Agricultural Development: Improving Resilience Through. Climate-Smart Agriculture, Agroecology and Conservation*. Routledge, London and New York, pp. 109–130.

Kerr, R. B., Madsen, S., Stüber, M., Liebert, J., Enloe, S., Borghino, N., Parros, P., Mutyambai, D. M., Prudhon, M. and Wezel, A. (2021). Can agroecology improve food security and nutrition? A review. *Global Food Security* 29, 100540.

Kerr, R. B., Postigo, J. C., Smith, P., Cowie, A., Singh, P. K., Rivera-Ferre, M., Tirado-von der Pahlen, M. C., Campbell, D. and Neufeldt, H. (2023). Agroecology as a transformative approach to tackle climatic, food, and ecosystemic crises. *Current Opinion in Environmental Sustainability* 62, 101275. 10.1016/j.cosust.2023.101275.

Kesavan, P. C. and Swaminathan, M. S. (2018). Modern technologies for sustainable food and nutrition security. *Current Science* 115(10), 1876–1883. DOI: 10.18520/cs/v115/i10/1876-1883.

Kesse-Guyot, E., Péneau, S., Méjean, C., Szabo de Edelenyi, F., Galan, P., Hercberg, S. and Lairon, D. (2013). Profiles of organic food consumers in a large sample of French adults: results from the Nutrinet-Santé cohort study. *PLoS ONE* 8(10), e76998.

Khadse, A., Rosset, P. M., Morales, H. and Ferguson, B. G. (2018). Taking agroecology to scale: the zero-budget natural farming peasant movement in Karnataka, India. *Journal of Peasant Studies* 45(1), 192–219. DOI: 10.1080/03066150.2016.1276450.

Khanal, S., Fulton, J., Klopfenstein, A., Douridas, N. and Shearer, S. (2018). Integration of high resolution remotely sensed data and machine learning techniques for spatial prediction of soil properties and corn yield. *Computers and Electronics in Agriculture* 153, 213–225.

Khangura, R., Ferris, D., Wagg, C. and Bowyer, J. (2023). Regenerative agriculture–a literature review on the practices and mechanisms used to improve soil health. *Sustainability* 15(3), 2338. DOI: 10.3390/su15032338.

Khin, M. M., Nair, A. S., Babu, V. J., Murugan, R. and Ramakrishna, S. (2012). A review on nanomaterials for environmental remediation. *Energy and Environmental Science* 5(8), 8075–8109.10.1039/c2ee21818f.

Khonje, M. G. and Qaim, M. (2019). Modernization of African food retailing and (un) healthy food consumption. *Sustainability* 11(16), 4306. DOI: 10.3390/su11164306.

Khonje, M. G., Ecker, O. and Qaim, M. (2020). Effects of modern food retailers on adult and child diets and nutrition. *Nutrients* 12(6), 1714.

Khoury, C. K., Bjorkman, A. D., Dempewolf, H., Ramirez-Villegas, J., Guarino, L., Jarvis, A., Rieseberg, L. H. and Struik, P. C. (2014). Increasing homogeneity in global food supplies and the implications for food security. *Proceedings of the National Academy of Sciences of the United States of America* 111(11), 4001–4006. DOI: 10.1073/pnas.1313490111.

Khurana, A. and Kumar, V. (2020). *State of Organic and Natural Farming: Challenges and Possibilities*. Centre for Science and Environment, New Delhi.

Kim, S. W., Lusk, L. and Brorsen, B. (2018). Look at me, I'm buying organic. The effects of social pressure on organic food purchases. *Journal of Agricultural and Resource Economics* 43(3), 364–387 ISSN 1068-5502. Copyright 2018 Western Agricultural Economics Association.

King, F. H. (1911). *Farmers of Forty Centuries; or Permanent Agriculture in China, Korea, and Japan*. King, Madison, WI.

Kirchmann, H. (2019). Why organic farming is not the way forward. *Outlook on Agriculture* 48(1), 22–27. DOI: 10.1177/0030727019831702.

Kirchmann, H. and Bergström, L. (2001). Do organic farming practices reduce nitrate leaching? *Communications in Soil Science and Plant Analysis* 32(7–8), 997–1028.

Kirchmann, H., Bergström, L., Kätterer, T., Andrén, O. and Andersson, R. (2008). Can organic crop production feed the world? In: Kirchmann, H. and Bergström, L. (Eds) *Organic Crop Production? Ambitions and Limitations*. Springer, Dordrecht, pp. 39–72. DOI: 10.1007/978-1-4020-9316-6_3.

Kirchmann, H., Bergström, L., Kätterer, T., Mattsson, L. and Gesslein, S. (2007). Comparison of long-term organic and conventional crop-livestock systems on a previously nutrient-depleted soil in Sweden. *Agronomy Journal* 99(4), 960–972.

Kirwan, J. (2006). The interpersonal world of direct marketing: examining conventions of quality at UK farmers' markets. *Journal of Rural Studies* 22(3), 301–312.

Kirwan, J., Ilbery, B., Maye, D. and Carey, J. (2013a). Grassroots social innovations and food localisation: an investigation of the local food programme in England. *Global Environmental Change* 23(5), 830–837.

Kirwan, J. and Maye, D. (2013). Food security framings within the UK and the integration of local food systems. *Journal of Rural Studies* 29, 91–100.

Kirwan, J., Maye, D. and Brunori, G. (2013b). Reflexive governance, incorporating ethics and changing understandings of food chain performance. *Sociologia Ruralis* 57(3), July 2017. DOI: 10.1111/soru.12169.

Klein, A.-M., Steffan-Dewenter, I. and Tscharntke, T. (2003). Pollination of *Coffea canephora* in relation to local and regional agroforestry management. *Journal of Applied Ecology* 40(5), 837–845.

Klein, J. (2021). Walmart digs into regenerative agriculture. *Greenbiz*. Available at: https://www.greenbiz.com/article/walmart-digs-regenerative-agriculture#:~:text=By%20working%20with%20the%20nonprofit,improve%20water%20qu. Accessed 27 September 2021.

Klerkx, L. and Rose, D. (2020). Dealing with the game-changing technologies of agriculture 4.0: how do we manage diversity and responsibility in food system transition pathways? *Global Food Security* 24, 100347.

Klima, K., Synowiec, A., Puła, J., Chowaniak, M., Püzyńska, K., Gala-Czekaj, D., Kliszcz, A., Galbas, P., Jop, B., Bkowska, T. D. and Lepiarczyk, A. (2020). Long-term productive, competitive, and economic aspects of spring cereal mixtures in integrated and organic crop rotations. *Agriculture* 10(6), 231. DOI: 10.3390/agriculture10060231.

Klitkou, A., Bolwig, S., Hansen, T. and Wessberg, N. (2015). The role of lock-in mechanisms in transition processes: the case of energy for road transport. *Environmental Innovation and Societal Transitions* 16, 22–37.

Kloppenburg, J. Jr., Hendrickson, J. and Stevenson, G. W. (1996). Coming in to the foodshed. *Agriculture and Human Values* 13(3), 33–42.

Klümper, W. and Qaim, M. (2014). A meta-analysis of the impacts of genetically modified crops. *PLoS ONE* 9(11), e111629. DOI: 10.1371/journal.pone.0111629.

Knapp, M. and Řezáč, M. (2015). Even the smallest non-crop habitat islands could be beneficial: distribution of carabid beetles and spiders in agricultural landscape. *PLoS ONE* 10(4), e0123052, 10.1371/journal.pone.0123052. Accessed 6 November 2020.

Knauer, K. and Bucheli, T. D. (2009). Nanomaterials: research needs in agriculture. *Revue Suisse d'Agriculture* 41(6), 337–341.

Kneafsey, M., Holloway, L., Cox, R., Dowler, E., Venn, L. and Tuomainen, H. (2008). *Reconnecting Consumers, Producers and Food: Exploring Alternatives*. Berg Publishers, Oxford.

Kneafsey, M., Venn, L., Schmutz, U., Balázs, B., Trenchard, L., Eyden-Wood, T., Bos, E., Sutton, G. and Blackett, M. (2014). *Short Food Supply Chains and Local Food Systems in the EU: A State of Play of Their Socio-Economic Characteristics*. European Commission, Brussels.

Kneen, B. (1995). *From Land to Mouth: Understanding the Food System*. NC Press, Toronto.

Knežević, B., Škrobot, P. and Žmuk, B. (2021). Position and role of social supermarkets in food supply chains. *Business Systems Research* 12(1), 179–196.

Knowler, D. and Bradshaw, B. (2007). Farmers' adoption of conservation agriculture: a review and synthesis of recent research. *Food Policy* 32(1), 25–48.

Koch, M. (2018). The naturalisation of growth: Marx, the Regulation Approach and Bourdieu. *Environmental Values* 27(1), 9–27.

Koepf, H. H. (2006). The biodynamic farm. Steiner Books, Dulles, VA.

Köhl, J. and Ravensberg, W. (Eds.) (2021). *Microbial Bioprotectants for Plant Disease Management*. Burleigh Dodds Science Publishing, Cambridge, UK.

Kolata, G. (1997). Study discounts DDT role in breast cancer. *The New York Times*, 10 October 1997, A26.

Kole, C., Kole, P., Randunu, K. M., Choudhary, P., Podila, R., Ke, P. C., Rao, A. M. and Marcus, R. K. (2013). Nanobiotechnology can boost crop production and quality:

first evidence from increased plant biomass, fruit yield and phytomedicine content in bitter melon (*Momordica charantia*). *BMC Biotechnology* 13, Article number 37.

Kolesnikova, M. (2011). Grantham says farmland will outperform all global assets. *Bloomberg.com*, 10 August. Available at: http://www.bloomberg.com/news/2011-08-10/grantham-says-farmlandwill-outperform-all-global-assets-1-.html.

Komorowicz, I., Gramowska, H. and Barałkiewicz, D. (2010). Estimation of the lake water pollution by determination of 18 elements using ICP-MS method and their statistical analysis. *Journal of Environmental Science and Health. Part A, Toxic/Hazardous Substances and Environmental Engineering* 45(3), 348–354. DOI: 10.1080/10934520903467873.

Koneswaran, G. and Nierenberg, D. (2008). Global farm animal production and global warming: impacting and mitigating climate change. *Environmental Health Perspectives* 116(5), 578–582. DOI: 10.1289/ehp.11034.

Konradsen, F., van der Hoek, W., Cole, D. C., Hutchinson, G., Daisley, H., Singh, S. and Eddleston, M. (2003). Reducing acute in developing countries: options for restricting the availability of pesticides. *Toxicology* 192(2–3), 249–261.

Konstantinis, A., Rozakis, S., Efpraxia-Aithra, M. and Shu, K. (2018). A definition of bioeconomy through the bibliometric networks of the scientific literature. *AgBioForum* 21(2), 64–85. ©2018 AgBioForum.

Kootstra, G., Wang, X., Blok, P. M., Hemming, J. and van Henten, E. (2021). Selective harvesting robotics: current research, trends, and future directions. *Current Robotics Reports* 2, 95–104.

Kottegoda, N., Munaweera, I., Madusanka, N. and Karunaratne, V. (2011). A green slow-release fertilizer composition based on urea-modified hydroxyapatite nanoparticles encapsulated wood. *Current Science* 101, 73–78.

Koutsos, T. and Menexes, G. (2019). Economic, agronomic, and environmental benefits from the adoption of precision agriculture technologies: a systematic review. *International Journal of Agricultural and Environmental Information Systems* 10(1), January–March.

Kovak, E., Blaustein-Rejto, D. and Qaim, M. (2022). Genetically modified crops support climate change mitigation. *Trends in Plant Science* 27(7), 627.

Kozai, T., Niu, G. and Takagaki, M. (2016). *Plant Factory: An Indoor Vertical Farming System for Efficient Quality Food Production*. Elsevier, New York. DOI: 10.1016/C2014-0-01039-8.

Kozai, T., Niu, G. and Takagaki, M. (2020). *Plant Factory, An Indoor Vertical Farming System for Efficient Quality Food Production*. Academic Press, New York. DOI: 10.1016/B978-0-12-816691-8.01001-3.

Kozicka, M., Havlík, P., Valin, H., Wollenberg, E., Deppermann, A., Leclère, D., Lauri, P., Moses, R., Boere, E., Frank, S., Davis, C., Park, E. and Gurwic, N. (2023). Feeding climate and biodiversity goals with novel plant-based meat and milk alternatives. *Nature Communications*. Available at: https://www.nature.com/articles/s41467-023-40899-2.

Krausmann, F. and Langthaler, E. (2019). Food regimes and their trade links: a socio-ecological perspective. *Ecological Economics* 160, 87–95.

Kremen, C., Iles, A. and Bacon, C. (2012). Diversified farming systems: an agroecological, systems-based alternative to modern industrial agriculture. *Ecology and Society* 17(4). DOI: 10.5751/ES-05103-170444.

Kremen, C. and Merenlender, A. M. (2018). Landscapes that work for biodiversity and people. *Science* 362(6412). 10.1126/science.aau6020.

Kremen, C. and Miles, A. (2012). Ecosystem services in biologically diversified versus conventional farming systems: benefits, externalities, and trade-offs. *Ecology and Society* 17(4). DOI: 10.5751/ES-05035-170440.

Krippner, G. (2005). The financialization of the American economy. *Socio-Economic Review* 3(2), 173–208.

Krippner, G. (2011). *Capitalizing on Crisis: The Political Origins of the Rise of Finance*. Harvard University Press, Cambridge.

Kristiansen, P., Taji, A. and Reganold, J. (Eds.) (2006). *Organic Agriculture: A Global Perspective*. CSIRO Publishing, Collingwood.

Krstić, B., Petrović, J., Stanišić, T. and Kahrović, E. (2017). Analysis of the organic agriculture level of development in the European Union Countries. *Economics of Agriculture* 3/2017 UDC: 330.341:631.147(4-672 EU).

Kuch, D., Kearnes, M. and Gulson, K. (2020). The promise of precision: datafication in medicine, agriculture, and education. *Policy Studies* 41(5), 527–546.

Kucukvar, M. and Samadi, H. (2015). Linking national food production to global supply chain impacts for the energy-climate challenge: the cases of the EU-27 and Turkey. *Journal of Cleaner Production* 108, 395–408.

Kudo, A. and Miseki, Y. (2009). Heterogeneous photocatalyst materials for water splitting. *Chemical Society Reviews* 38(1), 253–278.

Kumar, M., Shamsi, T. N., Parveen, R. and Fatima, S. (2017). Application of nanotechnology in enhancement of crop productivity and integrated pest management. In: Prasad, R., Kumar, M. and Kumar, V. (Eds) *Nanotechnology*. Springer, Singapore, pp. 361–371. DOI: 10.1007/978-981-10-4573-8_17.

Kumar, S. (2012). Biopesticides: a need for food and environmental safety. *Journal of Biofertilizers and Biopesticides* 3(4), e107. 10.4172/2155-6202.1000-107.

Kumar, S., Kumar, S., Kumar, D. and Dilbaghi, N. (2017). Preparation, characterization, and bio-efficacy evaluation of controlled release carbendazim loaded polymeric nanoparticles. *Environmental Science and Pollution Research* 24(1), 926–937. 10.1007/s11356-016-7774-y.

Kuo, H. J. and Peters, D. J. (2017). The socioeconomic geography of organic agriculture in the United States. *Agroecology and Sustainable Food Systems* 41(9–10), 1162–1184. DOI: 10.1080/21683565.2017.1359808.

Kuokkanen, A., Mikkila, M., Kuisma, M., Kahiluoto, H. and Linnanen, L. (2017). The need for policy to address the food system lock-in: a case study of the Finnish context. *Journal of Cleaner Production* 140(2), 933–944.

Kuokkanen, A., Nurmi, A., Mikkilä, M., Kuisma, M., Kahiluoto, H. and Linnanen, L. (2018). Agency in regime destabilization through the selection environment: the Finnish food system's sustainability transition. *Research Policy* 47(8), 1513–1522.

Kuzma, J. (2007). Moving forward responsibly: oversight for the nanotechnology-biology interface. *Journal of Nanoparticle Research* 9(1), 165–182.

La Via Campesina (2010). *Sustainable Peasant and Family Farm Agriculture Can Feed the World*. Via Campesina, Jakarta. Available at: http://www.foodmovementsunite.org/addenda/via-campesina.pdf. Accessed 24 July 2012.

LaCanne, C. E. and Lundgren, J. G. (2018). Regenerative agriculture: merging farming and natural resource conservation profitably. *PeerJ* 6, e4428. DOI: 10.7717/peerj.4428.

Lajoie-O'Malley, A., Bronson, K., van der Burg, S. and Klerkx, L. (2020). The future(s) of digital agriculture and sustainable food systems: an analysis of high-level policy documents. *Ecosystem Services* 45, 101183.

Lambek, N., Claeys, P., Wong, A. and Brilmayer, L. (Eds) (2014). *Rethinking Food Systems*. Springer Science & Business Media, Dordrecht. DOI: 10.1007/978-94-007-7778-1.

Lamine, C. (2005). Settling the shared uncertainties: local partnerships between producers and consumers. *Sociologia Ruralis* 45(4), 324–345.

Lamine, C. (2011). Transition pathways towards a robust ecologization of agriculture and the need for system redesign. Cases from organic farming and IPM. *Journal of Rural Studies* 27(2), 209–219.

Lamine, C. (2014). Sustainability and resilience in agrifood systems: reconnecting agriculture, food, and the environment. *Sociologia Ruralis* 55(1), January 2015.

Lamine, C. and Dawson, J. (2018). The agroecology of food systems: reconnecting agriculture, food, and the environment. *Agroecology and Sustainable Food Systems* 42(6), 629–636.

Lamine, C., Garçon, L. and Brunori, G. (2019c). Territorial agrifood systems: a Franco-Italian contribution to the debates over alternative food networks in rural areas. *Journal of Rural Studies* 68, 159–170.

Lamine, C., Magda, D. and Amiot, M. J. (2019a). Crossing sociological, ecological, and nutritional perspectives on agri-food systems transitions: towards a transdisciplinary territorial approach. *Sustainability* 11(5), 1284. DOI: 10.3390/su11051284.

Lamine, C., Darnhofer, I. and Marsden, T. K. (2019b). What enables just sustainability transitions in agrifood systems? An exploration of conceptual approaches using international comparative case studies. *Journal of Rural Studies* 68, 144–146.

Lamine, C., Renting, H., Rossi, A., Wiskerke, J. S. C. and Brunori, G. (2012). Agri-food systems and territorial development: innovations, new dynamics and changing governance mechanisms. In: Darnhofer, I., Gibbon, D. and Dedieu, B. (Eds) *Farming Systems Research into the 21st Century: The New Dynamic*. Springer, Dordrecht, pp. 229–256.

Lampayan, R. M., Rejesus, R. M., Singleton, G. R. and Bouman, B. A. M. (2015). Adoption and economics of alternate wetting and drying water management for irrigated lowland rice. *Field Crops Research* 170, 95–108.

Lampkin, N. (2002). *Organic Farming* (revised edition). Old Pond Publishing, Suffolk.

Lang, T. (2003). Food industrialisation and food power: implications for food governance. *Development Policy Review* 21(5–6), 555–568.

Lang, T. and Barling, D. (2012). Food security and food sustainability: reformulating the debate. *The Geographical Journal* 178(4), 313–326. DOI: 10.1111/j.1475-4959.2012.00480.x.

Larson, J. A., Velandia, M. M., Buschermohle, M. J. and Westlund, S. M. (2016). Effect of field geometry on profitability of automatic section control for chemical application equipment. *Precision Agriculture* 17(1), 18–35.

Latouche, S. (2003). *The World Downscaled* (English edition). Le Monde Diplomatique. Available at: http://mondediplo.com/2003/12/17growth.

Latouche, S. (2009). *Farewell to Growth*. Polity, Cambridge.

Latouche, S. (2012). *For a Frugal Plenty. (Per Un'abbondanza Frugale)*. Bollati Boringhieri, Torino (in Italian).

Latouche, S. (2016). *The Degrowth before the Degrowth (La Decrescita Prima Della Decrescita)*. Bollati Boringhieri, Torino (in Italian).

Lavicoli, I., Leso, V., Beezhold, D. H. and Shvedova, A. A. (2017). Nanotechnology in agriculture: opportunities, toxicological implications, and occupational risks. *Toxicology and Applied Pharmacology* 329, 96–111. 10.1016/j.taap.2017.05.025.

Laville, S. and Vidal, J. (2006). Supermarkets accused over organic foods. *The Guardian*, 5 October.

Lawniczak, A. E., Zbierska, J., Nowak, B., Achtenberg, K., Grześkowiak, A. and Kanas, K. (2016). Impact of agriculture and land use on nitrate contamination in groundwater and running waters in central-west Poland. *Environmental Monitoring and Assessment* 188(3), 172.

Le Grand, L. and Van Meekeren, M. (2008). Urban-rural relations: Dutch experiences of the leader+ network and rural innovation in areas under strong urban influences. Presented at the Conference Rurality near the City, Leuven, Belgium, 7–8 February.

Le Mire, G., Nguyen, M. L., Fassotte, B., du Jardin, P., Verheggen, F., Delaplace, P. and Jijakli, M. H. (2016). Implementing plant biostimulants and biocontrol strategies in the agroecological management of cultivated ecosystems. A review./intégrer les biostimulants et les stratégies de biocontrôle dans la gestion agroécologique des écosystèmes cultivés (synthèse bibliographique). *Biotechnology, Agronomy and Society and Environment* 20, 299.

Lee, C. W., Mahendra, S., Zodrow, K., Li, D., Tsai, Y. C., Braam, J. and Alvarez, P. J. (2010). Developmental phytotoxicity of metal oxide nanoparticles to *Arabidopsis thaliana*. *Environmental Toxicology and Chemistry* 29(3), 669–675.

Lee, J. C., Son, Y. O., Pratheeshkumar, P. and Shi, X. (2012). Oxidative stress and metal carcinogenesis. *Free Radical Biology and Medicine* 53(4), 742–757. DOI: 10.1016/j.freeradbiomed.2012.06.002.

Lee-Smith, D. (2010). Cities feeding people: an update on urban agriculture in equatorial Africa. *Environment and Urbanization* 22(2), 483–499.

Lefroy, E. C. (2009). *Agroforestry and the Functional Mimicry of Natural Ecosystems*. CSIRO, Melbourne, pp. 23–35. Agroforestry for Natural Resource Management.

Lefsrud, M. G., Kopsell, D. A., Kopsell, D. E. and Curran-Celentano, J. (2006). Irradiance levels affect growth parameters and carotenoid pigments in kale and spinach grown in a controlled environment. *Physiologia Plantarum* 127(4), 624–631.

Lefsrud, M. G., Kopsell, D. A. and Sams, C. E. (2008). Irradiance from distinct wavelength light-emitting diodes affect secondary metabolites in kale. *HortScience* 43(7), 2243–2244.

Léger, D. and Hervieu, B. (1979). *Le Retour à la Nature: Au Fond de la Forêt … l'État*. Éditions du Seuil, Paris, p. 240.

Lehnert, C., McCool, C., Sa, I. and Perez, T. (2020). Performance improvements of a sweet pepper harvesting robot in protected cropping environments. *Journal of Field Robotics*. DOI: 10.1002/rob.21973.

Lei, X. G. (2021). *Seaweed and Microalgae as Alternative Sources of Protein*. Burleigh Dodds Science Publishing, Cambridge.

Leippert, F., Darmaun, M., Bernoux, M. and Mpheshea, M. (2020). *The Potential of Agroecology to Build Climate-Resilient Livelihoods and Food Systems*. Food and Agriculture Organization, Rome and Biovision. DOI: 10.4060/cb0438en.

Lelieveld, J., Evans, J. S., Fnais, M., Giannadaki, D. and Pozzer, A. (2015). The contribution of out-door air pollution sources to premature mortality on a global scale. *Nature* 525(7569), 367–371. DOI: 10.1038/nature15371.

Letourneau, D. K. and Bothwell, S. G. (2008). Comparison of organic and conventional farms: challenging ecologists to make biodiversity functional. *Frontiers in Ecology and the Environment* 6(8), 430–438.

Levidow, L. (2015). European transitions towards a corporate-environmental food regime: agroecological incorporation or contestation? *Journal of Rural Studies* 40, 76–89.

Levidow, L. and Bijman, J. (2002). Farm inputs under pressure from the European food industry. *Food Policy* 27(1), 31–45.

Ley, S., Hamdy, O., Mohan, V. and Hu, F. B.(2014). *Prevention and Management of Type 2 Diabetes: Dietary Components and Nutritional Strategies*. Lancet Publishing Group, Cambridge, MA. DOI: 10.1016/S0140-6736(14)60613-9.

Liao, R., Zhang, S., Zhang, X., Wang, M., Wu, H. and Zhangzhong, L. (2021). Development of smart irrigation systems based on real-time soil moisture data in a greenhouse: proof of concept. *Agricultural Water Management* 245, 106632.

Lieblein, G., Breland, A., Østergard, E., Salomonsson, L. and Francis, C. (2007a). Educational perspectives in agroecology: steps on a dual learning ladder toward responsible action. *NACTA Journal* March, 37–44.

Lieblein, G., Østergaard, E. and Francis, C. (2007b). Becoming an agroecologist through action education. *International Journal of Agricultural Sustainability* 2(3), 147–153.

Lieblein, G., Francis, C., Barth Eide, W., Torjusen, H., Solber, S., Salomonsson, L., Lund, V., Ekblad, G., Persson, P., Helenius, J., Loiva, M., Sepannen, L., Kahiluoto, H., Porter, J., Olsen, H., Sriskandarajah, N., Mikk, M. and Flora, C. (2000). Future education in ecological agriculture and food systems: a student-faculty evaluation and planning process. *Journal of Sustainable Agriculture* 16(4), 49–69.

Ligutti, L. G. and Rawe, J. C. (1940). *Rural Roads to Security*. The Bruce Publishing Company, Milwaukee.

Lin, B. B. (2011). Resilience in agriculture through crop diversification: adaptive management for environmental change. *BioScience* 61(3), 183–193.

Lin, D. and Xing, B. (2007). Phytotoxicity of nanoparticles: inhibition of seed germination and root growth. *Environmental Pollution* 150(2), 243–250.

Lin, N., Wang, X., Zhang, Y., Hu, X. and Ruan, J. (2020). Fertigation management for sustainable precision agriculture based on Internet of Things. *Journal of Cleaner Production* 277, 124119.

Ling, X., Zhao, Y., Gong, L., Liu, C. and Wang, T. (2019). Dual-arm cooperation and implementing for robotic harvesting tomato using binocular vision. *Robotics and Autonomous Systems* 114, 134–143.

Linnerooth-Bayer, J. and Hochrainer-Stigler, S. (2015). Financial instruments for disaster risk management and climate change adaptation. *Climatic Change* 133(1), 85–100.

Lipinski, B., Hanson, C., Lomax, J., Kitinoja, L., Waite, R. and Searchinger, T. (2013). *Reducing Food Loss and Waste*. World Resources Institute, Washington, DC.

Lipper, L., Thornton, P., Campbell, B. M., Baedeker, T., Braimoh, A., Bwalya, M., Caron, P., Cattaneo, A., Garrity, D., Henry, K., Hottle, R., Jackson, L., Jarvis, A., Kossam, F., Mann, W., McCarthy, N., Meybeck, A., Neufeldt, H., Remington, T., Sen, P. T., Sessa, R., Shula, R., Tibu, A. and Torquebiau, E. F. (2014). Climate-smart agriculture for food security. *Nature Climate Change* 4(12), 1068–1072. Published online 26 November 2014; corrected after print 13 March 2015.

Little, A. (2021). *The Future of Vertical Farming Is Brighter than Once Thought*. Bloomberg, New York. Available at: https://www.livemint.com.

Liu, B., Tu, C., Hu, S., Gumpertz, M. and Ristaino, J. B. (2007). Effect of organic, sustainable, and conventional management strategies in grower fields on soil physical, chemical, and biological factors and the incidence of southern blight. *Applied Soil Ecology* 37(3), 202–214.

Liu, J., You, L., Amini, M., Obersteiner, M., Herrero, M., Zehnder, A. J. and Yang, H. (2010). A high-resolution assessment on global nitrogen flows in cropland. *PNAS* 107(17), 8035–8040. DOI: 10.1073/pnas.0913658107.

Liu, Q., Yang, F., Zhang, J., Liu, H., Rahman, S., Islam, S., Ma, W. and She, M. (2021). Application of CRISPR/Cas9 in crop quality improvement. *International Journal of Molecular Sciences* 22(8), 4206. DOI: 10.3390/ijms22084206.

Liu, R. Q. and Lal, R. (2014). Synthetic apatite nanoparticles as a phosphorus fertilizer for soybean (Glycine max). *Scientific Reports* 4, 5686.

Liu, R. Q. and Lal, R. (2015). Potentials of engineered nanoparticles as fertilizers for increasing agronomic productions. *Science of the Total Environment* 514, 131–139. DOI: 10.1016/j.scitotenv.2015.01.104.

Liu, Y., Langemeier, M. R., Small, I. M., Joseph, L. and Fry, W. E. (2017). Risk management strategies using PA technology to manage potato late blight. *Agronomy Journal* 109(2), 562–575. DOI: 10.2134/agronj2016.07.0418.

Liverani, M., Waage, J., Barnett, T., Pfeifer, D., Rushton, J., Rudge, J., Loevinsohn, M., Loboguerrero, A. M., Campbell, B. M., Cooper, P. J. M., Hansen, J. W., Rosenstock, T. and Wollenberg, E. (2019). Food and earth systems: priorities for climate change adaptation and mitigation for agriculture and food systems. *Sustainability* 11(5), 1372. DOI: 10.3390/su11051372.

Lockeretz, W. (2007). What explains the rise of organic farming? In: Lockeretz, W. (Ed.) *Organic Farming: An International History*. CAB International, Oxfordshire, pp. 1–8.

Lockeretz, W., Shearer, G. and Kohl, D. H. (1981). Organic farming in the Corn Belt. *Science* 211(4482), 540–547.

Lohrberg, F., Lička, L., Scazzosi, L. and Timpe, A. (Eds) (2016). *Urban Agriculture Europe*. Jovis, Berlin.

Loisel, J. P., François, M., Chiffoleau, Y., Hérault-Fournier, C., Sirieix, L. and Costa, D. (2013). La consommation alimentaire en circuits courts: equête nationale. Research Report. Gret-INC–INRA, Paris.

Long, T. (2017). *Permaculture - Bringing Together the Layers and the Zones*. Permaculture Research Institute. 20 April 2017. Available at: https://www.permaculturenews.org/2017/04/20/permaculture-bringing-together-layers-zones/. Accessed 25 September 2022.

Loos, J., Abson, D. J., Chappell, M. J., Hanspach, J., Mikulcak, F., Tichit, M. and Fischer, J. (2014). Putting meaning back into "sustainable intensification". *Frontiers in Ecology and the Environment* 12(6), 356–361.

Lopez, A. M., Loopstra, R., McKee, M. and Stuckler, D. (2017). Is trade liberalisation a vector for the spread of sugar-sweetened beverages? A cross-national longitudinal analysis of 44 low- and middle-income. *Social Science and Medicine* 172, 21–27. DOI: 10.1016/j.socscimed.2016.11.001.

Lotter, D. W. (2003). Organic agriculture. *Journal of Sustainable Agriculture* 21(4), 59–128.

Lotter, D. W., Seidel, R. and Liebhart, W. (2003). The performance of organic and conventional cropping systems in an extreme climate year. *American Journal of Alternative Agriculture* 18(3), 146–154.

Loures, L., Chamizo, A., Ferreira, P., Loures, A., Castanho, R. and Panagopoulos, T. (2020). Assessing the effectiveness of precision agriculture management systems in Mediterranean small farms. *Sustainability* 12(9), 3765.

Low, S. A., Adalja, A., Beaulieu, E., Key, N., Martinez, S., Melton, A. and Jablonski, B. B. R. (2015). *Trends in U.S. Local and Regional Food Systems (AP-068)*. U.S. Department of

Agriculture, Economic Research Service, Washington, DC. Available at: https://www.ers.usda.gov/publications/pub-details/?pubid=42807.

Lowder, S. K., Sánchez, M. V. and Bertini, R. (2021). Which farms feed the world and has farmland become more concentrated? *World Development* 142, 105455.

Lowe, P., Murdoch, J., Marsden, T., Munton, R. and Flynn, A. (1993). Regulating the new rural spaces, the uneven development of land. *Journal of Rural Studies* 9(3), 205-222.

Lu, C. M., Zhang, C. Y., Wen, J. Q., Wu, G. R. and Tao, M. X. (2002). Research of the effect of nanometer materials on germination and growth enhancement of glycine max and its mechanism. *Soybean Science* 21(3), 168-171.

Lu, Y. and Ozcan, S. (2015). Green nanomaterials: on track for a sustainable future. *Nano Today* 10(4), 417–420.

Ludwig, D. S. (2011). Technology, diet, and the burden of chronic disease. *JAMA* 305(13), 1352-1353.

Lumen (2021). The agricultural revolution. Available at: https://courses.lumenlearning.com/boundless-worldhistory/chapter/the-agricultural-revolution/. Accessed 11 May 2021.

Luna, A. G., Ferguson, B. G., Giraldo, O., Schmook, B. and Aldasoro Maya, E. M. (2019). Agroecology and restoration ecology: fertile ground for Mexican peasant territoriality? *Agroecology and Sustainable Food Systems* 43(10), 1174-1200. DOI: 10.1080/21683565.2019.1624284.

Luna-Gonzalez, D. V. and Sorensen, M. (2018). Higher agrobiodiversity is associated with improved dietary diversity, but not child anthropometric status, of Mayan achí people of Guatemala. *Proceedings of the International Astronomical Union* 21(11), 2128-2141. DOI: 10.1017/S1368980018000617.

Lundgren, J. G., McDonald, T. M., Rand, T. A. and Fausti, S. W. (2015). Spatial and numerical relationships of arthropod communities associated with key pests of maize. *Journal of Applied Entomology* 136, 446456. DOI: 10.1111/jen.12215.

Lusser, M., Parisi, C., Plan, D. and Rodríguez-Cerezo, E. (2011). New plant breeding techniques: state-of-the-art and prospects for commercial development. JRC Report EUR 24760 EN. European Commission Joint Research Centre (JRC), Seville. Available at: http://ftp.jrc.es/EURdoc/JRC63971.pdf.

Mabhaudhi, T., Chibarabada, T. P., Petrova Chimonyo, V. G., Murugani, V. G., Pereira, L. M., Sobratee, N., Govender, L., Slotow, R. and Thembinkosi Modi, A. (2019). Mainstreaming underutilized indigenous and traditional crops into food systems: a South African perspective. *Sustainability* 11, 172. DOI: 10.3390/su11010172.

Maciejczak, M. (2018). Quality as value-added bioeconomy: analysis of the EU policies and empirical evidence from polish agriculture. *AgBioForum* 21(2), 86-96.

Mäder, P., Fließbach, A., Dubois, D., Gunst, L., Fried, P. and Niggli, U. (2002a). Soil fertility and biodiversity in organic farming. *Science* 296(5573), 1694-1697.

MAFRA (2021). *Feed Efficiency in Feedlot Production*. Ministry of Agriculture, Food, and Rural Affairs, Guelph. Available at: http://www.omafra.gov.on.ca/english/livestock/beef/news/vbn0218a2.htm. Accessed 8 July 2021.

Maia, M. F. and Moore, S. J. (2011). Plant-based insect repellents: a review of their efficacy, development and testing. *Malaria Journal* 10 (Suppl. 1). DOI: 10.1186/1475-2875-10-S1-S11.

Mair, J. and Reischauer, G. (2017). Capturing the dynamics of the sharing economy: institutional research on the plural. *Technological Forecasting and Social Change* 125, 11-20.

Maity, S. (2018). *What is Digital Farming? Discussion with Tobias Menne, Head of Digital Farming at BASF*. Capgemini World, Paris. Available at: https://www.capgemini.com/2018/04/what-is-digital-farming-discussion-with-tobias-menne-head-of-digital-farming-basf/. Accessed 7 January 2020.

Makate, C. (2019). Effective scaling of climate smart agriculture innovations in African smallholder agriculture: a review of approaches, policy, and institutional strategy needs. *Environmental Science and Policy* 96, 37–51.

Makate, C., Makate, M. and Mango, N. (2018). Farm household typology and adoption of climate-smart agriculture practices in smallholder farming systems of southern Africa. *African Journal of Science, Technology, Innovation and Development* 10(4), 421–439.

Malak-Rawlikowska, A., Majewski, E., Was, A., Borgen, S. O., Csillag, P., Donati, M., Freeman, R., Hoàng, V., Lecoeur, J. L., Mancini, M. C., Nguyen, A., Saïdi, M., Tocco, B., Török, Á., Veneziani, M., Vittersø, G. and Wavresky, P. (2019). Measuring the economic, environmental, and social sustainability of short food supply chains. *Sustainability* 11(15), 4004.

Malik, Z., Ahmad, M., Abassi, G. H., Dawood, M., Hussain, A. and Jamil, M. (2017). Agrochemicals and soil microbes: interaction for soil health. In: Hashmi, M. Z., Kumar, V., Varma, A. (Eds) *Xenobiotics in the Soil Environment, Soil Biology*. Springer International Publishing, Cham, pp. 139–152. DOI: 10.1007/978-3-319-47744-2_11.

Malkanthi, S. H. P. (2021). Outlook of present organic agriculture policies and future needs in Sri Lanka. *Science Journal - Warsaw University of Life Sciences* 21(3), 55–72.

Malthus, T. R. (1798). *An Essay on the Principle of Population: Or a Vview of Its Past and Present Effects on Human Happiness*. J. Johnson, London.

Marcelis, L. (2019). *Vertical Farming: Plants in Control*. Wageningen University, Wageningen.

Marchetti, L., Cattivelli, V., Cocozza, C., Salbitano, F. and Marchetti, M. (2020). Beyond sustainability in food systems: perspectives from agroecology and social innovation. *Sustainability* 12(18), 7524. DOI: 10.3390/su12187524.

Mares, T. M. and Alkon, A. H. (2011). Mapping the food movement: addressing inequality and neoliberalism. *Environment and Society: Advances in Research* 2, 68–86.

Marinari, S., Mancinelli, R., Campiglia, E. and Grego, S. (2006). Chemical and biological indicators of soil quality in organic and conventional farming systems in Central Italy. *Ecological Indicators* 6(4), 701–711.

Markantonatou, M. (2016). Growth critique in the 1970s crisis and today: malthusianism, social mechanics, and labor discipline. *New Political Science* 38(1), 23–43.

Marquis, C. (2021). *rePlant Fuels Food System's Shift to Regenerative Agriculture with $2 Billion Soil Fund and Support for Farmers*. Forbes Business Council, Boston, MA. 5 October. Available at: https://www.forbes.com/sites/christophermarquis/2021/10/05/replant-fuels-food-systems-shift-to-regenerative-agriculture-with-2b-soil-fund-and-su...

Marriott, E. E. and Wander, M. (2006). Qualitative and quantitative differences in particulate organic matter fractions in organic and conventional farming systems. *Soil Biology and Biochemistry* 38(7), 1527–1536.

Mars, R. (2005). *The Basics of Permaculture Design*. Chelsea Green, White River Junction.

Marschner, M. P. (2012). *Mineral Nutrition of Higher Plants*. Elsevier, New York. DOI: 10.1016/C2009-0-63043-9.

Marsden, T. (2013). From post-productionism to reflexive governance: contested transitions in securing more sustainable food futures. *Journal of Rural Studies* 29, 123-134.

Marsden, T. (2016). Exploring the rural eco-economy: beyond neoliberalism. Sociologia Ruralis VC 2016 European society for rural sociology. *Sociologia Ruralis* 56(4), October. DOI: 10.1111/soru.12139.

Marsden, T., Banks, J. and Bristow, G. (2000). Food supply chain approaches: exploring their role in rural development. *Sociologia Ruralis* 40(4), 424-438. DOI: 10.1111/soru.2000.40.issue-4.

Marsden, T., Murdoch, J., Lowe, P., Munton, R. and Flynn, A. (1993). *Constructing the Countryside*. UCL Press, London.

Marteau, T. M., Hollands, G. J. and Fletcher, P. C. (2012). Changing human behaviour to prevent disease: the importance of targeting automatic processes. *Science* 337(6101), 1492-1495.

Martin, A., Coolsaet, B., Corbera, E., Dawson, N., Fisher, J., Franks, P., Mertz, O., Pascual, U., Rasmussen, L. and Ryan, C. (2018). Land use intensification: the promise of sustainability and the reality of trade-offs. In: Suich, H., Howe, C. and Mace, G. (Eds) *Ecosystem Services and Poverty Alleviation*. Routledge, London.

Martinez-Alier, J. (2011). The EROI of agriculture and its use of the via campesina. *Journal of Peasant Studies* 38(1), 145-160.

Martínez-Alier, J., Pascual, U., Vivien, F. D. and Zaccai, E. (2010). Sustainable de-growth: mapping the context, criticisms, and future prospects of an emergent paradigm. *Ecological Economics* 69(9), 1741-1747. DOI: 10.1016/j.ecolecon.2010.04.017.

Martinez-Torres, M. E. and Rosset, P. M. (2010). La Via Campesina: the birth and evolution of a transnational social movement. *Journal of Peasant Studies* 37(1), 149-175.

Martínez-Torres, M. E. and Rosset, P. M. (2014). Diálogo de saberes in la Vía campesina: food sovereignty and agroecology. *The Journal of Peasant Studies* 41(6), 979-997. DOI: 10.1080/03066150.2013.872632.

Martin-Guay, M. O., Paquette, A., Dupras, J. and Rivest, D. (2018). The new green revolution: sustainable intensification of agriculture by intercropping. *Science of the Total Environment* 615, 767-772. DOI: 10.1016/j.scitotenv.2017.10.024.

Mattick, C. S., Landis, A. E. and Allenby, B. R. (2015). A case for systemic environmental analysis of cultured meat. *Journal of Integrative Agriculture* 14(2), 249-254.

Maxwell, D., Levin, C. and Csete, J. (1998). Does urban agriculture help prevent malnutrition? Evidence from Kampala. *Food Policy* 23(5), 411-424.

May, J. (2017). Food security and nutrition: impure, complex and wicked? Food Security SA Working Paper Series No.002; DST-NRF Centre of Excellence in Food Security, Cape Town, South Africa. University of the Western Cape, Cape Town, South Africa.

Maye, D. (2018). Examining innovation for sustainability from the bottom UP: an analysis of the Permaculture Community in England. Sociologia Ruralis VC 2016 European Society for Rural Sociology. *Sociologia Ruralis* 58(2), April. DOI: 10.1111/soru.12141.

Maye, D. and Duncan, J. (2017). Understanding Sustainable Food System Transitions: Practice, Assessment and Governance. Sociologia Ruralis VC 2017 European Society for Rural Sociology. *Sociologia Ruralis* 57(3), July. DOI: 10.1111/soru.12177.

Mayer Labba, I. C., Steinhausen, H., Almius, L., Bach Knudsen, K. E. and Sandberg, A. S. (2022). Nutritional composition and estimated iron and zinc bioavailability of meat substitutes available on the Swedish market. *Nutrients* 14(19), 3903. DOI: 10.3390/nu14193903.

Maysinger, D. (2007). Nanoparticles and cells: good companions and doomed partnerships. *Organic and Biomolecular Chemistry* 5(15), 2335–2342.

Mbow, C., Rosenzweig, C., Barioni, L. G., Benton, T. G., Herrero, M., Krishnapillai, M., Liwenga, E., Pradhan, P., Rivera-Ferre, M. G., Sapkota, T., Tubiello, F. N. and Xu, Y. (2019). Food security, chapter 5. Special Report on Climate Change and Land. Intergovernmental Panel on Climate Change (IPCC). Available at: https://www.ipcc.ch/srccl/chapter/chapter-5/.

McCarthy, M., Rich, E., Smith, S., Mitchell, L. and Uren, S. (2020). *Growing Our Future: Scaling Regenerative Agriculture in the United States of America*. Forum for the Future, New York.

McClements, D. J., Decker, E. A., Park, Y. and Weiss, J. (2009). Structural design principles for delivery of bioactive components in nutraceuticals and functional foods. *Critical Reviews in Food Science and Nutrition* 49(6), 577–606. DOI: 10.1080/10408390902841529.

McClintock, N. (2014). Radical, reformist, and garden-variety neoliberal: coming to terms with urban agriculture's contradictions. *Local Environment* 19(2), 147–171.

McEntee, J. (2010). Contemporary and traditional localism: a conceptualization of rural local food. *Local Environment* 15(9-10), 785–803.

McEwen, F. and Stephenson, G. (1979). *The Use and Significance of Pesticides in the Environment*. John Wiley and Sons, New York, Chichester, Brisbane, and Toronto.

McFadden, S. (2004). The history of community supported agriculture. *Rodale Inst.* Available at: https://rodaleinstitute.org/blog/the-history-of-community-supported-agriculture/.

McFarlane, I. D., Jones, P. J., Park, J. R. and Tranter, R. B. (2018). Identifying GM crops for future cultivation in the EU through a Delphi forecasting exercise. *AgriculturistsBioForum* 21(1), 35-43.

McGuire, A. (2018). Regenerative agriculture: solid principles, extraordinary claims. Available at: http://csanr.wsu.edu/regen-ag-solid-principlesextraordinary-claims/.

McKeon, N. (2015). *Food Security Governance. Empowering Communities, Regulating Corporations*. Routledge, London and New York.

McMichael, P. (2005). Global development and the corporate food regime. In: Buttel, F. H. and McMichael, P. (Eds) *New Directions in the Sociology of Global Development*, Vol. 11. Elsevier, Oxford, pp. 229–267.

McMichael, P. (2009a). A food regime analysis of the 'world food crisis'. *Agriculture and Human Values* 26(4), 281–295. 10.1007/s10460-009-9218-5.

McMichael, P. (2009b). A food regime genealogy. *Journal of Peasant Studies* 36(1), 139–169.

McMichael, P. (2012). The land grab and corporate food regime restructuring. *The Journal of Peasant Studies* 39(3-4), 681–701. DOI: 10.1080/03066150.2012.661369.

McMichael, P. (2013). Land grabbing as security mercantilism in international relations. *Globalizations* 10(1), 47–64. DOI: 10.1080/14747731.2013.760925.

McMichael, P. (2014). Historicizing food sovereignty. *Journal of Peasant Studies* 41(6), 933–957. DOI: 10.1080/03066150.2013.876999.

McMichael, P. (2016). Commentary: food regime for thought. *The Journal of Peasant Studies* 43(3), 648–670. DOI: 10.1080/03066150.2016.1143816.

MDA (2004). Ministério do desenvolvimento agrário, secretaria de agricultura familiar, grupo de trabalho ater. Política Nacional de Assistência Técnica e Extensão Rural: Versão Final: 25/05/2004, Brasilia, 12 p.

Meadows, D. H., Meadows, G., Randers, J. and Behrens III, W. W. (1972). *The Limits to Growth*. Universe Books, New York.

Méda, Y. J. M., Egyir, I. S., Baptist, J. B. D. and Jatoe, D. (2017). The Green economics of conventional, organic and genetically modified crops farming: a review. *International Journal of Green Economics*, January. DOI: 10.1504/IJGE.2017.082714.

Meerow, S., Newell, J. P. and Stults, M. (2016). Defining urban resilience: a review. *Landscape and Urban Planning* 147, 38–49.

Meesterburrie, A. and Dupuy, L. (2018). *Spectaculaire Winst Voor GroenLinks in Groningen*. NRC, Amersfoort. Available at: https://www.nrc.nl/nieuws/2018/11/21/herindelingsverkie zingen-stembussen-in-37-fusiegemeenten-geslotena2756135.

Meier, T., Christen, O., Semler, E., Jahreis, G., Voget-Kleschin, L., Schrode, A. and Artmann, M. (2014). Balancing virtual land imports by a shift in the diet. Using a land balance approach to assess the sustainability of food consumption. Germany as an example. *Appetite* 74, 20–34. DOI: 10.1016/j.appet.2013.11.006.

Mendes, J., Pinho, T. M., Neves dos Santos, F., Sousa, J. J., Peres, E., Boaventura-Cunha, J., Cunha, M. and Morais, R. (2020). Smartphone applications targeting precision agriculture practices–a systematic review. *Agronomy* 10(6), 855. DOI: 10.3390/agronomy10060855.

Mendonça, R. D. D., Lopes, A. C. S., Pimenta, A. M., Gea, A., Martinez-Gonzalez, M. A. and Bes-Rastrollo, M. (2017). Ultra-processed food consumption and the incidence of hypertension in a Mediterranean cohort: the Seguimiento Universidad de Navarra Project. *American Journal of Hypertension* 30(4), 358–366.

Mendonça, R. D. D., Pimenta, A. M., Gea, A., de la Fuente-Arrillaga, C., Martinez-Gonzalez, M. A., Lopes, A. C. S. and Bes-Rastrollo, M. (2016). Ultra-processed food consumption and risk of overweight and obesity: the University of Navarra Follow-UP (SUN) cohort study. *American Journal of Clinical Nutrition* 104(5), 1433–1440.

Meng, F., Qiao, Y., Wu, W., Smith, P. and Scott, S. (2017). Environmental impacts and production performances of organic agriculture in China: a monetary valuation. *Journal of Environmental Management* 188, 49–57.

Mengistu, F. and Assefa, E. (2019). Farmers' decision to adopt watershed management practices in Gibe basin, southwest Ethiopia. *International Soil and Water Conservation Research* 7(4), 376–387. DOI: 10.1016/j.iswcr.2019.08.006.

Mercury News (2014). *As Organic Food Goes Mainstream, Consumers Can Expect Price Breaks*. Mercury News, California, 19 May.

Mercy Corps (2020). *Covid-19 Rapid Market Impact Report*. Mercy Corps, Portland, OR.

Merfield, C. N. (2019). *An Analysis and Overview of Regenerative Agriculture*. Report number 2-2019. The BHU Future Farming Centre, Lincoln.

Meticulous Research (2021). *Biopesticides Market Worth $9.6 Billion by 2028*. 27 September 2021. Available at: https://www.globenewswire.com/news-release/2021/09/27/2303450/0/en/Biopesticides-Market-Worth-9-6-Billion-by-2028-Exclusive-Report-by-Meticulous-Research.html.

Michalopoulos, S. (2015). Europe entering the era of 'precision agriculture'. *EurActiv.com*. Available at: http://www.euractiv.com/sections/innovation-feeding-world/europe-entering-era-precision-agriculture-318794. Accessed 15 November 2023.

Michelinia, L., Principato, L. and Iasevolia, G. (2018). Understanding food sharing models to tackle sustainability challenges. *Ecological Economics* 145, 205–217.

Migliore, G., Schifani, G., Guccione, G. D. and Cembalo, L. (2014). Food community networks as leverage for social embeddedness. *Journal of Agricultural and Environmental Ethics* 27(4), 549–567.

Migliore, G., Schifani, G. and Cembalo, L. (2015). Opening the black box of food quality in the short supply chain: effects of conventions of quality on consumer choice. *Food Quality and Preference* 39, 141–146.

Migliorini, P., Bàrberi, P., Bellon, S., Gaifami, T., Gkisakis, V. D., Peeters, A. and Wezel, A. (2020). Controversial topics in agroecology: a European perspective. *International Journal of Agriculture and Natural Resources* 47(3), 159–173. Available at: www.ijanr.cl.

Migliorini, P., Galioto, F., Chiorri, M. and Vazzana, C. (2018). An integrated sustainability score based on agro-ecological and socioeconomic indicators. A case study of stockless organic farming in Italy. *Agroecology and Sustainable Food Systems* 42(8), 859–884. DOI: 10.1080/21683565.2018.1432516.

Mignot, J. P. and Poncet, C. (2001). The industrialization of knowledge in life sciences: convergence between public research policies and industrial strategies. *Cahiers de Recherche du Creden*, 20 January .

Mijatović, D., Van Oudenhoven, F., Eyzaguirre, P. and Hodgkin, T. (2013). The role of agricultural biodiversity in strengthening resilience to climate change: towards an analytical framework. *International Journal of Agricultural Sustainability* 11(2), 95–107. DOI: 10.1080/14735903.2012.691221.

Mikler, J. (2014). *The Handbook of Global Companies*. John Wiley and Sons Inc., New York. Available at: http://public.eblib.com/EBLPublic/PublicView.do?ptiID=1158410.

Mikula, K., Izydorczyk, G., Skrzypczak, D., Mironiuk, M., Moustakas, K., Witek-Krowiak, A. and Chojnacka, K. (2020). Controlled release micronutrient fertilizers for precision agriculture–a review. *Science of the Total Environment* 712, 136365.

Miles, A., DeLonge, M. S. and Carlisle, L. (2017). Triggering a positive research and policy feedback cycle to support a transition to agroecology and sustainable food systems. *Agroecology and Sustainable Food Systems* 41(7), 855–879. DOI: 10.1080/21683565.2017.1331179.

Milinchuk, A. (2020). Is regenerative agriculture profitable? *Forbes*. Available at: https://www.forbes.com/sites/forbesfinancecouncil/2020/01/30/is-regenerative-agriculture-profitable/.

Miller, N. J., Griffin, T. W., Ciampitti, I. A. and Sharda, A. (2019). Farm adoption of embodied knowledge and information intensive precision agriculture technology bundles. *Precision Agriculture* 20(2), 348–361.

Mitchell, C. A. and Sheibani, F. (2020). In: Kozai, T., Niu, G. and Takagaki, M. (Eds) *Plant Factory*. Academic Press, Cambridge, MA, pp. 167–184. DOI: 10.1016/B978-0-12-816691-8.00010-8.

Mittal, A. (2009). The blame game: understanding structural causes of the food crisis. In: Clapp, J. and Cohen, M. (Eds) *The Global Food Crisis: Governance Challenges and Opportunities*. WLU Press, Waterloo, pp. 13–28.

Mizik, T. (2021). Climate-smart agriculture on small-scale farms: a systematic literature review. *Agronomy* 11(6), 1096. DOI: 10.3390/agronomy11061096.

Moaveni, P. and Kheiri, T. (2011). TiO nano particles affected on maize (*Zea mays* L). 2nd International Conference on Agricultural and Animal Science, Vol. 22, 25–27 November, 2011, Maldives. IACSIT Press, Singapore, pp. 160–163.

Mohr, S. and Evans, G. (2013). Projections of future phosphorus production. *Philica - The Instant, Open Access Journal of Everything*, Article number 380.
Molden, D. E. (2007). Water for food, water for life: a comprehensive assessment of water management in agriculture. EarthScan and International Water Management Institute, London and Colombo.
Mollison, B. (1988). *Permaculture: A Designer's Manual*. Tagari, Tasmania.
Mollison, B. (2003). The foundation yearbook of the permaculture academy. Available at: http://www.patriciamichaeldesign.com/texts/YearBook.pdf. Accessed 14 May 2013.
Mollison, B. and Holmgren, D. (1978). *Permaculture One: A Perennial Agricultural System for Human Settlements*. Tagari, Tyalgum.
Mollison, B. C. and Slay, R. M. (1997). *Introduction to Permaculture* (2nd edn). Tagari, Tasmania.
Monforti-Ferrario, F., Monforti, F., Dallemand, J. F., Pascua, I. P., Motola, V., Banja, M., Scarlat, N., Medarac, H., Castellazzi, L., Labanca, N., Bertoldi, P., Pennington, D., Goralczyk, M., Schau, E. M., Saouter, E., Sala, S., Notarnicola, B., Tassielli, G. and Renzulli, P. (2015). *Energy Use in the EU Food Sector: State of Play and Opportunities for Improvement*. Publications Office, Luxembourg.
Monier, V., Mudgal, S., Escalon, V., O'Connor, C., Gibon, T., Anderson, G. and Morton, G. (2010). *Preparatory Study on Food Waste across EU 27*. European Commission, Directorate-General for the Environment, Brussels.
Monteiro, A., Santos, S. and Gonçalves, P. (2021). Precision agriculture for crop and livestock farming–brief review. *Animals* 11(8), 2345. DOI: 10.3390/ani11082345.
Monteiro, C. A., Levy, R. B., Claro, R. M., de Castro, I. R. and Cannon, G. (2011). Increasing consumption of ultra-processed foods and likely impact on human health: evidence from Brazil. *Public Health Nutrition* 14(1), 5-13.
Monteiro, C. A. and Cannon, G. (2012). The impact of transnational "big food" companies on the South: a view from Brazil. *PLoS Medicine* 9(7), e1001252.
Monteiro, C. A., Moubarac, J. C., Cannon, G., Ng, S. W. and Popkin, B. (2013). Ultra-processed products are becoming dominant in the global food system. *Obesity Reviews* 14(Suppl. 2), 21-28.
Montes, F., Meinen, R., Dell, C., Rotz, A., Hristov, A. N., Oh, J., Waghorn, G., Gerber, P. J., Henderson, B., Makkar, H. P. S. and Dijkman, J. (2013). SPECIAL TOPICS - mitigation of methane and nitrous oxide emissions from animal operations: II. A review of manure management mitigation options. *Journal of Animal Science* 91(11), 5070-5094.
Montgomerie, J. (2008). Bridging the critical divide: global finance, financialisation and contemporary capitalism. *Contemporary Politics* 14(3), 233-252.
Montgomery, D. R. (2007). Soil erosion and agricultural sustainability. *Proceedings of the National Academy of Sciences of the United States of America* 104(33), 13268-13272.
Montgomery, D. R. and Biklé, A. (2021). Soil health and nutrient density: beyond organic vs. conventional farming. *Frontiers in Sustainable Food Systems* 5, 699147. DOI: 10.3389/fsufs.2021.699147.
Montgomery, D. R., Biklé, A., Archuleta, R., Brown, P. and Jordan, J. (2022). Soil health and nutrient density: preliminary comparison of regenerative and conventional farming. *PeerJ* 10, e12848. DOI: 10.7717/peerj.12848.
Moodie, R., Stuckler, D., Monteiro, C., Sheron, N., Neal, B., Thamarangsi, T., Lincoln, P. and Casswell, S. (2013). Profits and pandemics: prevention of harmful effects of tobacco, alcohol, and ultra-processed food and drink industries. *The Lancet* 381, 670-679. DOI: 10.1016/S0140-6736(12)62089-3.

Moore, J. W. (2010). Cheap food & bad money food, frontiers, and financialization in the rise and demise of neoliberalism. *Review* 33(2-3), 225-261.

Moore, J. W. (2014). Cheap food & bad climate: from surplus value to negative-value in the capitalist world-ecology. *Critical Historical Studies* 2(1), 1-42. University of Chicago Press.

Moorsom, T. L., Rao, S., Gengenbach, H. and Huggins, C. (2020). Food security and the contested visions of agrarian change in Africa. *Canadian Journal of Development Studies / Revue Canadienne d'études du Développement* 41(2), 212-223. DOI: 10.1080/02255189.2020.1786356.

Moragues-Faus, A. and Sonnino, R. (2018). Re-assembling sustainable food cities: an exploration of translocal governance and its multiple agencies. *Urban Studies* 56(4), 778-794.

Morée, A. L., Beusen, A. H. W., Bouwman, A. F. and Willems, W. J. (2013). Exploring global nitrogen and phosphorus flows in urban wastes during the twentieth century. *Global Biogeochemical Cycles* 25(27), 1-11. DOI: 10.1002/gbc.20072.

Moreira, P. V. L., Baraldi, L. G., Moubarac, J. C., Monteiro, C. A., Newton, A., Capewell, S. and O'Flaherty, M. (2015). Comparing different policy scenarios to reduce the consumption of ultra-processed foods in UK: impact on cardiovascular disease mortality using a modelling approach. *PLoS ONE* 10(2), e0118353. DOI: 10.1371/journal.pone.0118353.

Morel, K., San Cristobal, M. and Léger, F. G. (2017). Small can be beautiful for organic market gardens: an exploration of the economic viability of French microfarms using MERLIN. *Agricultural Systems* 158, 39-49.

Morgan, D. (1979). *Merchants of Grain: the Power and Profits of the Five Giant Companies at the Centre of the World's Food Supply*. Viking Press, New York.

Morgan, K. and Sonnino, R. (2010). The urban foodscape: world cities and the new food equation. *Cambridge Journal of Regions, Economy and Society* 3(2), 209-224.

Morone, P. (2019). Food waste: challenges and opportunities for enhancing the emerging bioeconomy. *Journal of Cleaner Production* 221, 10-16. DOI: 10.1016/j.jclepro.2019.02.258.

Morone, P., Falcone, P. M., Imbert, E. and Morone, A. (2018). Does food sharing leads to food waste reduction? An experimental analysis to assess challenges and opportunities of a new consumption model. *Journal of Cleaner Production* 185, 749-760. DOI: 10.1016/j.jclepro.2018.01.208.

Morris, D. (1987). Healthy cities: self-reliant cities. *Health Promotion International* 2(2), 169-176.

Morris, J. and Bate, R. (1999). *Fearing Food: Risk, Health, and Environment*. Butterworth-Heinemann, Oxford.

Moss, B. (2008). Water pollution by agriculture. *Philosophical Transactions of the Royal Society of London. Series B, Biological Sciences* 363(1491), 659-666.

Moss, M. (2013). *Salt, Sugar, Fat: How the Food Giants Hooked Us*. Random House, New York.

Mossa, A. H. (2016). Green pesticides: essential oils as biopesticides in insect-pest management. *Journal of Environmental Science and Technology* 9(5), 354-378.

Mostafalou, S. and Abdollahi, M. (2017). Pesticides: an update of human exposure and toxicity. *Archives of Toxicology* 91(2), 549-599. DOI: 10.1007/s00204-016-1849-x.

Mottet, A., de Haan, C., Falcucci, A., Tempio, G., Opio, C. and Gerber, P. (2017). Livestock: on our plates or eating at our table? A new analysis of the feed/food debate. *Global Food Security* 14, 1-8. DOI: 10.1016/j.gfs.2017.01.001.

Mouat, M. J., Prince, R. and Roche, M. M. (2019). Making value out of ethics: the emerging economic geography of lab-grown meat and other animal-free food products. *Economic Geography* 95(2), 136–158.

Moudrý, J., Bernas, J., Moudrý, J., Konvalina, P., Ujj, A., Manolov, I., Stoeva, A., Rembialkowska, E., Stalenga, J., Toncea, I., Fitiu, A., Bucur, D., Lacko-Bartošová, M. and Macák, M. (2018). Agroecology development in Eastern Europe–cases in Czech Republic, Bulgaria, Hungary, Poland, Romania, and Slovakia. *Sustainability* 10(5), 1311.

Mount, P. (2012). Growing local food: scale and local food system governance. *Agriculture and Human Values* 29(1), 107–121.

Mourad, M. (2016). Recycling, recovering and preventing "food waste": competing solutions for food systems sustainability in the United States and France. *Journal of Cleaner Production* 126, 461–477.

Moyer, J., Smith, A., Rui, Y. and Hayde, J. (2020). Regenerative agriculture and the soil carbon solution. Rodale Institute, Kutztown, PA. RodaleInstitute.org.

Muangprathu, J., Boonnam, N., Kajornkasirat, S., Lekbangpong, N., Wanichsombat, A. and Nillaor, P. (2019). IoT and agriculture data analysis for smart farm. *Computers and Electronics in Agriculture* 156, 467–474.

Mucheru-Muna, M., Mugendi, D., Pypers, P., Mugwe, J., Kung'u, J. Vanlauwe, B. and Merckx, R. (2014). Enhancing maize productivity and profitability using organic inputs and mineral fertiliser in Central Kenya small-holder farms. *Expl. Agriculturists* 50, 250–269.

Mueller, N. D., Gerber, J. S., Johnston, M., Ray, D. K., Ramankutty, N. and Foley, J. A. (2012). Closing yield gaps through nutrient and water management. *Nature* 490(7419), 254–257.

Mukhopadhyay, S. S. (2014). Nanotechnology in agriculture: prospects and constraints. *Nanotechnology, Science and Applications* 7, 63–71. DOI: 10.2147/NSA.S39409.

Muldowney, J., Mounsey, J. and Kinsella, L. (2013). Agriculture in the climate change negotiations; ensuring that food production is not threatened. *Animal* 7 (Suppl. 2), 206–211.

Muller, A., Schader, C., El-Hage Scialabba, N., Bruggemann, J., Isensee, A., Erb, K. H., Smith, P., Klocke, P., Leiber, F., Stolze, M. and Niggli, U. (2017). Strategies for feeding the world more sustainably with organic agriculture. *Nature Communications* 8(1), 1290. DOI: 10.1038/s41467-017-01410-w.

Mundler, P. and Laughrea, S. (2016). The contributions of short food supply chains to territorial development: a study of three Quebec territories. *Journal of Rural Studies* 45, 218–229.

Mundler, P. and Rumpus, L. (2012). The energy efficiency of local food systems: a comparison between different modes of distribution. *Food Policy* 37(6), 609–615.

Munné, A., Ginebreda, A. and Prat, N. (Eds) (2015). *Experiences from Ground, Coastal and Transitional Water Quality Monitoring: The EU Water Framework Directive Implementation in the Catalan River Basin District* (Vol. 43). Springer, New York.

Munns, D. P. D. (2014). The awe in which biologists hold physicists: Frits Went's first phytotron at Caltech, and an experimental definition of the biological environment. *History and Philosophy of the Life Sciences* 36(2), 209–231.

Muñoz, J. P. (2010). Constituyente, gobierno de transición y soberanía alimentaria en Ecuador. In: Gascón, J. and Montagut, X. (Eds) *Cambio de rumbo en las políticas agrarias latinoamericanas? Estado, movimientos sociales campesinos y soberanía alimentaria*. Icaria Editorial, Barcelona, pp. 151–168.

Murdoch, J., Marsden, T. and Banks, J. (2000). Quality, nature, and embeddedness: some theoretical considerations in the context of the food sector. *Economic Geography* 76(2), 107–125.

Murphy, S., Burch, D. and Clapp, J. (2012). *Cereal Secrets: The World's Largest Grain Traders and Global Agriculture*. Oxfam International, Nairobi. Available at: http://www.oxfam.org/en/grow/policy/cerealsecrets-worlds-largest-grain-traders-global-agriculture.

Murray, U., Gebremedhin, Z., Brychkova, G. and Spillane, C. (2016). Smallholder farmers and climate smart agriculture: technology and labour-productivity constraints amongst women smallholders in Malawi. *Gender, Technology and Development* 20(2), 117–148.

Mutenje, M. J., Farnworth, C. R., Stirling, C., Thierfelder, C., Mupangwa, W. and Nyagumbo, I. A. (2019). Cost-benefit analysis of climate-smart agriculture options in Southern Africa: Balancing gender and technology. *Ecological Economics* 163, 126–137.

Muthayya, S., Rah, J. H., Sugimoto, J. D., Roos, F. F., Kraemer, K. and Black, R. E. (2013). The global hidden hunger indices and maps: an advocacy tool for action. *PLoS ONE* 8(6), e67860. DOI: 10.1371/journal.pone.0067860.

MWA (2015). Massachusetts Workforce Alliance, Metropolitan Area Planning Council, Franklin Regional Council of Governments, Pioneer Valley Planning Commission. Massachusetts Local Food Action Plan.

Naderi, M. R. and Abedi, A. (2012). Application of nanotechnology in agriculture and refinement of environmental pollutants. *Journal of Nanotechnology* 11(1), 18–26. Available at: http://jm.birjand.ac.ir/archive_jou/escs/139261/4P-ESCS61-A91216.pdf.

Naderi, M. R. and Danesh-Shahraki, A. (2013). Nanofertilizers and their roles in sustainable agriculture. *International Journal of Agriculture and Crop Sciences* 5(19), 2229–2232.

Nadiminti, P. P., Dong, Y. D., Sayer, C., Hay, P., Rookes, J. E., Boyd, B. J. and Cahill, D. M. (2013). Nanostructured liquid crystalline particles as an alternative delivery vehicle for plant agrochemicals. *ACS Applied Materials and Interfaces* 5(5), 1818–1826.

Nair, P. K. R. (2014). Grand challenges in agroecology and land use systems. *Frontiers in Environmental Science*. Available at: http://journal.frontiersin.org/article/10.3389/fenvs.2014.00001/full.

Nair, R., Varghese, S. H., Nair, B. G., Maekawa, T., Yoshida, Y. and Kumar, D. S. (2010). Nanoparticulate material delivery to plants. *Plant Science* 179(3), 154–163. DOI: 10.1016/j.plantsci.2010.04.012.

Nanotechproject (2022). An inventory of nanotechnology-based consumer products introduced on the market. Available at: https://www.nanotechproject.tech/cpi/products/nanoceuticalstm-slim-shake-chocolate/.

Nature Editorial (2019). Counting the hidden $12-trillion cost of a broken food system. *Nature* 574(7778). DOI: 10.1038/d41586-019-03117-y.

Navrozidis, I., Alexandridis, T. K., Dimitrakos, A., Lagopodic, A. L., Moshoud, D. and Zalidis, G. (2018). Identification of purple spot disease on asparagus crops across spatial and spectral scales. *Computers and Electronics in Agriculture* 148, 322–329.

Nawar, S., Corstanje, R., Halcro, G., Mulla, D. and Mouazen, A. M. (2017). Delineation of soil management zones for variable-rate fertilization: a review. *Advances in Agronomy* 143, 175–245. DOI: 10.1016/bs.agron.2017.01.003.

NCD-RisC (2016). Worldwide trends in diabetes since 1980: a pooled analysis of 751 population-based studies with 4.4 million participants. *The Lancet* 387(10027), 1513–1530. DOI: 10.1016/S0140-6736(16)00618-8.

NCD Risk Factor Collaboration (2019). Rising rural body-mass index is the main driver of the global obesity epidemic in adults. *Nature* 569(7755), 260–264. DOI: 10.1038/s41586-019-1171-x.

Ndakidemi, B., Mtei, K. and Ndakidemi, P. A. (2016). Impacts of synthetic and botanical pesticides on beneficial insects. *Agricultural Sciences* 7(6), 364–372. DOI: 10.4236/as.2016.76038.

Neo, P. (2020). Stupid and ideological: New Zealand industry rejects anti-GMO group suggestion to establish new food standards authority. *FoodNavigator Asia*, 20 September. Available at: https://www.foodnavigator-asia.com/Article/2020/09/28/Stupid-and-ideological-New-Zealand-industry-rejects-anti-GMO-group-suggestion-to-establishnew-food-standards-authority.

Neset, T. S. and Cordell, D. (2012). Global phosphorus scarcity: identifying synergies for a sustainable future. *Journal of the Science of Food and Agriculture* 92(1), 2–6. DOI: 10.1002/jsfa.4650. Epub 2011 Oct 3. PMID: 21969145.

New Scientist (2002). Much ado about nothing, 18 May. Available at: http://www.biotech-info.net/much_ado.html.

New Scientist (2021). Are vegan meat alternatives putting our health on the line? Available at: https://www.newscientist.com/article/mg25133581-600-are-vegan-meat-alternatives-putting-our-health-on-the-line/.

New Scientist (2022). Protein from plant-based 'meat' may be less well absorbed by the body. 22 June 2022. Available at: https://www.newscientist.com/article/2325589-protein-from-plant-based-meat-may-be-less-well-absorbed-by-the-body/.

Newbold, T., Hudson, L. N., Hill, S. L., Contu, S., Lysenko, I., Senior, R. A., Börger, L., Bennett, D. J., Choimes, A., Collen, B., Day, J., De Palma, A., Díaz, S., Echeverria-Londoño, S., Edgar, M. J., Feldman, A., Garon, M., Harrison, M. L., Alhusseini, T., Ingram, D. J., Itescu, Y., Kattge, J., Kemp, V., Kirkpatrick, L., Kleyer, M., Correia, D. L., Martin, C. D., Meiri, S., Novosolov, M., Pan, Y., Phillips, H. R., Purves, D. W., Robinson, A., Simpson, J., Tuck, S. L., Weiher, E., White, H. J., Ewers, R. M., Mace, G. M., Scharlemann, J. P. and Purvis, A. (2015). Global effects of land use on local terrestrial biodiversity. *Nature* 520, 45–50.

Newell, P. and Taylor, O. (2018). Contested landscapes: the global political economy of climate-smart agriculture. *The Journal of Peasant Studies* 45(1), 108–129. DOI: 10.1080/03066150.2017.1324426.

Newton, P., Civita, N., Frankel-Goldwater, L., Bartel, K. and Johns, C. (2020). What is regenerative agriculture? A review of scholar and practitioner definitions based on processes and outcomes. *Frontiers in Sustainable Food Systems* 4, 577723. DOI: 10.3389/fsufs.2020.577723. Journal homepage: http://www.ijcmas.com.

Ng'ang'a, S. K., Jalang'o, D. A. and Girvetz, E. H. (2019). Adoption of technologies that enhance soil carbon sequestration in East Africa. What influence farmers' decision? *International Soil and Water Conservation Research*. DOI: 10.1016/j.iswcr.2019.11.001.

NGIN (2002). *Bayer/Monsanto/DuPont*. Norfolk Genetic Information Network (NGIN), London. Available at: Ngin.tripod.com/070402a.htm. Accessed 22 January 2004.

NGLS Roundup (1997). The World Food SummitUnited Nations Non-Governmental Liaison Service. Available at: http://www.un-ngls.org/orf/documents/text/roundup/11WFS.

NGO Forum (1996). Profit for few or food for all: food sovereignty and security to eliminate the globalisation of hunger. Available at: http://www.foodsovereignty.org/Portals/0/documenti%20sito/Resources/Archive/Forum/1996/wfs+5_NGO_FORUM96.pdf.

Nicholls, C. I. and Altieri, M. A. (2018). Pathways for the amplification of agroecology. *Agroecology and Sustainable Food Systems* 42(10), 1170–1193. DOI: 10.1080/21683565.2018.1499578.

Nieberg, H., Offermann, F. and Zander, K. (2007). Organic farms in a changing policy environment: impacts of support payments, EU-enlargement, and Luxembourg reforms. *Organic Farming European Commission: Economic Policy*, Vol. 13. University of Hohenheim, Department of Farm Economics, Hohenheim.

Nielson, D., Meng, Y. T., Buyvolova, A. and Hakobyan, A. (2018). *Unleashing the Power of Digital on Farms in Russia – and Seeking Opportunities for Small Farms*. The World Bank, Washington, DC.

Niggli, U., Earley, J. and Ogorzalek, K. (2007). Organic agriculture and environmental stability of the food supply. International Conference on Agriculture and Food Security. FAO, Rome. Available at: ftp://ftp.fao.org/docrep/fao/meeting/012/ah950e.pdf.

Niggli, U. and Riedel, J. (2020). Agroecology empowers a new, solution-oriented dialogue. *Landbauforschung* 70, 15–20.

Niggli, U., Slabe, A., Schmid, O., Halberg, N. and Schlüter, M. (2008). *Vision for an Organic Food and Farming Research Agenda to 2025*. IFOAM-EU Group, Brussels. Available at: http://www.organic-research.org/index.html; http://orgprints.org/13439/.

Nijdam, D., Rood, T. and Westhoek, H. (2012). The price of protein: review of land use and carbon footprints from life cycle assessments. *Food Policy* 37(6), 760–770. DOI: 10.1016/j.foodpol.2012.08.002.

Nirupama, K. V., Adlin Jino Nesalin, J. and Tamizh Mani, T. (2019). Preparation and evaluation of microparticles containing charantin by solvent evaporation technique. *SSRG International Journal of Pharmacy and Biomedical Engineering (SSRG-IJPBE)* 6(1) Jan–Apr.

Nnaoaham, K. E., Sacks, G., Rayner, M., Mytton, O. and Gray, A. (2009). Modelling income group differences in the health and economic impacts of targeted food taxes and subsidies. *International Journal of Epidemiology* 38(5), 1324–1333.

Norgaard, R. B. (1984). Traditional agricultural knowledge: past performance, future prospects, and institutional implications. *American Journal of Agricultural Economics* 66(5), 874–878.

Norton, L., Johnson, P., Joys, A., Stuart, R., Chamberlain, D., Feber, R., Firbank, L., Manley, W., Wolfe, M., Hart, B., Mathews, F., Macdonald, D. and Fuller, R. J. (2009). Consequences of organic and non-organic farming practices for field, farm, and landscape complexity. *Agriculture, Ecosystems and Environment* 129(1–3), 221–227.

Nowak, B. (2021). Precision agriculture: where do we stand? A review of the adoption of precision agriculture technologies on field crops farms in developed countries. *Agricultural Research* 10(4), 515–522 DOI: 10.1007/s40003-021-00539-x.

Nowak, B., Nesme, T., David, C. and Pellerin, S. (2013a). Disentangling the drivers of fertilizing material inflows in organic farming. *Nutrient Cycling in Agroecosystems* 96(1), 79–91.

Nowak, B., Nesme, T., David, C. and Pellerin, S. (2013b). To what extent does organic farming rely on nutrient inflows from conventional farming? *Environmental Research Letters* 8(4), 044045.

Nowak, B., Nesme, T., David, C. and Pellerin, S. (2015). Nutrient recycling in organic farming is related to diversity in farm types at the local level. *Agriculture, Ecosystems and Environment* 204, 17–26. Elsevier, London.

NRC (2010). *Toward Sustainable Agricultural Systems in the 21st Century*. National Research Council. National Academies Press, Washington, DC. DOI: 10.17226/12832.

NRC (2015). *A Framework for Assessing Effects of the Food System*. National Research Council. National Academies Press, Washington, DC.

NRDC (2007). *Food Miles: Health Facts*. Natural Resources Defence Council, Los Angeles, CA.

NTTC (2003). National Technology Transfer Center. Wheeling College Inc. Available at: Nttc.edu/products/guide/seca02.html. Accessed 15 January 2004.

Nugent, R. (2011). *Bringing Agriculture to the Table: How Agriculture Can Play a Role in Preventing Chronic Disease*. The Chicago Council on Global Affairs, Chicago.

Nunes, K. (2021). Nestle to invest $1.3 billion in regenerative agriculture. *Food Business News*. 16 September. Available at: https://www.foodbusinessnews.net/articles/19616-nestle-to-invest-13-billion-in-regenerative-agriculture.

Nuzzo, A., Madonna, E., Mazzei, P., Spaccini, R. and Piccolo, A. (2016). In situ photopolymerization of soil organic matter by heterogeneous nano-TiO_2 and biomimetic metal-porphyrin catalysts. *Biology and Fertility of Soils* 52(4), 585–593.

NXTaltfoods (2023). Plant-based $114 billion meat giant cargill partners with cubiq foods to scale plant-based fat. Available at: https://www.nxtaltfoods.com/news/articles/plant-based/114-billion-meat-giant-cargill-partners-with-cubiq-foods-to-scale-plant-based-fat/.

Nyasimi, M., Amwata, D., Hove, L., Kinyangi, J. and Wamukoya, G. (2014). Evidence of impact: climatesSmart agriculture in Africa. CCAFS Working Paper 86. CGIAR Research Program on Climate Change, Agriculture and Food Security (CCAFS), Copenhagen.

O'Brien, K. and Sygna, L. (2013). Responding to climate change: the three spheres of transformation. *Proceedings of the Transformation in a Changing Climate* (Oslo, Norway, 19–21 June), 16–23.

O'Neill, D. W. (2012). Measuring progress in the degrowth transition to a steady state economy. *Ecological Economics* 84, 221–231.

O'Riordan, T. (1981). *Environmentalism*. Pion, London.

Oakland Institute (2018). *Agroecology Case Studies*. Oakland Institute, Oakland, CA. Available at: https://www.oaklandinstitute.org/agroecology-case-studies. Accessed 2 November 2018.

Odegard, I. Y. R. and Van der Voet, E. (2014). The future of food–scenarios and the effect on natural resource use in agriculture in 2050. *Ecological Economics* 97, 51–59.

Odum, E. P. (1969). The strategy of ecosystem development. *Science* 164(3877), 262–270. DOI: 10.1126/science.164.3877.262.

Odum, H. T. (1994). *Ecological and General Systems: an Introduction to Systems Ecology*. University Press of Colorado, Niwot.

OECD (2002). *Measuring Material Flows and Resource Productivity. Volume 1. The OECD Guide Glossary*. OECD, Paris.

OECD (2010). *Sustainable Management of Water Resources in Agriculture*. OECD, Paris.

OECD (2012). *OECD Environmental Outlook to 2050*. OECD, Paris.

OECD (2017). *Agriculture Policy Monitoring and Evaluation 2017*. OECD, Paris).

OECD (2018). *Human Acceleration of the Nitrogen Cycle: Managing Risks and Uncertainty*. OECD, Paris. DOI: 10.1787/9789264307438-en.

OECD (2019a). *Agricultural Policy Monitoring and Evaluation 2019*. OECD, Paris. Available at: https://www.oecd-ilibrary.org/docserver/39bfe6f3-en.pdf?

OECD (2019b). Impacts of agricultural policies on productivity and sustainability performance in agriculture: a literature review. Available at: http://www.oecd.org/tad/agricultural-policies/innovation-foodagriculture.htm.

OECD (2020). *Towards Sustainable Land Use: Aligning Biodiversity, Climate and Food Policies*. OECD, Paris. Available at: https://www.oecd.org/environment/resources/towards-sustainable-land-usealigning- biodiversity-climate-and-food-policies.pdf.

OECD (2021). *Making Better Policies for Food Systems*. OECD, Paris. DOI: 10.1787/ddfba4de-en.

OECD-FAO (2017). *Agricultural Outlook 2017-2026*. OECD/FAO, Paris/France.

OECD/FAO (2019). *OECD-FAO Agricultural Outlook 2019-2028*. OECD, Paris/Food and Agriculture Organization of the United Nations, Rome. DOI: 10.1787/agr_outlook-2019-en.

Offer, A., Pechey, R. and Ulijaszek, S. (2010). Obesity under affluence varies by welfare regimes: the effect of fast food, insecurity, and inequality. *Economics and Human Biology* 8(3), 297–308.

Oggioni, C., Cena, H., Wells, J. C. K., Lara, J., Celis-Morales, C. and Siervo, M. (2015). Association between worldwide dietary and lifestyle patterns with total cholesterol concentrations and DALYs for infectious and cardiovascular diseases: an ecological analysis. *Journal of Epidemiology and Global Health* 5(4), 315–325. DOI: 10.1016/j.jegh.2015.02.002.

OilPrice (2023). $300 million fake meat fund aims to decarbonize the food industry. Available at: https://oilprice.com/Latest-Energy-News/World-News/300-Million-Fake-Meat-Fund-Aims-To-Decarbonize-The-Food-Industry.

Olwande, J. and Ayieko, M. (2020). *Impact of Covid-19 on Food Systems and Rural Livelihoods in Kenya*. Covid-19 Country Report 2 – December 2020.

Omotayo, O. E. and Chukwuka, K. S. (2009). Soil fertility restoration techniques in Sub-Saharan Africa using organic resources. Review. *African Journal of Agricultural Research* 4(3), 144–150.

Omran, A. R. (1971). The epidemiologic transition: a theory of the epidemiology of population change. *Milbank Memorial Fund Quarterly* 49(4), 509–538.

Onyango, C. M., Nyaga, J. M., Wetterlind, J., Söderström, M. and Piikki, K. (2021). Precision agriculture for resource use efficiency in smallholder farming systems in sub-Saharan Africa: a systematic review. *Sustainability* 13(3), 1158. DOI: 10.3390/su13031158.

Onyeneke, R. U., Igberi, C. O., Uwadoka, C. O. and Aligbe, J. O. (2018). Status of climate-smart agriculture in southeast Nigeria. *GeoJournal* 83(2), 333–346.

Oosterveer, P. and Sonnenfel, D. (2012). *Food, Globalization, and Sustainability*. Earthscan, London.

Opitz, I., Berges, R., Piorr, A. and Krikser, T. (2015). Contributing to food security in urban areas: differences between urban agriculture and peri-urban agriculture in the Global North. *Agriculture and Human Values* 33(2), 341–358.

Orhangazi, O. (2008). Financialization and capital accumulation in the non-financial corporate sector: a theoretical and empirical investigation on the US economy: 1973–2003. *Cambridge Journal of Economics* 32(6), 863–886.

Orsini, F., Pennisi, G., Zulfiqar, F. and Gianquinto, G. (2020). Sustainable use of resources in plant factories with artificial lighting (PFALs). *European Journal of Horticultural Science* 85(5), 297–309.

Osborne, H. (2014). A short history of food banks, a modern phenomenon. *International Business Times*. Available at: https://www.ibtimes.co.uk/short-history-food-banks-modern-phenomenon-1445071. Accessed 20 October 2022.

Osty, P. L. (2008). Raisonnement à partir de cas et agronomie des territoires. *Revue d'Anthropologie des Connaissances* 2(2), 169–193. DOI: 10.3917/rac.004.0169.

O'Sullivan, C. A., Fillery, I. R. P., Roper, M. M. and Richards, R. A. (2016). Identification of several wheat landraces with biological nitrification inhibition capacity. *Plant and Soil* 404(1–2), 61–74. 10.1007/s11104-016-2822-4.

O'Sullivan, L., Wall, D., Creamer, R., Bampa, F. and Schulte, R. P. O. (2018). Functional land management: bridging the think-do-gap using a multi-stakeholder science policy interface. *Ambio* 47(2), 216–230.

Otles, S. and Yalcin, B. (2010). Nano-biosensors as new tool for detection of food quality and safety. *LogForum* 6(4), 67–70.

Otte, J., Roland-Holst, D., Pfeifer, D., Soares- Magalhaes, R., Rushton, J., Graham, J. and Silbergeld, E. (2007). *Industrial Livestock Production and Global Health Risks*. FAO, Rome. Available at: http://www.fao.org/ag/againfo/programmes/en/pplpi/docarc/rep- hpai_industrialisationrisks.pdf.

Owen, G. (2020). What makes climate change adaptation effective? A systematic review of the literature. *Global Environmental Change* 62. DOI: 10.1016/j.gloenvcha.2020.102071.

Owolade, O. F., Ogunleti, D. O. and Adenekan, M. O. (2008). Titanium dioxide affects diseases, development and yield of edible cowpea. *EJEAFChe* 7(5), 2942–2947.

Page, K. L., Dang, Y. P. and Dalal, R. C. (2020). The ability of conservation agriculture to conserve soil organic carbon and the subsequent impact on soil physical, chemical, and biological properties and yield. *Frontiers in Sustainable Food Systems* 4(31). DOI: 10.3389/fsufs.2020.00031.

Pahpy, L. (2020). Viewpoint: organic food represents a 'reactionary' ideology that doesn't support health or sustainable farming–and should not be subsidized. Genetic Literacy Project. 10 July 2020. Available at: https://geneticliteracyproject.org/2020/07/10/viewpoint-organic-food-represents-a-reactionary-ideology-thatdoesnt-promote-health-or-sustainable-farming%E2%81%A0-and-should-not-be-subsidized.

Painter, J. A., Hoekstra, R. M., Ayers, T., Tauxe, R. V., Braden, C. R., Angulo, F. J. and Griin, P. M. (2013). Attribution of foodborne illnesses, hospitalizations, and deaths to food commodities by using outbreak data, United States, 1998-2008. *Emerging Infectious Diseases* 19(3), 407–415.

Palley, T. I. (2007). Financialization: what it is and why it matters. The Levy Economics Institute, Working Paper No. 525.

Pallottino, F., Antonucci, F., Costa, C., Bisaglia, C., Figorilli, S. and Menesatti, P. (2019). Optoelectronic proximal sensing vehicle-mounted technologies in precision agriculture: a review. *Computers and Electronics in Agriculture* 162, 859–873.

Palm, C., Blanco-Canqui, H., DeClerck, F., Gatere, L. and Grace, P. (2014). Conservation agriculture and ecosystem services: an overview. *Agriculture, Ecosystems and Environment* 187, 87–105.

Pampana, E. and Russell, P. (1955). *Malaria. A World Problem, Chronicle of the WHO 9*. WHO, Geneva.

PAN (2002). PAN. 28 October 2002. Available at: http://www.panna.org. Accessed 11 November 2003.

Pangaribowo, E. and Gerber, N. (2016). Innovations for food and nutrition security: impacts and trends. In: Gatzweiler, F. W. and von Braun, J. (Eds). *Technological and Institutional Innovations for Marginalized Smallholders in Agricultural Development*. Springer, Cham. DOI: 10.1007/978-3-319-25718-1_3.

Paoletti, M. G., Schweigl, U. and Favretto, M. R. (1995). Soil macroinvertebrates, heavy metals and organochlorines in low and high input apple orchards and a coppiced woodland. *Pedobiologia* 39(1), 20–33.

Paoletti, M. G., Sommaggio, D., Favretto, M. R., Petruzzelli, G., Pezzarossa, B. and Barbafieri, M. (1998). Earthworms as useful bioindicators of agroecosystem sustainability in different input orchards. *Applied Soil Ecology* 10(1–2), 137–150.

Paracchini, M. L., Petersen, J. E., Hoogeveen, Y., Bamps, C., Burfield, I. and van Swaay, C. (2008). *High Nature Value Farmland in Europe: An Estimate of the Distribution Patterns on the Basis of Land Cover and Biodiversity Data*. Office for Official Publications of the European Communities, Luxemburg.

Parchami, M., Ferreira, J. A. and Taherzadeh, M. J. (2021). Starch and protein recovery from brewer's spent grain using hydrothermal pre-treatment and their conversion to edible filamentous fungi - a brewery biorefinery concept. *Bioresource Technology* 337, 125409. DOI: 10.1016/j.biortech.2021.125409.

Parenteau, R. (2005). The late 1990s' bubble: financialization in the extreme. In: Epstein, G. (Ed.) *Financialization and the World Economy*. Edward Elgar Publishing, Cheltenham, pp. 111–148.

Parfitt, J., Barthel, M. and Macnaughton, S. (2010). Food waste within food supply chains: quantification and potential for change to 2050. *Philosophical Transactions of the Royal Society of London* 365(1554), 3065–3081.

Park, H. J., Kim, S. H., Kim, H. J. and Choi, S. H. (2006). A new composition of nanosized silica-silver for control of various plant diseases. *Plant Pathology Journal* 22(3), 295–302.

Parsons, K. and Hawkes, C. (2019). Rethinking food policy: a fresh approach to policy and practice. Brief 5: Policy Coherence in Food Systems.

Partel, V., Kakarla, S. C. and Ampatzidis, Y. (2019). Development and evaluation of a low-cost and smart technology for precision weed management utilizing artificial intelligence. *Computers and Electronics in Agriculture* 157, 339–350.

Pasitka, L., Cohen, M., Ehrlich, A., Gildor, B., Reuveni, E., Ayyash, M., Wissotsky, G., Herscovici, A., Kaminker, R., Niv, A., Bitcover, R., Dadia, O., Rudik, A., Voloschin, A., Shimoni, M., Cinnamon, Y. and Nahmias, Y. (2022). Spontaneous immortalization of chicken fibroblasts generates stable, high-yield cell lines for serum-free production of cultured meat. *Nature Food*. DOI: 10.1038/s43016-022-00658-w.

Patel, R., Holt-Giménez, E. and Shattuck, A. (2009). Ending Africa's hunger. *The Nation*. Available at: http://www.foodfirst.org/en/node/2556.

Pathak, H. S., Brown, P. and Best, T. (2019). A systematic literature review of the factors affecting the precision agriculture adoption process. *Precision Agriculture* 20(6), 1292–1316.

Pattison, P. M., Tsao, J. Y., Brainard, G. C. and Bugbee, B. (2018). LEDs for photons, physiology, and food. *Nature* 563(7732), 493–500.

Paull, J. (2014). Lord Northbourne, the man who invented organic farming, a biography. *Journal of Organic Systems* 9(1), 31–53.

Paulson, S. (2017). Degrowth: culture, power, and change. In: Gezon, L. L. and Paulson, S. (Eds) "Degrowth, culture and power". Special Section of the *Journal of Political Ecology* 24(1), 425–666.

Paustian, K., Chenu, C., Conant, R., Cotrufo, F., Lal, R., Smith, P. and Soussana, J. F. (2020). Climate mitigation potential of regenerative agriculture is significant! Available at: https://scholar.princeton.edu/sites/default/files/tsearchi/files/paustian_et_al._response_to_wri_soil_carbon_blog_.pdf.

Paustian, M. and Theuvsen, L. (2017). Adoption of precision agriculture technologies by German crop farmers. *Precision Agriculture* 18(5), 701–716. DOI: 10.1007/s11119-016-9482-5.

PBN (2022). 42% of global consumers believe plant-based food will replace meat within A decade. 27 November 2022. Available at: https://plantbasednews.org/news/environment/vegan-vegetarian-food-replace-meat-decade/.

PCFS (2007). People's coalition on food sovereignty (PCFS). *People's Coalition on Food Sovereignty*. Available at: http://www.archive.foodsov.org/html/aboutus.htm.

Peano, C., Massaglia, S., Ghisalberti, C. and Sottile, F. (2020). Pathways for the amplification of agroecology in African sustainable urban agriculture. *Sustainability* 12(7), 2718. DOI: 10.3390/su12072718.

Peano, C., Migliorini, P. and Sottile, F. (2014). A methodology for the sustainability assessment of agri-food systems: an application to the Slow Food Presidia project. *Ecology and Society* 19(4), 24. DOI: 10.5751/ES-06972-190424.

Pearson, C. J. (2007). Regenerative, semi-closed systems: a priority for twenty-first-century agriculture. *BioScience* 57(5), 409–418.

Peck, J. and Tickell, A. (2000). Searching for a new institutional fix. In: Amin, A. (Ed.). *Post-Fordism: A Reader*. Blackwell Publishers, Oxford.

Pecl, G. T., Araújo, M. B., Bell, J. D., Blanchard, J., Bonebrake, T. C., Chen, I. C., Clark, T. D., Colwell, R. K., Danielsen, F., Evengård, B., Falconi, L., Ferrier, S., Frusher, S., Garcia, R. A., Griffis, R. B., Hobday, A. J., Janion-Scheepers, C., Jarzyna, M. A., Jennings, S., Lenoir, J., Linnetved, H. I., Martin, V. Y., McCormack, P. C., McDonald, J., Mitchell, N. J., Mustonen, T., Pandolfi, J. M., Pettorelli, N., Popova, E., Robinson, S. A., Scheffers, B. R., Shaw, J. D., Sorte, C. J., Strugnell, J. M., Sunday, J. M., Tuanmu, M. N., Vergés, A., Villanueva, C., Wernberg, T., Wapstra, E. and Williams, S. E. (2017). Biodiversity redistribution under climate change: impacts on ecosystems and human well-being. *Science* 355(6332). DOI: 10.1126/science.aai9214.

Peet, J. R. (1969). The spatial expansion of commercial agriculture in the nineteenth century: a von Thunen interpretation. *Economic Geography* 45(4), 283–301.

Peeters, A., Dendoncker, N. and Jacobs, S. (2013). Chapter 22. Enhancing ecosystem services in Belgian agriculture through agroecology: a vision for a farming with a future. In: Jacobs, S., Dendoncker, N. and Keune, H. (Eds) *Ecosystem Services Global Issues, Local Practices*. Elsevier, Boston, MA, pp. 285–304.

Pendrill, F., Persson, U. M., Godar, J., Kastner, T., Moran, D., Schmidt, S. and Wood, R. (2019). Agricultural and forestry trade drives large share of tropical deforestation emissions. *Global Environmental Change* 56, 1–10. DOI: 10.1016/j.gloenvcha.2019.03.002.

Penker, M. (2006). Mapping and measuring the ecological embeddedness of food supply chains. *Geoforum* 37(3), 368–379.

Pereira, L., Wynberg, R. and Reis, Y. (2018). Agroecology: the future of sustainable farming? *Environment: Science and Policy for Sustainable Development* 60(4), 4–17. DOI: 10.1080/00139157.2018.1472507.

Pereira, L. M., Drimie, S., Maciejewski, K., Tonissen, P. B. and Biggs, R. O. (2020). Food system transformation: integrating a political-economy and social-ecological approach to regime shifts. *International Journal of Environmental Research and Public Health* 17(4), 1313. DOI: 10.3390/ijerph17041313.

Pérez-Marín, D., Paz, P., Guerrero, J.-E., Garrido-Varo, A. and Sánchez., M. T. (2010). Miniature handheld NIR sensor for the on-site non-destructive assessment of post-harvest quality and refrigerated storage behaviour in plums. *Journal of Food Engineering* 99(3), 294–302.

Perfecto, I. and Vandermeer, J. (2008). Biodiversity conservation in tropical agroecosystems. *Annals of the New York Academy of Sciences* 1134(1), 173–200. DOI: 10.1196/annals.1439.011.

Perfecto, I. and Vandermeer, J. (2010). The agroecological matrix as alternative to the land-sparing/agriculture intensification model. *Proceedings of the National Academy of Sciences of the United States of America* 107(13), 5786–5791. DOI: 10.1073/pnas.0905455107.

Perfecto, I., Vandermeer, J. and Wright, A. (2009). *Nature's Matrix: Linking Agriculture, Conservation, and Food Sovereignty*. Earthscan, Routledge, London. ISBN-13: 978-1844077823.

Perkins, J. H. (1997). *Geopolitics and the Green Revolution: Wheat, Genes, and the Cold War*. Oxford University Press, Oxford.

Permaculture Institute (2013). What is permaculture. *Permaculture Institute*. Available at: http://www.permaculture.org/nm/index.php/site/permaculture_design_course. Accessed 30 May 2013.

Perrott-White, Z. (2019). Hungry Europe: are food banks becoming the new supermarkets? *Europe and Me*. Available at: https://europeandme.eu/hungry-europe-are-food-banks-becoming-the-new-supermarkets/. Accessed 20 October 2022.

Perry, E. D., Ciliberto, F., Hennessy, D. A. and Moschini, G. C. (2016). Genetically engineered crops and pesticide use in U.S. maize and soybeans. *Science Advances* 2(8), e1600850.

Perryer, S. (2019). Cutting back: how the degrowth movement could save the planet. Available at: https://www.europeanceo.com/lifestyle/cutting-back-how-the-degrowth-movement-could-save-the-planet/. Accessed 13 November 2022.

Petrini, C. (2005). *Slow Food Nation; Why Our Food Should Be Good, Clean, and Fair*. Slow Food Editore, Bra.

Phalan, B., Green, R. E., Dicks, L. V., Dotta, G., Feniuk, C., Lamb, A., Strassburg, B. B., Williams, D. R., zu Ermgassen, E. K. and Balmford, A. (2016). How can higher-yield farming help to spare nature? *Science* 351(6272), 450–451.

Phalan, B., Onial, M., Balmford, A. and Green, R. E. (2011). Reconciling food production and biodiversity conservation: land sharing and land sparing compared. *Science* 333(6047), 1289–1291. DOI: 10.1126/science.1208742.

Phalan, B. T. (2018). What have we learned from the land sparing-sharing model? *Sustainability* 10(6), 1760. DOI: 10.3390/su10061760. Available at: www.mdpi.

Phelan, P. L. (2009). Ecology-based agriculture and the next green revolution. In: Bohlen, P. J. and House, G. (Eds) *Sustainable Agroecosystem Management*. CRC Press, Boca Raton, FL, pp. 97–135.

Phillips McDougall (2018). Evolution of the crop protection industry since 1960. Available at: https://croplife.org/wp-content/uploads/2018/11/Phillips-McDougall-Evolution-of-the-Crop-Protection-Industry-since-1960-FINAL.pdf. Accessed 11 November 2023.

Phillips McDougall (2019). Evolution of the crop protection industry since 1960. *Phillips McDougall AgriService*, April.

Pimbert, M. (2008). *Toward Food Sovereignty: Reclaiming Autonomous Food Systems*. IIED, London.

Pimbert, M. (2015). Agroecology as an alternative vision to conventional development and climate smart agriculture. *Development* 58(2-3), 286-298.

Pimbert, M. P. and Moeller, N. I. (2018). Absent agroecology aid: on UK agricultural development assistance since 2010. *Sustainability* 10(2), 505. DOI: 10.3390/su10020505.

Pimentel, D. (2005). Environmental and economic costs of the application of pesticides primarily in the United States. *Environment, Development and Sustainability* 7(2), 229-252. DOI: 10.1007/s10668-005-7314-2.

Pimentel, D., Hepperly, P., Hanson, J., Douds, D. and Seidel, R. (2005). Environmental, energetic, and economic comparisons of organic and conventional farming systems. *BioScience* 55(7), 573-582. DOI: 10.1007/BF01965614.

Pimentel, D. and Pimentel, M. (1979). *Food, Energy and Society*. Edward Arnold, London.

Pimm, S. L., Russell, G. J., Gittelman, J. L. and Brooks, T. M. (1995). The future of biodiversity. *Science* 269(5222), 347-350.

Pingali, P. (2006). Westernization of Asian diets and the transformation of food systems: implications for research and policy. *Food Policy* 32(3), 281-298.

Piorr, A., Ravetz, J. and Tosics, I. (2011). Peri-urbanisation in Europe: towards a European policy to sustain urban-rural futures. A Synthesis Report. *Life Sciences*. University of Copenhagen/Academic Books, Frederiksberg.

Pires, S. M., Vigre, H., Makela, P. and Hald, T. (2010). Using outbreak data for source attribution of human salmonellosis and campylobacteriosis in Europe. *Foodborne Pathogens and Disease* 7(11), 1351-1361. DOI: 10.1089/fpd.2010.0564.

Pittelkow, C. M., Liang, X., Linquist, B. A., van Groenigen, K. J., Lee, J., Lundy, M. E., van Gestel, N., Six, J., Venterea, R. T. and van Kessel, C. (2015). Productivity limits and potentials of the principles of conservation agriculture. *Nature* 517(7534), 365-368.

Pollan, M. (2013). *Cooked: A Natural History of Transformation*. Penguin Press, New York.

Ponisio, L. C., M'Gonigle, L. K., Mace, K. C., Palomino, J., de Valpine, P. and Kremen, C. (2015). Diversification practices reduce organic to conventional yield gap. *Proceedings of the Royal Society of Biological Sciences* 282(1799), 20141396. DOI: 10.1098/rspb.2014.1396.

Poore, J. and Nemecek, T. (2018). Reducing food's environmental impacts through producers and consumers. *Science* 360(6392), 987-992.

Poorter, H., Van Berkel, Y., Baxter, R., Den Hertog, J., Dijkstra, P., Gifford, R. M., Griffin, K. L., Roumet, C., Roy, J. and Wong, S. C. (1997). The effect of elevated CO_2 on the chemical composition and construction costs of leaves of 27 C3 species. *Plant, Cell and Environment* 20(4), 472-482.

Popkin, B. M. (1993). Nutritional patterns and transitions. *Population and Development Review* 19(1), 138-157.

Popkin, B. M. (1994). The nutrition transition in low-income countries: an emerging crisis. *Nutrition Reviews* 52(9), 285-298.

Popkin, B. M. (2002). Part II: What is unique about the experience in lower- and middle-income less industrialised countries compared with the very high-income countries? The shift in the stages of the nutrition transition differ from past experiences! *Public Health Nutrition* 5(1A), 205–214. DOI: 10.1079/PHN2001295.

Popkin, B. M. (2006). Global nutrition dynamics: the world is shifting rapidly toward a diet linked with noncommunicable diseases. *American Journal of Clinical Nutrition* 84(2), 289–298.

Popkin, B. M. (2009). Reducing meat consumption has multiple benefits for the World's health. *Archives of Internal Medicine* 169(6), 543–545.

Popkin, B. M. (2017). Relationship between shifts in food system dynamics and acceleration of the global nutrition transition. *Nutrition Reviews* 75(2), 73–82. DOI: 10.1093/nutrit/nuw064.

Popkin, B. M. (2019). *Ultra-Processed Foods' Impacts on Health*. Food and Agriculture Organization of the United Nations, Santiago.

Popkin, B. M., Adair, L. S. and Ng, S. W. (2012). Now and then: the global nutrition transition: the pandemic of obesity in developing countries. *Nutrition Reviews* 70(1), 3–21.

Popkin, B. M., Corvalan, C. and Grummer-Strawn, L. M. (2019). Dynamics of the double burden of malnutrition and the changing nutrition reality. *Lancet* 395(10217), 65–74. DOI: 10.1016/S0140-6736(19)32497-3.

Popkin, B. M. and Gordon-Larsen, P. (2004). The nutrition transition: worldwide obesity dynamics and their determinants. *International Journal of Obesity* 28(Suppl. 3), S2–S9. DOI: 10.1038/sj.ijo.0802804.

Popkin, B. M. and Hawkes, C. (2016). Sweetening of the global diet, particularly beverages: patterns, trends, and policy responses. *Lancet. Diabetes and Endocrinology* 4(2), 174–186. DOI: 10.1016/S2213-8587(15)00419-2-8587(15)00419-2.

Popkin, B. M. and Reardon, T. (2018). Obesity and the food system transformation in Latin America. *Obesity Reviews*. DOI: 10.1111/obr.12694.

Poppe, K., Wolfert, J., Verdouw, C. and Renwick, A. (2015). A European perspective on the economics of big data. *Farm Policy J.* 12, 11–19.

Popular Science (2021). Vertical farms are finally branching out.. Available at: https://www.popsci.com/science/vertical-farms-energy-use-photos/.

Porder, S. and Chadwick, O. A. (2009). Climate and soil-age constraints on nutrient uplift and retention by plants. *Ecology* 90(3), 623–636. DOI: 10.1890/07-1739.1.

Positive News (2022). Salvaging crops and tackling a broken food system: meet the 'gleaners'. Available at: https://www.positive.news/environment/salvaging-crops-and-tackling-a-broken-food-system-meet-the-gleaners/.

Pothukuchi, K. and Kaufman, J. L. (1999). Placing the food system on the urban agenda: the role of municipal institutions in food systems planning. *Agriculture and Human Values* 16(2), 213–224.

Poti, J. M., Mendez, M. A., Ng, S. W. and Popkin, B. M. (2015). Is the degree of food processing and convenience linked with the nutritional quality of foods purchased by US households? *American Journal of Clinical Nutrition* 101(6), 1251–1262.

Poulton, P., Johnston, J., Macdonald, A., White, R. and Powlson, D. (2018). Major limitations to achieving "4 per 1000" increases in soil organic carbon stock in temperate regions: evidence from long-term experiments at Rothamsted Research, United Kingdom. *Global Change Biology* 24(6), 2563–2584.

Powell, B., Thilsted, S. H., Ickowitz, A., Termote, C., Sunderland, T. and Herforth, A. (2015). Improving diets with wild and cultivated biodiversity from across the landscape. *Food Security* 7(3), 535–554. DOI: 10.1007/s12571-015-0466-5.

Powlson, D. S., Stirling, C. M., Jat, M. L., Gerard, B. G., Palm, C. A., Sanchez, P. A. and Cassman, K. G. (2014). Limited potential of no-till agriculture for climate change mitigation. *Nature Climate Change* 4(8), 678–683. DOI: 10.1038/nclimate2292.

Prasad, R., Bhattacharyya, A. and Nguyen, Q. D. (2017). Nanotechnology in sustainable agriculture: recent developments, challenges, and perspectives. *Frontiers in Microbiology* 8, 1014. DOI: 10.3389/fmicb.2017.01014.

Pratt, A. (2023). Food banks in the UK. Research briefing. 18 October. Available at: https://commonslibrary.parliament.uk/research-briefings/cbp-8585/.

Pretty, J. (1995). *Regenerating Agriculture: Policies and Practice for Sustainability and Self-Reliance*. Earthscan, London.

Pretty, J. (1998). *The Living Land: Agriculture, Food, and Community Regeneration in Rural Europe*. Earthscan, London.

Pretty, J. (2008). Agricultural sustainability: concepts, principles, and evidence. *Philosophical Transactions of the Royal Society of London. Series B, Biological Sciences* 363(1491), 447–465.

Pretty, J. and Bharucha, Z. P. (2013). Integrated pest management for sustainable intensification of agriculture in Asia and Africa. *Insects* 6(1), 152–182. DOI: 10.3390/insects6010152.

Pretty, J. and Hine, R. (2000). Feeding the world with sustainable agriculture: a summary of new evidence. Final Report from SAFE-World Research Project. University of Essex, Colchester.

Pretty, J. and Hine, R. (2001). Reducing food poverty with sustainable agriculture: a summary of new evidence. Final Report from the 'SAFE World' Research Project, University of Essex. Available at: http://www.essex.ac.uk/ces/esu/occasionalpapers/SAFE%20FINAL%20-%20Pages1-22.pdf.

Priefer, C., Jörissen, J. and Bräutigam, K. (2013). Technology options for feeding 10 billion people. Science and technology options assessment: options for cutting food waste. Directorate General for Internal Policies, European Parliament, p. 127.

Priester, J. H., Ge, Y., Mielke, R. E., Horst, A. M., Moritz, S. C., Espinosa, K., Gelb, J., Walker, S. L., Nisbet, R. M., An, Y. J., Schimel, J. P., Palmer, R. G., Hernandez-Viezcas, J. A., Zhao, L., Gardea-Torresdey, J. L. and Holden, P. A. (2012). Soybean susceptibility to manufactured nanomaterials with evidence for food quality and soil fertility interruption. *Proceedings of the National Academy of Sciences of the United States of America* 109(37), E2451–E2456.

Princen, T. (1997). The shading and distancing of commerce: when internalization is not enough. *Ecological Economics* 20(3), 235–253.

Princen, T. (2002). Distancing: consumption and the severing of feedback. In: Princen, T., Maniates, M. and Conca, K. (Eds) *Confronting Consumption*. MIT Press, Cambridge, MA, pp. 103–131.

Principato, L., Secondi, L. and Pratesi, C. A. (2015). Reducing food waste: an investigation on the behaviour of Italian youths. *British Food Journal* 117(2), 731–748. DOI: 10.1108/BFJ-10-2013-0314.

Pritchard, B. (2009). The long hangover from the second food regime: a world-historical interpretation of the collapse of the WTO Doha Round. *Agriculture and Human Values* 26(4), 297–307.

PRNewswire (2021). JBS is entering the cultivated protein market with the acquisition of BioTech Foods and the construction of a plant in Europe. Available at: https://www.prnewswire.com/news-releases/jbs-is-entering-the-cultivated-protein-market-with-the-acquisition-of-biotech-foods-and-the-construction-of-a-plant-in-europe-301428082.html.

Project Drawdown (2023). Conservation agriculture. Available at: https://drawdown.org/solutions/conservation-agriculture#.

Prové, C., Michiel, M. P. M. M., Dessein, J. and de Krom, J. (2019). Politics of scale in urban agriculture governance: a transatlantic comparison of food policy councils. *Journal of Rural Studies* 68, 171–181.

Pulido, M. D. and Parrish, A. R. (2003). Metal-induced apoptosis: mechanisms. *Mutation Research* 533(1-2), 227–241. DOI: 10.1016/j.mrfmmm.2003.07.015.

Puma, M. J., Bose, S., Chon, S. Y. and Cook, B. I. (2015). Assessing the evolving fragility of the global food system. *Environmental Research Letters* 10(2), 024007.

Purcell, M. and Brown, J. C. (2005). Against the local trap: scale and the study of environment and development. *Progress in Development Studies* 5(4), 279-297.

Qaim, M. (2020). Role of new plant breeding technologies for food security and sustainable agricultural development. *Applied Economic Perspectives and Policy* 42(2), 129–150.

Racuciu, M. and Creanga, D. E. (2007). TMA-OH coated magnetic nanoparticles internalized in vegetal tissue. *Romanian Journal of Physics* 52, 395–402.

Ragaei, M. and Sabry, A. K. (2014). Nanotechnology for insect pest control. *International Journal of Environmental Science and Technology* 3, 528–545.

Rai, M. and Ingle, A. (2012). Role of nanotechnology in agriculture with special reference to management of insect pests. *Applied Microbiology and Biotechnology* 94(2), 287–293. DOI: 10.1007/s00253-012-3969-4.

Raja, N. (2013). Biopesticides and biofertilizers: ecofriendly sources for sustainable agriculture. *Journal of Biofertilizers and Biopesticides* 4(1), e112. DOI: 10.4172/2155-6202.1000e112.

Ram, U. (2004). Glocommodification: how the global consumes the local – McDonald's in Israel. *Current Sociology* 52(1), 11–31.

Ramankutty, N., Evan, A. T., Monfreda, C. and Foley, J. A. (2008). Farming the planet: 1. Geographic distribution of global agricultural lands in the year 2000. *Global Biogeochemical Cycles* 22(1). DOI: 10.1029/2007GB002952.

Ramankutty, N., Ricciardi, V., Mehrabi, Z. and Seufert, V. (2019). Trade-offs in the performance of alternative farming systems. *Agricultural Economics* 50(S1), 97–105.

Ramirez-Villegas, J., Watson, J. and Challinor, A. J. (2015). Identifying traits for genotypic adaptation using crop models. *Journal of Experimental Botany* 66(12), 3451–3462.

Ranganathan, J., Vennard, D., Waite, R., Dumas, P., Lipinski, B., Searchinger, T. and GLOBAGRI Model Authors. (2016). Shifting diets for a sustainable food future. Working Paper, Instalment 11 of Creating a Sustainable Food Future. World Resources Institute, Washington, DC.

Ranganathan, J., Waite, R., Searchinger, T. and Zionts, J. (2020). Regenerative agriculture: good for soil health, but limited potential to mitigate climate change. Available at: https://www.wri.org/blog/2020/05/regenerativeagriculture-climate-change. Accessed 7 October 2020.

Rashid, M. I., Shahzad, T., Shahid, M., Imran, M., Dhavamani, J., Ismail, I. M. I., Basahi, J. M. and Almeelbi, T. (2017a). Toxicity of iron oxide nanoparticles to grass litter decomposition in a sandy soil. *Scientific Reports* 7, 41965.

Rashid, M. I., Shahzad, T., Shahid, M., Ismail, I. M. I., Shah, G. M. and Almeelbi, T. (2017b). Zinc oxide nanoparticles affect carbon and nitrogen mineralization of *Phoenix dactylifera* leaf litter in a sandy soil. *Journal of Hazardous Materials* 324(B), 298–305.

Ravishankar, E., Charles, M., Xiong, Y., Henry, R., Swift, J., Rech, J., Calero, J., Cho, S., Booth, R. E., Kim, T., Balzer, A. H., Qin, Y., Ho, C. H. Y., So, F., Stingelin, N., Amassian, A., Saravitz, C., You, W., Ade, H., Sederoff, H. and O'Connor, B. T. (2021). Balancing crop production and energy harvesting in organic solar-powered greenhouses. Cell Reports Physical Science 2, 100381.

Raworth, K. (2017). A doughnut for the anthropocene: humanity's compass in the 21st century. *The Lancet. Planetary Health* 1(2), e48–e49. DOI: 10.1016/S2542-5196(17)30028-1.

Reardon, T., Barrett, C. B., Berdegué, J. A.. and Swinnen, J. F. M. (2009). Agrifood industry transformation and small farmers in developing countries. *World Development* 37(11), 1717–1727.

Reardon, T. and Berdegué, J. A. (2002). The rapid rise of supermarkets in Latin America: challenges and opportunities for development. *Development Policy Review* 20(4), 371–388.

Reardon, T., Echeverria, R., Berdegué, J., Minten, B., Liverpool-Tasie, S., Tschirley, D. and Zilberman, D. (2019). Rapid transformation of food systems in developing regions: highlighting the role of agricultural research & innovations. *Agricultural Systems* 172, 47–59. DOI: 10.1016/j.agsy.2018.01.022.

Reardon, T. and Hopkins, R. (2006). The supermarket revolution in developing countries: policies to address emerging tensions among supermarkets, suppliers and traditional retailers. *European Journal of Development Research* 18(4), 522–545.

Reardon, T., Timmer, C. P., Barrett, C. B. and Berdegué, J. (2003). The rise of supermarkets in Africa, Asia, and Latin America. *American Journal of Agricultural Economics* 85(5), 1140–1146. DOI: 10.1111/j.0092-5853.2003.00520.x.

Reardon, T., Tschirley, D., Liverpool-Tasie, L. S. O., Awokuse, T., Fanzo, J., Minten, B., Vos, R., Dolislager, M., Sauer, C., Dhar, R., Vargas, C., Lartey, A., Raza, A. and Popkin, B. M. (2021). The processed food revolution in African food systems and the double burden of malnutrition. *Global Food Security* 28, 100466.

Recanati, F., Maughan, C., Pedrotti, M., Dembska, K. and Antonelli, M. (2019). Assessing the role of CAP for more sustainable and healthier food systems in Europe: a literature review. *Science of the Total Environment* 653, 908–919.

Reeder, J. D. and Schuman, G. E. (2002). Influence of livestock grazing on C sequestration in semi-arid mixed-grass and short-grass rangelands'. *Environmental Pollution* 116(3), 457–463. DOI: 10.1016/S0269-7491(01)00223-8.

Rees, W. E. (2019). Why place-based food systems? Food security in a chaotic world. *Journal of Agriculture, Food Systems, and Community Development* 9(Suppl. 1), 5–13. DOI: 10.5304/jafscd.2019.091.014.

Reeve, J. R., Hoagland, L. A., Villalba, J. J., Carr, P. M., Atucha, A., Cambardella, C., Davis, D. R. and Delate, K.(2016). Chapter six – organic farming, soil health, and food quality: considering possible links. In: Sparks, D. L. (Ed.) *Advances in Agronomy* 137, 319-367. DOI: 10.1016/bs.agron.2015.12.003.

Reeves, D. W. (1997). The role of soil organic matter in maintaining soil quality in continuous cropping systems. *Soil and Tillage Research* 43(1–2), 131–167.

Reganold, J. P. (1995). Soil quality and profitability of biodynamic and conventional farming systems: a review. *American Journal of Alternative Agriculture* 10(1), 36–45.

Reganold, J. P., Elliott, L. F. and Unger, Y. L. (1987). Long-term effects of organic and conventional farming on soil erosion. *Nature* 330(6146), 370–372.

Reganold, J. P., Jackson-Smith, D., Batie, S. S., Harwood, R. R., Kornegay, J. L., Bucks, D., Flora, C. B., Hanson, J. C., Jury, W. A., Meyer, D., Schumacher, A., Sehmsdorf, H., Shennan, C., Thrupp, L. A. and Willis, P. (2011). Transforming U.S. agriculture. *Science* 332(6030), 670–671.

Reganold, J. P. and Wachter, J. M. (2016). Organic agriculture in the twenty-first century. *Nature Plants* 2, 15221. DOI: 10.1038/nplants.2015.221.

Reich, L. (2010). Why I'm not a permaculturist. *Garden Rant*. Available at: http://gardenrant.com/2010/03/why-im-not-a-permaculturist.html. Accessed 28 May 2012.

Reichardt, M. and Jürgens, C. (2009). Adoption and future perspective of precision farming in Germany: results of several surveys among different agricultural target groups. *Precision Agriculture* 10(1), 73–94. DOI: 10.1007/s11119-008-9101-1.

Renting, H., Marsden, T. K. and Banks, J. (2003). Understanding alternative food networks: exploring the role of short food supply chains in rural development. *Environment and Planning A* 35(3), 393–411. DOI: 10.1068/a3510.

Research and Markets (2022). Nutraceuticals: global markets to 2026. Available at: https://www.researchandmarkets.com/reports/4584273/nutraceuticals-global-markets-to-2026?utm_source=GNOM&utm_medium=PressRelease&utm.

Retzbach, A. and Maier, M. (2015). Communicating scientific uncertainty: media effects on public engagement with science. *Communication Research* 42(3), 429–456. DOI: 10.1177/0093650214534967.

Reuters (2022a). Beyond meat cuts revenue view, jobs as inflation hits plant protein demand. *Reuters*, 15 October.

Reuters (2022b). COP27: farm climate innovation commitments double to $8 billion. *Reuters*, 11 November. Available at: https://www.reuters.com/business/cop/cop27-farm-climate-innovation-commitments-double-8-billion. Accessed 02 July 2023.

Reuters (2023). Lab-grown meat moves closer to American dinner plates. 23 January 2023. Available at: https://www.reuters.com/business/retail-consumer/lab-grown-meat-moves-closer-american-dinner-plates-2023-01-23/.

Reynolds, B. (2009). Feeding a world city: the London food strategy. *International Planning Studies* 14(4), 417–424.

Reynolds, C., Goucher, L., Quested, T., Bromley, S., Gillick, S., Wells, V. K., Evans, D., Koh, L., Carlsson Kanyama, A., Katzeff, C., Svenfelt, Å. and Jackson, P. (2019). Review: consumption stage food waste reduction interventions – what works and how to design better interventions. *Food Policy* 83, 7–27. DOI: 10.1016/j.foodpol.2019.01.009.

Rhodes, C. J. (2012). Feeding and healing the world: through regenerative agriculture and permaculture. *Science Progress* 95(4), 345–446.

Rhodes, C. J. (2017). The imperative for regenerative agriculture. *Science Progress* 100(1), 80–129.

Ricciardi, V. (2019). The role of small-scale farms in the global food system. PhD dissertation. The University of British Columbia, Vancouver, Canada.

Ricciardi, V., Ramankutty, N., Mehrabi, Z., Jarvis, L. and Chookolingo, B. (2018). An open-access dataset of crop production by farm size from agricultural censuses and surveys. *Data in Brief* 19, 1970–1988.

Rice, A. M., Einbinder, N. and Calderón, C. I. (2023). 'With agroecology, we can defend ourselves': examining campesino resilience and economic solidarity during pandemic-era economic shock in Guatemala. *Agroecology and Sustainable Food Systems* 47(2), 273-305. DOI: 10.1080/21683565.2022.2140378.

Richardson, K., Steffen, W., Lucht, W., Bendtsen, J., Cornell, S. E., Donges, J. F., Drüke, M., Fetzer, I., Bala, G., von Bloh, W., Feulner, G., Fiedler, S., Gerten, D., Gleeson, T., Hofmann, M., Huiskamp, W., Kummu, M., Mohan, C., Nogués-Bravo, D., Petri, S., Porkka, M., Rahmstorf, S., Schaphoff, S., Thonicke, K., Tobian, A., Virkki, V., Wang-Erlandsson, L., Weber, L. and Rockström, J. (2023). Earth beyond six of nine planetary boundaries. *Science Advances* 9(37), eadh2458, 13 September.

Rickerl, D. and Francis, C. (Eds) (2004). *Agroecosystems Analysis. Monograph Series* 43. American Society of Agronomy, Madison, WI.

Ricketts, T. H., Daily, G. C., Ehrlich, P. R. and Michener, C. D. (2004). Economic value of tropical forest to coffee production. *Proceedings of the National Academy of Sciences of the United States of America* 101(34), 12579–12582.

Rico, C. M., Majumdar, S., Duarte-Gardea, M., Peralta-Videa, J. R. and Gardea-Torresdey, J. L. (2011). Interaction of nanoparticles with edible plants and their possible implications in the food chain. *Journal of Agricultural and Food Chemistry* 59(8), 3485–3498. DOI: 10.1021/jf104517j.

Rico-Campà, M., Martínez-González, A., Alvarez-Alvarez, I., de Deus Mendonça, R. D., de la Fuente-Arrillaga, C., Gómez-Donoso, C. and Bes-Rastrollo, M. (2019). Association between consumption of ultra-processed foods and all-cause mortality: SUN prospective cohort study. *British Medical Journal*, May. DOI: 10.1136/bmj.l1949.

Rigby, D. and Cáceres, D. M. (2001). Organic farming and the sustainability of agricultural systems. *Agricultural Systems* 68(1), 21–40.

Risku-Norja, H., Kurppa, S. and Helenius, J. (2009). Dietary choices and greenhouse gas emissions – assessment of impact of vegetarian and organic options at national scale. Progress in Industrial Ecology: An International Journal 6, 340–354.

Risner, D., Kim, Y., Nguyen, C., Siegel, J. B. and Spang, E. S. (2023). Environmental impacts of cultured meat: a cradle-to-gate life cycle assessment. *bioRxiv*. DOI: 10.1101/2023.04.21.537778.

Ro, D. K., Paradise, E. M., Ouellet, M., Fisher, K. J., Newman, K. L., Ndungu, J. M., Ho, K. A., Eachus, R. A., Ham, T. S., Kirby, J., Chang, M. C., Withers, S. T., Shiba, Y., Sarpong, R. and Keasling, J. D. (2006). Production of the antimalarial drug precursor artemisinic acid in engineered yeast. *Nature* 440(7086), 940–943.

Robbins, M. J. (2015). Exploring the 'localisation' dimension of food sovereignty. *Third World Quarterly* 36(3), 449–468.

Robertson, M. J., Llewellyn, R. S., Mandel, R., Lawes, R., Bramley, R. G. V., Swift, L., Metz, N. and O'Callaghan, C. (2012). Adoption of variable rate fertiliser application in the Australian grains industry: status, issues, and prospects. *Precision Agriculture* 13(2), 181–199.

Robles, M., Torrero, M. and von Braun, J. (2009). When speculation matters. *IFPRI Issue Brief* 57, February. IFPRI, Washington, DC. Available at: http://www.ifpri.org/pubs/ib/ib57.pdf. Accessed 29 July 2009.

Robotics and Automation News (2019). John Deere showcases autonomous electric tractor and other new tech. 19 November 2019. Available at: https://roboticsandautomationnews.com/2019/11/19/john-deere-showcases-autonomous-electric-tractor-and-other-new-tech/26774/. Accessed 21 September 2021.

Robra, B. and Heikkurinen, P. (2019). Degrowth and the sustainable development goals. In: Leal Filho, W. (Eds) *Decent Work and Economic Growth*. DOI: 10.1007/978-3-319-71058-7_37-1.

Rockström, J., Gaffney, O., Rogelj, J., Meinshausen, M., Nakicenovic, N. and Schellnhuber, H. J. (2017). A roadmap for rapid decarbonization. *Science* 355(6331), 1269–1271.

Rockström, J., Karlberg, L., Wani, S. P., Barron, J., Hatibu, N., Oweis, T., Bruggeman, A., Farahani, J. and Qiang, Z. (2010). Managing water in rainfed agriculture: the need for a paradigm shift. *Agricultural Water Management* 97(4), 543–550. DOI: 10.1016/j.agwat.2009.09.009.

Rockström, J., Steffen, W., Noone, K., Persson, Å., Chapin, F. S., III, Lambin, E., Lenton, T. M., Scheffer, M., Folke, C., Schellnhuber, H. J., Nykvist, B., de Wit, C. A., Hughes, T., van der Leeuw, S., Rodhe, H., Sörlin, S., Snyder, P. K., Costanza, R., Svedin, U., Falkenmark, M., Karlberg, L., Corell, R. W., Fabry, V. J., Hansen, J., Walker, B., Liverman, D., Richardson, K., Crutzen, P. and Foley, J. (2009a). Planetary boundaries: exploring the safe operating space for humanity. *Ecology and Society* 14(2), 32.

Rockström, J., Steffen, W., Noone, K., Persson, A., Chapin, F. S., Lambin, E. F., Lenton, T. M., Scheffer, M., Folke, C., Schellnhuber, H. J., Nykvist, B., de Wit, C. A., Hughes, T., van der Leeuw, S., Rodhe, H., Sörlin, S., Snyder, P. K., Costanza, R., Svedin, U., Falkenmark, M., Karlberg, L., Corell, R. W., Fabry, V. J., Hansen, J., Walker, B., Liverman, D., Richardson, K., Crutzen, P. and Foley, J. A. (2009b). A safe operating space for humanity. *Nature* 461(7263), 472–475. DOI: 10.1038/461472a.

Rodale Institute (2014). *Regenerative Organic Agriculture and Climate Change: A Down-to-Earth Solution to Global Warming*. Rodale Institute, Kutztown, PA.

Rodale Institute (2020). *Regenerative Organic Agriculture and Climate Change: A Down-to-Earth Solution to Global Warming*. Rodale Institute, Kutztown, PA. Available at: https://rodaleinstitute.org/wp-content/uploads/rodale-white-paper.pdf.

Rodale Institute (2022). Regenerative agriculture identified as a major investment opportunity. Available at: https://rodaleinstitute.org/blog/regenerative-agriculture-identified-as-major-investment-opportunity/.

Rodale, R. (1983). Breaking new ground: the search for a sustainable agriculture. *The Futurist* 1, 15–20.

Rogers, K. (2022). Lab-grown meat could make strides in 2022 as start-ups push for U.S. approval. CNBC food & beverage. Available at: https://www.cnbc.com/2022/01/23/lab-grown-meat-start-ups-hope-to-make-strides-in-2022.html.

Rogovska, N., Laird, D. A., Chiou, C. P. and Bond, L. J. (2019). Development of field mobile soil nitrate sensor technology to facilitate precision fertilizer management. *Precision Agriculture* 20(1), 40–55. DOI: 10.1007/s11119-018-9579-0.

Rohatgi, K. W., Tinius, R. A., Cade, W. T., Steele, E. M., Cahill, A. G. and Parra, D. C. (2017). Relationships between consumption of ultra-processed foods, gestational weight gain and neonatal outcomes in a sample of US pregnant women. *PeerJ* 5, e4091.

Röös, E., Bajželj, B., Smith, P., Patel, M., Little, D. and Garnett, T. (2017). Greedy or needy? Land use and climate impacts of food in 2050 under different livestock futures. *Global Environmental Change* 47, 1–12. DOI: 10.1016/j.gloenvcha.2017.09.001.

Rosa-Schleich, J., Loos, J., Mußhoff, O. and Tscharntke, T. (2019). Ecological-economic trade-offs of Diversified Farming Systems – a review. *Ecological Economics* 160, 251–263.

Rose, D. C. and Chilvers, J. (2018). Agriculture 4.0: broadening responsible innovation in an era of smart farming. *Frontiers in Sustainable Food Systems* 2, 87. DOI: 10.3389/fsufs.2018.00087.

Rosegrant, M. W., Ringler, C. and ,Zhu, T. (2009). Water for agriculture: maintaining food security under growing scarcity. *Annual Review of Environment and Resources*, 205-222. DOI: 10.1146/annurev.environ.030308.090351.

Rosenstock, T. S., Mpanda, M., Rioux, J., Aynekulua, E., Kimaro, A. A., Neufeldt, H., Shepherd, K. D. and Luedeling, E. (2014). Targeting conservation agriculture in the context of livelihoods and landscapes. *Agriculture, Ecosystems and Environment* 187, 47-51.

Rosenstock, T. S., Wilkes, A., Jallo, C., Namoi, N., Bulusu, M., Suber, M., Mboi, D., Mulia, R., Simelton, E., Richards, M., Gurwick, N. and Wollenberg, E. (2019). Making trees count: measurement and reporting of agroforestry in UNFCCC national communications of non-Annex I countries. *Agriculture, Ecosystems and Environment* 284, 106569.

Rosin, C. (2013). Food security and the justification of productivism in New Zealand. *Journal of Rural Studies* 29, 50–58.

Rosin, C. and Campbell, H. (2009). Beyond bifurcation: examining the conventions of organic agriculture in New Zealand. *Journal of Rural Studies* 25(1), 35–47.

Ross, R. (1971). The smooth muscle cell: II. Growth of smooth muscle in culture and formation of elastic fibers. *The Journal of Cell Biology* 50(1), 172–186.

Rosset, P., Patel, R. and Courville, M. (2006). *Promised Land: Competing Visions of Agrarian Reform*. Food First Books, Oakland, CA.

Rosset, P. M. and Altieri, M. A. (1997). Agroecology versus input substitution: a fundamental contradiction in sustainable agriculture. *Society and Natural Resources* 10(3), 283–295.

Rosset, P. M. and Martínez-Torres, M. E. (2012). Rural social movements and agroecology: context, theory, and process. *Ecology and Society* 17(3), 17.

Rosset, P. M., Sosa, B. M., Jaime, A. M. and Lozano, D. R. (2011). The Campesino-to-Campesino agroecology movement of ANAP in Cuba: social process methodology in the construction of sustainable peasant agriculture and food sovereignty. *Journal of Peasant Studies* 38(1), 161–191.

Rothamsted (2023). The history of Rothamsted research. Available at: https://www.rothamsted.ac.uk/history-rothamsted-research. Accessed 9 November 2023.

Rotz, S., Duncan, E., Small, M., Botschner, J., Dara, R., Mosby, I., Reed, M. and Fraser, E. D. G. (2019). The politics of digital agricultural technologies: a preliminary review. *Sociologia Ruralis* 59(2), 203–229.

Roulac, J. W. (2021). Making America's rivers blue again: connecting the dots between regenerative agriculture and healthy waterways. *Common Dreams*, 17 February. Available at: https://www.commondreams.org/views/2021/02/17/making-americas-rivers-blue-again-connecting-dots-between-regenerative-agriculture.

Rouphael, Y., du Jardin, P., Brown, P. H., De Pascale, S. and Colla, G. (Eds) (2020). *Biostimulants for Sustainable Crop Production*. Burleigh Dodds Science Publishing, Cambridge.

Rouphael, Y., Kyriacou, M. C., Petropoulos, S. A., De Pascale, S. and Colla, G. (2018). Improving vegetable quality in controlled environments. *Scientia Horticulturae* 234, 275–289.

Rousidou, C., Papadopoulou, E. S., Kortsinidou, M., Giannakou, I. O., Singh, B. K., Menkissoglu-Spiroudi, U. and Karpouzas, D. G. (2013). Bio-pesticides: harmful or harmless to ammonia oxidizing microorganisms? The case of a Paecilomyces lilacinus based nematicide. *Soil Biology and Biochemistry* 67, 98–105. DOI: 10.1016/j.soilbio.2013.08.014.

Rovira-Más, F., Zhang, Q. and Saiz-Rubio, V. (2020). Mechatronics and intelligent systems in agricultural machinery. In: Holden, N. M., Wolfe, M. L., Ogejo, J. A. and Cummins, E. J. (Eds) *Introduction to Biosystems Engineering*. American Society of Agricultural and Biological Engineers (ASABE) and Virginia Tech Publishing, St. Joseph, MI.

Ruben, R., Cavatassi, R., Lipper, L., Smaling, E. and Winters, P. (2021). Towards food systems transformation–five paradigm shifts for healthy, inclusive, and sustainable food systems. *Food Security* 13(6), 1423–1430. DOI: 10.1007/s12571-021-01221-4.

Runck, B., Streed, A., Wang, D. R., Ewing, P. M., Kantar, M. B. and Raghavan, B. (2023). State spaces for agriculture: a meta-systematic design automation framework. *PNAS Nexus* 2(4), 1–8. DOI: 10.1093/pnasnexus/pgad084.

Russi, L. (2013). *Hungry Capital: The Financialization of Food*. Zero Books, Winchester.

Sacchi, G., Cei, L., Stefani, G., Lombardi, G. V., Rocchi, B., Belletti, G., Padel, S., Sellars, A., Gagliardi, E., Nocella, G., Cardey, S., Mikkola, M., Ala-Karvia, U., Macken-Walsh, À., McIntyre, B., Hyland, J., Henchion, M., Bocci, R., Bussi, B., De Santis, G., Rodriguez y Hurtado, I., de Kochko, P., Riviere, P., Carrascosa-García, M., Martínez, I., Pearce, B., Lampkin, N., Vindras, C., Rey, F., Chable, V., Cormery, A. and Vasvari, G. (2018). A multi-actor literature review on alternative and sustainable food systems for the promotion of cereal biodiversity. *Agriculture* 8(11), 173. DOI: 10.3390/agriculture8110173.

Sachs, J. D., Schmidt-Traub, G., Mazzucato, M., Messner, D., Nakicenovic, N. and Rockström, J. (2019). Transformations to achieve the sustainable development goals. *Nature Sustainability* 2(9), 805–814.

Sage, C. (2003). Social embeddedness and relations of regard: alternative 'good food' networks in south-west Ireland. *Journal of Rural Studies* 19(1), 47–60.

Said-Al Ahl, H. A. H., Hikal, W. M. and Tkachenko, K. G. (2017). Essential oils with potential as insecticidal agents: a review. *International Journal of Environmental Planning and Management* 3(4), 23–33. Available at: http://www.aiscience.org/journal/ijepm.

Saj, S., Torquebiau, E., Hainzelin, E., Pages, J. and Maraux, F. (2017). The way forward: an agroecological perspective for climate-smart agriculture. *Agriculture, Ecosystems and Environment* 250, 20–24.

Salamanca-Buentello, F., Persad, D. L., Court, E. B., Martin, D. K., Daar, A. S. and Singer, P. A. (2005). Nanotechnology and the developing world. *PLoS Medicine* 2(5), e97. DOI: 10.1371/journal.pmed.0020097.

Salbitano, F., Fini, A., Borelli, S. and Konijnendijk, C. C. (2019). Editorial–Urban Food Forestry: current state and future perspectives. *Urban Forestry and Urban Greening* 45, 126482.

Salgadoe, A. S. A., Robson, A. J., Lamb, D. W., Dann, E. K. and Searle, C. (2018). Quantifying the severity of phytophthora root rot disease in avocado trees using image analysis. *Remote Sensing* 10(2), 226.

Salles, J. M., Teillard, F., Tichit, M. and Zanella, M. (2017). Land sparing versus land sharing: an economist's perspective. *Regional Environmental Change* 17(5), 1455–1465. DOI: 10.1007/s10113-017-1142-4ff.

SAM (2019). *A Scoping Review of Major Works Relevant to Scientific Advice Towards an EU Sustainable Food System*. The Scientific Advice Mechanism Unit of the European Commission, Brussels, 26 p.

Samuel, S. (2019). Del Taco's newest 'meat' taco is 100% meatless. *Vox*, 15 April. Available at: vox.com.

Sanchez, P. A. (2020). Viewpoint: time to increase production of nutrient-rich foods. *Food Policy* 91, 101843.

Sánchez-Bayo, F. and Wyckhuys, K. A. G. (2019). Worldwide decline of the entomofauna: a review of its drivers. *Biological Conservation* 232, 8-27, DOI: 10.1016/j.biocon.2019.01.020.

Sanders, J., Stolze, M. and Padel, S. (2011). *Use and Efficiency of Public Support Measures Addressing Organic Farming*. Thunen-Institute of Farm Economics, Braunschweig.

Sanderson, S. E. (1986). *The Transformation of Mexican Agriculture: International Structure and the Politics of Rural Change*. Princeton University Press, Princeton, NJ.

Santos, P. Z. F., Crouzeilles, R. and Sansevero, J. B. B. (2019). Can agroforestry systems enhance biodiversity and ecosystem service provision in agricultural landscapes? A meta-analysis for the Brazilian Atlantic Forest. *Forest Ecology and Management* 433, 140-145. DOI: 10.1016/j.foreco.2018.10.064.

SAPEA (2020). *A Sustainable Food System for the European Union*. Science Advice for Policy by European Academies, Berlin. DOI: 10.26356/sustainablefood.

Sarandón, S. J. and Flores, C. C. (2014). *Agroecología: Bases Teóricas para el Diseño y Manejo de Agroecosistemas Sustentables. Primera*. Editorial de la Universidad de La Plata, Buenos Aires.

Sarri, D., Lombardo, S., Pagliai, A., Perna, C., Lisci, R., De Pascale, V., Rimediotti, M., Cencini, G. and Vieri, M. (2020). Smart farming introduction in wine farms: a systematic review and a new proposal. *Sustainability* 12(17), 7191.

Sauer, M. (1990). Fordist modernization of German agriculture and the future of family farms. *Sociologia Ruralis* 30(3-4), 260-279.

Saxe, H., Larsen, T. M. and Mogensen, L. (2012). The global warming potential of two healthy Nordic diets compared with the average Danish diet. *Climatic Change* 116(2), 249-262.

Schader, C., Muller, A. and El-Hage Scialabba, N. (2015). Impacts of feeding less food competing feedstuffs to livestock on global food system sustainability. *Journal of the Royal Society Interface* 12, 20150891.

Scharf, N., Wachtel, T., Reddy, S. E. and Säumel, I. (2019). Urban commons for the edible city–first insights for future sustainable urban food systems from Berlin, Germany. *Sustainability* 11(4), 966. DOI: 10.3390/su11040966.

Schimmelpfennig, D. (2016). Farm profits and adoption of precision agriculture, economic research. Report No. ERR-217. USDA, Washington, DC. Available at: https://www.ers.usda.gov/publications/pub-details/?pubid580325. Accessed 8 December 2017.

Schlosser, E. (2001). *Fast Food Nation: The Dark Side of the All-American Meal*. Houghton Mi in Harcourt, New York.

Schmelzkopf, K. (1995). Urban community gardens as contested space. *Geographical Review* 85(3), 364-381.

Schmidhuber, J. and Shetty, P. (2005). The nutrition transition to 2030. Why developing countries are likely to bear the major burden. *Acta Agriculturae Scandinavica, Section C* 2(3-4), 150-166.

Schmit, T. M., Jablonski, B. B. R. and Mansury, Y. (2016). Assessing the economic impacts of local food system producers by scale: a case study from New York. *Economic Development Quarterly* 30(4), 316-328.

Schmitt, E., Galli, F., Menozzi, D., Maye, D., Touzard, J. M., Marescotti, A., Six, J. and Brunori, G. (2017). Comparing the sustainability of local and global food products in Europe. *Journal of Cleaner Production* 165, 346-359. ISSN 09596526.

Schneider, A., Friedl, M. A. and Potere, D. (2010). Mapping global urban areas using MODIS 500-m data: new methods and datasets based on urban ecoregions. *Remote Sensing of Environment* 114(8), 1733–1746. DOI: 10.1016/j.rse.2010.03.003.

Schneider, F. and Scherhaufer, S. (2015). Advancing social supermarkets across Europe. *WP4 - Testing Social Innovation-Feasibility Study Final Report*. Available at: https://www.researchgate.net/publication/299367202.

Schonfield, A., Anderson, W. and Moore, M. (1995). PAN's dirty dozen campaign. *Global Pesticide Campaigner* 5(3). Available at: panna.org/resources/pestis/pestis.1995.36.html.

Schreefel, L., Schulte, R. P. O., de Boer, I. J. M., Pas Schrijver, A. P. and van Zanten, H. H. E. (2020). Regenerative agriculture - the soil is the base. *Global Food Security* 26, 100404.

Schumacher, E. F. (1973). *Small Is Beautiful*. Harper & Row, New York.

Schupp, J. L. and Sharp, J. S. (2012). Exploring the social bases of home gardening. *Agriculture and Human Values* 29(1), 93–105.

Schupp, J. L., Som Castellano, R. L., Sharp, J. S. and Bean, M. (2016). Exploring barriers to home gardening in Ohio households. *Local Environment* 21(6), 752–767.

Scognamiglio, V. (2013). Nanotechnology in glucose monitoring: advances and challenges in the last 10 years. *Biosensors and Bioelectronics* 47, 12–25.

Scoones, I., Smith, R., Cooper, B., White, L., Goh, S., Horby, P., Wren, B. and Gundogdu, Scott, N. and Chen, H. (2013). Nanoscale science and engineering for agriculture and food systems. *Industrial Biotechnology* 9(1), 17–18.

Scott, R. (2010). A critical review of permaculture in the United States–a critical review of permaculture. Available at: http://robscott.net/; http://robscott.net/2010/comments/.

Searchinger, T., Waite, R., Hanson, C. and Ranganathan, J. (2019). *Creating a Sustainable Food Future: A Menu of Solutions to Feed Nearly 10 Billion People by 2050*. World Resources Institute, Washington (DC), USA. ISBN: 978-1-56973-963-1. Library of Congress Control Number: 2019907466.

Seelan, S. K., Laguette, S., Casady, G. M. and Seielstad, G. A. (2003). Remote sensing applications for precision agriculture: a learning community approach. *Remote Sensing of Environment* 88(1–2), 157–169.

Segerson, K. (2013). Voluntary approaches to environmental protection and resource management. *Annual Review of Resource Economics* 5(1), 161–180.

Seibold, S., Gossner, M. M., Simons, N. K., Blüthgen, N., Müller, J., Ambarlı, D., Ammer, C., Bauhus, J., Fischer, M., Habel, J. C., Linsenmair, K. E., Nauss, T., Penone, C., Prati, D., Schall, P., Schulze, E. D., Vogt, J., Wöllauer, S. and Weisser, W. W. (2019). Arthropod decline in grasslands and forests is associated with landscape-level drivers. *Nature* 574(7780), 671–674. DOI: 10.1038/s41586-019-1684-3.

Seitzinger, S. P., Mayorga, E., Bouwman, A. F., Kroeze, C., Muscanescu, B., Sen, A. H. W., Billen, G., Van Drecht, G., Dumont, E., Fekete, B. M., Garnier, J. and Harrison, J. A. (2010). Global River nutrient export: a scenario analysis of past and future trends. *Global Biogeochemical Cycles* 24, GB0A08. DOI: 10.1029/2009gb003587.

Sekhon, B. S. (2014). Nanotechnology in agri-food production: an overview. *Nanotechnology, Science and Applications* 2014, 31–53.

Sekulova, F., Kallis, G., Rodríguez-Labajos, B. and Schneider, F. (2013). Degrowth: from theory to practice. *Journal of Cleaner Production* 38, 1–6. DOI: 10.1016/j.jclepro.2012.06.022.

Selman, P. (2000). *Environmental Planning*, 2nd edn. Sage Publications, London.
Servin, A. D. and White, J. C. (2016). Nanotechnology in agriculture: next steps for understanding engineered nanoparticle exposure and risk. *NanoImpact* 1, 9-12.
Seto, K. and Ramankutty, N. (2016). Hidden linkages between urbanization and food systems. *Science*. American Association for the Advancement of Science. DOI: 10.1126/science.aaf7439.
Setshedi, K. and Modirwa, S. (2020). Socio-economic characteristics influencing small-scale farmers' level of knowledge on climate-smart agriculture in mahikeng local municipality, Northwest Province, South Africa. *South African Journal of Agricultural Extension* 48, 139-152.
Seufert, V., Mehrabi, Z., Gabriel, D. and Benton, T. G. (2019). Current and potential contributions of organic agriculture to diversification of the food production system. In: Lemaire, G., De Faccio Carvalho, P. C., Kronberg, S. and Recous, S. (Eds) *Agroecosystem Diversity*. Academic Press, Cambridge, MA, pp. 435-452. DOI: 10.1016/B978-0-12-811050-8.00028-5.
Seufert, V. and Ramankutty, N. (2017). Many shades of gray—the context-dependent performance of organic agriculture. *Science Advances* 3(3), e1602638. DOI: 10.1126/sciadv.1602638.
Seufert, V., Ramankutty, N. and Foley, J. A. (2012). Comparing the yields of organic and conventional agriculture. *Nature* 485(7397), 229-232. DOI: 10.1038/nature11069.
Seufert, V., Ramankutty, N. and Mayerhofer, T. (2017). What is this thing called organic? – How organic farming is codified in regulations. *Food Policy* 68, 10-20.
Seychell, M. (2016). Towards better prevention and management of chronic diseases. *Health-EU Newsletter* 169. Available at: http://ec.europa.eu/health/newsletter/169/focus_newsletter_en.htm.
Seyfang, G. and Smith, A. (2007). Grassroots innovations for sustainable development: towards a new research and policy agenda. *Environmental Politics* 16(4), 584-603.
Seyran, E. and Craig, W. (2018). New breeding techniques and their possible regulation. *AgBioForum* 21(1), 1-12. ©2018 AgBioForum.
Shannon, K. L., Kim, B. F., McKenzie, S. E. and Lawrence, R. S. (2015). Food system policy, public health, and human rights in the United States. *Annual Review of Public Health* 36, 151-173. DOI: 10.1146/annurev-publhealth-031914-122621.
Shao, H., Xi, N. and Zhang, Y. (2018). Microemulsion formulation of a new biopesticide to control the diamondback moth (Lepidoptera: Plutellidae). *Scientific Reports* 8(1), 10565. DOI: 10.1038/s41598-018-28626-0.
SharathKumar, M., Heuvelink, E. and Marcelis, L. F. M. (2020). Vertical farming: moving from genetic to environmental modification. *Trends in Plant Science* 25(8), 724-727.
Shaw, B. T., Eyck, T. T., Coppin, D., Konefal, J., Oliver, C. and Fairweather, J. (2004). *External Review of the Collaborative Research Agreement between Novartis and the University of California*. Institute for Food and Agricultural Standards, Michigan State University, East Lansing, MI.
Shaw, D. J. (2007). *World Food Security: A History Since 1945*. Palgrave Macmillan, New York.
Sheffield City Council (2014). The Sheffield Food Strategy 2014-2017. Available at: http://democracy.sheffield.gov.uk/documents/s13929/FoodStrategy3.pdf. Accessed 10 May 2019.
Shepard, M. (2013). *Restoration Agriculture: Real World Permaculture for Farmers*. Acres U.S.A., Austin.

Shepon, A., Eshel, G., Noor, E. and Milo, R. (2016). Energy and protein feed-to-food conversion efficiencies in the US and potential food security gains from dietary changes. *Environmental Research Letters* 11(10), 105002.

Shepon, A., Eshel, G., Noor, E. and Milo, R. (2018). The opportunity cost of animal-based diets exceeds all food losses. *Proceedings of the National Academy of Sciences of the United States of America* 115(15), 3804–3809. DOI: 10.1073/pnas.1713820115.

Sherwood, S. and Uphoff, N. (2000). Soil health: research, practice, and policy for a more regenerative agriculture. *Applied Soil Ecology* 15(1), 85–97.

Sherwood, S., Arce, A. and Paredes, M. (2017). *Food, Agriculture, and Social Change: The Everyday Vitality of Latin America*. Routledge, Earthscan, New York.

Shi, H., Magaye, R., Castranova, V. and Zhao, J. (2013). Titanium dioxide nanoparticles: a review of current toxicological data. *Particle and Fibre Toxicology* 10(15). DOI: 10.1186/1743-8977-10-15.

Shi, J., Ye, J., Fang, H., Zhang, S. and Xu, C. (2018). Effects of copper oxide nanoparticles on paddy soil properties and components. *Nanomaterials* 8(10), 839.

Shroff, R., Ramos Cortés, C. and Bessol, M. (2021). *Bill Gates and His Fake Solutions to Climate Change*. Navdanya International, Rome. Available at: www.navdanyainternational.org.

Shumba, A., Chikowo, R., Thierfelder, C., Corbeels, M., Six, J. and Cardinael, R. (2023). Conservation agriculture increases soil organic carbon stocks but not soil CO_2 efflux in two 8-year-old experiments in Zimbabwe. *EGUsphere*. DOI: 10.5194/egusphere-2023-1233.

Siedenburg, K. (1991). Demise of the drins. *GPC*, January.

Siegel, K. R., Ali, M. K., Srinivasiah, A., Nugent, R. A. and Narayan, K. M. (2014). Do we produce enough fruits and vegetables to meet global health need? *PLoS ONE* 9(8), 1–7.

Siegrist, S., Staub, D., Pfiffner, L. and Mäder, P. (1998). Does organic agriculture reduce soil erodibility? The results of a long-term field study on loess in Switzerland. *Agriculture, Ecosystems and Environment* 69(3), 253–264.

Sifted (2022). Can vertical farming survive the downturn? 1 September 2022. Available at: https://sifted.eu/articles/vertical-farming-recession-profitability.

Sileshi, G., Akinnifesi, F. K., Debusho, L. K., Beedy, T., Ajayi, O. C. and Mong'omba, S. (2010). Variation in maize yield gaps with plant nutrient inputs, soil type and climate across Sub-Saharan Africa Field Crops. *Research* 116, 1–13. DOI: 10.1016/j.fcr.2009.11.014.

Silici, L. (2014). *Agroecology. What It Is and What It Has to Offer*. IIED, London.

Simberloff, D. and Dayan, T. (1991). The guild concept and the structure of ecological communities. *Annual Review of Ecology and Systematics* 22(1), 115–143. DOI: 10.1146/annurev.es.22.110191.000555.

Simone, A., Andrea, L. and Federica, R. (2009). Introduction. Managing the uncertainty of nanotechnology across boundaries and scales. *Stand Alone*, 1–6. DOI: 10.3233/978-1-60750-022-3-1.

Simonin, M., Guyonnet, J. P., Martins, J. M. F., Ginot, M. and Richaume, A. (2015). Influence of soil properties on the toxicity of TiO_2 nanoparticles on carbon mineralization and bacterial abundance. *Journal of Hazardous Materials* 283, 529–535.

Singh, R., Singh, Y., Xalaxo, S., Verulkar, S., Yadav, N., Singh, S., Singh, N., Prasad, K. S. N., Kondayya, K., Rao, P. V. R., Rani, M. G., Anuradha, T., Suraynarayana, Y., Sharma, P. C., Krishnamurthy, S. L., Sharma, S. K., Dwivedi, J. L., Singh, A. K., Singh, P. K., Nilanjay, Singh, N. K., Kumar, R., Chetia, S. K., Ahmad, T., Rai, M., Perraju, P., Pande, A., Singh,

D. N., Mandal, N. P., Reddy, J. N., Singh, O. N., Katara, J. L., Marandi, B., Swain, P., Sarkar, R. K., Singh, D. P., Mohapatra, T., Padmawathi, G., Ram, T., Kathiresan, R. M., Paramsivam, K., Nadarajan, S., Thirumeni, S., Nagarajan, M., Singh, A. K., Vikram, P., Kumar, A., Septiningshih, E., Singh, U. S., Ismail, A. M., Mackill, D. and Singh, N. K. (2016). From QTL to variety-harnessing the benefits of QTLs for drought, flood, and salt tolerance in mega rice varieties of India through a multi-institutional network. *Plant Science* 242, 278–287.

Singh, R. P., Handa, R. and Manchanda, G. (2021). Nanoparticles in sustainable agriculture: an emerging opportunity. *Journal of Controlled Release* 329, 1234–1248.

Sinke, P., Swartz, E., Sanctorum, H., van der Giesen, C. and Odegard, I. (2023). Exante life cycle assessment of commercialscale cultivated meat production in 2030. *The International Journal of Life Cycle Assessment* 28(3), 234–254. DOI: 10.1007/s11367-022-02128-8.

Sirieix, L., Kledal, P. R. and Sulitang, T. (2011). Organic food consumers' trade-offs between local or imported, conventional or organic products: a qualitative study in Shanghai. *International Journal of Consumer Studies* 35(6), 670–678.

Sirinathsinghji, E. (2020). Why genome edited organisms are not excluded from the Cartagena Protocol on Biosafety. *Biosafety Briefing*. December. TWN-THIRD WORLD NETWORK and GeneWatch UK. Available at: www.twn.my; www.genewatch.org.

Sirohi, S., Mago, P., Gunwal, I. and Singh, L. (2014). Genetic pollution and biodiversity. *International Journal of Recent Scientific Research* 5(9), 1639–1642.

Sishodia, R. P., Ray, R. L. and Singh, S. K. (2020). Applications of remote sensing in precision agriculture: a review. *Remote Sensing* 12(19), 3136. DOI: 10.3390/rs12193136.

SkyGreens (2010). About sky greens. Available at: https://www.skygreens.com/about-skygreens.

Slow Food (1989). Slow Food Manifesto. Available at: https://slowfood.com/filemanager/Convivium%20Leader%20Area/Manifesto_ENG.pdf.

Slow Food (2021). Slow Food Annual Report 2021. Published June 2022. Available at: www.slowfood.com.

Slow Food (2022). Good, clean and fair food for all. Available at: https://www.slowfood.com/. Accessed 15 October 2022.

Slow Food International (2016). About Us. Available at: http://www.slowfood.com/about-us/.

Smetana, S., Mathys, A., Knoch, A. and Heinz, V. (2015). Meat alternatives: life cycle analysis of most known meat alternatives. *International Journal of Life Cycle Assessment* 20(9), 1254–1267.

Smil, V. (2000). *Feeding the World: A Challenge for the Twenty-First Century*. MIT Press, Cambridge, MA.

Smith, A. and Raven, R. (2012). What is protective space? Reconsidering niches in transitions to sustainability. *Research Policy* 41(6), 1025–1036.

Smith, A. and Stirling, A. (2010). The politics of social-ecological resilience and sustainable socio-technical transitions. *Ecology and Society* 15(1), 11.

Smith, A., Vos, J.-P. and Grin, J. (2010). Innovation studies and sustainability transitions: the allure of the multi-level perspective and its challenges. *Research Policy* 39(4), 435–448.

Smith, B. G. (2008). Developing sustainable food supply chains. *Philosophical Transactions of the Royal Society of London. Series B, Biological Sciences* 363(1492), 849–861. DOI: 10.1098/rstb.2007.2187.

Smith, J. R. (1929). *Tree Crops: A Permanent Agriculture*. Brace and Co., New York.
Smith, L. G., Jones, P. J., Kirk, G. J. D., Pearce, B. D. and Williams, A. G. (2018). Modelling the production impacts of a widespread conversion to organic agriculture in England and Wales. *Land Use Policy* 76, 391–404.
Smith, P., Bustamante, M., Ahammad, H., Clark, H., Dong, H., Elsiddig, E. A., Haberl, H., Harper, R., House, J., Jafari, M., Masera, O., Mbow, C., Ravindranath, N. H., Rice, C. W., Robledo Abad, C., Romanovskaya, A., Sperling, F. and Tubiello, F. (2014). Agriculture, forestry and other land use (AFOLU). In: Edenhofer, O., Pichs-Madruga, Y., Sokona, E., Farahani, S., Kadner, K., Seyboth, A., Adler, I., Baum, S., Brunner, P., Eickemeier, B., Kriemann, J., Schlömer, S., von Stechow, C., Zwickel, T. and Minx, J.C. (Eds) *Climate Change 2014: Mitigation of Climate Change*. Contribution of Working Group III to the Fifth Assessment Report of the Intergovernmental Panel on Climate Change. Cambridge University Press, Cambridge and New York.
Smolik, J. D., Dobbs, T. L. and Rickerl, D. H. (1995). The relative sustainability of alternative, conventional and reduced-till farming system. *American Journal of Alternative Agriculture* 10(1), 25–35.
Snapp, S., Kebede, Y., Wollenberg, E., Dittmer, K. M., Brickman, S., Egler, C. and Shelton, S. (2021). *Agroecology and Climate Change Rapid Evidence Review: Performance of Agroecological Approaches in Low- and Middle- Income Countries*. CGIAR Research Program on Climate Change, Agriculture and Food Security (CCAFS), Wageningen.
Soares, F. V., Knowles, M., Daidone, S. and Tirivayi, N. (2016). *Combined Effects and Synergies Between Agricultural and Social Protection Interventions: What Is the Evidence So Far*. FAO, Rome.
Solazzo, R., Donati, M., Tomasi, L. and Arfini, F. (2016). How effective is greening policy in reducing GHG emissions from agriculture? Evidence from Italy. *Science of the Total Environment* 573, 1115–1124. DOI: 10.1016/J.SCITOTENV.2016.08.066.
Soloviev, E. R. and Landua, G. (2016). *Levels of Regenerative Agriculture*. Terra Genesis International, Driggs, ID.
Song, C., Zhou, Z., Zang, Y., Zhao, L., Yang, W., Luo, X., Jiang, R., Ming, R., Zang, Y., Zi, L. and Zhu, Q. (2021). Variable-rate control system for UAV-based granular fertilizer spreader. *Computers and Electronics in Agriculture* 180, 105832.
Sonka, S. (2015). Big Data: from hype to agricultural tool. *Farm Policy Journal* 12, 1–9.
Sonnino, R. (2014). The new geography of food security: exploring the potential of urban food strategies: the new geography of food security. *The Geographical Journal*. DOI: 10.1111/geoj.12129.
Soubry, B. and Sherren, K. (2022). 'You keep using that word . . .': disjointed definitions of resilience in food systems adaptation. *Land Use Policy* 114, 105954. DOI: 10.1016/j.landusepol.2021.105954.
Soussana, J. F., Tichit, T., Lecomte, P. and Dumont, B. (2015). Agroecology: integration with livestock, in agroecology for Food Security and Nutrition. In: Proceedings of the FAO International Symposium, 18–19 September 2014. FAO, Rome, pp. 225–249.
Sperling, F., Havlík, P., Denis, M., Valin, H., Palazzo, A., Gaupp, F. and Visconti, P. (2020). IIASA-ISC consultative science platform: resilient food systems. Thematic Report of the International Institute for Applied Systems Analysis (IIASA). Laxenburg, and the International Science Council (ISC), Paris.
Spratt, S. (2013). Food price volatility and financial speculation. Future Agricultures Consortium, Working Paper 047.

Springmann, M., Clark, M., Mason-D'Croz, D., Wiebe, K., Bodirsky, B. L., Lassaletta, L., Vermeulen, S. J., Herrero, M., Carlson, K. M., Jonell, M., Troell, M., DeClerck, F., Gordon, L. J., Zurayk, R., Scarborough, P., Rayner, M., Loken, B., Fanzo, J., Godfray, H. C. J., Tilman, D., Rockström, J. and Willett, W. (2018). Options for keeping the food system within environmental limits. *Nature* 562(7728), 519–525. DOI: 10.1038/s41586-018-0594-0.

Spyrou, I. M., Karpouzas, D. G. and Menkissoglu-Spiroudi, U. (2009). Do botanical pesticides alter the structure of the soil microbial community? *Microbial Ecology* 58(4), 715–727. DOI: 10.1007/s00248-009-9522-z.

Średnicka-Tober, D., Barański, M., Seal, C. J., Sanderson, R., Benbrook, C., Steinshamn, H., Gromadz-ka-Ostrowska, J., Rembiałkowska, E., Skwarło-Sońta, K., Eyre, M., Cozzi, G., Larsen, M. K., Jordon, T., Niggli, U., Sakowski, T., Calder, P. C., Burdge, G. C., Sotiraki, S., Stefanakis, A., Stergiadis, S., Yolcu, H., Chatzidimitriou, E., Butler, G., Stewart, G. and Leifert, C. (2016a). Higher PUFA and n-3 PUFA, conjugated linoleic acid, α-tocopherol and iron, but lower iodine and selenium concentrations in organic milk: a systematic literature review and meta- and redundancy analyses. *British Journal of Nutrition* 115, 1043–1060. DOI: 10.1017/S0007114516000349.

Średnicka-Tober, D., Barański, M., Seal, C., Sanderson, R., Benbrook, C., Steinshamn, H., Gromadz-ka-Ostrowska, J., Rembiałkowska, E., SkwarłoSońta, K., Eyre, M., Cozzi, G., Krogh Larsen, M., Jordon, T., Niggli, U., Sakowski, T., Calder, P. C., Burdge, G. C., Sotiraki, S., Stefanakis, A., Yolcu, H., Stergiadis, S., Chatzidimitriou, E., Butler, G., Stewart, G., and Leifert, C. (2016b). Composition differences between organic and conventional meat: a systematic literature review and meta-analysis. *British Journal of Nutrition* 115, 994–1011. doi: 10.1017/S0007114515005073.

Stafford, J. (Eds) (2018). *Precision Agriculture and Sustainability*. Burleigh Dodds Science Publishing, Cambridge.

Stamatiadis, S., Schepers, J. S., Evangelou, E., Glampedakis, A., Glampedakis, M., Dercas, N., Tsadilas, C., Tserlikakis, N. and Tsadila, E. (2020). Variable-rate application of high spatial resolution can improve cotton N-use efficiency and profitability. *Precision Agriculture* 21(3), 695–712.

Stanhill, G. (1990). The comparative productivity of organic agriculture. *Agriculture, Ecosystems and Environment* 30(1–2), 1–26.

Statista (2023a). The growing global hunger for meat. 3 July . Available at: https://www.statista.com/chart/28251/global-meat-production/.

Statista (2023b). Projected vertical farming market worldwide from 2022 to 2032. Available at: https://www.statista.com/statistics/487666/projection-vertical-farming-market-worldwide/. Accessed 1 December 2023.

Stefanini, M., Larson, J. A., Lambert, D. M., Yin, X., Boyer, C. N., Scharf, P., Tubaña, B. S., Varco, J. J., Dunn, D., Savoy, H. J. and Buschermohle, M. J. (2019). Effects of optical sensing based variable rate nitrogen management on yields, nitrogen use and profitability for cotton. *Precision Agriculture* 20(3), 591–610.

Stefanovic, L., Freytag-Leyer, B. and Kahl, J. (2020). Food system outcomes: an overview and the contribution to food systems transformation. *Frontiers in Sustainable Food Systems* 4, 546167. DOI: 10.3389/fsufs.2020.546167.

Steffen, W., Richardson, K., Rockström, J., Cornell, S. E., Fetzer, I., Bennett, E. M., Biggs, R., Carpenter, S. R., De Vries, W., De Wit, C. A., Folke, C., Geertsen, D., Heinke, J., Mace, G. M., Persson, L. M., Ramanathan, V., Reyers, B. and Sörlin, S. (2015a). Planetary boundaries: guiding human development on a changing planet. *Science* 347(6223). DOI: 10.1126/science.1259855.

Steffen, W., Richardson, K., Rockstrom, J., Cornell, S. E., Fetzer, I., Bennett, E. M., Stehle, S. and Schulz, R. (2015b). Agricultural insecticides threaten surface waters at the global scale. *PNAS* 112, 5750–5755.

Steinbach, D., Wood, R. G., Kaur, N., D'Errico, S., Choudhary, J., Sharma, S., Rahar, V. and Jhajharia, V. (2016). *Aligning Social Protection and Climate Resilience*. IIED, London. Available at: http://pubs.iied.org/pdfs/10157IIED.pdf.

Steiner, A., Aguilar, G., Bomba, K., Bonilla, J. P., Campbell, A., Echeverria, R., Gandhi, R., Hedegaard, C., Holdorf, D., Ishii, N., Quinn, K., Ruter, B., Sunga, I., Sukhdev, P., Verghese, S., Voegele, J., Winters, P., Campbell, B., Dinesh, D., Huyer, S., Jarvis, A., Loboguerrero Rodriguez, A. M., Millan, A., Thornton, P., Wollenberg, L. and Zebiak, S. (2020). *Actions to Transform Food Systems under Climate Change*. CGIAR Research Program on Climate Change, Agriculture and Food Security (CCAFS), Wageningen.

Stender, S., Astrup, A. and Dyerberg, J. (2016). Artificial transfat in popular foods in 2012 and in 2014: a market basket investigation in six European countries. *BMJ Open* 6(3). DOI: 10.1136/bmjopen-2015-010673.

Stenmarck, Å., Jensen, C., Quested, T. and Moates, G. (2016). Estimates of European food waste levels. Available at: https://www.eu-fusions.org/phocadownload/Publications/Estimates%20of%20European%20food%20waste%20levels.pdf.

Stephens, N., Di Silvio, L., Dunsford, I., Ellis, M., Glencross, A. and Sexton, A. (2018). Bringing cultured meat to the market: technical, socio-political, and regulatory challenges in cellular agriculture. *Trends in Food Science and Technology* 78, 155–166.

Stephenson, G., Larry, L. and Brewer, L. (2008). I'm getting desperate: what we know about farmers' markets that fail. *Renewable Agriculture and Food Systems* 23(3), 188–199.

Stevenson, J. R., Serraj, R. and Cassman, K. G. (2014). Evaluating conservation agriculture for small-scale farmers in Sub-Saharan Africa and South Asia. *Agriculture, Ecosystems and Environment* 187, 1–10.

Stockdale, E. A., Lampkin, N. H., Hovi, M., Keatinge, R., Lennartsson, E. K. M., Macdonald, D. W., Padel, S., Tattersall, F. H., Wolfe, M. S. and Watson, C. A. (2001). Agronomic and environmental implications for organic farming systems. *Advances in Agronomy* 70, 261–327.

Stokes, E. (2013). Demand for command: responding to technological risks and scientific uncertainties. *Medical Law Review* 21(1), 11–38. DOI: 10.1093/medlaw/fws042.

Stölze, M., Piorr, A., Häring, A. and Dabbert, S. (Eds) (2000). The environmental impact of organic farming in Europe. In: *Organic Farming in Europe: Economics and Policy*. University of Hohenheim, Hohenheim. Available at: http://orgprints.org/2366/02/Volume6.pdf.

Stombaugh, T. (2018). Satellite-based positioning systems for precision agriculture. In: *Precision Agriculture Basics*. American Society of Agronomy; Crop Science Society of America; Soil Science Society of America, Madison, WI, pp. 25–35.

Stout, A. J., Arnett, M. J., Chai, K., Guo, T., Liao, L., Mirliani, A. B., Rittenberg, M. L., Shub, M., White, E. C., Yuen Jr., J. S. K., Zhang, X. and Kaplan, D. L. (2023). Immortalized bovine satellite cells for cultured meat applications. *ACS Synthetic Biology* 12(5), 1567–1573. DOI: 10.1021/acssynbio.3c00216.

Strand, R. and Kjølberg, K. L. (2011). Regulating nanoparticles: the problem of uncertainty. *European Journal of Law and Technology* 2(3), 1–10.

Stuart, T. (2009). *Waste – Uncovering the Global Food Scandal*. Penguin Books, London.

Stuckler, D., McKee, M., Ebrahim, S. and Basu, S. (2012). Manufacturing epidemics: the role of global producers in increased consumption of unhealthy commodities including processed foods, alcohol, and tobacco. *PLoS Medicine* 9(6), e1001235.

Sullivan, D. G., Shaw, J. N. and Rickman, D. (2005). IKONOS imagery to estimate surface soil property variability in two Alabama physiographies. *Soil Science Society of America Journal* 69(6), 1789–1798.

Sullivan, P. (2002). *Drought Resistant Soil*. ATTRA, National Center for Appropriate Technology USDA, Butte, MT. Available at: http://attra.ncat.org/attra-pub/PDF/drought.pdf.

Sun, W., Canadell, J. G., Yu, L., Yu, L., Zhang, W., Smith, P., Fischer, T. and Huang, Y. (2020). Climate drives global soil carbon sequestration and crop yield changes under conservation agriculture. *Global Change Biology* 26(6), 3325–3335.

Sundkvist, Å., Milestad, R. and Jansson, A. (2005). On the importance of tightening feedback loops for sustainable development of food systems. *Food Policy* 30(2), 224–239. DOI:10.1016/j.foodpol.2005.02.003.

SuperMeat (2022). Home page. Available at: https://supermeat.com/our-meat/. Accessed 27 January 2022.

Sussman, L. and Bassarab, K. (2017). Food policy council report 2016. Retrieved from the Johns Hopkins Bloomberg School of Public Health website. Available at: https://assets.jhsph.edu/clf/mod_clfResource/doc/FPC%20Report%202016_Final.pdf.

Sutton, M. A., Bleeker, A., Howard, C. M., Bekunda, M., Grizzetti, B., de Vries, W., van Grinsven, H. J. M., Abrol, Y. P., Adhya, T. K., Billen, G., Davidson, E. A., Datta, A., Diaz, R., Erisman, J. W., Liu, X. J., Oenema, O., Palm, C., Raghuram, N., Reis, S., Scholz, R. W., Sims, T., Westhoek, H., Zhang, F. S., with contributions from Ayyappan, S., Bouwman, A. F., Bustamante, M., Fowler, D., Galloway, J. N., Gavito, M. E., Garnier, J., Greenwood, S., Hellums, D. T., Holland, M., Hoysall, C., Jaramillo, V. J., Klimont, Z., Ometto, J. P., Pathak, H., Plocq Fichelet, V., Powlson, D., Ramakrishna, K., Roy, A., Sanders, K., Sharma, C., Singh, B., Singh, U., Yan, X. Y. and Zhang, Y. (2013). *Our Nutrient World: the Challenge to Produce More Food and Energy with Less Pollution*. Global Overview of Nutrient Management. Centre for Ecology and Hydrology, Edinburgh. (Edinburgh on behalf of the Global Partnership on Nutrient Management and the International Nitrogen Initiative.)

Swagemakers, P., Domínguez García, M. D., Milone, P., Ventura, F. and Wiskerke, J. S. C. (2019). Exploring cooperative place-based approaches to restorative agriculture. *Journal of Rural Studies* 68, 191–199.

Swift, A. (2020). Has the debate on gene editing missed the question? *The Grocer*, 23 September.

Swinburn, B. A., Caterson, I., Seidell, J. C. and James, W. P. T. (2007). Diet, nutrition and the prevention of excess weight gain and obesity. *Public Health Nutrition* 7(1a), 123–146.

Swinnen, J. (2010). The political economy of the most radical reform of the common agricultural policy. Journal of International Agricultural Trade and Development 59, 37–48.

Symes, D. and Marsden, T. K. (1985). Industrialisation of agriculture: intensive livestock farming in Humberside. In: Healey, M. J. and Ilbery, B. W. (Eds) *The Industrialization of the Countryside*. Geo Books, Norwich, pp. 99–120.

Tacoli, C. (2019). Editorial: The urbanization of food insecurity and malnutrition. *Environment and Urbanization* 31(2), 371-374. DOI: 10.1177/0956247819867255.

Tall, A., Hansen, J., Jay, A., Campbell, B., Kinyangi, J., Aggarwal, P. K. and Zougmoré, R. (2014). Scaling up climate services for farmers: mission possible. Learning from good practice in Africa and South Asia. CCAFS Report No. 13. CGIAR Research Program on Climate Change, Agriculture and Food Security (CCAFS), Copenhagen , p. 44.

Tarjan, R. and Kemeny, T. (1969). Multi-generation studies on DDT in mice. *Food and Cosmetics Toxicology* 7, 214–222.

Taulavuori, K., Hyöky, V., Oksanen, J., Taulavuori, E. and Julkunen-Tiitto, R. (2016). Species-specific differences in synthesis of flavonoids and phenolic acids under increasing periods of enhanced blue light. *Environmental and Experimental Botany* 121, 145–150.

Tavares, E. T., Schramm, F. R., Tavares, E. T. and Schramm, F. R. (2015). The principle of precaution and the nano-techno-sciences. *Revista Bioética* 23, 244–255. DOI: 10.1590/1983-80422015232063.

Taylor, J. B. and Uhlig, H. (2016). *Handbook of Macroeconomics* (Vol. 2). Elsevier, Amsterdam.

Teasdale, J. R., Rosecrance, R. C., Coffman, C. B., Starr, J. L., Paltineanu, I. C., Lu, Y. C. and Watkins, B. K. (2000). Performance of reduced tillage cropping systems for sustainable grain production in Maryland. *American Journal of Alternative Agriculture* 15(2), 79–87.

TedxUniversity of Groningen (2015). Free Cafe Groningen: Ivanka Annott. Available at: https://www.youtube.com/watch?v=NWgJKWkWvow.

Tefft, J., Jonasova, M., Adjao, R. and Morgan, A. (2017). *Food Systems for an Urbanising World*. World Bank and FAO, Amsterdam.

Templer, N., Hauser, M., Owamani, A., Kamusingize, D., Ogwali, H., Mulumba, L., Onwonga, R., Adugna, B. T. and Probst, L. (2018). Does certified organic agriculture increase agroecosystem health? Evidence from four farming systems in Uganda. *International Journal of Agricultural Sustainability* 16(2), 150–166. DOI: 10.1080/14735903.2018.1440465.

Termeer, C. J. A. M., Dewulf, A. and Biesbroek, G. R. (2017). Transformational change: governance interventions for climate change adaptation from a continuous change perspective. *Journal of Environmental Planning and Management* 60(4), 558–576.

Teron, A. C. and Tarasuk, V. S. (1999). Charitable food assistance: what are food bank users receiving? *Canadian Journal of Public Health* 90(6), 382–384.

Thaler, E. A., Kwang, J. S., Quirk, B. J., Quarrier, C. L. and Larsen, I. J. (2022). Rates of historical anthropogenic soil erosion in the Midwestern United States. *Earth's Future* 10(3), EF002396. DOI: 10.1029/2021EF002396.

The Conversation (2022). Nanotechnology could make our food tastier and healthier – but can we stomach it? 22 June 2016. Available at: https://theconversation.com/nanotechnology-could-make-our-food-tastier-and-healthier-but-can-we-stomach-it-60349. Accessed 13 February 2022.

The Market Herald (2021). *How a Vancouver Based Company Will Dominate the Vertical Farming Market*. The Market Herald, Vancouver. 23 November 2021.

The Optimist Daily (2022). Fish poop feeds world's largest vertical farm. 21 February 2022. Available at: https://www.optimistdaily.com/2022/01/fish-poop-feeds-worlds-largest-vertical-farm.

The Telegraph (2023). The futuristic farming technique that could solve our food crisis. 22 March 2023. Available at: https://www.telegraph.co.uk/food-and-drink/features/could-vertical-farming-solve-food-crisis/.

Therond, O., Debril, T., Duru, M., Magrini, M. B., Plumecocq, G. and Sarthou, J. P. (2019). Socio-economic characterisation of agriculture models. In: Bergez, J. E., Audouin, E. and Therond, O. (Eds) *Agroecological Transitions: From Theory to Practice in Local Participatory Design*. Springer, Cham, pp. 21-43.

Thompson, E. P. (1971). The moral economy of the English crowd in the 18th century. *Past and Present* 50(1), 76-136.

Thornton, P. K., Whitbread, A., Baedeker, T., Cairns, J., Claessens, L., Baethgen, W., Bunn, C., Friedmann, M., Giller, K. E., Herrero, M., Howden, M., Kilcline, K., Nangia, V., Ramirez-Villegas, J., Kumar, S., West, P. C. and Keating, B. (2018). A framework for priority-setting in climate smart agriculture research. *Agricultural Systems* 167, 161-175.

Thow, A. M., Downs, S. and Jan, S. (2014). A systematic review of the effectiveness of food taxes and subsidies to improve diets: understanding the recent evidence. *Nutrition Reviews* 72(9), 551-565.

Tilman, D. (1999). Global environmental impacts of agricultural expansion: the need for sustainable and efficient practices. *Proceedings of the National Academy of Sciences of the United States of America* 96(11), 5995-6000.

Tilman, D., Balzer, C., Hill, J. and Befort, B. L. (2011). Global food demand and the sustainable intensification of agriculture. *Proceedings of the National Academy of Sciences of the United States of America* 108(50), 20260-20264.

Tilman, D. and Clark, M. (2014). Global diets link environmental sustainability and human health. *Nature* 515(7528), 518-522. DOI: 10.1038/NATURE13959. Available at: http://www.nature.com/nature/journal/v515/n7528/full/nature13959.html.

Tilman, D., Clark, M., Williams, D. R., Kimmel, K., Polasky, S. and Packer, C. (2017). Future threats to biodiversity and pathways to their prevention. *Nature* 546(7656), 73-81.

Tipraqsa, P., Craswell, E. T., Noble, A. D. and Schmidt-Vogt, D. (2007). Resource integration for multiple benefits: multifunctionality of integrated farming systems in Northeast Thailand. *Agricultural Systems* 94(3), 694-703. DOI: 10.1016/j.agsy.2007.02.009.

Tischler, W. (1965). *Agrarökologie*. Gustav Fischer Verlag, Jena, 499 p.

Titonell, P. (2014). Ecological intensification of agriculture - sustainable by nature. *Current Opinion in Environmental Sustainability* 8(Oct), 53-61. DOI: 10.1016/j.cosust.2014.08.006.

Tittonell, P. (2015). Food security and ecosystem services in a changing world: it is time for agroecology. In: Agroecology for Food Security and Nutrition: Proceedings of the FAO International Symposium, Rome, Italy, 18-19 September 2014, pp. 16-31.

Tittonell, P. (2020). Assessing resilience and adaptability in agroecological transitions. *Agricultural Systems* 184, 102862.

Tittonell, P. and Giller, K. E. (2013). When yield gaps are poverty traps: the paradigm of ecological intensification in African smallholder agriculture. *Field Crops Research* 143, 76-90. DOI: 10.1016/j.fcr.2012.10.007.

Tittonell, P., Klerkx, L., Baudron, F., Félix, G. F., Ruggia, A., van Apeldoorn, D., Dogliotti, S., Mapfumo, P. and Rossing, W. A. H. (2016). Ecological intensification: local innovation to address global challenges. In: Lichtfouse, E. (Ed.) *Sustainable Agriculture Reviews* 19, 1-34.

Toledo, V. M. (1990). The ecological rationality of peasant production. In: Altieri, M. and Hecht, S. (Eds). *Agroecology and Small Farm Development*. CRC Press, Boca Raton, FL, pp. 51-58.

Tomlinson, I. J. (2010). Acting discursively: the development of UK organic food and farming policy networks. *Public Administration* 88(4), 1045-1062.

Tonitto, C., David, M. B. and Drinkwater, L. E. (2006). Replacing bare fallows with cover crops in fertilizer-intensive cropping systems: a meta-analysis of crop yield and N dynamics. *Agriculture, Ecosystems and Environment* 112(1), 58–72.

Toogoodtogo (2022). Let's start saving food. Available at: https://toogoodtogo.com/en-us. Accessed 1 November 2022.

Torbett, J. C., Roberts, R. K., Larson, J. A. and English, B. C. (2007). Perceived importance of precision farming technologies in improving phosphorus and potassium efficiency in cotton production. *Precision Agriculture* 8(3), 127–137.

Torney, F., Trewyn, B. G., Lin, V. S. and Wang, K. (2007). Mesoporous silica nanoparticles deliver DNA and chemicals into plants. *Nature Nanotechnology* 2(5), 295–300. DOI: 10.1038/nnano.2007.108.

Totin, E., Segnon, A. C., Schut, M., Affognon, H., Zougmoré, R. B., Rosenstock, T. and Thornton, P. K. (2018). Institutional perspectives of climate-smart agriculture: a systematic literature review. *Sustainability* 10(6), 1990. DOI: 10.3390/su10061990. Available at: www.mdpi.com/journal/sustainability.

Trainer, T. (2020). De-growth: some suggestions from the Simpler Way perspective. *Economics* 167, 106436.

Trasande, L., Zoeller, R. T., Hass, U., Kortenkamp, A., Grandjean, P., Myers, J.P., DiGangi, J., Hunt, P. M., Rudel, R., Sathyanarayana, S., Bellanger, M., Hauser, R., Legler, J., Skakkebaek, N. E. and Hein-del, J. J. (2016). Burden of disease and costs of exposure to endocrine disrupting chemicals in the European Union: an updated analysis. *Andrology* 4(4), 565–572. DOI: 10.1111/andr.12178.

Tregear, A. (2011). Progressing knowledge in alternative and local food networks: critical reflections and a research agenda. *Journal of Rural Studies* 27(4), 419–430.

Treu, H., Nordborg, M., Cederberg, C., Heuer, T., Claupein, E., Hoffmann, H. and Berndes, G. (2017). Carbon footprints and land use of conventional and organic diets in Germany. *Journal of Cleaner Production* 161, 127–142. DOI: 10.1016/j.jclepro.2017.05.041.

Trewavas, A. J. (2001a). The population/biodiversity paradox. Agricultural efficiency to save wilderness. *Plant Physiology* 125(1), 174–179. DOI: 10.1104/pp.125.1.174.

Trewavas, A. (2001b). Urban myths of organic farming. *Nature* 410, 409–410.

TropoGo (2022). Agricultural drones – application of drones in agriculture in India. 6 February. Available at: https://tropogo.com/blogs/application-of-drones-in-agriculture-in-india.

Tscharntke, T., Clough, Y., Wanger, T. C., Jackson, L., Motzke, I., Perfecto, I., Vandermeer, J. and Whitbread, A. (2012). Global food security, biodiversity conservation and the future of agricultural intensification. *Biological Conservation* 151(1), 53–59. DOI: 10.1016/j.biocon.2012.01.068.

Tscharntke, T., Klein, A. M., Kruess, A., Steffan-Dewenter, I. and Thies, C. (2005). Landscape perspectives on agricultural intensification and biodiversity–ecosystem service management. *Ecology Letters* 8(8), 857–874.

Tsiafouli, M. A., Thébault, E., Sgardelis, S. P., de Ruiter, P. C., van der Putten, W. H., Birkhofer, K., Hemerik, L., de Vries, F. T., Bardgett, R. D., Brady, M. V., Bjornlund, L., Jørgensen, H. B., Christensen, S., Hertefeldt, T. D., Hotes, S., Gera Hol, W. H., Frouz, J., Liiri, M., Mortimer, S. R., Setälä, H., Tzanopoulos, J., Uteseny, K., Pižl, V., Stary, J., Wolters, V. and Hedlund, K. (2014). Intensive agriculture reduces soil biodiversity across Europe. *Global Change Biology* 21(2), 973–985.

Tsouros, D. C., Bibi, S. and Sarigiannidis, P. G. (2019). A review on UAV-based applications for precision agriculture. *Information* 10(11), 349.

Tu, C., Louws, F. J., Creamer, N. G., Paul Mueller, J., Brownie, C., Fager, K., Bell, M. and Hu, S. (2006). Responses of soil microbial biomass and N availability to transition strategies from conventional to organic farming systems. *Agriculture, Ecosystems and Environment* 113(1-4), 206-215.

Tuomisto, H. L., Hodge, I. D., Riordan, P. and Macdonald, D. W. (2012). Does organic farming reduce environmental impacts? A meta-analysis of European research. *Journal of Environmental Management* 112, 309-320.

Tuong, T. P. and Bouman, B. A. M. (2003). Rice production in water scarce environments. In: Kijne, J. W., Barker, R. and Molden, D. (Eds) *Water Productivity in Agriculture: Limits and Opportunities for Improvement*. CABI Publishing, Wallingford, pp. 53-67.

Turnbull, C., Lillemo, M. and Hvoslef-Eide, T. A. K. (2021). Global regulation of genetically modified crops amid the gene edited crop boom - a review. *Frontiers in Plant Science* 12, 630396. DOI: 10.3389/fpls.2021.630396.

Ulbrich, R. and Pahl-Wostl, C. (2019). The German permaculture community from a community of practice perspective. *Sustainability* 11(5), 1241. DOI: 10.3390/su11051241.

Uldrich, J. (2021). *Regenerative Agriculture: The Next Trend in Food Retailing*. Forbes Business Council, Boston, MA. 19 August. Available at: https://www.forbes.com/sites/forbesbusinesscouncil/2021/08/19/regenerative-agriculture-the-next-trend-in-food-retailing/?sh=10a9db4e2153.

Ullah, A., Nawaz, A., Farooq, M. and Siddique, K. H. M. (2021). Agricultural innovation and sustainable development: a case study of rice-wheat cropping systems in South Asia. *Sustainability* 13(4), 1965.

Ulrichs, C., Mewis, I. and Goswami, A. (2005). Crop diversification aiming nutritional security in West Bengal: biotechnology of stinging capsules in nature's water-blooms. *Annals of the Technical Issue of State Agricultural Technologists Service Association*, 1-18.

UN (2019a). Population. Available at: https://www.un.org/en/sections/issues-depth/population/index.html.

UN (2019b). World population prospects 2019. Available at: https://www.un.org/development/desa/publications/worldpopulation-prospects-2019-highlights.html.

UN (2020). World population is expected to reach 9.7 billion in 2050. Available at: https://www.un.org/development/desa/en/news/population/world-population-prospects-2019.html. Accessed 10 July 2020.

UN Water (2015). *The United Nations World Water Development Report 2015: Water for a Sustainable World*. UNESCO, Paris. Available at. https://sustainabledevelopment.un.org/content/documents/1711Water%20for%20a%20Sustainable%20World.pdf.

UNCCD (2012). *Zero Net Land Degradation: A Sustainable Development Goal for Rio +20*. United Nations Convention to Combat Desertification, Bonn.

UNCTAD (2009). *The Global Economic Crisis: Systemic Failures and Multilateral Remedies*.United Nations Conference on Trade and Development, Geneva.

UNCTAD (2011). Price formation in financialized commodity markets: the role of information. Available at: http://www.unctad.org/en/docs/gds20111_en.pdf.

UNEP (1992). *Saving Our Planet, Challenges and Hopes, the State of the Environment (1972-92)*. United Nations Environment Program, Nairobi.

UNEP (2014). Assessing global land use: balancing consumption with sustainable supply. In: Bringezu, S., Schütz, H., Pengue, W., OBrien, M., Garcia, F., Sims, R., Howarth, R., Kauppi, L., Swilling, M. and Herrick, J. (Eds). *A Report of the Working Group on*

Land and Soils of the International Resource Panel. United Nations Environment Programme (UNEP), Nairobi / Paris.

UNEP (2016). Food Systems and Natural Resources. A Report of the Working Group on Food Systems of the International Resource Panel. (Westhoek, H., Ingram, J., Van Berkum, S., Özay, L. and Hajer, M.) Job Number: DTI/1982/PA. ISBN: 978-92-807-3560-4.

UNEP (2021a). Worldwide food waste. Available at: https://www.unep.org/thinkeatsave/get-informed/worldwide-food-waste. Accessed 07 June 2021.

UNEP (2021b). *Food Waste Index Report 2021*. UNEP, Nairobi.

UNFPA (2023). World population dashboard. United Nations Population Fund, New York. Available at: https://www.unfpa.org/data/world-population-dashboard. Accessed 29 November 2023.

UN-Habitat and IHS-Erasmus University Rotterdam (2018). *The State of African Cities 2018: The Geography of African Investment*. United Nations Human Settlements Programme. UN-Habitat, Nairobi.

University of Barcelona (2003). Available at: http://bus-web.ad.uab.edu/studentdrive/danthony/Sept%2030/Calgene.pdf.

Unruh, G. C. (2000). Understanding carbon lock-in. *Energy Policy* 28(12), 817–830.

Uphoff, N. (2002). *Agroecological Innovations: Increasing Food Production with Participatory Development*. Earthscan, London.

Upside Foods (2022). Home page. Available at: https://upsidefoods.com/our-foods/. Accessed 27 January 2022.

USC (2023). *Computational Agroecology – The Future of Farming*. University of Southern California, Los Angeles, CA. 5 July.

USDA (2014a). New data reflects the continued demand for farmers markets. WWW Document. Available at: http://www.usda.gov/wps/portal/usda/usdahome?-contentid=2014/08/0167.xml. Accessed 4 April 2016.

USDA (2014b). Agricultural marketing service–farmers market growth. United States Department of Agriculture, Washington, DC. Available at: http://www.ams.usda.gov/AMSv1.0/ams.fetchTemplateData.do?template=TemplateS&leftNav=WholesaleandFarmersMarkets&page=WFMFarmersMarketGrowth&description=Farmers+Market+Growth.

USDA (2021). Food and Nutrition Service. FNS-101: Supplemental Nutrition Assistance Program. Available at: https://www.fns.usda.gov/resources?f%5B0%5D=program%3A2&f%5B1%5D=resource_type%3A155. Accessed 10 July 2021.

Usman, M., Byrne, J. M., Chaudhary, A., Orsetti, S., Hanna, K., Ruby, C., Kappler, A. and Haderlein, S. B. (2018). Magnetite and green rust: synthesis, properties, and environmental applications of mixed-valent iron minerals. *Chemical Reviews* 118(7), 3251–3304.

Usman, M., Farooq, M., Wakeel, A., Nawaz, A., Cheema, S. A., Rehman, H. U., Ashraf, I. and Sanaullah, M. (2020). Nanotechnology in agriculture: current status, challenges, and future opportunities. *Science of the Total Environment* 721, 137778.

Vaarst, M., Escudero, A. G., Chappell, M. J., Brinkley, C., Nijbroek, R., Arraes, N. A. M., Andreasen, L., Gattinger, A., Almeida, G. F. D., Bossio, D. and Halberg, N. (2017). Exploring the concept of agroecological food systems in a city-region context. *Agroecology and Sustainable Food Systems*42 686–711. DOI: 10.1080/21683565.2017.1365321.

Vaarst, M., Escudero, A. G., Chappell, M. J., Brinkley, C., Nijbroek, R., Arraes, N. A. M., Andreasen, L., Gattinger, A., De Almeida, G. F., Bossio, D. and Halberg, N.

(2018). Exploring the concept of agroecological food systems in a city-region context. *Agroecology and Sustainable Food Systems* 42(6), 686–711. DOI: 10.1080/21683565.2017.1365321.

Valin, H., Havlik, P., Mosnier, A., Herrero, M., Schmid, E. and Obersteiner, M. (2013). Agricultural productivity and greenhouse gas emissions: trade-offs or synergies between mitigation and food security? *Environmental Research Letters* 8(3), 035019.

Valko, M., Morris, H. and Cronin, M. T. D. (2005). Metals, toxicity and oxidative stress. *Current Medicinal Chemistry* 12(10), 1161–1208. DOI: 10.2174/0929867053764635.

Vallgårda, S. (2015). Governing obesity policies from England, France, Germany, and Scotland. *Social Science and Medicine* 147, 317–323.

Van de Poel, I. (2000). On the role of outsiders in technical development. *Technology Analysis and Strategic Management* 12(3), 383–397.

van Delden, S. H., Sharath Kuma, M., Butturini, M., Graamans, L. J. A., Heuvelink, E., Kacira, M., Kaiser, E., Klamer, R. S., Klerkx, L., Kootstra, G., Loeber, A., Schouten, R. E., Stanghellini, C., van Ieperen, W., Verdonk, J. C., Vialet-Chabrand, S., Woltering, E. J., van de Zedde, R., Zhang, Y. and Marcelis, L. F. M. (2021). Current status and future challenges in implementing and upscaling vertical farming. *Nature Food* 2(12), 944–956. DOI: 10.1038/s43016-021-00402-w.

van der Esch, S., Brink, B. and Stehfest, E. (2017). *Exploring Future Changes in Land Use and Land Condition and the Impacts on Food, Water, Climate Change and Biodiversity: Scenarios for the UNCCD Global Land Outlook*. PBL Netherlands Environmental Assessment Agency, The Hague.

van der Ploeg, J. D. (2014). Peasant-driven agricultural growth and food sovereignty. *Journal of Peasant Studies* 41(6), 999–1030. DOI: 10.1080/03066150.2013.876997.

van der Ploeg, J. D., Barjolle, D., Bruil, J., Brunori, G., Costa Madureira, L. M., Dessein, J., Drąg, Z., Fink-Kessler, A., Gasselin, P., Gonzalez de Molina, M., Gorlach, K., Jürgens, K., Kinsella, J., Kirwan, J., Knickel, K., Lucas, V., Marsden, T., Maye, D., Migliorini, P., Milone, P., Noe, E., Nowak, P., Parrott, N., Peeters, A., Rossi, A., Schermer, M., Venturar, F., Visser, M. and Wezel, A. (2019). The economic potential of agroecology: empirical evidence from Europe. *Journal of Rural Studies* 71, 46–61.

van der Weele, C., Feindt, P., Jan van der Goot, A., van Mierloa, B. and van Boekel, M. (2019). Meat alternatives: an integrative comparison. *Trends in Food Science and Technology* 88, 505–512.

Van Dooren, C., Marinussen, M., Blonk, H., Aiking, H. and Vellinga, P. (2014). Exploring dietary guidelines based on ecological and nutritional values: a comparison of six dietary patterns. *Food Policy* 44, 36–46.

Van Drecht, G., Bouwman, A. F., Harrison, J. and Knoop, J. M. (2009). Global nitrogen and phosphate in urban wastewater for the period 1970 to 2050. *Global Biogeochemical Cycles* 23(4). DOI: 10.1029/2009GB003458.

Van Driel, H. and Schot, J. (2005). Radical innovation as a multi-level process: introducing floating grain elevators in the port of Rotterdam. *Technology and Culture* 46(1), 51–76.

Van Hal, O., De Boer, I. J. M., Muller, A., De Vries, S., Erb, K.-H., Schader, C., Gerrits, W. J. and Van Zanten, H. H. E. (2019). Upcycling food leftovers and grass resources through livestock: impact of livestock system and productivity. *Journal of Cleaner Production* 219, 485–496.

Van Henten, E. J., Van Tuijl, B. A. J., Hoogakker, G.-J., Van Der Weerd, M. J., Hemming, J., Kornet, J. G. and Bontsema, J. (2006). An autonomous robot for de-leafing cucumber plants grown in a high-wire cultivation system. *Biosystems Engineering* 94(3), 317–323.

Van Kernebeek, H. R. J., Oosting, S. J., Van Ittersum, M. K., Bikker, P. and De Boer, I. J. M. (2016). Saving land to feed a growing population: consequences for consumption of crop and livestock products. *International Journal of Life Cycle Assessment* 21(5), 677–687.

van Klink, R., Bowler, D. E., Gongalsky, K. B., Swengel, A. B., Gentile, A. and Chase, J. M. (2020). Meta-analysis reveals declines in terrestrial but increases in freshwater insect abundances. *Science* 368(6489), 417–420.

Van Vuuren, D. P., Bouwman, A. F. and Beusen, A. H. W. (2010). Phosphorus demand for the 1970–2100 period: a scenario analysis of resource depletion. *Global Environmental Change* 20(3), 428–439.

van Walsum, E., van den Berg, L., Bruil, J. and Gubbels, P. (2014). From vulnerability to resilience: agroecology for sustainable dryland management. *Planet@Risk* 2(1), 62–69.

Van Zanten, H. H. E., Herrero, M., Van Hal, O., Roos, E., Muller, A., Garnett, T., Gerber, P. J., Schader, C. and De Boer, I. J. M. (2018). Defining a land boundary for sustainable livestock consumption. *Global Change Biology* 24(9), 4185–4194. DOI: 10.1111/gcb.14321.

Van Zanten, H. H. E., Van Ittersum, M. K. and De Boer, I. J. M. (2019). The role of farm animals in a circular food system. *Global Food Security* 21, 18–22.

Vandermeer, J. (2010). *The Ecology of Agroecosystems*. Jones & Bartlett, Burlington, MA.

Vanderroost, M., Ragaert, P., Devlieghere, F. and De Meulenaer, B. (2014). Intelligent food packaging: the next generation. *Trends in Food Science and Technology* 39(1), 47–62. DOI: 10.1016/j.tifs.2014.06.009.

Vandevijvere, S., Jaacks, L. M., Monteiro, C. A., Moubarac, J. C., Girling-Butcher, M., Lee, A. C., Pan, A., Bentham, J. and Swinburn, B. (2019). Global trends in ultra-processed food and drink product sales and their association with adult body mass index trajectories. *Obesity Reviews* 20 (Suppl. 2), 10–19.

Vanheukelom, J., Mackie, J. and Ronceray, M. (2018). Good enough coherence. In: *Policy Coherence for Sustainable Development 2018*. OECD, Paris. DOI: 10.1787/9789264301061-en.

Vanloqueren, G. and Baret, P. V. (2009). How agricultural research systems shape a technological regime that develops genetic engineering but locks out agroecological innovations. *Research Policy* 38(6), 971–983.

Veen, E. J., Bock, B. B., Van den Berg, W., Visser, A. J. and Wiskerke, J. S. C. (2015). Community gardening and social cohesion: different designs, different motivations. *Local Environment* 21(10), 1271–1287.

Veldkamp, A., Kok, K., De Koning, G. H. J., Schoorl, J. M., Sonneveld, M. P. W. and Verburg, P. H. (2001). Multi-scale system approaches in agronomic research at the landscape level. *Soil and Tillage Research* 58(3–4), 129–140. DOI: 10.1016/S0167-1987(00)00163-X.

Venn, L., Kneafsey, M., Holloway, L., Cox, R., Dowler, E. and Tuomainen, H. (2006). Researching European "alternative" food networks: some methodological considerations. *Area* 38(3), 248–258. DOI: 10.1111/j.1475-4762.2006.00694.x.

Verhoeckx, K. C. M., Vissers, Y. M., Baumert, J. L., Faludi, R., Feys, M., Flanagan, S., Herouet-Guicheney, C., Holzhauser, T., Shimojo, R., van der Bolt, N., Wichers, H. and Kimber,

I. (2015). Food processing and allergenicity. *Food and Chemical Toxicology* 80, 223-240. DOI: 10.1016/j.fct.2015.03.005.

Verhulst, N., Govaerts, B., Verachtert, E., Castellanos-Navarrete, A., Mezzalama, M., Wall, P., Deckers, J. and Sayre, K. D. (2010). Conservation agriculture, improving soil quality for sustainable production systems? In: Lal, R. and Stewart, B. A. (Eds) *Advances in Soil Science: Food Security and Soil Quality*. CRC Press, Boca Raton, FL, pp. 137-208.

Verma, K. K., Song, X. P., Joshi, A., Tian, D. D., Rajput, V. D., Singh, M., Arora, J., Minkina, T. and Li, Y. R. (2022). Recent trends in nano-fertilizers for sustainable agriculture under climate change for global food security. *Nanomaterials* 12(1), 173. DOI: 10.3390/nano12010173.

Verma, M. L. (2017). Enzymatic nanobiosensors in the agricultural and food industry. In: Ranjan, S., Dasgupta, N. and Lichtfouse, E. (Eds) *Nanoscience in Food and Agriculture 4*. Springer International Publishing, Cham, pp. 229–245. DOI: 10.1007/978-3-319-53112-0_7.

Vermeulen, S. J., Campbell, B. M. and Ingram, J. S. I. (2012). Climate change and food systems. *Annual Review of Environment and Resources* 37(1), 195–222. DOI: 10.1146/annurev-environ-020411-130608.

Vertical Farming Planet (2023). Can vertical farming be organic? Available at: https://verticalfarmingplanet.com/can-vertical-farming-be-organic/. Accessed 2 July 2023.

Villar, J. L. and Freese, B. (2008). *Who Benefits from GM Crops?* Friends of the Earth International, Amsterdam.

Vittersø, G., Torjusen, H., Laitala, K., Tocco, B., Biasini, B., Csillag, P., de Labarre, M. D., Lecoeur, J. L., Maj, A., Majewski, E., Malak-Rawlikowska, A., Menozzi, D., Török, Á. and Wavresky, P. (2019). Short food supply chains and their contributions to sustainability: participants' views and perceptions from 12 European cases. *Sustainability* 11(17), 4800. DOI: 10.3390/su11174800.

von Koerber, K., Bader, N. and Leitzmann, C. (2017). Wholesome nutrition: an example for a sustainable diet. *Proceedings of the Nutrition Society* 76(1), 34–41.

Voogt, W., Holwerda, H. T. and Khodabaks, R. (2010). Biofortification of lettuce (*Lactuca sativa* L.) with iodine: the effect of iodine form and concentration in the nutrient solution on growth, development and iodine uptake of lettuce grown in water culture. *Journal of the Science of Food and Agriculture* 90(5), 906-913.

Wachter, J. M., Painter, K. M., Carpenter-Boggs, L. A., Huggins, D. R. and Reganold, J. P. (2019). Productivity, economic performance, and soil quality of conventional, mixed, and organic dryland farming systems in eastern Washington State. *Agriculture, Ecosystems and Environment* 286, 106665.

Wackernagel, M., Kitzes, J., Moran, D., Goldfinger, S. and Thomas, M. (2006). The ecological footprint of cities and regions: comparing resource availability with resource demand. *Environment and Urbanization* 18(1), 103–112.

Wada, Y., Van Beek, L. P. H. and Bierkens, M. F. P. (2011). Modelling global water stress of the recent past: on the relative importance of trends in water demand and climate variability. *Hydrology and Earth System Sciences* 15(12), 3785–3808.

Waggoner, P. E. (1996). How much land can ten billion people spare for nature? *Daedalus*, 73-93. Available at: http://www.jstor.org/stable/20027371.

Wagner, K. H. and Brath, H. (2012). A global view on the development of non-communicable diseases. *Preventive Medicine* 54, S38–S41. DOI: 10.1016/j.ypmed.2011.11.012.

Wagstaff, A. and van Doorslaer, E. (2000). Income inequality and health: what does the literature tell us? *Annual Review of Public Health* 21, 543-567.

Wahl, P. (2009). *Food Speculation: The Main Factor of the Price Bubble in 2008*. Briefing Paper. World Economy, Ecology and Development, Berlin.

Wakefield, S., Yeudall, F., Taron, C., Reynolds, J. and Skinner, A. (2007). Growing urban health: community gardening in South-East Toronto. *Health Promotion International* 22(2), 92–101.

Waldman, K. B. and Kerr, J. M. (2014). Limitations of certification and supply chain standards for environmental protection in commodity crop production. *Annual Review of Resource Economics* 6(1), 429–449.

Walingo, M. K. and Ekesa, B. N. (2013). Nutrient intake, morbidity and nutritional status of preschool children are influenced by agricultural and dietary diversity in western Kenya. *Pakistan Journal of Nutrition* 12(9), 854–859.

Walker, B., Holling, C. S., Carpenter, S. R. and Kinzig, A. P. (2004). Resilience, adaptability, and transformability in social-ecological systems. *Ecology and Society* 9(2). Available at: http://www.ecologyandsociety.org/vol9/iss2/art5/.

Wallace, I. (1985). Towards a geography of agribusiness. *Progress in Human Geography* 9, 401–414.

Wallgren, C. (2006). Local or global food markets: a comparison of energy use for transport. *Local Environment* 11(2), 233–251. DOI: 10.1080/13549830600558598.

Wang, C. Y., Wang, S. Y., Yin, J. J., Parry, J. and Yu, L. L. (2007). Enhancing antioxidant, antiproliferation, and free radical scavenging activities in strawberries with essential oils. *Journal of Agricultural and Food Chemistry* 55(16), 6527–6532.

Wang, N., Zhang, N. and Wang, M. (2006). Wireless sensors in agriculture and food industry–recent development and future perspective. *Computers and Electronics in Agriculture* 50(1), 1–14.

Wang, Y. C., McPherson, K., Marsh, T., Gortmaker, S. L. and Brown, M. (2011). Health and economic burden of the projected obesity trends in the USA and the UK. *The Lancet* 378(9793), 815–825. DOI: 10.1016/S0140-6736(11)60814-3.

Wanger, T. C., DeClerck, F., Garibaldi, L. A., Ghazoul, J., Kleijn, D., Klein, A. M., Kremen, C., Mooney, H., Perfecto, I., Powell, L. L., Settele, J., Solé, M., Tscharntke, T. and Weisser, W. (2020). Integrating agroecological production in a robust post-2020 Global biodiversity framework. *Nature Ecology and Evolution*. DOI: 10.1038/s41559-020-1262-y.

Wanyama, R., Gödecke, T., Chege, C. G. K. and Qaim, M. (2019). How important are supermarkets for the diets of the urban poor in Africa? *Food Security* 11(6), 1339–1353.

Ward, M. H., Jones, R. R., Brender, J. D., de Kok, T. M., Weyer, P. J., Nolan, B. T., Villanueva, C. M. and van Breda, S. G. (2018). Drinking water nitrate and human health: an updated review. *International Journal of Environmental Research and Public Health* 15(7), 1507–1557.

Ward, N. (1993). The agricultural treadmill and the rural environment in the post-productivist era. *Sociologia Ruralis* 33(3–4), 348–364.

Warner, K. (2006). *Agroecology in Action: Extending Alternative Agriculture through Social Networks*. The MIT Press, Cambridge, MA.

Watson, D. J. (2018). *Pesticides and Agriculture: Profit, Politics, and Policy*. Burleigh Dodds Science Publishing Limited, Cambridge. Available at: www.bdspublishing.com.

Watson, D. (2019a). Adaption to climate change: climate adaptive breeding of maize, wheat and rice. In: Sarkar, A., Sensarma, S. R. and vanLoon, G. W. (Eds) *Sustainable*

Solutions for Food Security Combating Climate Change by Adaptation. Springer, New York.

Watson, D. (2019b). Adaptation to climate change through adaptive crop management. In: Sarkar, A., Sensarma, S. R. and vanLoon, G. W. (Eds) *Sustainable Solutions for Food Security Combating Climate Change by Adaptation*. Springer.

Watson, D. J. (2021). *Secondary Impacts of COVID-19 on Food Systems Resilience in SSA*. Alliance for a Green Revolution in Africa (AGRA), Nairobi.

Watson, D. (2022). *Food, Fuel and Fertiliser Crisis. Russia Ukraine Crisis Series*. AGRA, Westend Towers, Nairobi.

Watson, J. L. and Caldwell, M. L. (2017). *The Cultural Politics of Food and Eating: A Reader*. Harvard University, Cambridge.

Watts, C. W. and Dexter, A. R. (1997). The influence of organic matter in reducing the destabilization of soil by simulated tillage. *Soil and Tillage Research* 42(4), 253–275.

Watts, M. and Williamson, S. (2015). *Replacing Chemicals with Biology: Phasing Out Highly Hazardous Pesticides with Agroecology*. PAN Asia and the Pacific, Penang.

Watts, D. C. H., Ilbery, B. and Maye, D. (2005). Making reconnections in agro-food geography: alternative systems of food provision. *Progress in Human Geography* 29(1), 22–40. DOI: 10.1191/0309132505ph526oa.

Weatherspoon, D. R. D. and Reardon, T. (2003). Supermarkets in Africa: implications for agrifood systems and the rural poor. *Development Policy Review* 2, 1–17.

Webb, A. and Hessel, A. (2022). *The Genesis Machine: Our Quest to Rewrite Life in the Age of Synthetic Biology*. Public Affairs, Hachette Book Group, Inc, New York.

Weber, H., Poeggel, K., Eakin, H., Fischer, D., Lang, D. J., Von Wehrden, H. V. and Wiek, A. (2020). What are the ingredients for food systems change towards sustainability? – insights from the literature. *Environmental Research Letters* 15(11), 113001.

WEF (2012). *What If the World's Soil Runs Out?* World Economic Forum. https://world.time.com/2012/12/14/what-if-the-worlds-soil-runs-out/.

Wegerif, M. C. A. and Hebinck, P. (2016). The symbiotic food system: an 'alternative' agri-food system already working at scale. *Agriculture* 6(3), 40. DOI: 10.3390/agriculture6030040.

Wei, A. and Cacho, J. (2000). Competition among foreign and Chinese agro-food enterprises in the process of globalization. *Int. Food Agribus Man* 2(3), 437–451.

Weiss, M., Jacob, F. and Duveiller, G. (2020). Remote sensing for agricultural applications: a meta-review. *Remote Sensing of Environment* 236, 111402.

Werdigier, J. (2009). British supermarket chain places bets on banking. *The New York Times*, 30 April.

West, G. H. and Kovacs, K. (2017). Addressing groundwater declines with precision agriculture: an economic comparison of monitoring methods for variable-rate irrigation. *Water* 9(1), 28. DOI: 10.3390/w9010028.

West, P. C., Gerber, J. S., Engstrom, P. M., Mueller, N. D., Brauman, K. A., Carlson, K. M., Cassidy, E. S., Johnston, M., MacDonald, G. K., Ray, D. K. and Siebert, S. (2014). Leverage points for improving global food security and the environment. *Science* 345(6194), 325–328.

Westhoek, H., Lesschen, J. P., Rood, T., Wagner, S., De Marco, A., Murphy-Bokern, D., Leip, A., van Grinsven, H., Suttons, M. A. and Oenema, O. (2014). Food choices, health, and environment: effects of cutting Europe's meat and dairy intake. *Global Environmental Change* 26, 196–205.

Wezel, A. and Silva, E. (2017). Agroecology and agroecological cropping practices. In: Wezel, A. (Ed.) *Agroecological Practices for Sustainable Agriculture: Principles, Applications, and Making the Transition*. World Scientific, Hackensack, NJ, pp. 19-51.

Wezel, A., Bellon, S., Doré, T., Francis, C., Vallod, D. and David, C. (2009). Agroecology as a science, a movement, and a practice: a review. *Agronomy for Sustainable Development*. INRA, EDP Sciences. DOI: 10.1051/agro/2009004.

Wezel, A., Casagrande, M., Celette, F., Vian, J. F., Ferrer, A. and Peigné, J. (2013). Agroecological practices for sustainable agriculture: a Review. *Agronomy for Sustainable Development* 34, 1-20.

Wezel, A., Brives, H., Casagrande, M., Clément, C., Dufour, A. and Vandenbroucke, P. (2016). Agroecology territories: places for sustainable agricultural and food systems and biodiversity conservation. *Agroecology and Sustainable Food Systems* 40(2), 132-144. DOI: 10.1080/21683565.2015.1115799.

Wezel, A., Goette, J., Lagneaux, E., Passuello, G., Reisman, E., Rodier, C. and Turpin, G. (2018a). Agroecology in Europe: research, education, collective action networks, and alternative food systems. *Sustainability* 10(4), 1214. DOI: 10.3390/su10041214.

Wezel, A., Goris, M., Bruil, J., Félix, G. F., Peeters, A., Bàrberi, P., Bellon, S. and Migliorini, P. (2018b). Challenges and action points to amplify agroecology in Europe. *Sustainability* 10(5), 1598. DOI: 10.3390/su10051598.

Wezel, A., Herren, B. G., Kerr, R. B., Barrios, E., Gonçalves, A. L. R. and Sinclair, F. (2020). Agroecological principles and elements and their implications for transitioning to sustainable food systems: a review. *Agronomy for Sustainable Development* 40(6): 40.

WFP (2020). *Update on the Impact of Covid-19 on Food and Nutrition Security in West and Central AfDrica*. World Food Programme, Regional Bureau Dakar, Research, Assessments and Monitoring, Emergency Preparedness, Nutrition and Supply Chain Units, Dakar. 17 April.

Wheaton, B. and Kiernan, W. J. (2012). Farmland: an untapped asset class? quantifying the opportunity to invest in agriculture. *Global Ag Investing: Research and Insight*, December.

Wheeler, D. C., Nolan, B. T., Flory, A. R., DellaValle, C. T. and Ward, M. H. (2015). Modelling groundwater nitrate concentrations in private wells in Iowa. *Science of the Total Environment* 536, 481-488.

Wheeler, T. and von Braun, J. (2013). Climate change impacts on global food security. *Science* 341(6145), 508-513.

White, B., Borras, S. Jr., Hall, R., Scoones, I. and Wolford, W. (2012). The new enclosures: critical perspectives on corporate land deals. *Journal of Peasant Studies* 39(3-4), 619-647.

White, B. and Dasgupta, A. (2010). Agrofuels capitalism: a view from political economy. *The Journal of Peasant Studies* 37(4), 593-607.

Whitefield, P. (2004). *The Earth Care Manual: A Permaculture Handbook for Britain & Other Temperate Climates*. Permanent, Portsmouth.

Whitney, C. W., Luedeling, E., Hensel, O., Tabuti, J. R. S., Krawinkel, M., Gebauer, J. and Kehlenbeck, K. (2018). The role of home gardens for food and nutrition security in Uganda. *Human Ecology* 46(4), 497-514. DOI: 10.1007/s10745-018-0008-9.

WHO (1990). *Public Health Impact of Pesticides Used in Agriculture*. WHO, Geneva.

WHO (2003). World Health Organization. Diet, Nutrition, and the Prevention of Chronic Diseases: Report of a Joint WHO/FAO Expert Consultation. Technical Report Series 916. World Health Organization, Geneva.

WHO (2010). Exposure to highly hazardous pesticides: a major public health concern. Available at: https://www.who.int/ipcs/features/hazardous_pesticides.pdf?ua=1. Accessed 1 August 2019.

WHO (2014). *Global Status Report on Noncommunicable Diseases 2014*. World Health Organization, Geneva.

WHO (2018). Healthy diet - key facts. Available at: http://www.who.int/en/news-room/fact-sheets/detail/healthy-diet.

WHO/FAO (2002). *Joint WHO/FAO Expert Consultation on Diet, Nutrition, and the Prevention of Chronic Diseases*. World Health Organization/Food and Agriculture Organization of the United Nations, Geneva.

Whyte, W. (1987). *Our American Land: 1987 Yearbook of Agriculture*. U.S. Government Printing Office, Washington, DC.

Wiebe, M. G. (2002). Myco-protein from *Fusarium venenatum*: a well-established product for human consumption. *Applied Microbiology and Biotechnology* 58(4), 421–427.

Wikipedia (2021a). World population. Available at: wikipedia.org/wiki/World_population. Accessed 1 July 2021.

Wikipedia (2021b). Cultured meat. Available at: https://en.wikipedia.org/wiki/Cultured_meat#mw-head. Accessed 05 April 2021.

Wilcox, M. (2021). For food companies, shifting to regenerative ag practices is complicated. *GreenBiz*, 18 March. Available at: https://www.greenbiz.com/article/food-companies-shifting-regenerative-ag-practices-complicated.

Wild, F., Czerny, M., Janssen, A. M., Kole, A. P. W., Zunabovic, M. and Domig, K. J. (2014). The evolution of a plant-based alternative to meat. From niche markets to widely accepted meat alternatives. *Agro Food Industry Hi-Tech* 25, 45–49.

Wilkinson, J. (2009). The globalization of agribusiness and developing world food systems. *Mon Reve* 61(04), 38–50.

Willer, H. and Lernoud, J. (2016). *The World of Organic Agriculture, Statistics, and Emerging Trends* (1st edn. Handbook). FIBL and IFOAM, Bonn. ISBN 978-3-03736-306-5.

Willer, H., Schlatter, B. and Trávníček, J. (Eds) (2023). *The World of Organic Agriculture. Statistics and Emerging Trends*. Research Institute of Organic Agriculture FiBL, Frick, and IFOAM – Organics International, Bonn. Online Version 2 of 23 February 2023.

Willer, H., Trávníček, J., Meier, C. and Schlatter, B. (Eds) (2022). *The World of Organic Agriculture. Statistics and Emerging Trends*. Research Institute of Organic Agriculture FiBL, Frick, and IFOAM – Organics International, Bonn.

Willett, W., Rockström, J., Loken, B., Springmann, M., Lang, T., Vermeulen, S., Garnett, T., Tilman, D., DeClerck, F., Wood, A., Jonell, M., Clark, M., Gordon, L. J., Fanzo, J., Hawkes, C., Zurayk, R., Rivera, J. A., De Vries, W., Sibanda, L. M., Afshin, A., Chaudhary, A., Herrero, M., Agustina, R., Branca, F., Lartey, A., Fan, S., Crona, B., Fox, E., Bignet, V., Troell, M., Lindahl, T., Singh, S., Cornell, R., K. S., Narain, S., Nishtar, S. and Murray, C. J. L. (2019). Food in the Anthropocene: the EAT-Lancet Commission on healthy diets from sustainable food systems. *Lancet* 393(10170), 447–492. Accessed 16 January 2019. Available at: http://dx.doi.org/10.1016/S0140-6736(18)31788-4.

Williams, C., Fenton, A. and Huq, S. (2015). Knowledge and adaptive capacity. *Nature Climate Change* 5(2), 82–83.

Williams, D. R., Clark, M., Buchanan, G. M., Ficetola, G. F., Rondinini, C. and Tilman, D. (2020). Proactive conservation to prevent habitat losses to agricultural expansion. *Nature Sustainability*, December. Available at: https://doi.org/10.1038/s41893-020-00656-5.

Williamson, S. (2019). *First Steps by British Farmers Towards Agroecological Systems*. PAN, UK. Available at: www.panna.org. Accessed 21 April 2021.

Wilson, J. (2011). Irrepressibly toward food sovereignty. In: Holt-Giménez, E. (Ed.) *Food Movements Unite! Strategies to Transform Our Food Systems*. Food First Books, Oakland, CA, pp. 71-92.

Winne, M. (2008). *Closing the Food Gap*. Beacon Press, Boston, MA.

Winter, M. (2003). Embeddedness, the new food economy, and defensive localism. *Journal of Rural Studies* 19(1), 23-32.

Wirsenius, S., Azar, C. and Berndes, G. (2010). How much land is needed for global food production under scenarios of dietary changes and livestock productivity increases in 2030? *Agricultural Systems* 103(9), 621-638.

Wise, R. M., Fazey, I., Stafford Smith, M., Park, S. E., Eakin, H. C., Archer Van Garderen, E. R. M. and Campbell, B. (2014). Reconceptualising adaptation to climate change as part of pathways of change and response. *Global Environmental Change* 28, 325-336.

Wiskerke, J. S. C. (2009). On places lost and places regained: reflections on the alternative food geography and sustainable regional development. *International Planning Studies* 14(4), 369-387.

Wissuwa, M., Wegner, J., Ae, N. and Yano, M. (2002). Substitution mapping of the Pup1: a major QTL increasing phosphorus uptake of rice from a phosphorus deficient soil. *Theoretical and Applied Genetics* 105(6-7), 890-897.

Withers, P., Neal, C., Jarvie, H. and Doody, D. (2014). Agriculture, and eutrophication: where do we go from here? *Sustainability* 6(9), 5853-5875.

Wittman, H., Beckie, M. and Hergesheimer, C. (2012). Linking local food systems and the social economy? Future roles for farmers' markets in Alberta and British Columbia. *Rural Sociology* 77(1), 36-61.

Wolfert, S., Ge, L., Verdouw, C. and Bogaardt, M. J. (2017). Big data in smart farming–a review. *Agricultural Systems* 153, 69-80.

Woodcock, T., Fagan, C. C., O'Donnell, C. P. and Downey, G. (2007). Application of near and mid-infrared spectroscopy to determine cheese quality and authenticity. *Food Bioprocess Technology* 1(2), 117-129.

Woomer, P. L. and Swift, M. J. (1992). The biological management of tropical soil fertility. In: Greenland, D. J. and Szabolcs, I. (Eds) *Soil Resilience and Sustainable Land Use*. CAB International, Wallingford.

World Bank (2000). *Entering the 21st Century: World Development Report 1999/2000*. Oxford University Press, New York.

World Bank (2007). *World Development Report 2008: Agriculture for Development*. The World Bank, Washington, DC.

World Bank (2010). *Rising Global Interest in Farmland: Can It Yield Sustainable and Equitable Benefits?* The World Bank, Washington, DC.

World Grain (2023a). *Companies Betting on Plant-Based Protein*, 15 February. Available at: https://www.world-grain.com/articles/18111-companies-betting-on-plant-based-protein.

World Grain (2023b). *ADM Expands Regenerative ag Program*, 18 July. Available at: https://www.agriculturedive.com/news/adm-expands-regenerative-agriculture-program/.

Worthy, M. (2011). *Broken Markets: How Financial Market Regulation Can Help Prevent Another Global Food Crisis*. World Development Movement, London.

Wrangham, R. (2013). The evolution of human nutrition. *Current Biology* 23(9), R354-R355.

Wrathall, J. E. (1988). Recent changes in arable crop production in England and Wales. *Land Use Policy* 5(2), 219–231.

WRI (1994). *World Resources 1994–95*. World Resources Institute, Washington DC and Oxford University Press, Oxford.

WRI (2023). *The Global Benefits of Reducing Food Loss and Waste, and How to Do It*. World Resources Institute, Washington, DC. Accessed 20 April 2023 Available at: https://www.wri.org/insights/reducing-food-loss-and-food-waste.

WWF (2020). *Communal Gardening Could Help Solve Our Food Crisis in This COVID-19 World*. WWF, Manila.

Xia, L., Cao, L., Yang, Y., Ti, C., Liu, Y., Smith, P., van Groenigen, K. J., Lehmann, J., Lal, R., Butterbach-Bahl, K., Kiese, R., Zhuang, M., Lu, X. and Yan, X. (2023). Integrated biochar solutions can achieve carbon-neutral staple crop production. *Nature Food*. DOI: 10.1038/s43016-023-00694-0.

Xiong, Y., Ge, Y., Grimstad, L. and From, P. J. (2020). An autonomous strawberry harvesting robot: design, development, integration, and field evaluation. *Journal of Field Robotics* 37(2), 202–224.

Yadvinder-Singh, K., Kukal, S. S., Jat, M. L. and Sidhu, H. S. (2014). Improving water productivity of wheat-based cropping systems in South Asia for sustained productivity. *Advances in Agronomy* 127. ISSN 0065-2113. DOI: 10.1016/B978-0-12-800131-8.00004-2.

Yang, F., Hong, F., You, W., Liu, C., Gao, F., Wu, C. and Yang, P. (2006). Influence of nano-anatase TiO_2 on the nitrogen metabolism of growing spinach. *Biological Trace Element Research* 110(2), 179–190.

Yang, G., Pu, R., Zhao, C. and Xue, X. (2014). Estimating high spatiotemporal resolution evapotranspiration over a winter wheat field using an IKONOS image based complementary relationship and lysimeter observation. *Agricultural Water Management* 133, 34–43.

Yara (2021). Partnership for a green energy transition. Available at: https://www.yara.com/news-and-media/green-ammonia-project-press-conference/. Accessed 21 September 2021.

Yatribi, T. (2020). Factors affecting precision agriculture adoption: a systematic litterature review. *Economics* 8(2). ISSN 2303-5005.

Yeomans, P. A. (1954). *The Keyline Plan*. P.A. Yeomans, Sydney.

Yeomans, P. A. (1958). *The Challenge of Landscape: The Development and Practice of Keyline*. Keyline Publishing Pty, Sydney.

Yeomans, P. A. (1971). *The City Forest: The Keyline Plan for the Human Environment Revolution*. Keyline Publishing Pty., Sydney.

Yeomans, P. A. (1981). *Water for Every Farm: Using the Keyline Plan*. Second Back Row, Katoomba.

Young, D. L., Kwon, T. J., Smith, E. G. and Young, F. L. (2003). Site-specific herbicide decision model to maximize profit in winter wheat. *Precision Agriculture* 4(2), 227–238.

Yu, H. and Huang, Q. (2013). Bioavailability and delivery of nutraceuticals and functional foods using nanotechnology. In: Bagchi, D., Bagchi, M., Moriyama, H. and Shahidi, F. (Eds) *Bio-nanotechnology*. Blackwell Publishing Ltd, Hoboken, NJ, pp. 593–604. Available at: http://onlinelibrary.wiley.com/doi:10.1002/9781118451915.ch35/summary.

Zabel, F., Delzeit, R., Schneider, J. M., Seppelt, R., Mauser, W. and Václavík, T. (2019). Global impacts of future cropland expansion and intensification on agricultural markets and biodiversity. *Nature Communications* 10(1), 2844. DOI: 10.1038/s41467-019-10775-z.

Zamora-Ros, R., Rabassa, M., Cherubini, A., Urpí-Sardà, M., Bandinelli, S., Ferrucci, L. and Andres-Lacueva, C. (2013). High concentrations of a urinary biomarker of polyphenol intake are associated with decreased mortality in older adults. *Journal of Nutrition* 143(9), 1445–1450. 10.3945/jn.113.177121.

Zasada, I. (2012). *Peri-urban Agriculture and Multifunctionality: Urban Influence, Farm Adaptation Behaviour and Development Perspectives*. TU München, Fakultät Wissenschaftszentrum Weihenstephan, Munich.

Zasada, I., Schmutz, U., Wascher, D., Kneafsey, M., Corsi, S., Mazzocchi, C., Monaco, F., Boyce, P., Doernberg, A., Sali, G. and Piorr, A. (2019). Food beyond the city – analysing foodsheds and self-sufficiency for different food system scenarios in European metropolitan regions. *City, Culture and Society* 16, 25–35.

Zazo-Moratalla, A., Troncoso-González, I. and Moreira-Muñoz, A. (2019). Regenerative food systems to restore urban-rural relationships: insights from the concepción metropolitan area foodshed (Chile). *Sustainability* 11(10), 2892. DOI: 10.3390/su11102892.

Zhai, J. B., Xia, J. F., Zhou, Y. and Zhang, S. (2014). Design and experimental study of the control system for precision seed-metering device. *International Journal of Agricultural and Biological Engineering* 7, 13–18.

Zhang, N., Wang, M. and Wang, N. (2002). Precision agriculture: a worldwide overview. *Computers and Electronics in Agriculture* 36(2–3), 113–132. DOI: 10.1016/S0168-1699(02)00096-0.

Zhang, X., Zou, T., Lassaletta, L., Mueller, N. D., Tubiello, F. N., Lisk, M. D., Lu, C., Conant, R. T., Dorich, C. D., Gerber, J., Tian, H., Bruulsema, T., Maaz, T. M., Nishina, K., Bodirsky, B. L., Popp, A., Bouwman, L., Beusen, A., Chang, J., Havlík, P., Leclère, D., Canadell, J. G., Jackson, R. B., Heffer, P., Wanner, N., Zhang, W. and Davidson, E. A. (2021). Quantification of global and national nitrogen budgets for crop production. *Nature Food* 2(7), 529–540. Available at: https://www.nature.com/articles/s43016-021-00318-5.

Zhang, Y., Min, Q., Li, H., He, L., Zhang, C. and Yang, L. (2017). A conservation approach of Globally Important Agricultural Heritage Systems (GIAHS): improving traditional agricultural patterns and promoting scale production. *Sustainability* 9(2), 295.

Zhang, Y. H. P. (2010). Production of bio-commodities and bioelectricity by cell-free synthetic enzymatic pathway bio-transformations: challenges and opportunities. *Biotechnology and Bioengineering* 105(4), 663–677.

Zhao, L., Wu, L., Dong, C. and Li, Y. (2010). Rice yield, nitrogen utilization and ammonia volatilization as influenced by modified rice cultivation at varying nitrogen rates. *Agricultural Sciences* 1(1), 10–16.

Zheng, L., Hong, F., Lu, S. and Liu, C. (2005). Effect of Nano-TiO_2 on strength of naturally aged seeds and growth of spinach. *Biological Trace Element Research* 104(1), 83–92.

Zheng, M., Huang, Z., Ji, H., Qiu, F., Zhao, D., Bredar, A. R. C. and Farnum, B. H. (2020). Simultaneous control of soil erosion and arsenic leaching at disturbed land using polyacrylamide modified magnetite nanoparticles. *Science of the Total Environment* 702, 134997.

Zhou, Y., Du, S., Su, C., Zhang, B., Wang, H. and Popkin, B. M. (2015). The food retail revolution in China and its association with diet and health. *Food Policy* 55, 92–100.

Zhu, Y., Chen, H., Fan, J., Wang, Y., Li, Y., Chen, J., Fan, J., Yang, S., Hu, L., Leung, H., Mew, T. W., Teng, P. S., Wang, Z. and Mundt, C. C. (2000). Genetic diversity and disease control in rice. *Nature* 406(6797), 718–722.

Zou, Z., Ye, J., Sayama, K. and Arakawa, H. (2001). Direct splitting of water under visible light irradiation with an oxide semiconductor photocatalyst. *Nature* 414(6864), 625-627.

Zurek, M., Hebinck, A., Leip, A., Vervoort, J., Kuiper, M., Garrone, M., Havlík, P., Heckelei, T., Hornborg, S., Ingram, J., Kuijsten, A., Shutes, L., Geleijnse, J. M., Terluin, I., van 't Veer, P., Wijnands, J., Zimmermann, A. and Achter Bosch, T. (2018). Assessing sustainable food and nutrition security of the EU food system—an integrated approach. *Sustainability* 10, 4271. DOI: 10.3390/su10114271.

Index

Milton Keynes UK
Ingram Content Group UK Ltd.
UKHW051652010724
444645UK00001BB/3